IN SILICO IMMUNOLOGY

IN SILICO IMMUNOLOGY

edited by

Darren Flower
The Jenner Institute, University of Oxford, UK

and

Jon Timmis
University of York, Heslington, UK

 Springer

Library of Congress Control Number: 2006931791

ISBN-10: 0-387-39238-6
ISBN-13: 978-0-387-39238-7

Printed on acid-free paper.

9 8 7 6 5 4 3 2 1

springer.com

Darren Flower should like to dedicate this book to the two most important people in his life: his wonderful, if long suffering, wife Christine and his beautiful daughter Isobel.

Jon Timmis should like to dedicate this book to his wife, Gráinne, without whom life would not be complete, and to his as yet unborn child: welcome to the mad house.

Contents

Part I Introducing In Silico Immunology

Part II The Nature of Natural and Artificial Immune Systems

Preface

Whatever its final readership and impact, we, the Editors, feel this book is important. It addresses the realisation that there is a deep and abiding synergy, albeit one only now being properly explored and exploited, between immunology and computational science. This area of intersection we christen *in silico* immunology. Immunology is an inspiration for computational scientists seeking practical and philosophical metaphors for their work; but, at the same time, it is itself a biological discipline of such discombobulating complexity that only computational help as different as simulation and data warehousing can make its modern study tractable. Thus immunology both inspires but also requires computational science. This book deals in detail with the three main areas of *in silico* immunology: theoretical immunology, immunoinformatics, and artificial immune systems. While all of these are now well-established the interactions between the three are only beginning to be developed. It is a truly exciting time to be working in *in silicio* immunology. We are reaching a critical mass that will enable great strides to be taken and significant achievements to be made. Like David Hume, we may yet come to regret that this book falls still born from the press but we hope not. Hopefully it will instead strike a cord and tap into a burgeoning zeitgeist ready to capitalise on the remarkable potential that is *in silico* immunology.

It is a rare book indeed that is the product of one mind, and this multi-author tome is certainly not an example of that, and it is never the case that the creation, production, and dissemination of a book lies solely in the hands of a single individual. So, we have many people to thank for helping us to bring this book before the public:

All the authors, without whose contributions this book would not be possible.

Springer staff, in particular Frank Holwarth who provided Latex support and Marcia Kidston our Editorial Assistant.

Referees of our original book proposal, who recommended that we pursue this venture, and provided excellent feedback to us.

Susan Stepney for assistance with the formatting of the bibliography and Jools Greensmith for helping with some of the more nasty Latex tables.

We also have some more personal thanks to give.

DRF would like offer his thanks to the following. Firstly, Prof Peter Beverley, former scientific head of the Edward Jenner Institute for Vaccine Research (EJIVR), for introducing me to the world of Immunology and for having the moral courage to support bioinformatics in immunology by recruiting me to my present post. Secondly, members of his research group at the EJIVR and latterly the Jenner Institute, University of Oxford. Of these, One person who must be singled out is Dr Irini Doytchinova as her contribution to the group has been without equal. The others are Valerie Walshe, Martin Blythe, Christianna Zygouri, Debra Clayton (nee Taylor), Shelley Hemsley, Christopher P Toseland, Kelly Paine, Dr Pingping Guan, Dr Paul Taylor, Dr Helen McSparron, Dr Matthew N Davies, Dr Channa Hattotuwagama, and Dr Shunzhou Wan. DRF should also like to thank other staff members at the EJIVR for their help and for stimulating discussions: Dr Persephone Borrow, Dr Shirley Ellis, Dr Simon Wong, Dr Helen Bodmer, Dr Sam Hou, Dr Elma Tchillian, and Dr Josef Walker. With a few ineffable exceptions, DRF should also like to thank all his other colleagues and co-workers at the EJIVR and the Institute for Animal Health (IAH), Compton for their close, nurturing, and supportive collaboration. Finally, DRF would like to thank all his colleagues, friends, and colleagues in science, but particularly those working in immunoinformatics, including those too busy to contribute to this book. In particular, he would like to thank the following: Dr Tongbin Li, Dr Pedro Reche, Dr G.P.S. Raghava, Prof Vladimir Brusic, Prof Shoba Ranganathan, Dr Anne DeGroot, and many, many more.

JT would like to offer his thanks to the following: First, Prof Susan Stepney and Prof Andy Tyrrell who encouraged me to apply for the position at York and provided me excellent support through the process. York is a great place to live and work and I am very pleased to have the opportunity to work in two excellent departments. Second, people I have worked with over the past few years in AIS who have provided me the opportunity to learn about new topics and challenge the way I think. In particular, Dr Mark Neal and Dr Rogério de Lemos who never let me get away with anything (and long may this continue). I also need to mention Dr Andy Hone, Dr Alex Freitas, Dr Emma Hart, Dr Thomas Stibor and Dr Guissepe Nicosia, Dr Miguel Mendao and Dr Andy Greensted, all of whom are great collaborators, and working with you all is great fun: the way it should be. Third, my research students (past and present), without whom the vast majority of work would not get done: Dr. Andrew Watkins, Dr. Tom Knight, Dr. Andy Secker, Dr Modupe Ayara, Pete May, Phil Mohr, Paul Andrews, Ed Clark and Adam Knowles. Fourth, I would like to thank DRF for suggesting the idea of undertaking this book project, it has been great fun and I have learnt a great deal. Finally, I would like to thank Prof Keith Mander from the University of Kent, who took the decision to appoint me at Kent some years ago. He was incredibly supportive of me in my early years, and I firmly believe if I had not had his support I would not be in the position I am now, thank you Keith.

Compton, July 2006   *Darren Flower*
York, July 2006   *Jon Timmis*

List of Contributors

Uwe Aickelin
School of Computer Science (ASAP),
University of Nottingham,
Nottingham, NG8 1BB. UK.
uwe.aickelin@nottingham.ac.uk

Paul Andrews
Department of Computer Science,
University of York,
Heslington, York,
YO10 5DD, UK.
psa@cs.york.ac.uk

Jóse A. M. Borghans
Department of Immunology,
UMCU, Lundlaan 6,
3584 EA Utrecht, The Netherlands
and
Center for Biological Sequence
Analysis,
BioCentrum-DTU,
Technical University of Denmark.
j.borghans@umcutrecht.nl

Persephone Borrow
The Jenner Institute,
University of Oxford,
Compton, Berkshire, UK.
RG20 7NN.
Persephone.Borrow@jenner.ac.uk

Gastone C. Castellani
C.I.G. "L. Galvani" Interdepartmental Centre,
University of Bologna,
40126 Bologna, Italy.
gastone@vet.unibo.it

Melvin Cohn
Conceptual Immunology Group,
The Salk Institute for Biological
Studies,
10010 N. Torrey Pines Rd.,
La Jolla, CA 92037.
and
Instituto Gulbenkian de Cincia,
6 Rua da Quinta Grande,
2780-156 Oeiras,
Portugal.
cohn@salk.edu

Vincenzo Cutello
Department of Mathematics and
Computer Science,
University of Catania,
V.le A. Doria, 6 - 95125 Catania
Italy.
cutello@dmi.unict.it

Matthew Davies
The Jenner Institute,
University of Oxford,

Compton, Berkshire,
RG20 7NN. UK.
m.davies@mail.cryst.bbk.ac.uk

Rob J. De Boer
Theoretical Biology,
Utrecht University Padualaan 8,
3584 CH Utrecht, The Netherlands.
R.J.DeBoer@bio.uu.nl

Irini A Doytchinova
The Jenner Institute,
University of Oxford,
Compton, Berkshire,
RG20 7NN. UK.
Irini.A.Doytchinova@jenner.ac.uk

Claudio Francesci
Department of Experimental
Pathology,
University of Bologna,
40126 Bologna, Italy
and
C.I.G. 'L. Galvani' Interdepartmental
Centre,
University of Bologna,
40126 Bologna, Italy.
claudio.franceschi@unibo.it

Darren R Flower
The Jenner Institute,
University of Oxford,
Compton, Berkshire,.
RG20 7NN. UK.
darren.flower@jenner.ac.uk

Simon Garrett
Computational Biology Group,
Department of Computer Science,
University of Wales, Aberystwyth,
SY23 3PG. UK.
smg@aber.ac.uk

Pingping Guan
The Jenner Institute,
University of Oxford,
Compton, Berkshire,
RG20 7NN. UK.
Pingping.Guan@bbsrc.ac.uk

Channa K Hattotuwagama
The Jenner Institute,
University of Oxford,
Compton, Berkshire,
RG20 7NN. UK.
Channa.Hattotuwagama@jenner.ac.uk

Shelley L Hemsley
The Jenner Institute,
University of Oxford,
Compton, Berkshire,
RG20 7NN. UK.
shelley.hemsley@jenner.ac.uk

Denise Kirschner
Department of Microbiology and
Immunology,
University of Michigan Medical
School,
Ann Arbor, MI 48109-0620. USA.
kirschne@umich.edu

Can Kesmir
Theoretical Biology,
Utrecht University Padualaan 8,
3584 CH Utrecht, Netherlands.
and
The Center for Biological Sequence
Analysis,
BioCentrum-DTU, Technical
University of Denmark,
C.Kesmir@bio.uu.nl

Ha Youn Lee
Theoretical Biology and Biophysics,
Los Alamos National
Laboratory, Los Alamos, NM 87545.
USA.
hayoun@lanl.gov

Andrew Hone
Institute Mathematics, Statistics
and Actuarial Sciences,
University of Kent,
Canterbury, Kent. UK.
a.n.w.hone@kent.ac.uk

Jonathan Loroni
Department of Experimental
Pathology,
University of Bologna,
40126 Bologna, Italy.
and
C.I.G. 'L. Galvani' Interdepartmental
Centre,
University of Bologna,
40126 Bologna, Italy.
jonathan.loroni@gmail.com

Mark Neal
Department of Computer Science,
University of Wales, Aberystwyth,
SY23 3PG. UK. mjn@aber.ac.uk*

Giuseppe Nicosia
Department of Mathematics and
Computer Science,
University of Catania,
V.le A. Doria,
6 - 95125 Catania Italy.
nicosia@dmi.unict.it

Alan Perelson
Theoretical Biology and Biophysics,
Los Alamos National Laboratory,
Los Alamos, NM 87545. USA.
alan_perelson@att.net

Daniel Remondini
C.I.G. "L. Galvani" Interdepartmental Centre,
University of Bologna,
40126 Bologna, Italy.
daniel.remondini2@unibo.it

Adrian Robins
Division of Immunology,
Nottingham University Medical
School,
Queens Medical Centre, Nottingham,
NG7 2UH. UK.
adrain.robins@nottingham.ac.uk

Martin Robbins
Computational Biology Group,
Department of Computer Science,
University of Wales, Aberystwyth,
SY23 3PG.
mjr00@aber.ac.uk

Stefano Salvio
Department of Experimental
Pathology,
University of Bologna,
40126 Bologna, Italy.
and
C.I.G. 'L. Galvani' Interdepartmental
Centre,
University of Bologna,
40126 Bologna, Italy.
stefano.salvioli2@unibo.it

Susan Stepney
Department of Computer Science,
University of York,
Heslington, York,
YO10 5DD, UK.
susan@cs.york.ac.uk

Paul Taylor
The Jenner Institute,
University of Oxford,
Compton, Berkshire,
RG20 7NN. UK.
paul.taylor@jenner.ac.uk

Debra J Taylor
The Jenner Institute,
University of Oxford,
Compton, Berkshire,
RG20 7NN. UK.
debra.clayton@bbsrc.ac.uk

Paolo Tieri
Department of Experimental
Pathology,
University of Bologna,
40126 Bologna, Italy.
and
C.I.G. 'L. Galvani' Interdepartmental
Centre,
University of Bologna,
40126 Bologna, Italy.
p.tieri@unibo.it

Jon Timmis
Department of Computer Science
and Department of Electronics,
University of York,
Heslington, York.
YO10 5DD. UK.
jtimmis@cs.york.ac.uk

Christopher Toseland
The Jenner Institute,
University of Oxford,
Compton, Berkshire,
RG20 7NN. UK.
Christopher.Toseland@jenner.ac.uk

Cole Trapnell
University of Maryland, U.S.A.
btrapnel@gmail.com

Silvana Valensi
Department of Experimental
Pathology,
University of Bologna,
40126 Bologna, Italy.
and
C.I.G. 'L. Galvani' Interdepartmental
Centre,

University of Bologna,
40126 Bologna, Italy.
silvana.valensin@unibo.it

Hugo van den Berg
Warwick Systems Biology,
University of Warwick, UK.
hugo@maths.warwick.ac.uk

Joanne Walker
Computational Biology Group,
Department of Computer Science,
University of Wales, Aberystwyth,
SY23 3PG. UK.
jw@aber.ac.uk

Andrew Watkins
Department of Computer Science
and Engineering,
Mississippi State University, Box
9637,
Mississippi State, MS 39762 USA.
andrew@cse.msstate.edu

Valerie Walshe
The Jenner Institute,
University of Oxford,
Compton, Berkshire,
RG20 7NN. UK.
Valerie.Walshe@jenner.ac.uk

William Wilson
School of Computer Science (ASAP),
University of Nottingham,
Nottingham, NG8 1BB. England,
UK.
wow@cs.nott.ac.uk

Overview of the book

Immunology is old: at least half a billion years for adaptive immunology and far longer for innate immunity. The study of immunology is, however, more of a Johnny-come-lately, with a history of scientific study stretching back only a few hundred years, although an empirical or phenomenological understanding of it probably reaches far into antiquity. Accounts litter the historical record. An example, that is often quoted, can found in the works of Thucydides. He was the principal historian of the Peloponnesian War, which raged between the Spartan Alliance and Greek city-state of Athens. The plague struck in the early summer of 430 BC and continued through the following year; after abating greatly, the plague returned again in 427 BC. Its initial effects were devastating: over a third of Athens population succumbed. Thucydides, who was himself a victim, described the disease, its symptoms and effects, in graphic detail, but also noted that:

> it was with those who had recovered from the disease that the sick and the dying found most compassion. These knew what it was from experience, and had now no fear for themselves; for the same man was never attacked twice - never at least fatally.

Today, science, manifest in the discipline of Immunology, can describe such phenomena in rather greater detail and within a presumed precision undreamt of by our ancestors. Immunology is intimately connected with disease: infectious, most obviously, but also autoimmune disease, inherited and multifactorial genetic disease, cancer, and allergy. We now believe we understand diseases at the most fundamental level and through the vaccine and the rational discovery and targeting of drugs can alleviate the effects of infection and even prevent the spread of contagious diseases. Thus immunology is often viewed as a science of paramount importance. Immunology is well regarded in the wider scientific community. It is a large and well funded discipline. Despite all that, Immunology nonetheless finds itself at a key point in its history. After centuries of hypothesis-driven research, it teeters on the brink of reinventing itself as a quantitative, genomic, data-driven science. It will need to engage with computational science in a way inconceivable less than a generation ago. The complexity of datasets will render if it inefficient, even impossible, to analyse them by hand. Simulations will be absolutely vital if any coherence is expected from emergent experimental data sets. Yet, immunology has much to teach computer science as well. The computer algorithms that it inspires are amongst the most insightful and exciting of the modern era. The era of in silico immunology is truly upon us.

In Silico Immunology covers a range of topics that is both diverting and extremely diverse. When reading this book you will encounter ideas taken from 'wet' immunology, theoretical immunology, mathematics, computational modeling, computer science and immunoinformatics. You will discover how the role of computing has radically altered the world of immunology, in terms of producing models to help further understand the immune system, to automated techniques which assist in the analysis of vast amounts of data collected by experiments. You will also learn about how immunology is informing the development of novel computational techniques in computer science and engineering to help build more effective tools, this field is known as Artificial Immune Systems (AIS).

In order to get through such a diverse range of topics, we have divided the book into 3 major sections: Part I sets the scene, covering the basics of most topics in the book; Part II goes onto explore the role of modeling the immune system (mathematically and computationally), and this is intertwined with topics connected with both the real immune system, and artificial immune systems; contributions in Part 3 are concerned with the interaction of both natural and artificial systems with their environment, and in many different ways discuss the often forgotten notion that a system is embodied in an environment, and that environment effects how the system operates.

However, we begin our journey with some context: the immune system. Chapter 1 by Robins provides a gentle introduction to the vast topic of immunology. Clearly, this is not a complete overview, as whole books have been written on the topic, and have often hardly scratch its surface, but what this chapter concentrates on is the interplay between the innate and adaptive immune systems. Robins highlights the intricate interaction between these two arms of immune systems, one that is only now be properly explored within immunological research. This has implications also for Artificial Immune Systems. The chapter by Neal and Trapnel (Chapter 14) discusses this issue at great length, and attempts to draw new inspiration for novel computational systems inspired by the innate immune system.

Chapter 2 provides a detailed introduction to the second area of this book: immunoinformatics. This is a discipline which has emerged in recent years within the wider world of bioinformatics. It addresses the problems particular to immunology, such as the prediction of immunogenicity, be that identification of epitopes or the prediction of whole protein immunogens. Chapter 2 introduces the scope of immuninformatics and highlights both its importance and potential while not shying away from its many remaining problems.

The final chapter in this section, Chapter 3 presents an introduction to the area of Artificial Immune Systems (AIS). This area of research is concerned with developing computer systems (or engineered systems) that in some abstract way mimic the properties of the natural immune system. One such example that we have in this book, is that the immune system learns about new antigen, in that during an immune response (amongst many complex interactions), new antibodies are generated that have in some way generalised so that they can recognise antigen that the immune system may not have encountered before. This observation has led to the development of machine learning algorithms capable of classification of previously unseen data items. However, it would seem that there is a great deal to be done in terms of developing new AIS, which somehow better capture the complexity of the immune system.

Now the scene is set, we move into Part II. Our first contribution here is by Lee and Perelson (Chapter 4). In this chapter, the authors present a review of computational models of B and T Cells receptors. Models range from simple binary based models, to more complex shape spaces, and include details of various affinity measures that can be used to simulate the binding of receptors. Attention is also given to the use of random energy models which are derived from the actual physics of protein interactions. The chapter then discusses the relevance and impact of using such models in the context of influenza modeling. The notion of shape space has influenced greatly the field of AIS, with the more basic models being adapted as a standard. However, consideration into how the shape space paradigm may have an impact on the area of AIS is discussed somewhat in the chapter by Stepney (Chapter 12), and who argues that clearly great care has to be taken when defining the actual shape space employed in an AIS, and that maybe, practitioners should be making much more use of richer shape spaces in their AIS.

We then move from modelling receptors, to modelling immune memory. Chapter 5 by Garrett et al. first discusses various theories in the immunological literature regarding how the immune system maintains a record or memory of past encounters. It soon becomes clear that there is no single clear message from the literature on how the immune system does achieve this. In order to begin addressing this, the authors argue that more powerful modelling tools are needed to be able to cope with the complexity of addressing such an issue.

The Sentinel system is described, which is a new tool capable of performing large scale simulations of the immune system. This tool is then used to simulate a theory of immunological memory, achieving insightful results, showing that such modelling tools can be of great use to understanding immunological processes.

Chapter 6 by Tieri et al. continues the theme of modelling, but now attention is drawn to the nature of receptors themselves, and the notion that receptors are degenerate i.e. some immune receptors bind to many types of ligands (rather than the more traditional view that the binding is more a one-to-one nature. The main argument of this chapter is that it is the degenerate nature of receptors that is a primary contributor the overall robustness of the immune system itself. The authors discuss the notion of degeneracy i the context of a class of network models where the idea of degeneracy is very natural, and how various degrees of specificity and selectivity to emerge from a such degenerate system. This chapter is complimented by the following chapter by Andrews and Timmis (Chapter 7) who take this idea of degeneracy even further in relation to the development of novel immune inspired algorithms. Their suggestion is that incorporating the degenerate nature may lead to more robust, scalable and better performing immune inspired systems. As with the previous chapter, the authors take the argument made in Chapter 3 that AIS are potentially suffering from a lack of novel biological input and that much could be gained from revisiting alternative theories of immunology. The authors focus on the cognitive theory by Cohen, that the immune system is a cognitive system capable of recognition, decision making and actions, much like other cognitive systems. Amongst many ideas of Cohen, is degeneracy. The chapter goes onto to discuss the role that degeneracy may play in the future development of AIS.

Chapters 8, 9 and 10 all address issues that are specific and molecular, rather than general and conceptual, in nature. Chapter 8 addresses the issue of predicting immunogenicity: the ability of the unprimed immune system to mount a response to some foreign protein or more general antigen. The ability to predict immunogencity lies at the heart of computer assisted vaccine design (CAVD). CAVD represents a set of techniques of tremendous potential importance in the search for the next set of life saving vaccines. Immunoinformatics has tried to deal with the crucial problem of immunogenicity prediction by developing a range of techniques that allow for the accurate identification of T cell mediated epitopes. Epitopes are small fragments of proteins which constitute the smallest quantum of immune recognition. Most T cell epitopes are presented by proteins called Major Histocompatibility Complexes (MHC), which are the subject of Chapters 9 and 10.

In chapter 9, Can Kesmir and colleagues use a potent combination of theoretical immunology and immunoinformatics to address the question of the origin of diversity within the human MHC and Chapter 10 addresses some techniques that can be used to master this bewildering variety in the context of vaccines. MHC-encoding genes are among the most polymorphic of human genes. Since they mediate such an important role in the immune responses, within populations such polymorphism may arise from selection for increased protection of hosts against pathogens. However, each individual expresses only a limited number of different MHC molecules. Chapter 9 seeks to square the circle in our understanding of this apparent paradox. Chapter 10 is concerned not with explaining this phenomenon but in trying to deal with its practical outcome: the vast range - which numbers in the thousands - of MHC alleles present in the human population. Guan, Doytchinova, and Flower describe attempts to reduce the huge number of alleles down to a more manageable handful of so-called supertypes, which each exhibit the same, or nearly the same, peptide specificity. If we can target a small set of supertypes of convincing provenance we have a chance of formulating vaccines and other immunotherapeutics.

The final chapter in this section by Cutello and Nicosia (Chapter 11) is our first implementation and application of an AIS. In this chapter the authors develop a clonal selection based algorithm that is capable of performing biomolecular structure prediction. The authors take the approach of casting the problem as an optimisation problem (hence the use of a clonal selection type algorithm, as this is their typical, but not exclusive, use. Casting the problem as such is not in itself novel, however, that happens in this chapter is that the problem is recast as a multi-objective optimisation problem for the structure prediction of medium sized proteins sequences $(46 - 70$ residues). Different solutions (for 3D conformations) may involve a trade-off between different objectives. An optimum solution with respect to one objective may not be optimum with respect to another objective. As a consequence, one cannot choose a solution which is optimal with respect to only one objective and the algorithm must search the landscape for the best compromise. The algorithm proposed appears to be very effective, creating a nice circle between the extraction of biological inspiration and its application to a biological problem.

The third and final part of our book turns attention to the environment in which the immune system operates and takes a closer look at interactions between the environment and the immune system, but also importantly the interactions within the immune system. The first of our contribution by Stepney (Chapter 12) takes a unique look at the consequences of considering the immune system as an embodied system i.e. operating within a certain environment. Stepney explores this concept describing the need for complex dynamics between the embodied system and its environment and the coupling between the two. It is argued that when one thinks carefully about the notion of embodiment, it has serious implications not only in a biological context, but importantly in an artificial context. Stepney proposes a number of design principles that can be extracted from the immune system, in the context of embodiment, for use in the construction of AIS. Following on from this, Kirshner in Chapter 13 discusses in detail the immune response in the context of *Mycobacterium tuberculosis*: this is real embodiment in a real immune system. The authors have made attempts to explore the complex system of immunity by studying the immune response to a specific pathogen. They present studies of the interaction of the immune system with the intracellular pathogen *Mycobacterium tuberculosis* at a number of biological and spatial scales. They highlight both the biology that they are addressing and mathematical approaches that have been adopted as a means to understand the integrated, multi-scale complex system know as the immune response.

Continuing the theme of interacting with the environment, but adding a second perspective of interactions within the system, Chapter 14 by Neal and Trapnel, explores the almost untapped area of innate immunity and how that interacts with the adaptive immune system, in the context of AIS. Whilst the chapter by Timmis and Andrews discusses the further exploration of the adaptive immune system for AIS, Neal and Trapnel identify a significant gap in the understanding of the role the complex set of interactions that make up the innate and adaptive immune system. Amongst many actors, the chapter discusses the role of dendritic cells, mast cells, PAMPS and many others, ultimately drawing a number of implications for people in the AIS community for consideration when developing new AIS.

Chapter 15 by Watkins explores the use of embodiment for the development of an AIS. Here Watkins makes use of distributed processing systems to develop an immune inspired classifier (a system capable of identifying unseen data items into various classes). In this chapter, Watkins details how a network of computers can be exploited to make significant gains in computational power for an immune inspired system: this is an excellent demonstration of making use of your environment and exploiting it within your computational system. The AIS in question is called AIRS (Artificial Immune Recognition System) and has received a great deal of attention in the AIS literature. Here, further advances on AIRS are proposed, and the role of distributed learning is discussed as possible future direction for making fuller use of the immune metaphor.

Concluding our discussions on AIS is our penultimate chapter (Chapter 16) by Hone and van den Burgh. In this chapter the authors bring together both theoretical immunology and AIS through the use mathematical techniques used in theoretical immunology for the analysis of immune inspired systems. The authors outline the role that certain mathematical techniques can play in the understanding of AIS (such as non-linear dynamics and ideas from optimal control theory), and go onto propose a novel cytokine network model that may be further exploited in future AIS.

Our final chapter (Chapter 17) is reserved for Melvin Cohn who has had a major impact on the theoretical immunological community over the years. In this chapter, Cohn revisits the fundamental nature of self and non-self of the immune system and provides a detailed argument as to the conceptual nature of the immune self. Cohn describes a principled approach to the understanding of the very nature of self/non-self discrimination, and supports his arguments from 5 different experiments. He concludes that, sadly, the world of theoretical immunology has not had a great impact on experimental immunology: this is a challenge to the theoretical community to make more of an impact, as there is much insight to be offered.

Of course, this book is finite, and thus limited, and there are several areas of in silico immunology that we would have loved to include. The fact that we were unable to do so is no reflection on these disciplines within a discipline or practitioners thereof. Something we would have liked to do was to include more about practical applications to real world problems. Such problems, and the application of techniques drawn from some area of in silico immunology to address such problems, are clearly legion. These range from methods drawn from immunoinformatics applied to clinical immunology through to the shipment of freighter cargo using AIS algorithms. Another obvious example of an omission is the mathematical modelling of epidemiology and infection disease progression and the efficacy of vaccination. The links to both theoretical immunology and to immuninformatics of such epidemiological modelling is both clear and profound.

The three main disciplines we discuss here theoretical immunology, immunoinformatics, and artificial immune systems can and do act synergistically. However, not all interactions between them are equally strong nor of the same nature. Likewise, their interactions with immunology itself are not all equally strong nor of the same nature. Immunoinformatics can be of direct benefit to immunology by addressing specific problems, such as the prediction of epitopes, which are more efficiently dealt with in silico than they are in vitro. By exploring concepts using theoretical immunology, valuable and testable hypothesises can be generated within immunology. AIS uses metaphors from immunology, as filtered through theoretical immunology, to design new and efficient computer algorithms which can be then be deployed in immunoinformatics. The issue now is to exploit the potential latent within all these disciplines both for their own sake and for the sake of immunology itself. As we know, and will see time and again in the pages that follow, by growing together, rather than apart, all of these areas can and will benefit.

We hope you enjoy reading in silico immunology. There is clearly a great deal that can be learnt in many disciplines from each other, and we feel that the future of in silico immunology is bright: working at the true intersection of many disciplines, aiming to break through in each. Whether you choose to read the book cover to cover, or dive in at various chapters, we hope that the eclectic mix that we have assembled here proves useful to you and provides a great deal of food for thought.

Darren Flower
Jon Timmis

Part I

Introducing In Silico Immunology

Innate and Adaptive Immunity

Adrian Robins

Division of Immunology, Nottingham University Medical School,Queen's Medical
Centre, Nottingham NG7 2UH. UK. adrian.robins@nottingham.ac.uk

Summary. Innate immune responses recognise generic targets on pathogens using germline encoded receptors, whereas adaptive immune responses recognise specific targets using randomly generated receptors which have an essentially unlimited
recognition repertoire. Interactions between innate and adaptive forms of immune
recognition are increasingly being recognised as essential for the effective functioning of the immune response. Examples given here demonstrate the advantages of
integrating pre-programmed recognition (rapid response using widely distributed
receptors) with random repertoire recognition (open repertoire for specific recognition of novel targets, with memory). The randomly generated repertoire brings
problems of self/non-self discrimination, which is solved at multiple levels in the human immune system, including shaping of the naïve repertoire, stringent control of
naïve cell activation by innate immune receptor recognition, and control by regulatory cells in the periphery. The interactions between innate and adaptive immunity
are many, complex, and bidirectional, with innate mechanisms being instrumental
in the initiation of adaptive responses, and controlling the type of adaptive response
induced; innate effector mechanisms are also recruited in the effector phase of adaptive responses. The challenge is now to abstract the essential components of the
innate-adaptive interaction in order to utilise this concept in alternative contexts.

1.1 Introduction

For many years, innate and adaptive immunity were separate areas of study:
adaptive immunologists were engaged in understanding the antibodies produced by B cells and the cell-based recognition structures of the T cell. Antibodies have a variable binding site which allows them to bind specifically
to target molecules (antigens) with a seemingly unlimited repertoire. T cell
recognition specificities are also diverse, but the target for recognition is a
complex between a self molecule (coded for by the major histocompatibility complex, MHC) and short peptide sequences derived from their target

antigen. The MHC antigens that present peptides to conventional T cells are very variable, meaning that different individuals present different peptide sequences from the same pathogen, depending on the peptide 'fit' to their MHC molecules. Antibodies recognise native antigens on the cell surface or in the extracellular compartment, whereas T cells recognise peptide sequences from the antigen. This happens in two main ways: antigens synthesised within a cell bind to class I MHC, whereas antigens taken up from outside the cell bind to class II MHC. Class I MHC/peptide recognition by cytotoxic T cells allows monitoring of the intracellular compartment by continuous sampling of peptides from proteins being synthesised within the cell which are transported to the cell surface complexed with class I MHC molecules. Class II MHC presents peptides to helper T cells, which participate in both cytotoxic T cell responses and antibody responses. Innate immunologists studied cellular and humoral innate mediators, such as macrophages, neutrophils, complement and interferons. In the last ten years, as more has been learned about the receptors and molecular mechanisms in the innate immune system [Germain 2004], the complex interactions between the innate and adaptive immunity have become more apparent, and their importance recognised [Hoebe *et al.* 2004]. In this chapter, these interactions will be exemplified, to illustrate the richness of the interplay between innate and adaptive immunity, and in particular to highlight the role of innate immunity in controlling the adaptive immune response.

1.2 Innate and Adaptive Recognition Structures

The recognition structures used by the immune system can be divided into two categories depending on the way in which they are generated. Firstly, those that are encoded for by germ line genes, and whose structure and thus specificity are inherited. Secondly, those encoded for by randomly modified genes, with specificities which arise by chance in clonally distributed cells of the individual. These two categories of receptor can be considered to be a basis for distinguishing innate and adaptive immunity respectively, although as we will see, this categorisation is a simplification, with a range of additional recognition structures with limited variability forming a grey area between classical innate and adaptive receptors. However, this broad distinction between modes of receptor genesis is useful, and also gives a logical basis for the familiar characteristics of the innate and adaptive arms of the immune response.

Innate receptors are expressed by large numbers of cells before exposure to the potential pathogen, and are thus available for immediate recognition and rapid effective first line defence against invaders. The need for a germ line code for the specificity of each receptor limits the recognition repertoire, and does not allow for recognition of specificities not previously encountered by the species. The direct link between germline sequence and receptor specificity means that these recognition specificities are selected by evolutionary pressure. The limited germ line defined repertoire can easily evolve to not recognise accessible self antigens, allowing robust self/non-self discrimination by the innate immune system.

The use of randomly coded clonally distributed receptors by the adaptive immune system has the major advantage of an open repertoire, capable of recognising, in principle, *any* target molecule (antigen). This allows for recognition of threats not previously encountered, and for memory – the ability to respond rapidly and effectively on second contact with antigen. Because of the diversity of the adaptive repertoire, the frequency of cells with receptors specific for a particular new antigen is low, requiring the expansion of these antigen specific cells before an effective specific response can be mounted. This means that adaptive responses are less rapidly mobilised than innate responses. The randomly encoded repertoire also presents a major challenge for self/non-self discrimination. A randomly generated antibody or T cell receptor does not 'know' what it is reacting with, because the specificity of recognition is not predetermined by the genes generating the receptor. The adaptive immune system therefore has to have a major investment in mechanisms to control potentially damaging self reactivity.

1.3 Controlling Self-reactivity

Initially, mechanisms that remove potentially damaging self reactive adaptive immune cells during their development were described, including positive and negative selection of developing T cells in the thymus. This ensures that emerging naïve T cells interact with self MHC molecules (positive selection), but lack strong reactivity with self-peptide/MHC complexes (negative selection) . Issues such as tissue specific self molecules, and molecules only expressed at specific developmental stages (such as puberty) were seen as problematic for this mechanism, suggesting that the purging of the self-reactive specificities from the repertoire must be very incomplete. This and other factors stimulated the development of alternative views of self/non-self discrimination (see below), and further discussions in the chapter by Cohn and the chapter by Andrews and Timmis in this book. However, recent studies of gene expression in the thymus have reinforced the view that repertoire selection is a critical component of the control of self-reactivity [Kyewski & Derbinski 2004]. Thus tissue specific and developmental stage specific antigens are expressed 'promiscuously' in the thymus, and mutations or allelic variations which result in altered levels of expression of these antigens are associated with an increased risk of autoimmunity [Kyewski & Derbinski 2004]. It is sobering to think that the immune system has such a detailed knowledge of present and future self.

A second level of control of the adaptive response involves a stringent activation threshold for naïve T cells; once this threshold has been achieved, a programme of activation occurs, generating memory and effector T cells. The requirements for activating T cells at each of these stages is illustrated schematically in Figure 1.1. The initial triggering of a naïve T cell requires the simultaneous triggering of the T cell receptor (often termed signal 1) and receptors responsive to ligands generated by the innate immune response; these signals unrelated to antigen specificity are termed signal 2. This requirement for signal 2 (also called costimulatory signal) for activation is also found to a lesser extent with early and intermediate memory cells, so that for example, increasing the costimulatory stimulus recruits more helper cells into cytokine production [Waldrop et al. 1998]. On the right of Figure 1.1, fully differentiated effector cells are exquisitely sensitive to the expression of their antigen, being able to respond to 1 or 2 molecules on a target cell [George et al. 2005]; they do not require costimulation for activation. Another important characteristic of late effector cells is that they have limited proliferative potential, and are susceptible to apoptosis, or programmed cell death. These characteristics are important for the involution of the specific adaptive response once an infection has been cleared.

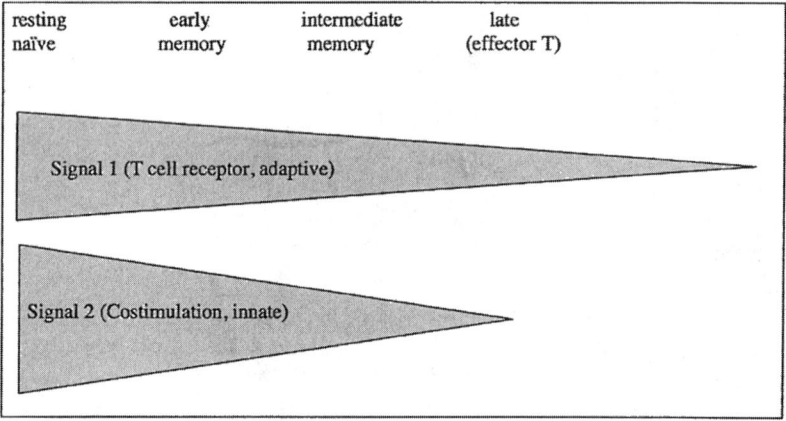

Fig. 1.1. Requirements for activation at different stages of T cell maturation.

The level of costimulation is controlled by signalling via innate immune receptors (see below). A possible representation of this relationship for naïve

T cell activation is shown in Figure 1.2, making the point that a threshold level of signal 1 is required for a response, and that below this threshold immunological ignorance occurs, and costimulatory signals have no effect. The precise shape of the response envelope is not known, but there is an inverse relationship between signal 1 and signal 2 requirements above the signal 1 threshold.

A third level of control of adaptive immunity is termed peripheral tolerance, which may involve rendering T cells refractory to stimulation by their cognate antigen (anergy), or the development of immunosuppressive regulatory T cells. Anergy may arise when the T cell receptor is activated in the absence of costimulation as illustrated in figure 1.2; these conditions may render T cells unresponsive to an antigenic stimulus to which they would previously have responded. Under some conditions, T cell receptor stimulation without costimulation may induce the death of the T cell, and/or induce regulatory T cells which actively control T cell responses. This points to a critical aspect of the innate/adaptive interaction: in addition to the respond/don't respond decision, the innate response controls the type and magnitude of the adaptive response. This polarisation or differentiation of the adaptive response is discussed in more detail below.

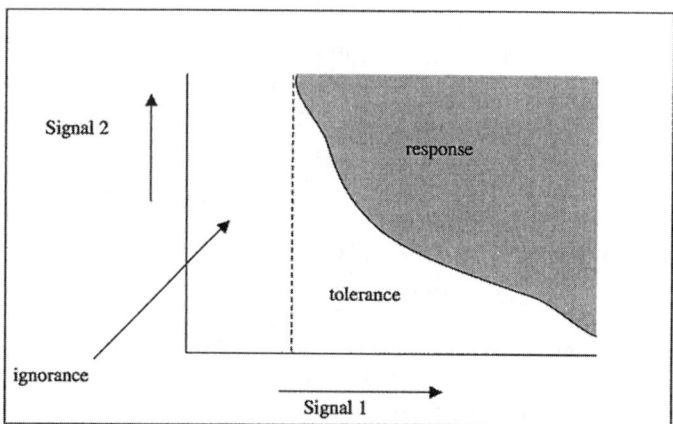

Fig. 1.2. Integration of innate and adaptive signals for stimulation of naïve T cell responses. Signal 1 is via the T cell receptor (adaptive) and signal 2 is costimulation, induced by innate receptors.

1.4 Infection and Danger

The importance of innate immunity in the control of adaptive immune responses was highlighted by Janeway, describing adjuvants required to make foreign proteins stimulate a strong immune response as '...immunologists' dirty little secret'. He suggested that the immune system has evolved to distinguish infectious non-self from non-infectious self [Janeway 1992]. This is achieved through the recognition of 'pathogen associated molecular patterns' (PAMPs) by pattern recognition receptors (PRRs) of the innate immune system. These pattern recognition receptors respond to classes of molecular structures that are found on pathogens, but not normal mammalian cells. Typically these pathogen associated molecules are shared by whole classed of organisms, for example lipopolysaccharide (lps) found on gram negative bacteria.

A different slant on this concept was developed by Matzinger, who also highlighted the role of the innate immune system in controlling adaptive responses, but suggested that innate immune receptors evolved to detect damage to self (Danger) rather than recognition of pathogens *per se* [Matzinger 1994b]. In this view, pathogens are seen as utilising receptors for endogenous danger signals, much in the way that morphine evolved in the opium poppy to utilise the receptor for endogenous ligands, the endorphins. Examples of danger signals might be intracellular components such as DNA and RNA which are released when cells undergo accidental (necrotic) cell death, and it is clear that necrotic cells induce an inflammatory response conducive to the development of adaptive immune responses [Gallucci *et al.* 1999].

1.5 Compartments and Anatomy

The immune system has sophisticated anatomical and cellular mechanisms to allow the effective exposure of the naïve repertoire of receptors, clonally distributed on rare cells, to invading pathogens. The specialised location for this interaction is the lymph nodes, which are connected to the tissues of the body by a drainage system (the afferent lymphatics). After passing through the lymph node, fluid and cells from the tissues rejoins the blood circulation via the efferent lymphatics and thoracic duct. Naïve lymphocytes are only able to leave the blood circulation in the lymph node, so they continuously revisit lymph nodes by recirculating via the bloodstream. Memory cells can extravasate into the tissues, allowing them to seek out and respond to invading pathogens. The lymph node is the place where adaptive immune responses are initiated: it is a crossroads between the naïve repertoire of lymphocytes continuously recirculating between lymph nodes and blood, and material from pathogenic invaders arriving from the tissues via the afferent lymphatics. There is also a subset of early memory cells which can recirculate via the lymph node. These early memory cells have a high proliferative potential, and will be the source of rapid expansion of specific adaptive T cell response on reexposure to a pathogen.

1.6 Dendritic cells

In this anatomical setting, the dendritic cell is the key cellular link between innate and adaptive responses [Rossi & Young 2005]. This innate immune cell has elegantly developed functionality to suit it for this role. In the tissues, dendritic cells sample their environment continuously, using a variety of mechanisms. These include macropinocytosis, in which surrounding medium is ingested, and receptor mediated uptake, where for example the mannose receptor recognises abnormal carbohydrate structures. On encountering a pathogen, the dendritic cell will respond to the PAMPs it bears in a generic way, triggering an activation/maturation response. Pathogen molecules ingested will be digested into short peptide fragments, which are transported to the cell surface in the groove of MHC molecules. As part of the triggering response, dendritic cells will leave the tissues and migrate via the afferent lymphatics to the lymph node, where they can present pathogen antigens to the naïve lymphocytes. Recent studies indicate that this process is made more efficient by changes to the blood vessels and lymphatics of lymph nodes induced by innate immune responses [Soderberg *et al.* 2005].

In addition to the antigen presentation function (generating signal 1), another crucial aspect of dendritic cell function is integration of innate signals that will control the extent and nature of the adaptive response. This is achieved by the upregulation of cell surface signalling (costimulatory) molecules and production of soluble signalling molecules (cytokines) in response to triggering of their PRRs by pathogen derives molecules. It is becoming clearer that the type of immune response generated (eg antibody, cytotoxic T cell, regulatory T cell) is controlled by innate signals integrated by the dendritic cells and translated into the appropriate costimulatory and cytokine response [Pulendran 2005]. A fascinating recent observation suggests that the magnitude of the response generated may also be controlled by synergy of signalling via different PRRs [Napolitani *et al.* 2005].

1.7 CD4 T cells

1.7.1 Helper T cells

A key means by which the nature of the adaptive response is controlled is by the function of the helper T cells induced following stimulation of naïve CD4 T cells by DC. This was originally described in terms of two subsets: T helper 1 (Th1) that secrete interferon, and help cellular immunity against intracellular invaders; and T helper 2 (Th2) that synthesize interleukin-4 and help antibody responses, particularly IgE responses against parasites [Mosmann & Coffman 1989]. Interestingly, Th1 cytokines inhibit Th2 differentiation, and vice versa, meaning that once directed by the innate signals from the initiation of the response, the type of the adaptive response tends to be locked in. The Th1/Th2 model has proved a valuable paradigm, although it has become apparent that additional CD4 functional subsets exist, and their activation determined by innate responses. Regulatory T cells were first added to the list [Kapsenberg 2003], and more recently, a subset producing interleukin 17 (IL-17), that have been dubbed Th17 [McKenzie *et al.* 2006]. These relationships are illustrated in Figure 1.3.

1.7.2 Regulatory T cells

Regulation of T cell responses by other T cells has recently become an intense area of study [Jiang & Chess 2006], after many years in the wilderness. Originally named suppressor T cells, the cellular and molecular basis of their function became unclear, and the field was largely discredited [Moller 1988]. Under the banner of regulatory T cells (Treg), much detailed analysis of the mechanisms responsible for the suppressive activity has been undertaken, and their potential importance in the control of immunologically mediated diseases recognised [Jiang & Chess 2006]. Regulatory T cells have been difficult to identify unequivocally, as markers used to identify them, such as CD25 (a constituent of the receptor for IL-2), are also expressed by activated T cells. However, recently the transcription factor Foxp3 has been shown to be an important feature of regulatory T cells [Fontenot & Rudensky 2005]. The detailed mechanisms by which regulatory T cells control adaptive responses are still under investigation, but key suppressive cytokines IL-10 and transforming growth factor beta (TGFβ) are involved.

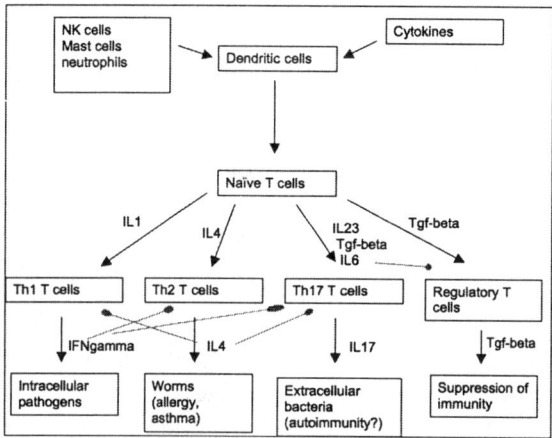

Fig. 1.3. Integration of innate immunity and variegation of the adaptive response by dendritic cells. Some activating (black arrows) and inhibitory (red lines) activities are shown.

1.7.3 Th17 cells

Th17 cells are the most recent to be added to the interacting subsets of helper cells induced by innate responses and controlling adaptive responses; they may not be the last, however[Tato et al. 2006]. Th17 cells were originally related to Th1 cells because of their proinflammatory properties, and because of the close relationship between IL23 (which is involved in Th17 differentiation) and IL12 (which induces Th1 differentiation). IL12 and IL23 share the same p40 subunit, which is partnered by p35 in the case of IL12, and p19 in the case of IL23. An alternative cytokine combination, TGFβ + IL6, has be shown to be a powerful inducer of Th17 cells. This illustrates the potential of individual cytokines to have contrasting effects, depending on other cytokines that may be present. Thus in this case, TGFβ is a suppressive cytokine, but in the presence of IL6, is key to induction of the proinflammatory Th17 subset of helper cells. IL6 is itself a multifunctional cytokine, being involved in the involution of innate (inflammatory) responses, and the activation of adaptive responses [Jones 2005].

1.8 Natural Killer (NK) cells

As illustrated in figure 1.3, cellular interactions are also important in influencing the activities of dendritic cells in controlling the adaptive response.

Of particular interest are Natural Killer (NK) cells, which have a recognition mechanism that is complementary to that of T cells, and also have important interactions with dendritic cells [Degli-Eposti & Smyth 2005], influencing the outcome of the adaptive immune response.

NK cells are a subset of lymphoid cells found mainly in the blood and spleen which were initially recognised functionally by their ability to kill cancer cells *in vitro*. They are a distinct morphological set of large lymphocytes, with granular cytoplasm. The means by which tumour cells are recognised and killed by NK cells remained mysterious for many years, until the elegant insight of Karre and colleagues, who realised that NK cells recognised the *absence* of self molecules [Karre *et al.* 1986]. The self molecules involved are the MHC molecules that present peptide antigens to T cells, thus covering an escape route for pathogens avoiding recognition by preventing MHC molecules bearing their antigens getting to the cell surface. Negative recognition is achieved by inhibitory receptors: an NK cell interacting with a self cell that has normal expression of MHC molecules will not kill that cell because the self MHC triggers the inhibitory receptors on the NK cell. It is now known that there is a range of inhibitory receptors interacting with different specific MHC molecules, which requires a 'licensing' mechanism for maturing NK cells to acquire full function [Kim *et al.* 2005].

There is even a receptor that monitors overall MHC expression by an ingenious strategy which uses a peptide sequence in the 'tail' of most MHC molecules. This peptide fits into the groove of an invariant MHC molecule, HLA-E, and has to be present for HLA-E to get transported to the cell surface. Any interference with the synthesis of MHC molecules reduces the presence of the tail peptide, reducing HLA-E expression, and thus reducing the inhibitory signal that HLA-E delivers to its own inhibitory receptor [Orange *et al.* 2002].

It is also now clear that NK cells have activating as well as inhibitory receptors. The activating receptors recognise molecules upregulated by stressed cells, perhaps an example of the detection of 'danger'. The recognition structures involved are again related to classical MHC molecules that present peptides to T cells, but as with HLA-E, they have limited variability, and can be considered an essential link between innate and adaptive immunity [Rodgers & Cook 2002]. As indicated above, using these receptors as an innate detectors of damage, activated NK cells are another cellular link between innate and adaptive immunity, producing cytokines and having powerful effects on dendritic cell function, influencing the outcome of adaptive immune response [Degli-Eposti & Smyth 2005].

1.9 Innate immunity and antibody responses

The role of innate immunity in activating appropriate helper T cells is a major controlling mechanism for antibody responses, but other aspects of innate immunity can also have a profound effect. A good example is the complement system, which was first identified as an effector mechanism activated by antibodies (the classical pathway of activation). The complement system is a complex cascade of enzymatically activated molecules, which culminates in the assembly of membrane attack complexes, which effectively punch holes in the membranes of target cells. Complement can also be activated by purely innate mechanisms, by the alternate pathway, and the lectin pathway. Complement activation by these routes can potentiate the adaptive immune system, as has been elegantly demonstrated by the construction of synthetic antigen-C3d complexes [Dempsey *et al.* 1996]. The dose of antigen required to induce an antibody response was reduced by four orders of magnitude when 3 copies of C3d were linked to the antigen.

1.10 Summary

Elements of the innate and adaptive immune system exemplified here demonstrate the advantages of integrating pre-programmed recognition (rapid response using widely distributed receptors recognising generic targets) with random repertoire recognition (open repertoire for specific recognition of novel target, with memory). The randomly generated repertoire brings problems of self/non-self discrimination, which is solved at multiple levels in the human immune system, including shaping of the naïve repertoire, stringent control of naïve cell activation by innate immune receptor recognition, and control by regulatory cells in the periphery. The interactions between innate and adaptive immunity are many and complex, with innate mechanisms being instrumental in the induction and differentiation of the adaptive response, as well as being recruited in the effector phase of adaptive responses. The challenge is now to abstract the essential components of the innate-adaptive interaction in order to utilise this concept in alternative contexts.

Immunoinformatics and Computational Vaccinology: A Brief Introduction

Paul D Taylor and Darren R Flower

The Jenner Institute, University of Oxford, Compton, Berkshire, RG20 7NN, UK
Darren.flower@jenner.ac.uk

Summary. Immunoinformatics has recently emerged as a buoyant and dynamic sub-discipline within the wider field of bioinformatics. Immunoinformatics is the application of bioinformatic methods to the unique problems of immunology and vaccinology. Immunoinformatics, as a principal component of incipient immunomic technologies, is beginning to foment important changes within immunology, as this key discipline tries to free itself from the empirical straight jacket that has characterised its development and attempts to grapple with the post-genomic revolution. Immunoinformatics is, importantly, also beginning to establish itself as a pivotal tool within vaccine discovery.

2.1 Introduction

Have you ever had a bout of the common cold? Do you suffer from Hay Fever or a nut allergy or Asthma? Have you ever had a more serious infectious disease? Are you a victim of a chronic autoimmune disease or even cancer? Now answer a seemingly distinct and unrelated question: have you ever used a computer? If you answered yes to either group of questions, have you ever thought of combining the two? The use of computers to fight infectious disease and other acute and chronic disease states may seem far fetched to many, but computers have long been used in the design of small molecule drugs, and now they are beginning to impact on the design and discovery of immunotherapeutics and prophylactic vaccines. We see this dramatic synergy made manifest through the discipline of immunoinformatics, a profound and exciting new computational science able to greatly accelerate the speed and effectiveness of vaccine and immunotherapeutic discovery.

The domain of infectious disease - allergy, in all its forms; autoimmune disease, such as rheumatoid arthritis; and even cancer - is the domain of Immunology. Immunology is, amongst other many other things, the study of how the body defends itself against infection, and from the standpoint of human disease, a proper appreciation of innate and adaptive immunity is crucial, as the immune system has evolved, at least in part, to combat infectious disease, the greatest source of preventable human mortality and morbidity. Immunology is thus a very broad branch of the biosciences which has led, directly or indirectly, to many pivotal advances in modern bio-medicine. Moreover, our knowledge of the molecular and cellular mechanisms which underlie immunity has also allowed for the development of new clinical and non-clinical technologies, which have an equally broad range of applications. While much of its focus remains strongly anthrocentric, or, at least, centred on the adaptive immune system of vertebrates, its societal importance can not be gainsaid, as it deals with the physiological function of the immune system in both health and disease; the malfunctioning of immunity in immunological disorders (autoimmune diseases, allograph rejection, hypersensitivities, and immune deficiency); and the in vitro, in vivo, and in situ, chemical and physiological properties of immunological components of the immune system.

However, immunology, and all its attendant disciplines, now find themselves at a turning point, whether or not practitioners realize it. After a hundred years of empirical research, immunology is increasingly poised to reinvent itself as a quantitative, genome-based science. Like most bioscience disciplines, immunology is increasingly facing the challenge of capitalizing on a potentially overwhelming cascade of new information delivered by high-throughput, post-genomic technologies. This data is both bamboozlingly complex and on a scale which has never been encountered before. It is also clear, at least to some, that such high throughput approaches will engender a paradigm-shift from traditional hypothesis-driven research to a new data-driven, information-focused approach, with new understanding emerging from the analysis of complex, intricate, multifaceted datasets.

In response to pressures such as these, there has been much recent interest in the development and deployment of informatics tools, which can analyze the data that arises from immunological research of all kinds. In turn, this has lead to the growth of two flavours of computational support for immunology. The first is straightforward bioinformatics support, technically indistinguishable from support given to other areas of biology, and includes genome annotation of both the human genome and diverse microbial species. For example, well in excess of 150 bacterial genomes have now been sequenced, and hundreds more are nearing completion [Paine & Flower 2002]. Another area of growth is immunotranscriptomics, or immunologically targeted Microarray analysis [Walker et al. 2002].

The other kind of support is more focussed: immunoinformatics. This is an exciting and dynamic specialism, which has emerged in recent years within the wider world of bioinformatics. It addresses the particular problems which arise within immunology, including the accurate prediction of immunogenicity, be that manifest as the identification of epitopes or the prediction of whole protein immunogens; this endeavour stands as the principal short- to medium-term goal of immunoinformatic investigation. The theoretical or mathematical modeling of the immune systems seeks, as a discipline, to address what some are wont to call *important scientific questions*: how might immunity work? What is the nature of host-pathogen interactions? Work of this sort is described in various Chapters elsewhere in this book: Chapter 4 by Lee and Perelson, which describes computational modeling of the function of B cell and T cell receptors; Chapter 13 by Denise Kirschner, which describes multi-scale modeling of the immune system in response to pathogens; and Chapter 17 by Melvin Cohn, which describes the self/non-self paradox. Immunoinformatics, on the other hand, is concerned with prosaic, nitty-gritty, nuts-and-bolts issues: is this particular amino acid sub-sequence of a protein an epitope? Is this protein within a viral genome more antigenic than another? Can we identify common virulence factors in the genomes of a distinct phylogenetic grouping of bacteria? It is with questions of this sort that immunoinformatics concerns itself; it is a discipline which rolls up its sleeves and gets on with the job.

Like bioinformatics, immunoinformatics is grounded in computer science. Increasingly, however, immunoinformatics integrates a whole array of cross-disciplinary techniques from physical biochemistry and biophysics; computational, medicinal, and analytical chemistry; structural biology and protein homology modeling, as well as many other branches of biological, physical, and computational science. Traditionally, it has emphasized problem solving and focused on data classification into discrete sets rather than predicting continuous, quantitative data, leading to the use of *black-box* neural networks for prediction and to databases such as SYFPEITHI [Rammensee *et al.* 1999]. Increasingly, however, approaches are turning towards more quantitative models, familiar from decades of QSAR analysis of drug molecules, which predict continuous binding measures. This approach is more overtly physico-chemical in nature, with a greater implicit emphasis on the explanation of underlying atomistic molecular mechanisms. These different points of view are highly complementary. Remaining conflicts between these differing perspectives are easily reconciled by methods from Drug Design. Such methods meet both objectives: seeking to explain and understand without sacrificing efficiency or loosing sight of the pragmatic and utilitarian purpose of the undertaking.

It is perhaps a cliché, or at least a truism, to say that the immune system is complex, complicated, and hierarchical, exhibiting considerable emergent behaviour at every level from subcellular to organismal. Yet, for all that, this aphorism retains an essential veracity. If it were not true, then the book you are reading, *in silico immunology,* would not need to be written. The complexity of the immune systems is confounding, and, though many might wish to deny it, our ignorance of it remains profound. Yet, at the heart of the immune system lie straightforward molecular recognition events: the coming together of two or more molecules to form stable complexes of measurable duration. In terms of atomistic interactions, these events are indistinguishable from the binding phenomena experienced by any macromolecule. The binding of an epitope to a major histocompatibility complex protein (MHC), or T Cell Receptor (TCR) to a peptide-MHC complex is, thus, in terms of underlying physico-chemical phenomena, identical in nature to any other molecular interaction in any other area of bioscience. It is only at higher levels - when tens, or thousands, or millions of different molecules come together – that immune systems exhibit, in time and space, complex and confusing emergent behavior. In seeking to understand immunology and address its problems, immunoinformatics can exploit the observation that the immune system is based on simple, understandable molecular events, and, within the biological context of the subject, it does seek, in the broadest sense, to model such phenomena. Much of the rest of this chapter will explore how.

2.2 Immunogenicity: A Brief Primer

Immunogenicity is that property of a molecular, or supramolecular, moiety that allows it to induce a significant response from the immune system. Here a molecular moiety may be a protein, lipid, carbohydrate, or some combination thereof. A supramolecular moiety may be a virus, bacteria or protozoan parasite. An immunogen - a moiety exhibiting immunogenicity - is a substance which can elicit a specific immune response, while an antigen - a moiety exhibiting antigenicity - is a substance recognized, in a recall response, by the extant machinery of the adaptive immune response, such as T cells or antibodies. Thus, antigenicity is the capacity, exhibited by an antigen, for recognition by one or several parts of the antibody or TCR immune repertoire. Immunogenicity, on the other hand, is the ability of an immunogen to induce a specific immune response when it is exposed to initial surveillance by the immune system. These two properties are clearly coupled but properly understanding how they are inter-related is by no means facile.

Predicting actual antigenicity and/or immunogenicity of a complex protein remains problematic. It depends simultaneously upon the context in which it is presented and also the nature of the immune repertoire that recognizes it. Either or both of these components may be critical. For example, the immune response in many immunogens or antigens is focused to a handful of immunodominant structures, while much of the rest of the molecule may be unable to engender a response. Mutating an antigen may eliminate, reduce, or even enhance its inherent immunogenicity, or, of course, it might move it to other regions of the molecule. In seeking to assess immunogenicity, we must consider properties of the host and the pathogenic organism of origin, and not just the intrinsic properties of the antigen itself. The composition of the available immune repertoire will affect its response to a given epitope and alter its recognition of a particular target. When mounting a response *in vivo*, those elements of an immune repertoire capable of participating, in a given response, might have been deleted through their cross-reactivity with host antigens. Moreover, fundamental restrictions on the antibody repertoire, imposed by the limited number of V genes that encode the antigen-binding site of the antibody, may also limit responses. Overall, it is clear that antigenicity and immunogenicity have many interlinked causes. The induction of immune responses requires critical interaction between parts of the innate immune system, which respond rapidly and in a relatively nonspecific manner, and other, more specific components of the adaptive immune system, which can recognize individual epitopes.

Immunogenicity is currently the most important and interesting property for analysis and prediction by immunoinformatics. Immunogenicity can manifest itself through both arms of the adaptive immune response: humoral (mediated through the binding of whole protein antigens by antibodies) and cellular immunology (mediated by the recognition of proteolytically cleaved peptides by T cells). Humoral Immunogenicity, as mediated by soluble or membrane-bound cell surface antibodies, can be measured in several ways. Methods such as enzyme-linked immunosorbent assay (ELISA) or competitive inhibition assays yield values for the Antibody Titre, the concentration at which the ability of antibodies in the blood to bind an antigen has reached its half maximal value. One can also measure directly the affinity of antibody and antigen, using, for example, equilibrium dialysis. Likewise, measurements of cellular immunity through T cell responses have become legion. For class I presentation, arguably the most direct approach is to measure T cell killing. Cytotoxic T lymphocytes or CTL, can induce lysis in target cells. This can be measured using a radioisotope of chromium, which is taken into target cells and released during CTL lysis. For class II presentation, the proliferative response of CD4+ T cells, which acts indirectly by activating B cells or macrophages, can be measured using the incorporation of tritiated thymidine into T cell DNA during cell division. Alternatively, Enzyme-Linked ImmunoSpot, or ELISpot, assays measure production of cytokines or other molecules by class I and /or class II T-cells when exposed to antigen. More recently attention has migrated towards tetramers as tools for detecting T cell responses [Doherty et al. 2000]. MHC:peptide tetramers are formed from four biotinylated peptide-MHC complexes (pMHCs) bound to tetrameric avidin or streptavidin. These tetramers bind to TCRs with a proportionately higher affinity allowing antigen-T cell interactions to be assessed with greatly enhanced specificity.

Much of immunogenicity is determined by the presence of epitopes, the principal chemical moieties recognized by the immune system. Consequently, the accurate prediction of B cell and T cell epitopes is the pivotal challenge for immunoinformatics. Despite a growing appreciation of the role played by non-peptide epitopes, such as carbohydrates and lipids, nonetheless peptidic B cell and T cell epitopes (as mediated by the humoral or cellular immune systems respectively) remain the principal tools by which the intricacy of immune responses can be surveyed and manipulated, since it is the recognition of epitopes by T cells, B cells, and soluble antibodies that lies at the heart of the adaptive immune response. Such initial responses lead, in turn, to the activation of the cellular and humoral immune systems and, ultimately, to the effective destruction of pathogenic organisms.

While the prediction of B cell epitopes remains primitive and largely unsuccessful [Blythe & Flower 2005], a multitude of sophisticated, and successful, methods for the prediction of T cell epitopes have been developed [Flower et al. 2002]. These began with early motif methods [Doytchinova et al. 2004c], and have grown to exploit both qualitative and semi-quantitative approaches, typified by neural network classification methods, and a variety of more quantitative techniques [Doytchinova & Flower 2002c]. Most modern methods for T cell epitope prediction rely on predicting the affinity of peptides binding to MHCs. The T cell, a specialized type of immune cell mediating cellular immunity, constantly patrols the body seeking out foreign proteins derived from microbial pathogens. T cells express a particular receptor: the T cell receptor or TCR, which exhibits a wide range of selectivities and affinities. TCRs bind MHCs, which are presented on the surfaces of other cells. These proteins in turn bind small peptide fragments (epitopes) originating from both endogenous, or self, and exogenous, including pathogen-derived, protein sources. It is, as we have said, the recognition of such complexes that lies at the heart of both the adaptive, and memory, cellular immune response.

2.2.1 T cell and Antibody Repertoire

Until recently, the immune system was thought to discriminate rigidly between "self" and "non-self". This discrimination was believed to form the basis of protection of the host against pathogens. Such views are changing as studies indicate that determinants of self do not always induce absolute immune tolerance in the host. Under certain conditions peptides from self-antigens can be processed and displayed by MHC as targets for immune surveillance. This provides a rationale for the investigation of, say, self-epitopes as mediators of autoimmunity, or epitopes from cancer antigens as targets for immunotherapy or targeting epitopes from proteins which induce allergic reactions.

As we have said, MHCs bind peptides. These are themselves derived through the degradation, by protelotytic enzymes, of foreign and self proteins. Foreign epitopes originate from benign or pathogenic microbes, such as viruses and bacteria. Self epitopes originate from host proteins that find their way into the degradation pathway as part of the cell's intrinsic quality control procedures. The proteolytic pathway by which peptides become available to MHCs is very complex and many, many important details and molecular components remain to be elucidated. Yet, it is the complexity and degeneracy of the T cell presentation pathway that allows peptides with diverse post-translational modifications, such as phosphorylation or glycosylation, to form pMHCs, and thus, ultimately, to be recognized by TCRs. Moreover, MHCs are very catholic in terms of the molecules they bind and are not restricted to peptides. Chemically modified peptides and peptidomimetics are also bound by MHCs. It is also well known that many drug-like molecules bind to MHCs [Pichler 2002].

As we shall see below, the overall presentation process is long, complicated, and involves many subsidiary steps. There are several alternative processing pathways, but the principal ones seem linked to the two major types of MHC: Class I and Class II. Class I MHCs are expressed by almost all nucleated cells in the body. They are recognized by T cells whose surfaces are rich in CD8 co-receptor protein. Class II MHCs are only really expressed on so-called *professional antigen presenting cells* and are recognized by T cells whose surfaces are rich in CD4 co-receptors. MHCs are polymorphic. Generally, most humans have six classic MHCs: 3 Class I (HLA-A, B, and C) and 3 class II (HLA-DR, DP, and DQ), these proteins will have different sequences, or different HLA alleles, in different individuals. Different MHC alleles, both class I and Class II, have different peptide specificities. A simple way to look at this phenomenon is to say that MHCs bind peptides which exhibit certain particular sequence patterns and not others. Within the human population there are an enormous number of different, variant genes coding for MHC proteins, each of which exhibits discernibly different peptide-binding sequence selectivities. T cell receptors, in their turn, also exhibit different and typically weaker affinities for different peptide-MHC complexes. The combination of MHC and TCR selectivities thus determines the power of peptide recognition in the immune system and thus the recognition of foreign proteins and pathogens. This will be discussed more thoroughly in accompanying chapters. Whatever dyed-in-the-wool immunologists may say, such interactions form the quintessential nucleus of immune recognition, and thus the principal point of intervention by immunotherapeutics.

2.3 Epitopes and Epitology

The word *epitope* is widely used amongst biological scientists. Etymologically speaking, its roots are Greek, and, like most words, its meanings are diverse and in a state of constant flux. It is most often used to refer to any region of a biomacromolecule which is recognized, or bound, by another biomacromolecule. For an immunologist, the meaning is more restricted and refers to particular structures recognized by the immune system in particular ways. The region on a macromolecule, which undertakes the recognition of an epitope, is called a paratope. In terms of the physical chemistry of binding, then we need think only of equal partners in a binding reaction. B cell epitopes are regions of a protein recognized by Antibody molecules. T cell epitopes are short peptides which are bound by major histocompatibility complexes (MHC) and subsequently recognized by T cells.

A B cell epitope is a region of a protein, or other biomacromolecule, recognized by soluble or membrane-bound Antibodies. B-cell epitopes are classified as either linear or discontinuous epitopes. Linear epitopes comprise a single continuous stretch of amino acids within a protein sequence, while an epitope whose residues are distantly separated in the sequence and are brought into physical proximity by protein folding is called a discontinuous epitope. Although most epitopes are, in all likelihood, discontinuous, experimental epitope detection has focused on linear epitopes. Linear epitopes are believed to be able to elicit antibodies that can subsequently cross-react with its parent protein.

A T cell epitope is a short peptide bound, in turn, by MHC and TCR, to form a ternary complex. The formation of such a complex is the primary, but not sole, molecular recognition event in the activation of T cells. Many other co-receptors and accessory molecules, in addition to CD4 and CD8 molecules, are also involved in T cell recognition. The recognition process is not simple and remains poorly understood. However, it has emerged that the process involves the creation of the immunological synapse, a highly organised, spatio-temporal arrangement of receptors and accessory molecules of many types. The involvement of these accessory molecules, although essential, is not properly understood, at least from a quantitative perspective. Ultimately, the accurate modelling of all these complex processes will be required to gain full and complete insight into the process of epitope presentation.

We will explore how epitopes arise rather more fully below. The peptides presented by class I and class II MHCs differ, principally in terms of their length. Class I peptides are primarily derived from intracellular proteins, such as viruses. These proteins are targeted to the proteasome, which cuts them into short peptides. Subsidiary enzymes also cleave these peptides, producing a range of peptide lengths, of which the distribution used to be believed to fall neatly into the range: 8 to 11 amino acids. More recently, however, this has been shown that much longer peptides, currently up to 15 amino acids, can also be bound by MHCs and recognized by TCRs [Probst-Kepper *et al.* 2001]. For Class II, the receptor mediated intake of extracellular protein derived from a pathogen is targeted to an endosomal compartment, where such proteins are cleaved by cathepsins, a particular class of protease, to produce peptides which are typically somewhat longer than Class I. These, again, exhibit a considerable distribution of lengths, centred on a range of 15-20 amino acids. However, longer and shorter peptides can also be presented, via Class II MHCs, to immune surveillance. Peptide cleavage specificity exhibited by Cathepsins has also been investigated and some insight has been gained into cleavage motifs[Chapman 1998]. However, considerably more work is required before truly efficient predictive methods can be realized.

It is now generally accepted that only peptides that bind to MHC at an affinity above a certain threshold will act as T cell epitopes and that, to some extent at least, peptide affinity for the MHC correlates with T cell response. This particular issue is somewhat complicated and obscured by hearsay and dogma: as with many questions important to immunoinformatics, the key, systematic studies remain to be done. The behaviour of heteroclitic peptides, where synthetic enhancements to binding affinity are often reflected in enhanced T cell reactivity, seems compelling evidence of the relationship between affinity and immunogenicity. However, and whatever people may say, affinity of binding is an important component of recognition and thus of the overall process leading to the generation of an immune response. Not the only, or, necessarily, the most important part, but a key part nonetheless. Its importance is debated, particularly by people critical of the immunoinformatic endeavour. Nevertheless, its utility in this context is clear. Experimental immunologists and vaccinologists are constantly using nascent immunoinformatic approaches to select, filter, or prune lists of candidate peptides in order to identify functional epitopes.

Epitopes, whether B cell or T cell, are, as we have mentioned several times above, short continuous or discontinuous sequences or strings of amino acids. These may be of different length and exist in different contexts, but they remain sequences. As such they can be stored in functional immunological databases, much as whole sequences are stored in GenBank or Swiss-Prot. As we shall see in the next section, there are many of such resources, most available via the Internet.

2.4 Databases

For some time, the database has been the *lingua franca* or, more prosaically, the common language of bioinformatics. The creation, and the manipulation of databases, which contain biologically relevant information, is the most critical feature of contemporary bioinformatics. The same is true of immunoinformatics. This is manifest through its support for post-genomic bioscience and as a discipline in its own right. Functional data, as housed in databases, will rapidly become the principal currency in the dynamic information economy of 21st Century immunology. Having said all that, there is nothing particularly new about immunological databases, at least in the sense that they do no more than apply standard data warehousing techniques in an immunological context. Nonetheless, the continuing development of an expanding variety of immunoinformatic database systems indicates that the application of bioinformatics to immunology is beginning to broaden and mature.

Example databases, such as IMGT [Robinson *et al.* 2003] or Kabat [Wu & Kabat 1970], have made the sequence analysis of important immunological macromolecules their focus for many years [Brusic *et al.* 2002]. Functional, or epitope-orientated, databases are somewhat newer, but their provenance is now well established. Generally speaking, such databases record data on T cell epitopes or peptide-MHC binding affinity. Arguably, the highest quality database currently available is the HIV Molecular Immunology database [Korber *et al.* 2001b], which focuses on the sequence and the sequence variations of a single virus, albeit one of unique medical important. However, the scope of the database is, in terms of the kinds of data it archives, less restricted than others, containing information on both cellular immunology (T cell epitopes and MHC binding motifs) and humoral immunology (linear and conformational B cell epitopes).

An early, and widely used, database is SYFPEITHI [Rammensee *et al.* 1999], another high quality development, which contains an up-to-date and useful compendium of T cell epitopes. SYFPEITHI also contains much data on MHC peptide ligands, peptides isolated from cell surface MHC proteins *ex vivo*, but purposely excludes binding data on synthetic peptide. MHCPEP [Brusic *et al.* 1998], a now defunct database, pooled both T cell epitope and MHC binding data in a flat file, introducing a widely used conceptual simplification, which combines together the bewildering variety of binding measures, reclassifying peptides as either *Binders* or *Non-Binders*. Binders are further sub-divided as High-binders, Medium binders, and Low binders. More recently, Brusic and coworkers have developed a much more complex and sophisticated database: FIMM [Schonbach *et al.* 2005]. This system integrates a variety of data on MHC-peptide interactions: in addition to T cell epitopes and MHC-peptide binding data, it also archives a wide variety of other data, including sequence data on MHCs themselves together with data on the disease associations of particular MHC alleles .

More recently, related, yet distinct, databases have begun to emerge, each addressing data on different aspects of molecular immunology. Kangueane and coworkers have developed a database that focuses solely on X-ray crystal structures of MHC-peptide complexes [Govindarajan *et al.* 2003], while Middleton et al. describe the Allele

Frequency Database which lists population frequencies of particular MHC alleles [Middleton *et al.* 2003]. All these databases are available via the Internet.

AntiJen [Toseland *et al.* 2005], formerly known as JenPep[Blythe *et al.* 2002, McSparron *et al.* 2003], is a database developed recently, which brings together a variety of kinetic, thermodynamic, functional, and cellular data within immunobiology. While it retains a focus on both T cell and B cell epitopes, AntiJen is the first functional database in immunology to contain continuous quantitative binding data on a variety of immunological molecular interactions, rather than the kind of subjective classifications described above. Data archived includes thermodynamic and kinetic measures of peptide binding to TAP and MHC, peptide-MHC complex binding to T cell receptors, and general immunological protein-protein interactions, such as the interaction of co-receptors, interactions with superantigens, etc. Although the nature of the data within AntiJen sets it apart from other immunology databases, there is, nonetheless, considerable overlap between other systems and AntiJen.

Moreover, AntiJen shares characteristics with several other newly-emergent non-immunological databases: thermodynamic binding databases, such as BindingDB [Chen *et al.* 2002], and a variety of other databases of different sorts, of which BIND [Bader & Hogue 2000] and Brenda [Schomburg *et al.* 2002] are prime examples. Such databases, which contain experimental measured binding affinities, are a relatively recent development. The focus of these databases is rigorously measured thermodynamic properties derived from experimental protocols such as Isothermal Titration Calorimetry (ITC), which can return not only free energies of binding, but also equivalent enthalpies, entropies, and heat capacities. As these protocols are well standardized, databases, such as BindingDB, can easily record precise experimental conditions.

In the domain of immunological experiments, AntiJen records IC_{50}, BL_{50}, $t1/2$ measurements, etc. For each such measurement, it also archives standard experimental details, such as pH, temperature, the concentration range over which the experiment was conducted, the sequence and concentration of the reference radiolabeled peptide competed against, together with their standard deviations. As it is rare to find a paper which records all such data in a reliable way, thus standardization remains a significant issue. It is also unclear how much more data remains to be collated.

Server Name	URL	
BIMAS	http://bimas.dcrt.nih.gov/molbio/hla_bind/	bind MHC class I ligands
EpiGenomix	http://epigenomix.com/pls/hiv/!www_user.front.end	
HLA-DR4 binding	http://www-dcs.nci.nih.gov/branches/surgery/sbprog.html	
LpPep	http://reiner.bu.edu/zhiping/lppep.html	
MHCPred	http://www.jenner.ac.uk/MHCPred	
MHC-THREAD	http://www.csd.abdn.ac.uk/~gjlk/MHC-Thread/	
NetMHC	http://www.cbs.dtu.dk/services/NetMHC/	
PREDEP	http://bioinfo.md.huji.ac.il/marg/Teppred/mhc-bind/	
PREDICT	http://sdmc.krdl.org.sg:8080/predict/	
ProPred	http://www.imtech.res.in/raghava/propred/	
RankPep	http://www.mifoundation.org/Tools/rankpep.html	
SVMHC	http://www.sbc.su.se/svmhc/	
SYFPEITHI	http://syfpeithi.bmi-heidelberg.com/Scripts/MHCServer.dll/EpiPredict.htm	
BIMAS	http://bimas.dcrt.nih.gov/molbio/hla_	MHC class I and proteasome cleavage
MAPPP	http://www.mpiib-berlin.mpg.de/MAPPP	Proteasome cleavage
NetChop	http://www.cbs.dtu.dk	HLA-A2 and H-2Kk
NetMHC	http://www.cbs.dtu.dk/services/NetMHC	Proteasome cleavage
PAProC	http://www.paproc.de	HLA-DR
ProPred	http://www.imtech.res.in/raghava/propred	MHC class I ligands
ProPred1	http://www.imtech.res.in/raghava/propred1	MHC class I and II ligands
SYFPEITHI	http://www.syfpeithi.de	MHC class I & II ligands
RANKPEP	http://www.mifoundation.org/Tools/rankpep.html	MHC class I ligands
SVMHC	http://www.sbc.su.se/svmhc	MHC class I ligands
Lib Score	http://www.ddbj.nig.ac.jp/analysesp-e.html	

Table 2.1. URLs of Prediction Servers

2.5 Immunoinformatic Datamining

A useful simplification of biological computation is to split methods between the areas of datamining and simulation. In truth, of course, there is a continuous spectrum of techniques stretching from one extreme to the other. Within immunology, a key example of data mining is the identification of peptide binding motifs, which seeks to characterize the peptide specificity of different MHC alleles in terms of dominant anchor positions with a strong preference for certain amino acids [Sette et al. 1989]. Such motifs are undoubtedly popular amongst immunologists, as they are simple to understand and just as simple to implement either visually or computationally. For example, human Class I allele HLA-A*0201, probably the best-studied allele, has anchor residues at peptide positions P2 and P9. At P2, acceptable amino acids would be L and M, and V and L at position P9. Secondary anchors, which are residues that are favourable, but not essential, for binding, may also be present. A seemingly uncountable number of papers have, over the past 15 years or so, successfully extended this to include the specificity patterns of many other alleles, both human and animal. However, despite this success, there are many fundamental problems with the motif approach.

The most significant of problem with motifs is that they are deterministic: a peptide either is, or is not, a binder. A brief reading of the literature shows that motif matches produce many false positives, and probably also produces an equal number of false negatives, although such negative results are seldom screened. Thus being motif-positive, as the jargon can put it, is neither necessary not sufficient for affinity for an MHC. Although it is clear that so-called primary anchors do often dominate binding, it is well known, that binding motifs, as descriptions of the process, are fundamentally flawed. Not hopeless, not useless, but partial and incomplete. In the sense that motifs are widely used and widely understood, they are indeed most useful, but as accurate predictors of binding they leave much to be desired. Such shortcomings have led many to seek other data mining solutions to the peptide-MHC affinity problem.

The development of data driven predictive methods in immunoinformatics is now two decades old. Early methods attempted to predict epitopes directly, and, in the absence of knowledge of the peptide preferences of MHC restriction, enjoyed limited success [Deavin et al. 1996, Flower 2003]. As described in chapter 8, several groups have used techniques from artificial intelligence research, such as artificial neural networks (ANNs) and hidden Markov models (HMMs), to tackle the problem of predicting peptide-MHC affinity [Brusic & Flower 2004]. ANNs and HMMs, are, for slightly different applications, particular favourites among bioinformaticians when looking for tools to build predictive models. However, the development of ANNs is often complicated by their preponderance for problems of interpretation, and also for overtraining and over-fitting. Of course, many other methods - indeed, in all probability, all methods - suffer similar or equivalent problems. Indeed, over-fitting is the curse of all data driven methods. Support vector machines are currently flavour-of-the-month. Whether this method, or indeed any other AIS-based approach, will ever escape the traps which have caught-out other techniques remains to be seen. A number of prediction servers are available over the web. See table 2.1.

In the prediction of MHC-binding, the main issues are the quality, quantity, ability to represent available data, the complexity of the selected predictive model relative to the natural complexity of the peptide-MHC interaction, and the training and testing of the predictive model. A good quality data set is critical to the creation of an accurate prediction system. Available data contains significant biases, as peptides are often pre-selected for experimental testing using binding motifs. Data is often intrinsically poor and requires data cleaning. Data quantity also has important implications for the selection of appropriate prediction methods. Guidelines have been given based on a recent comparative study of algorithm performance [Brusic & Flower 2004, Yu *et al.* 2002] and were suggested in the context of Artificial Intelligence techniques, which have well defined data requirements:

1. If there is no binding data at all, then speculative molecular modeling is the only option. Here, supertype analysis, as described later, can be useful.
2. When the number of available peptides is below 50, binding motifs are the most pragmatic solution.
3. With 50-100 peptides, quantitative matrices or SVMs can be used.
4. With data sets comprising over 100 peptides, HMMs or ANNs can be used.
5. With very large data sets, only really available for HLA-A*0201, ANNs can provide high specificity predictions, albeit at the price of slightly lowered sensitivity.

Our own QSAR methods have slightly different data requirements. For more information on these approaches see Chapter 8. The minimum set size is about 20 peptides, though models only begin to gain statistical significance at 40 peptides and above. When sets reach 200 or above, then it becomes possible to introduce reliable cross-terms: 1-2 and 1-3 side chain-side chain interactions in our case.

However, as we have explained above, it is not just quantity, but data diversity, that is an issue. As diversity in peptide sequence and binding affinity increases, so does the predictivity and generality of the models. Highly degenerate data or data with a very narrow affinity range often prove difficult. Predictive models should be tested before use, using internal cross-validation and the splitting of data into training and test sets.

2.6 Modelling T cell Mediated Antigen Presentation and Recognition

One of the most challenging problems in modern computational vaccinology is the effective modeling of the cellular presentation of antigenic epitopes. Professional antigen-presenting cells (APCs), such as dendritic cells or macrophages, endocytose and process protein antigens into peptides, which are subsequently presented on the cell surface associated with MHC molecules. This presentation can then result in the stimulation of cytotoxic or helper T cells. Conceptually, the phenomenon of

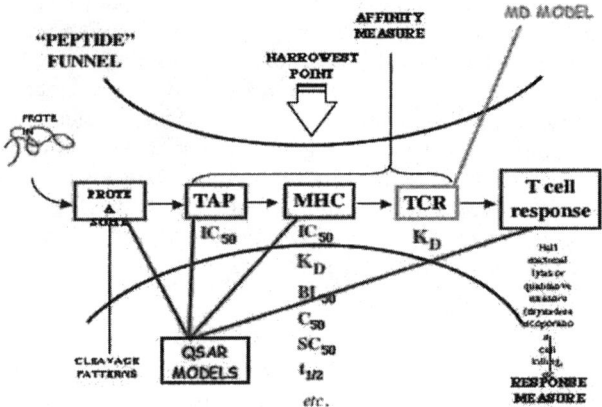

Fig. 2.1. The 'Simple' Class I Processing Pathway: A schematic showing a simplified view of class I antigen presentation. Peptides are generated initially from whole proteins via cleavage in the proteasome, followed by transport, into the endoplasmic reticulum (ER), by the transporter associated with processing (TAP). ERAAP trims peptides prior to binding by MHC molecules. MHCs are transported to the cell surface where they are recognized by T cells. The kind of measurable quantities, such as affinities or cleavage patterns, available for each step on the pathway are shown. This process approximates to a funnel with the principal bottleneck being binding by the MHC. The kind of model we have worked on (QSAR or MD model) is also indicated.

antigen presentation can be divided into three mechanistic stages. Firstly, antigen uptake: the recognition of antigen proteins by cell surface-receptors and the subsequent internalization of soluble, extracellular antigens. Secondly, antigen processing: intracellular enzymatic degradation and transport of endocytosed and cytoplasmic endogenous and exogenous proteins followed by peptide loading of MHC molecules. Thirdly, the exocytosis of MHC complexes, containing self and exogenous and endogenous antigenic peptides, and CD1, which presents potentially antigenic lipids. Put at its simplest, fragments of extracellular proteins are presented by class II MHCs and fragments of intracellular proteins are presented by class I MHCs; so-called cross-presentation refers to the presentation of extracellular antigens in the context of class I MHCs and vice-versa.

Much of what we adumbrate below will focus on class I antigen presentation; a somewhat simplified description of the many subsidiary steps involved in class I presentation is shown in Figure 2.1. Significant advances have been made recently in the modeling of class I presentation, particularly the prediction of proteasomal cleavage patterns and peptide binding to TAP. Together studies on proteasomal cleavage and TAP transport represent a good first attempt to produce useful predictive tools for the processing aspect of Class I restricted epitope presentation.

Cytosolic proteins, after labeling with ubiquitin, are transported to the proteasome, a multimeric protease responsible for most protein digestion within the cytosol, where they are cleaved into short peptides, typically 15 or fewer amino acids in length. Several methods have been development for predicting semi-stochastic proteasomal protein cleavage [Brusic & Flower 2004]. All perform statistical analysis of digested fragments from a small set of proteins, principally enolase-1, and augmented this sparse data-set with signals apparent in the termini of peptides eluted from cell surface MHCs. This developed the work of [Altuvia & Margalit 2000], who showed that the termini of peptides eluted from cell surface MHCs exhibit distinct sequence motifs at the C, but not the N, terminus, consistent with peptides undergoing N terminal trimming by other proteases subsequent to digestion by the proteasome. Several of these methods are available via the Internet. The predictive power shown by different prediction methods is only beginning to be evaluated objectively. [Saxova et al. 2003] evaluated three publicly available methods for proteasomal cleavage prediction, and found that the best method gave an accuracy value of 70% at the C-termini. Clearly, considerable progress is still required.

Peptides generated by the proteasome are subsequently bound by the transmembrane peptide transporter TAP, which translocates them from the cytoplasm to the endoplasmic reticulum (ER). In the ER peptides are bound by MHCs. A number of studies have been conducted into the peptide substrate specificities exhibited by the TAP transporter [Doytchinova et al. 2004a], leading to the development of several predictive models for the determination of peptides that bind to TAP. Most identify strong sequence patterns at the C-termini. This feature, also present in proteasome cleavage patterns, is consistent with a role for ERAAP in N-terminal *trimming* of peptides within the lumen of the endoplasmic reticulum.

So far, so good: a reasonably straightforward and uncomplicated linear pathway has been modeled with some success. However, there are, in reality, many other processing components and, indeed, whole presentation pathways, which greatly complicate the simple picture sketched out above. The growing complexity of antigen presentation is best exemplified by the class I processing pathway. See Figure 2.2. As well as the proteasome, peptides are cleaved by other cytosolic proteases, such as Tripeptidyl peptidase II (TPPII), currently the only well characterized protolytic enzyme known to be involved in presenting epitopes, although it is most probable that many others are involved. Peptides cleaved by the proteasome or TPPII are degraded by cytosolic proteases such as LAP and TOP. Peptides transported into the ER by TAP then bind to MHCs. This process is catalysed by a variety of chaperones, including Tapasin, calnexin, and ERp57. Peptides in the ER pool are trimmed by ERAAP and other proteases, such as L-RAP. Other anterograde and retrograde routes operate between the cytosol and the ER, by which means protein fragments can access different proteases, including puromycin resistant aminopeptidease.

At the other end of the process, extracellular proteins undergo antigen capture mediated by receptor-mediated endocytosis, entering the class I pathway through mechanisms of cross-presentation. The exact nature and number of such receptors remains obscure. Accurate modeling of this process is complicated by the observation

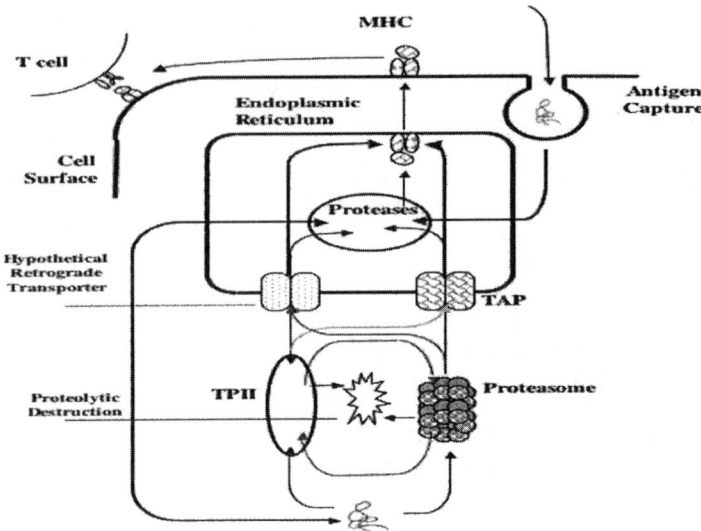

Fig. 2.2. Complex Antigen Presentation Pathway: A more realistic schematic of the class I antigen presentation pathway compared to Figure 2.1. This incorporates several proteases as well as different mechanisms of cross-presentation. Peptides are created or bound at indicated stage. The route taken by peptides is shown by arrows. Two points are of note: first, the synergistic interaction of TPPII and proteasome in the proteolytic creation of peptides and second the retrograde access to the cytosol.

that all cellular compartments or pools, whether conceptual or membrane bound, are leaky.

So far, not so good; at least from the view-point of someone trying to model the process. The accurate prediction of antigen processing and presentation depends on a proper understanding of the molecular mechanisms underlying the overall pathway. In order to develop a general model of cell surface epitope presentation, each of these steps would require its own predictive model for both the thermodynamics of peptide specificity and substrate-dependent peptide kinetics. The process requires decomposition into a set of peptide cleavage and peptide binding steps, each of which would then be open to modeling. This would not, in itself, account for the complexities of antigen presentation. Rather, we will need to supplement individual bioinformatic models with well understood mathematical models, such as those prototyped on reaction kinetics within multifurcating, multi-enzyme pathways: so called "metabolic control theory" [Fell 2005], which can account for substrate flux within multi-step, multi-component metabolic pathways, and allow for the ready incorporation of quantitative aspects of individual bioinformatic models. An effective model of this type, however hard to realize in practice, would, in all probability, better help us to understand why certain peptides come to dominate presentation: the apparently intractable problem of epitope immunodominance.

The presentation of peptides by the MHC is often viewed as the most discriminatory step of the presentation process. However, peptide recognition by the T cell is also vitally important. If we define recognition as the interaction of TCR and pMHC, then many complex subsequent steps are involved in the actual activation of T cells. Recognition is not an isolated event and the context in which an antigen is encountered by a T cell will determine if TCR engagement leads to full activation or tolerance. In the presence of costimulation, antigen presentation by an activated APC will lead to full activation. However, antigen presentation by a resting APC will lead to tolerance. Moreover, molecular recognition of toll-like receptor ligands by other receptors also has a key role in activating APCs and promoting the activation of T cells.

2.7 Immunoinformatics and Systems Biology

At the level of the antigen, and more specifically the protein antigen, immunogenicity is contingent upon properties of the molecule itself, as well as properties of both the host and the pathogen, be that microbe or cancer cell. It is, therefore, a collective property of the entire system of interacting cells and organisms. The response of the host is mediated by the recognition of T cell and B cell epitopes, as well as the recognition of more mechanistically-generic danger signals. The level of expression of the antigen and its subcellular location within the pathogenic organism are also potentially key arbiters of immunogenicity. The argument runs thus: a poorly expressed, under-represented protein in an inaccessible compartment of a microbial cell is unlikely to be an important antigen, however potent its individual epitopes may be. How such antigens interact with components of the presentation pathway is also important: both viral and bacterial proteins are known to interfere with processes of antigen presentation: some down regulate MHC production, for example, others interfere with peptide transport.

Immunonomics is a newly coined term which subsumes both the theoretical and experimental study of immunology and related disciplines in a post-genomic context. We have already described how the complex process of antigen presentation and subsequent T Cell recognition is beginning to be modeled. Such attempts, while noteworthy, are still floundering due to lack of relevant data. There is still an obvious need for experiments which directly support the development of useful and accurate *in silico* models. Immunoinformaticians need quality data to work from; existing data is seldom satisfactory. Informaticians can no longer exist solely on the crumbs dropped from the experimentalist's table. Instead, there is a clear and palpable requirement for experiments specifically addressing the kind of predictions that immunoinformaticans need to make. Antigen Presentation is being addressed by experiment as well as through the development of theoretical methods. Such experiments are, typically, still operating at the phenomenological level: describing the phenomenon but not dissecting it.

Another aspect of immunogenicity prediction, focuses on system properties of individual gene products from pathogenic micro-organisms. These seek to predict post-translational modifications, subcellular localization, and expression levels. Together these appear to be important factors in the identification of potential antigens. Cell surface proteins, or ones secreted into the extracellular milieu, are more directly open to surveillance by the immune system. Poorly expressed genes are unlikely to be potent antigens because, again, they will not be seen by the immune system. The presence of post-translational modifications can often act as danger signals or are potent immunomodulators either as components of T cell epitopes or through their binding to other receptors. The identification of pathogen proteins which are highly expressed and/or found or outside the pathogen cell and/or contain post-translational modifications is a highly complementary approach to the detection of epitopes, which can be used to select potential antigens with or without knowledge of T cell or B cell responses. An alternative to this approach is to attempt the direct identification of antigens without reference to any mechanistic detail. Here one might endeavor to discriminate between sets of known antigens and sets of known non-antigens or random sets of proteins. While conceptually simple and straightforward, this approach is untested: at present, neither large sets of antigens nor appropriate descriptors are forthcoming.

The effective prediction of protein expression levels in a pathogenic microbe is a potentially important indicator of putative immunogenicity. However, there are inherent difficulties in both the process of prediction itself and in even knowing what an appropriate expression level is. Clearly, under certain conditions, such as starvation, patterns of expression will change dramatically, being up-regulated or down-regulated significantly. Generally, we can assume that the successful surveillance of a microbial protein by the immune system will be linked, in part at least, to its presence in sufficient quantities. There are many ways to predict expression levels but the best studied is codon usage [Karlin & Mrazek 2000]. Different organisms display different codon biases: the preference for one codon rather than another when coding for amino acids. Moreover, there is also a correlation between the choice of which codon is used and the level and rate at which a protein is expressed. The ability to predict different expression levels under different conditions is difficult and requires at least a partial understanding of the whole hierarchy of immune regulation: transcription factors and their binding sites, operons, promotors, mulitcomponent regulatory networks, etc. To address this pivotal challenge will require the combined ingenuity and imagination of experimentalists, theorists, immunoinformaticans, computer scientists, and mathematicians.

Another important aspect of the prediction of immunogenicity is the accurate identification of Post-Translational Modifications (PTMs). These can take many forms and many potentially contribute to the molecular basis of immunogenicity. PTMs can include glycosylation and lipidation. Glycosylated proteins can be targets for binding by cell surface receptors based on sugar binding leptin domains. Glycosylated epitopes can also be bound by TCRs and antibodies. Lipids can act as epitopes directly through their presentation by CD1. PTMs can also be transitory, such as phosphorylation, or more permanent, such as modified amino acids. Many of these can be part of functional epitopes recognized by the immune system. Glycosylation of a protein, for example, is dependent on the presence of sequence patterns or motifs (Ser/Thr-X-Asn for N-linked glycosylation and Ser/Thr for O-linked) but this is not enough to correctly predict them. If these motifs are present at solvent inaccessible regions of a protein rather on the surface then they will not be glycosylated. More-over, the other residues which surround these patterns will also affect the specificity of the glyscosylating enzymes: Pro as the X in the Ser/Thr-X-Asn motif for N-linked glycosylation will essentially prevent glycosylation. Glycosylation, in particular, is also very dependent on context, and it is thus a system property of an organism, and can vary considerably in terms of the nature and extent of the different sugars that can become attached to proteins, at least in eukaryotic systems.

Arguably, the most useful, and thus the best studied, of what we might broadly term system approaches to the identification of immunogens, has been the prediction of subcellular location. There are two basic types of prediction method: the manual construction of rules based on knowledge of what determines subcellular location and the application of data-driven machine learning methods, which automatically identify factors determining subcellular location by discriminating between proteins from different known location. Accuracy differs markedly between different methods and different compartments, due to a paucity of data or the inherent complexity that determines protein location. Such methods are often classified according to the input data required and how the prediction rules are constructed. Input data which is used to discriminate between compartments include: the amino acid composition of the whole protein; sequence derived features of the protein, such as hydrophobic regions; the presence of certain specific motifs; or a combination thereof. Phyloge-netic profiles can also be used to predict protein location, as the location of close protein homologues can be assumed to be similar.

Signal complexity is a more complicated problem. A very complex signal will require considerable data so that one might be confident in the model. A simple signal, on the other hand, may prove difficult because many, otherwise unrelated, proteins may posses a sorting signal which appears similar, but only by chance. For example, the SWISS-PROT sequence database contains about twice as many non-perioxisomal proteins with a PTS1 sorting signal than real perioxisome-located proteins.

Another challenge is the difference in locations evinced by different organisms. PSORT, a knowledge-based, multi-category program for the prediction of subcellular location, and often regarded as the gold standard for such predictions, is composed of several different programs. Of special interest in this context is PSORT-B, which generates predictions for subcellular location in bacteria. It reports precision values of 96.5% and recall values of 74.8%. PSORT-B is a multi-category method which makes use of six algorithms: SCL-BLAST, which uses protein homology to identify location; PROSITE, which detects motifs; HMMTOP, which predicts membrane proteins; outer membrane β-barrel proteins are identified using specific sequence patterns; SubLocC, is an SVM that uses protein amino acid composition to assign a cytoplasmic or non-cytoplasmic location; and a Hidden Markov Model trained to identify signal peptide cleavage sites. The results of these 6 methods are combined using a Bayesian Network.

Another well known method of interest here is SignalP, which is based on neural networks and predicts N-terminal Spase-I-cleaved secretion signal sequences and their cleavage site. The signal predicted is the type-II signal peptide common to both eukaryotic and prokaryotic organisms, for which there is wealth of data, in terms of both quality and quantity. A recent enhancement of SignalP is a Hidden Markov Model version which is also able to discriminate uncleaved signal anchors from cleaved signal peptides. One of the limitations of SignalP is over-prediction, as it is unable to discriminate between several very similar signal sequences, regularly predicting membrane proteins and lipoproteins as type-II signals. Many other kinds of signal sequence exist. A number of methods have been developed to predict lipoproteins, for example. The prediction of proteins that are translocated via the TAT-dependent pathway is also important but has not been addressed in any depth.

2.8 Immunoinformatics and Vaccinology

Vaccines are molecular entities which can, in effect, mimic infectious organisms so that such microbes can later be recognised and destroyed by the human body or other host, without harm to itself, during subsequent infection. Based on sound, experimental data, immunoinformaticians are using statistical and artificial intelligence methods to identify computationally antigenic proteins and epitopes from pathogenic micro-organisms – bacteria, virus, parasites, or fungi – which the immune system can then recognize, tagging these invading microbes for eventual destruction. However, in order to realise the burgeoning power of these advances still requires much effort.

Vaccines can provide both therapeutic and prophylatic treatments of autoimmune diseases, allergy, and cancer, as well as infectious disease. In light of the many perceived threats to human health, views about infectious disease, in particular, are altering rapidly, leading to a radical reappraisal of the role of vaccines in the fight against pathogenic micro-organisms. Immunovaccinology is the name given to a rational form of vaccinology based very firmly upon our increasing understanding of the fundamental mechanisms which underpin immunology. It must also exploit the potential power of post-genomic technologies. Humanity has sought to address infection through the systematic use of biological and chemical entities: small molecule drugs and supramolecular vaccines. It is now generally accepted that mass vaccination, taking account, as it does, of the principal of herd immunity, is amongst the most effective prophylactic approaches to the treatment, or rather, pretreatment, of infectious disease.

The discovery of vaccination is generally attributed to Edward Jenner, who noted that milkmaids, who had contracted cowpox, a virus related to smallpox, seemed immune to the disease. On 14th May 1796, he used the fluid from a cowpox pustule to build protective immunity against smallpox in James Phipps, an 8 year old boy. Jenner then infected him with smallpox. The boy did not become ill. Later, *Vaccination* - the word Jenner had invented for his treatment (from the Latin *vacca*, a cow) – was adopted for immunization against any disease. In 1980, the World Health Organisation was able to announce the total eradication of smallpox through worldwide vaccination.

However, vaccination has, until relatively recently, been a highly empirical science, relying on poorly understood, non-mechanistic approaches to the development of new vaccines. As a consequence of this, relatively few effective vaccines were developed, and deployed, during most of the two centuries that have elapsed since Jenner's work in the closing years of the Eighteenth century. In the post war years, when antibiotics were king, the threat posed by serious infectious disease, at least in First World countries, seemed to all but vanish, as vaccines and antimicrobial drugs combined to almost eliminate it. The present era is characterized by worries over a variety of burgeoning threats to human well-being: bio-terrorism, climate change, antibiotic resistance, etc. These changes have led, amongst other things, to the re-emergence of diseases such as TB, and exotic emergent diseases, e.g. SARS or avian flu.

A vaccine is a molecular, or super-molecular, agent which elicits specific, protective immunity against pathogenic microbes and the diseases they cause. Protective immunity is an enhanced adaptive immune response to re-infection, as potentiated by immune memory, which, ultimately, mitigates the effect of subsequent infection. Historically, vaccines have been attenuated whole pathogen vaccines such as Sabin's Polio vaccine or BCG for TB. Recently, safety concerns have led to the development of other strategies, focusing separately on subunit/antigen and epitope vaccines (see Figure 2.3). Hepatitis B vaccine is an example of an antigen - or subunit - vaccine, and many epitope-based vaccines have now entered clinical trials. Nevertheless, despite much effort, both publicly and commercially funded, efficacious vaccines are not yet available for many major pathogens such as Shigella, H. pylori, or Meningococcus B.

WHOLE ORGANISM   **SUBUNIT VACCINE**   **EPITOPE VACCINE**

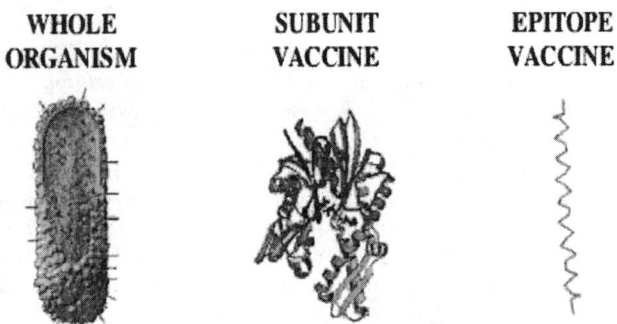

Fig. 2.3. Types of Vaccine: The three main kinds of vaccine component: whole organism attenuated pathogen; subunit whole protein vaccine; polyepitope vaccine. These three are the principal kind of core components of modern vaccines. Epitopes can, potentially, also be carbohydrate or lipid based or a mixture. Modern vaccines are delivered in a variety of ways, such as DNA or as part of a viral vector. Vaccines are also often delivered with adjuvants, molecules which can exacerbate an initial immune response.

Ultimately, the utilitarian value of epitope and immunogenicity prediction will need to be demonstrated through their usefulness in experimental vaccine discovery programmes. All of the methods, we have adduced, focus primarily on the discovery of T cell epitopes, which can prove useful, amongst other things, as diagnostic markers of microbial infection and as the potential basis of epitope vaccines. Many workers have, in recent years, used computational methods as part of their strategy for the identification of both Class I and Class II restricted T cell epitopes. However, it is certainly encouraging that many experimental immunologists are now beginning to see the need for informatics techniques. Computer-based data and knowledge management is essential if this data deluge is not to overwhelm the post-genomic vaccinologist.

There is a clear need to produce more accurate prediction algorithms, which cover more Class I and Class II alleles in more species. Yet, for these improved methodologies to be ultimately effective, i.e. that they are taken up and used routinely by experimental immunologists, these methods must also be tested rigorously for a sufficiently large number of peptides that their accuracy can be shown to work to statistical significance. To do this requires more than new algorithms and software, it requires the confidence of experimentalists to exploit the methodology and to commit laboratory experimentation. Yet most of these tools remain daunting for laboratory-based immunologists. The use of these methods should be routine. It is not only a matter of training and education. These methods must, ultimately, be made more accessible and robust.

2.9 Discussion

From a societal standpoint, immunology is rightly viewed as an important - even a paramount - science. Immunologists are sometimes regarded as a discipline apart. Immunology has a high standing in the wider scientific community: its journals have high impact factors, and it is a large and, generally speaking, a well funded discipline. Immunology is intimately connected with disease: infectious, most obviously, but also autoimmune disease, inherited and multi-factorial genetic disease, cancer, and allergy. Yet, for all its prestige, immunology finds itself at a pivotal point in its history. After more than a century of empirical research, it is on the brink of reinventing itself as a post-genomic science. How will it cope? One obvious way is through embracing computational science.

Immunoinformatics is an amalgam of many different disciplines. Operationally, it has grown from bioinformatics and much of immunoinformatics is ostensibly the application of standard bioinformatic techniques, such as MicroArray analysis or comparative genomics, to the context of immunology. There are, however, several areas which are unique to immunology. Amongst these, the accurate prediction of immunogenicity, be that manifest as the identification of epitopes or the prediction of whole protein antigenicity. It can be fairly described as both the high frontier of immunoinformatic investigation and a grand scientific challenge: it is difficult, yet exciting, and, as a central tool in the drive to develop improved vaccines and diagnostics, is also of true practical value. It requires not only an understanding of immunology but also the integration of many other disciplines, both experimental (physical biochemistry, cell biology, *etc.*) and theoretical (computer science, *etc.*).

We have discussed several distinct areas of immunoinformatic research, yet, there are many others, such as predicting B cell epitopes and adjuvant discovery among them. Immunoinformatics is changing quickly, with many groups trying to improve databases and algorithms. However, despite the steady increase in studies reporting the real-world use of prediction algorithms, there is still an on-going need for truly convincing validations of the underlying approach. Why should this be? As we have seen, predicting T-cell epitopes remains a daunting challenge. We still need to understand the underlying cell biology and model accurately the complexities of the class I and class II antigen presentation pathway. We also still need to understand and accurately model the underlying physical chemistry, in terms of both thermodynamics and kinetics, of peptide binding to MHCs and of TCRs binding to pMHCs.

We have come to a turning point, where a number of technologies have obtained the necessary level of maturity: post-genomic strategies on the one hand and predictive computational methods on the other. Progress will occur in two ways. One will involve closer connections between immunoinformaticians and experimentalists seeking to discover new vaccines. In such a situation, work would progress through a cyclical process of using and refining models and experiments, at each stage moving closer towards a common goal of effective, cost-efficient vaccine development. The other way is the devolved model, where methods are made accessible and used remotely via the web and the GRID.

However, when deprived of direct collaboration, there is still a clear and obvious need for experimental work to be conducted in support of the development of accurate *in silico* methods. Recent work from our laboratory shows the way. Peptides, as reported in literature binding experiments and epitope identification exercises, have heavily biased sequence compositions, resulting from a process of pre-selection which leads to spiraling self-reinforcement. Since only part of a given selection will bind, this rapidly converges to a very limited, and thus incomplete, model of binding dominated by the selection criteria used. These problems would be resolved by a properly designed training set. We have addressed this experimentally, beginning by correlating 90 literature peptide IC_{50}s with cell surface BL_{50} measurements [Doytchinova *et al.* 2004c]. Using models derived from these values, we predicted super-binders with pico-molar affinities much greater than reported values. Using analogues of super binders with modified anchor positions, we then evaluated the relative dominance of anchor positions in a fully systematic manner. Our ability to combine *in vitro* and *in silico* analysis allows us to improve both the scope and power of our predictions in a way that would be impossible using only data from the literature. To ensure we produce useful, quality *in silico* models, rather than worthless and unusable methods, we need to value the predictions generated by immunoinformatics for themselves and conduct experiments appropriately.

The innate and inherent complexity of the immune system is confounding at all levels. Nevertheless, the work of many skilled immunoinformaticains has attempted and nonetheless clearly succeeded in producing useful, if doubtless imperfect, models with true utilitarian value. Progress is, and will continue, being made. We should feel confident that the great synergy arising within this discipline will be of true benefit to Immunology, leading to clear improvements in vaccine candidates, diagnostics, and laboratory reagents. Methods able to predict immunogenicity accurately will become landmark tools for the immunologist and vaccinologist working in the world of tomorrow.

Acknowledgements

We should like to thank the following for the help. Firstly, members of the bioinformatics group at the Jenner Institute, University of Oxford, both past and present. Of these, we should like to single out Dr Irini Doytchinova for special mention, as her contribution to the group has been without equal. The others include, in no particular order: Valerie Walshe, Kelly Paine, Christopher P Toseland, Shelley L Hemslay, Christianna Zygouri, Dr Pingping Guan, Dr Helen McSparron, Dr Matt Davies, and Dr Channa Hattotuwagama, Dr Shunzhou Wan, Martin Blythe, and Debra Clayton (nee Taylor). We should also like to thank the following for their help and stimulating discussions: Prof Peter Beverley, Prof Vladimir Brusic, Dr Persephone Borrow, Dr Shirley Ellis, Dr Kevin Rigley, Dr Simon Wong, Dr Helen Bodmer, Dr Sam Hou, Dr Elma Tchillian, and Dr Josef Walker. Finally, we should also like to thank all our other colleagues and co-workers at the Edward Jenner Institute for Vaccine Research and the Institute for Animal Health, Compton for their close, nurturing, and supportive collaboration.

A Beginners Guide to Artificial Immune Systems

Jon Timmis[1,2] and Paul Andrews[2]

[1] Department of Electronics, University of York, Heslington, York, UK.
[2] Department of Computer Science, University of York, Heslington, York, UK.
{jtimmis,psa}@cs.york.ac.uk

Summary. Artificial Immune Systems (AIS) have recently emerged as a computational intelligence approach that show great promise. Inspired by the complexity of the immune system, computer scientists and engineers have created systems that in some way mimic or capture certain computationally appealing properties of the immune system, with the aim of building more robust and adaptable solutions. In this chapter, we will explore the basics of AIS, charting their brief history, and outlining what type of immunology has served as inspiration. We will see that different immune processes and ideas have been captured within simple artificial systems, each with their own dynamics and application niches. As a final note, we then outline considerations that need to be borne in mind when building your own AIS.

3.1 Introduction

As we have seen in chapter 1 (and you will see in chapter 14), the immune system is a very complex system that undertakes a myriad of tasks. The abilities of the immune system have helped to inspire computer scientists to build systems that *mimic*, in some way, various properties of the immune system. We have already read in chapter 2, how people are using computers to help solve problems in immunology, but now we are concerned with the opposite: using immunology to help solve problems in computing.

Over the years, biology has provided a rich source of inspiration for many different scientists in many different domains, ranging from the design of aircraft wings to bulletproof vests. In computing, there has been an extensive amount of work undertaken on the use of biological metaphors, for example neural networks [Haykin 1999], swarm systems [Kennedy & Eberhart 2001], genetic algorithms [Holland 1975] and genetic programming [Banzhaf et al. 1998]. Recently, there has been increasing interest in using the natural immune system as a metaphor for computation in a variety of domains [de Castro & Timmis 2002a]. This field of research, Artificial Immune Systems (AIS), has seen the application of immune inspired algorithms to problems such as robotic control [Krohling et al. 2002], network intrusion detection [Forrest et al. 1997, Kim 2002], fault tolerance [Canham & Tyrrell 2002, Ayara 2005], bioinformatics [Cutello et al. 2004, Nicosia 2004] and machine learning [Kim & Bentley 2002a, Knight & Timmis 2003, Watkins et al. 2004], to name a few. To many, trying to mimic how the immune system operates in a computer may seem an unusual thing to do, why then would people in computing wish to do this? The answer is that, from a computational point of view, the immune system has many desirable properties that they would like their computer systems to possess. These properties are such things as robustness, adaptability, diversity, scalability, multiple interactions on a variety of timescales and so on. There is a real challenge in the world of computer science (and engineering) to build systems that can cope with increasingly complex problems, and are thus more scalable and robust (i.e. they break less!). Indeed, there is the notion of a *Grand Challenge* in computer science to try and do this very thing [Stepney et al. 2005a].

When working in the world of biologically inspired computing, a word of caution should be given. It is essential that metaphors are adopted carefully. Just because the immune system has the *desirable property* , it does not mean that it is necessarily suitable to solve your problem with, therefore, careful thought has to be given to the applicability of any technique [Freitas & Timmis 2003]. We will explore this later in the chapter. First, we will discuss the area of AIS, providing a history of its development, a summary of what AIS are, and a simple outline of the immunology that has been used to date to develop the basic principles underlying AIS. Chapter 7 later in the book, explores a different approach to the inspiration that may be explored in AIS, so no discussion on that matter will be given here. This is meant as an introductory chapter, so many technical details have been omitted, as has an extensive overview of current state of the art. However, the reader is encouraged to follow given references if interested in these technical details.

3.2 A Brief History of Artificial Immune Systems

3.2.1 The Humble Beginnings

The origins of AIS has its roots in the early theoretical immunology work of Farmer, Perelson and Varela [Farmer et al. 1986, Perelson 1989, Varela et al. 1988]. These works investigated a number of theoretical immune network models proposed to describe the maintenance of immune memory (as opposed to the more widely held view presented in Chapter 1). Whilst controversial from an immunological perspective, these models began to give rise to an interest from the computing community. The most influential people at crossing the divide between computing and immunology in the early days were Hugues Bersini and Stephanie Forrest. In the case of Bersini, after attending a talk by Francisco Varela in 1985, he made the decision there and then to begin working with Varela[3]. In the case of Forrest, she happened to be car sharing on the way to work with Alan Perelson, thus their working relationship began there. It is fair to say that some of the early work by Bersini [Bersini 1991, Bersini 1992] was very well rooted in immunology, and this is also true of the early work by Forrest [Forrest *et al.* 1994, Hightower *et al.* 1995]. It was these works that formed the basis of a solid foundation for the area of AIS . In the case of Bersini, he concentrated on the immune network theory, examining how the immune system maintained its memory and how one might build models and algorithms mimicing that property. With regards to Forrest, her work was focussed on computer security (in particular network intrusion detection) [Forrest *et al.* 1997, Hofmeyr & Forrest 2000] and formed the basis of a great deal of further research by the community on the application of immune inspired techniques to computer security.

3.2.2 Starting to Gain Pace

At about the same time as Forrest was undertaking her work, researchers in the UK started to investigate the nature of learning in the immune system and how that might by used to create *machine learning* algorithms [Cooke & Hunt 1995]. The term machine learning is used to cover a wide range of topics [Mitchell 1997], but essentially, machine learning techniques are computational methods applied to data in order to learn or discover something new about that data, or alternatively, to predict answers based on previous knowledge. The work of [Cooke & Hunt 1995] came about from the collaboration of Denise Cook, a biologist working at the University of Wales, Aberystwyth, with her husband John Hunt, a computer scientist working at the same institution. They had the idea that it might be possible to exploit mechanisms of the immune system (in particular the immune network) in learning systems, so they set about doing a proof of concept [Cooke & Hunt 1995]. Initial results were very encouraging, and they built on their success by applying the immune ideas to the classification of DNA sequences as either promoter or non-promoter classes, [Hunt & Cooke 1996] and the detection of potentially fraudulent mortgage applications [Hunt *et al.* 1998].

The work of Hunt and Cook spawned more work in the area of immune network based machine learning over the next few years, notably in [Timmis 2000] where the Hunt and Cook system was totally rewritten, simplified and applied to unsupervised learning (very similar to cluster analysis). Concurrently, similar work was carried out by [de Castro & Von Zuben 2002, de Castro & Von Zuben 2001], who developed algorithms for use in function optimisation and data clustering (the details of these are described in more details later in the chapter). The work of Timmis on machine learning spawned yet more work in the unsupervised learning domain, in trying to perform dynamic clustering (where the patterns in the input data move over time). This was met with some success in works such as [Wierzchon & Kuzelewska 2002, Neal 2002]. At the same time, using ideas other than the immune network

---

[3] Personal communication with Hugues Bersini

theory, work by [Hart & Ross 2002] used immune inspired associative memory ideas to track moving targets in databases.

In the supervised learning domain, very little happened until work by [Watkins 2001] (later augmented in [Watkins *et al.* 2004]) developed an immune based classifier known as AIRS. The system developed by Watkins was then adapted into a parallel and distributed learning system in [Watkins 2005], and has shown itself to be one of the real success stories of immune inspired learning [Goodman *et al.* 2003, Goodman *et al.* 2002, Watkins *et al.* 2003]. More information on AIRS, can be found in Chapter 15 of this book.

In addition to the work on machine learning, there has been plenty of other activity in AIS over the years. To outline all the applications of AIS and developments over the past 10 years would take a long time, and there are some good review papers in the literature, thus the reader is directed those [Dasgupta 1999, Timmis & Knight 2001, de Castro & Timmis 2002a, Garrett 2005]. In addition to these works, [Hart & Timmis 2005] investigated the application areas AIS have been applied to, and considered the contribution AIS have made to these areas. Their survey of AIS is not exhaustive, but attempts to produce a picture of the general areas to which they have been applied. Of the 97 papers reviewed, 12 categories were identified to reflect the natural groupings of the papers. These were, in the order of most papers first: clustering/classification, anomaly detection (e.g. detecting faults in engineering systems), computer security, numerical function optimisation, combinatoric optimisation (e.g. scheduling), learning, bio-informatics, image processing, robotics (e.g. control and navigation), adaptive control systems, virus detection and web mining. Hart and Timmis go on to note that these categories can be summarised into three general application areas of learning, anomaly detection and optimisation.

Due to a growing amount of work conducted on AIS, the International Conference on Artificial Immune Systems (ICARIS) conference series was started in 2002[4] and has operated in subsequent years [Timmis & Bentley 2002, Timmis *et al.* 2003, Nicosia *et al.* 2004, Jacob *et al.* 2005]. This is the best source of reference material to read in order to grasp the variety of application areas of AIS, and also the developments in algorithms and the more theoretical side of AIS. It would be impossible, and indeed this is not the place, to review all of these advances, so the reader is encouraged to pursue those references if they desire.

3.2.3 What Motivated Them?

The AIS work we have highlighted above was motivated by a variety of factors. However, if one reads the early AIS literature, there are clear reasons why people were attracted to the immune system in the first place:

- *Self-organisation.* The immune system does not appear to have a central controller telling immune agents (cells and molecules) what to do under different

---

[4] http:/www.artificial-immune-systems.org

circumstances. The observed behavior of the system is a result of many local interactions, giving rise to a complex, self organizing system. Computationally, algorithms that can self organize can be attractive in many situations, such as where outside control is not possible or desirable.

- *Learning.* B cells and their associated antibodies are, in effect, a record of the type of antigens the immune system has been exposed to. The primary and secondary immune responses of the adaptive immune system, in response to continual exposure to antigens, produce new cells and antibodies to combat the infection. This has lead to the development of systems that can learn from repeated exposure, based loosely on the primary and secondary immune responses.
- *Adaptation and diversity.* Some B cell clones undergo somatic hypermutation. This is an attempt by the immune system to develop a set of B cells and antibodies that cannot only remove the specific antigen, but also similar antigens. By using the idea of mutation a more diverse representation of the data being learnt is gained than a simple mapping of the data could achieve.
- *Classification.* The immune system is able to classify antigens into those that are self and non-self via the use of antigen receptors. Computationally, these receptors can act as detectors monitoring a system to determine when something anomolous (i.e. non-self) has occurred.

This very limited set of ideas, motivated the AIS practitioner to investigate the *computational properties* of the immune system, and to be honest, *use and abuse* the immune system as inspiration. We will now provide an overview of how the AIS practitioner has viewed the immune system in the past, and outline their attempts at building simple AIS. These AIS typically attempt to capture a tiny part of what the immune system has to offer – as you will see, it is very limited, but can be surprisingly successfully.

3.3 What is an Artificial Immune System?

AIS have been defined by [de Castro & Timmis 2002a] as:

> "adaptive systems, inspired by theoretical immunology and observed immune functions, principle and models, which are applied to problem solving"

In an attempt to create a common basis for AIS, work in [de Castro & Timmis 2002a] proposed the idea of a framework for engineering AIS. They argued the case for such a framework as the existance of similar frameworks in other biologically inspired approaches, such as artificial neural networks (ANN) and evolutionary algorithms (EAs), has helped considerably with the understanding and construction of such systems. For example, de Castro and Timmis [de Castro & Timmis 2002a] consider a set of artificial neurons, which can be arranged together to form an artificial neural network. In order to acquire knowledge, these neural networks undergo an adaptive process, known as learning or training, which alters (some of) the parameters within the network. Therefore, they argued that in a simplified form, a framework to design an ANN is composed of: a set of artificial neurons, a pattern of interconnection for these neurons, and a learning algorithm. Similarly, they argued that in evolutionary algorithms, there is a set of artificial chromosomes representing a population of individuals that iteratively suffer a process of reproduction, genetic variation, and selection. As a result of this process, a population of evolved artificial individuals arises. A framework, in this case, would correspond to the genetic representation of the individuals of the population, plus the procedures for reproduction, genetic variation, and selection. Therefore, they proposed that a framework to design a biologically inspired algorithm requires, at least, the following basic elements:

- A representation for the components of the system
- A set of mechanisms to evaluate the interaction of individuals with the environment and each other. The environment is usually simulated by a set of input stimuli, one or more fitness function(s), or other means
- Procedures of adaptation that govern the dynamics of the system, i.e., how its behavior varies over time

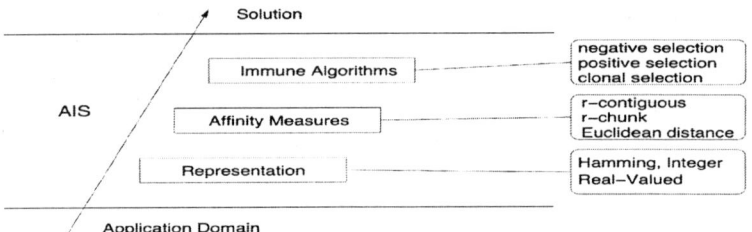

Fig. 3.1. AIS Layered Framework adapted from [de Castro & Timmis 2002a]

This framework can be thought of as a layered approach such as the specific framework for engineering AIS of [de Castro & Timmis 2002a] shown in figure 3.1. This framework follows the three basic elements for designing a biologically inspired algorithm just described, where the set of mechanisms for evaluation are the affinity measures and the procedures of adaptation are the immune algorithms. In order to build a system such as an AIS, one typically requires an application domain or target function. From this basis, the way in which the components of the system will be represented is considered. For example, the representation of network traffic may well be different than the representation of a real time embedded system. In AIS, the way in which something is represented is known as *shape space*. There are many kinds of shape space, such as Hamming, real valued and so on, each of which carries it own bias and should be selected with care [Freitas & Timmis 2003]. Once the representation has been chosen, one or more affinity measures are used to quantify the interactions of the elements of the system. There are many possible affinity measures (which are partially dependent upon the representation adopted), such as Hamming and Euclidean distance metrics. Again, each of these has its own bias, and the affinity function must be selected with great care, as it can affect the overall performance (and ultimately the result) of the system [Freitas & Timmis 2003]. This was also recently shown experimentally in the case of immune networks, where the affinity function affected the overall outcome of the shape of the network [Hart & Ross 2004, Hart 2005] The final layer involves the use of algorithms, which govern the behavior (dynamics) of the system. Such algorithms include those based on the following immune processes: negative and positive selection, clonal selection, bone marrow, and immune network algorithms.

3.4 Current Artificial Immune Systems Biology and Basic Algorithms

The main developments within AIS, have focussed on three main immunological theories: clonal selection, immune networks and negative selection. Researchers in AIS have concentrated, for the most part, on the *learning* and *memory* mechanisms of the immune system inherent in clonal selection and immune networks, and the negative selection principle for the generation of *detectors* that are capable of classifying changes in *self*. In this section, we review the immunology that has been capitalised on by the AIS community. We outline the three main immunological theories noted above that have acted as a source of inspiration. At each stage, we review a simple AIS approach that has extracted some feature from that theory. It is worth noting that, although not covered here, a large effort is currently being made in the AIS community into exploring other immune ideas and mechanisms such as danger theory and innate immunity. For more details see section 3.6 and Chapter 14 and Chapter 7 in this book.

3.4.1 Immunity

The vertebrate immune system (the one which has been used to inspire the vast majority of AIS to date) is composed of diverse sets of cells and molecules. These work in collaboration with other systems, such as the neural and endocrine, to maintain a steady state of operation within the host: this is termed *homeostasis*. The role of the immune system is typically viewed as one of protection from infectious agents such as viruses, bacteria, fungi and other parasites. On the surface of these agents are antigens that allow the identification of the invading agents (pathogens) by the immune cells and molecules, which in turn provoke an immune response. There are two basic types of immunity, innate and adaptive. Innate immunity is not directed towards specific pathogens, but against any pathogen that enter the body. The innate immune system plays a vital role in the initiation and regulation of immune responses, including adaptive immune responses. Specialized cells of the innate immune system evolved so as to recognize and bind to common molecular patterns found only in microorganisms. However, the innate immune system is by no means a complete solution to protecting the body.

Adaptive, or acquired immunity, is directed against specific invaders, with adaptive immune cells being modified by exposure to such invaders. The adaptive immune system mainly consists of lymphocytes, which are white blood cells, more specifically B and T cells. These cells aid in the process of recognizing and destroying specific substances. Any substance that is capable of generating such a response from the lymphocytes is called an antigen or immunogen. Antigens are not the invading microorganisms themselves; they are substances such as toxins or enzymes in the microorganisms that the immune system considers foreign. Adaptive immune responses are normally directed against the antigen that provoked them and are said to be antigen-specific.

3.4.2 Clonal Selection

The clonal selection theory (CST) [Burnet 1959] is the theory used to explain the basic response of the adaptive immune system to an antigenic stimulus. It establishes the idea that only those cells capable of recognizing an antigenic stimulus will proliferate, thus being selected against those that do not. Clonal selection operates on both T cells and B cells. In the case of B cells, when their antigen receptors (antibodies) bind with an antigen, the B cell becomes activated and begins to proliferate producing new B cell clones that are an exact copy of the parent B cell. The clones then undergo somatic hypermutation and produce antibodies that are specific to the invading antigen [Berek & Ziegner 1993]. After proliferation, B cells differentiate into *plasma cells* or long-lived B *memory cells*. Plasma cells produce large amounts of *antibodies* which will attach themselves to the antigen and act as a type of *tag* for other immune cells to pick up on and remove from the system. This whole process is known as *affinity maturation*.

Memory cells help the immune system to be protective over periods of time. In the normal course of the evolution of the immune system, an organism would be expected to encounter a given antigen repeatedly during its lifetime. The initial exposure to an antigen that stimulates an adaptive immune response is handled by a small number of B cells, each producing antibodies of different affinity. Storing some high affinity antibody producing cells (memory cells) from the first infection, so as to form a large initial specific B cell sub-population for subsequent encounters, considerably enhances the effectiveness of the immune response to secondary encounters. Rather than starting from a *tabula rosa*, such a strategy ensures that both the speed and accuracy of the immune response becomes successively stronger after each infection.

Autoimmunity is the term used to describe the existence of antigen receptors that recognise the body's own molecules, or self-antigens. According to the CST, immune specificity is a property of immune receptors. When a non-self antigen is detected, a suitable immune response is elicited and the antigen is destroyed. Thus, the recognition of self-antigen is forbidden, and self-reacting receptors must be deleted.

Artificial Clonal Selection

Work in [de Castro & Von Zuben 2000, de Castro & Von Zuben 2002] proposes an optimisation algorithm, known as CLONALG, inspired by the clonal selection process, as outlined in the previous section. Given a function F, a population of candidate solutions (antibodies) are evolved to either minimize or maximise the function. Each member of this population is a vector, in a certain shape space, which maps values to the parameters of the function F. Figure 3.2 provides a simple flowchart of the CLONALG algorithm. CLONALG exploits the cloning, mutation and selection mechanisms of clonal selection, to effectively evolve a set of memory cells that contain candidate solutions to the function F.

CLONALG operates via the following procedure. A population P is initialized with random vectors, where P is set of candidate solutions for the given function. Each member of P is evaluated against the function, and the highest affinity n number are selected for cloning, where affinity can be measured as the distance to the optimal value. Clones are produced at a rate proportional to the affinity (so the better the affinity, the more clones are produced). Each clone is subject to a mutation rate, which is inversely proportional to the affinity. These clones are added to P and then the n highest affinity are selected to remain in the population. A number of low affinity members are then removed from the population and replaced with the same number of randomly generated members. This process is repeated until some convergence criteria is satisfied, or a fixed number of iterations has been performed.

Experimentally, CLONALG has been shown to perform reasonably on standard benchmark tests for optimisation problems [de Castro & Von Zuben 2002]. However, it has not been reported in the literature that CLONALG itself outperforms any well known technique. Other algorithms similar to CLONALG exist in the literature, such as [Kelsey & Timmis 2003] and [Cutello et al. 2004], with comparative studies showing that whilst CLONALG is effective, better results can be obtained with more specialised versions of the algorithm [Cutello et al. 2004, Nicosia 2004]. Clonal selection based algorithms have also been developed for dynamic environments, reporting good performance [Gaspar & Hirsbrunner 2002, Kim & Bentley 2002a, Kelsey et al. 2003]. CLONALG has also been adapted for simple pattern recognition problems, but the results from that work are less conclusive [Whitesides & Boncheva 2002]. It has also been adapted for more sophisticated learning systems where results are very encouraging indeed for static learning [Goodman et al. 2002, Watkins et al. 2003, Watkins & Timmis 2004] and for dynamic learning [Secker et al. 2003a]. In this book, we have two applications of the basic clonal selection idea. The first is in Chapter 11 where we see the clonal selection algorithm applied to the problem of protein structure prediction. The second, in Chapter 15, describes a clonal selection based distributed supervised machine learning algorithm

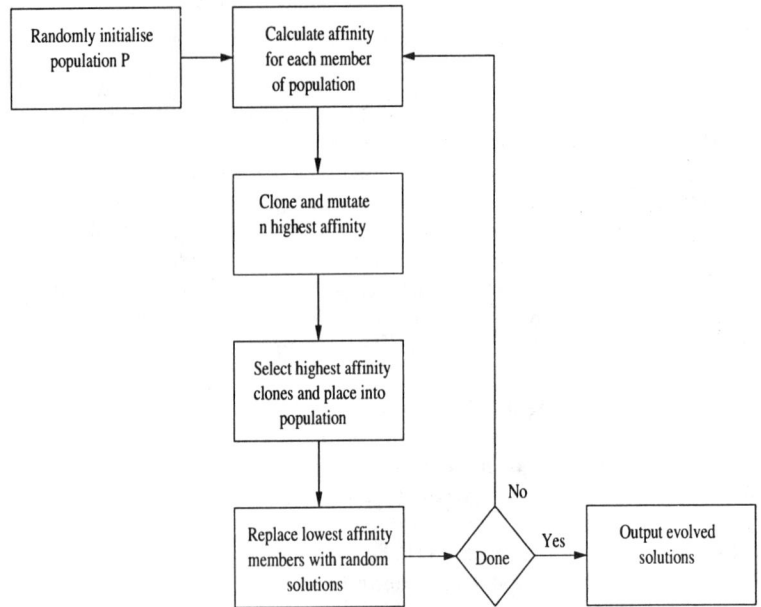

Fig. 3.2. Flowchart of CLONALG

3.4.3 Immune Networks

In a landmark paper for its time, [Jerne 1974] proposed that the immune system is capable of achieving immunological memory by the existence of a mutually reinforcing network of B cells. This network of B cells occurs due to the ability of paratopes (molecular portions of an antibody) located on B cells, to match against idiotopes (other molecular portions of an antibody) on other B cells. The binding between idiotopes and paratopes has the effect of stimulating the B cells. This is because the paratopes on B cells react to the idiotopes on similar B cells, as it would an antigen. However, to counter the reaction there is a certain amount of suppression between B cells which acts as a regulatory mechanism. This interaction of B cells due to the network, was said to contribute to a *stable* memory structure, and account for the retainment of memory cells, even in the absence of antigen. This theory was refined and formalised in successive works by [Farmer *et al.* 1986, Perelson 1989] and combined with work by [Bersini & Varela 1994] was very influential in development of the immune network based AIS such as [Hunt & Cooke 1996, Timmis *et al.* 2000, Timmis & Neal 2001, Neal 2002]. Indeed, in Chapter 16 in this book, some of this theory is revisited to help understand the dynamics of AIS today. Whilst acknowledging that the immune network theory (by the time some AIS people had read it!), was out of favor with the majority of the immunological community, it still played a key role in the inspiration to develop new immune inspired algorithms. Rather than review all of the AIS immune network systems, we will focus on one that is very similar to the CLONALG system described in section 3.4.2, but with a few simple additions.

Artificial Immune Networks

Based on the work of CLONALG, an algorithm known as aiNet was proposed in [de Castro & Von Zuben 2001]. As we have already said, aiNet is a simple extension of CLONALG, but exploits interactions between B cells according to the immune network theory. Figure 3.3 provides a flowchart for aiNet. As can be seen, the main difference between the two approaches, is that after new clones are integrated into the population, a network suppression function is employed throughout the population to remove cells that have similar affinities[5].

aiNet was initially designed for data clustering, but has been extended over the years, most recently as a hierarchical clustering tool in [de Castro & Timmis 2002b] and through hybridization with fuzzy systems methods by [Bezerra *et al.* 2005]. In the last paper, aiNet was augmented to take into account an adaptive radius measure instead of a fixed radius for B cell matching. This lead to a much improved version of aiNet, being able to achieve better separation of the data, forming clusters in less time. Work by [Castro & Timmis 2002] adapted aiNet for multi-modal function optimisation. In that paper, aiNet was also modified to be applied to the same optimisation problems as CLONALG, and was shown to have greatly improved performance over CLONAG, but this is not as comparable to other clonal selection based systems [Timmis & Edmonds 2004, Timmis *et al.* 2004]. However, it was recently identified that if careful thought was given to the optimisation problem, the basic aiNet algorithm can be augmented to give significant gains in performance [Andrews & Timmis 2005].

3.4.4 Negative Selection

Negative selection is a process of *selection* that takes place in the thymus gland. T cells are produced in the bone marrow and before they are released into the lymphatic system, undergo a maturation process in the thymus gland. The maturation of the T cells is conceptually very simple. T cells are exposed to self-proteins in a binding process. If this binding activates the T cell, then the T cell is killed, otherwise it is allowed into the lymphatic system. This process of *censoring* prevents cells that are reactive to *self* from entering the lymph system, thus endowing (in part) the host's immune system with the ability to distinguish between self and non-self agents. However, as discussed in Chapter 7, this distinction is very contentious, and that debate will not be entered into here.

---

[5] It should be noted that this is a slight departure from the immune network theory, where both suppression and stimulation occur between cells

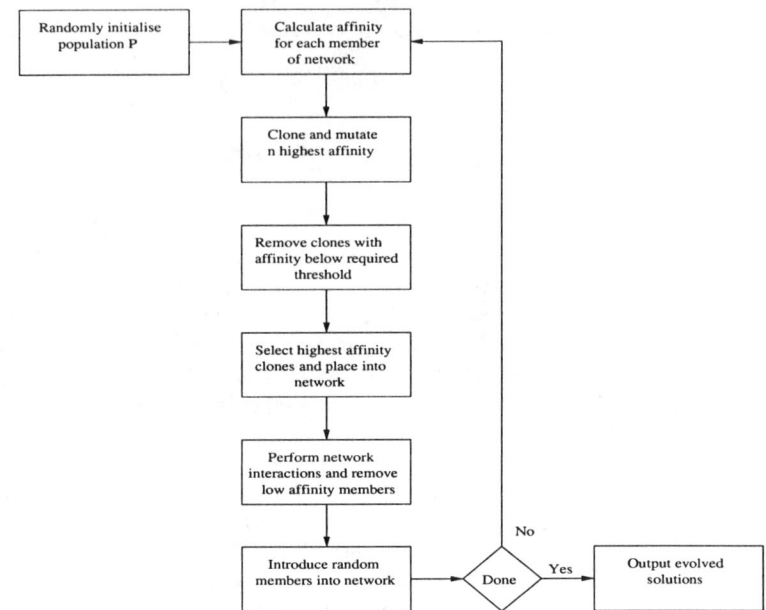

Fig. 3.3. Flowchart of aiNET

Artificial Negative Selection

The negative selection principle inspired [Forrest *et al.* 1994] to propose a negative selection algorithm to detect data manipulation caused by computer viruses. The basic idea is to generate a number of detectors in the complementary space and then to apply these detectors to classify new (unseen) data as self (no data manipulation) or non-self (data manipulation). The negative selection algorithm proposed by Forrest *et al.* is illustrated in Figure 3.4 and summarized in the following steps. We can define self as a set **S** of elements of length l in shape-space. Then generate a set **D** of detectors, such that each fails to match any element in **S**. With these detectors, monitor a continual data stream for any changes, by continually matching the detectors in **D** against the stream. This work spawned a great deal of investigations into the use of negative selection for intrusion detection, with early work meeting with some success [Forrest *et al.* 1997], but later works showing many limitations of the approach [Stibor *et al.* 2004, Stibor *et al.* 2005a].

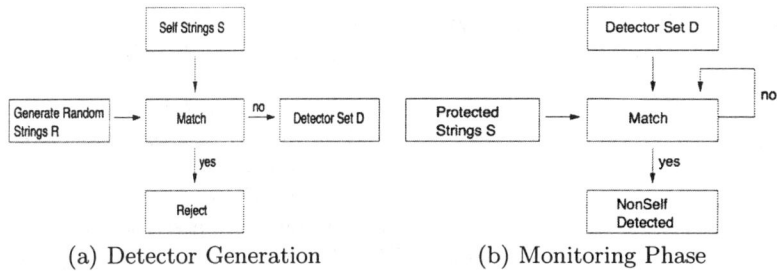

(a) Detector Generation   (b) Monitoring Phase

Fig. 3.4. Negative Selection Algorithm by Forrest *et al.*

3.5 Building Artificial Immune Systems

When constructing an AIS, there are many computational and practical issues to consider. The first is computational complexity of the approach. This relates to the time and space required to generate the suitable number of detectors (members of a population) that are required for the job [Timmis *et al.* 2002]. For example, there are a number of works that outline the unacceptable computational complexity of the negative selection approach [Kim & Bentley 2002b, Stibor *et al.* 2004, Stibor *et al.* 2005a] as there is an exponential relationship between the size of the data set to be used, and the number of detectors that it is possible to generate. However, other approaches within AIS, such as clonal selection and immune networks, seem not to suffer quite the same problem. The second aspect to consider is the data to be used. In the context of embedded systems for example, if one abstracts away from the system components and uses state machines, then one has to be careful that there is an accurate mapping between the state machine and the actual system, and ensure that the state machine adequately scopes the space to be immunised [Timmis *et al.* 2002]. Consideration here also has to be given to the way in which data is represented. The shape space paradigm proposes varying ways of data representation and interaction. However, when dealing with discrete values, such as those found in embedded systems, the method of defining affinity (i.e. seeing how similar one item is to another) is not as clear-cut as it may seem. This is coupled with the fact that mutation, even what might be thought of as a small amount, could have a huge impact on the meaning of the data. Should a binary shape space be employed, the mere flipping of one bit could indicate a huge shift in meaning of the state, rather than the small shift that may be desired. In both of these situations, domain knowledge can play a pivotal role in the success or failure of such as system [Timmis *et al.* 2002]. This type of problem is not unique to the design of AIS, but hinders many other biologically inspired approaches such as evolutionary algorithms.

3.5.1 Consider the Application Area

[Freitas & Timmis 2003] outline the need to consider carefully the application domain when developing AIS. They review the role AIS have played in the development of a number of machine learning tasks, including that of classification. However, Freitas and Timmis point out that there is a lack of appreciation for possible inductive bias within algorithms and positional bias within the choice of representation and affinity measures. For example, recent studies by [Hart & Ross 2004], point out that the main effect on immune network algorithms may well be the way in which interaction is defined. Through the development of a simple model Hart and Ross demonstrate the evolution of various immune network structures which are considerably affected by the choice of affinity measure between two B cells, which in turn effects how B cells interact with each other. Whilst no concrete conclusions are drawn here, the message is clear: think before you design. This may be facilitated by the development of more theoretical aspects of AIS, which will help us to understand how, when and where to apply various AIS techniques.

3.5.2 Design Principles

There have been some previous attempts at providing *design principles* for immune systems, such as work by [Cohen & Segal 2001] and [Bersini & Varela 1994]. However, work by Segal, whilst extremely interesting, focussed primarily on network signalling, and did not provide a comprehensive set of general design principles, or provide any test application areas for those principles. Work by Bersini, focussed on the immune network and *self assertion* ideas of the immune system to create design principles. Whilst being more concrete, these are still quite high level:

- Principle 1: The control of any process is distributed around many operators in a network structure. This allows for the development of a self-organising system that can display emerging properties.
- Principle 2: The controller should maintain the viability of the process being controlled. This is keeping the system within certain limits and preventing the system from being driven in one particular way.
- Principle 3: While there may be perturbations that can affect the process, the controller learns to maintain the viability of the process through adaptation. This learning and adaptation requires two kinds of plasticity: a parametric plasticity, which keeps a constant population of operators in the process, but modifies parameters associated with them; and a structural plasticity which is based on the recruitment mechanism which can modify the current population of operators.
- Principle 4: The learning and adaptation are achieved by using a reinforcement mechanism between operators. Operators interact to support common operations or controls.
- Principle 5: The dynamics and metadynamics of the system can be affected by the sensitivity of the population.

- Principle 6: The system retains a population-based memory, which can maintain a stable level in a changing environment.

These are potentially useful principles, that should be refined in light of immunological advances and possibly taken on board (to some degree) by the community. These need to be tested in various application areas, and refined to allow for the creation of not only a generic set of AIS design principles that are useful to the community, but also specific ones for specific application areas. With this, may come a better understanding of how to apply AIS, and avoid falling into the traps highlighted by [Freitas & Timmis 2003].

3.6 Future Directions

In section 3.2.3 we described how the original AIS researchers were motivated by a number of computationally appealing properties present in the vertebrate immune system. It is true that today, current AIS researchers are still motivated by the same properties, as it apparent that all AIS to date fail to fully capture the complex operation of the immune system. What has changed is the increased scope of immunological theories that those working with AIS take inspiration from. For example, in recent years there has been a growing interest in the mechanisms of innate immune system in immunology [Germain 2004]. This has filtered down into the AIS community, resulting in AIS inspired by theories such as danger theory [Secker et al. 2003b, Aickelin et al. 2003] and innate immunity [Greensmith et al. 2005, Bentley et al. 2005]. In their summaries of the future for AIS, both [Garrett 2005] and [Hart & Timmis 2005] point towards an increased emphasis on the innate and homeostatic functions of the immune system as possible areas for AIS exploitation. In addition to the increased scope of AIS, there has been a recent and healthy rise in investigating the theoretical workings of various immune algorithms [Clark et al. 2005a, Stibor et al. 2005b]. The way in which AIS are built has also been addressed by work in [Stepney et al. 2005b]. This paper proposes a conceptual framework that allows for the development of more biologically grounded AIS, through the adoption of an interdisciplinary approach. Metaphors employed have typically been simple, but somewhat effective. However, as proposed in [Stepney et al. 2005b], through greater interaction between computer scientists, engineers, biologists and mathematicians, better insights into the workings of the immune system, and the applicability (or otherwise) of the AIS paradigm will be gained. These interactions should be rooted in a sound methodology in order to fully exploit the synergy.

3.7 Summary

We have presented in this chapter a brief, informal tour of AIS and their development over the years. Our aim has been to provide a starting place for the uninitiated researcher to explore the world of AIS (please follow any interesting references!) and to appreciate the AIS chapters that follow in this book. As a final point, it is clear that AIS is still a young field of research, especially when compared to other biologically inspired paradigms such as evolutionary algorithms and artificial neural networks. The AIS field, however, is starting to mature at an increasing rate, with work presented both here in Chapters 11, 14, 15 and 7 elsewhere [Timmis & Bentley 2002, Timmis *et al.* 2003, Nicosia *et al.* 2004, Jacob *et al.* 2005], exploring alternative immunological ideas, design principles for AIS, and the theoretical aspects underpinning AIS approaches.

Part II

The Nature of Natural and Artificial Immune Systems

Computational Models of B cell and T cell Receptors

Ha Youn Lee[1] and Alan S. Perelson[2]

Theoretical Biology and Biophysics,Los Alamos National Laboratory, Los Alamos, NM 87545. USA. hayoun@lanl.gov

Summary. We review computational models of B cell and T cell receptors. We first consider string models where both antigen specific receptors on immune cells and antigens are represented by binary strings, or more generally, digit strings with a given number of letters. Various match rules are presented to describe the binding interaction between receptors and antigens. A second class of models is geometric models where receptors and antigens are represented as geometric shapes. A rule of interaction among receptors in shape space is introduced. Lastly, the random energy model is introduced where the binding interaction between receptors and antigens is quantified with an energy function derived from the physics of protein interaction. To compare these approaches, we explicitly calculate how receptor-ligand affinity is affected by a point mutation in different models. These calculations are relevant to understanding the correlation between the change in the sequence and the change in the binding strength. We finally review a method for connecting string models and shape space models in the context of analyzing antibody binding assay data relevant to the immune response against the influenza virus.

4.1 Introduction

The adaptive immune system, which can provide protection against a huge variety of pathogens, accomplishes this task by generating B cells and T cells that, to a first approximation, have different specificity receptors on their cell surface. The receptor diversity is called the repertoire size. Based on the size of the genes family used to encode B and T cell receptors, it has been estimated that the mouse genome encodes information for making at least 10^{10} different B cell receptors and of order 10^{18} T cell receptors (TCR).

In order to build realistic computational models of the immune system it is desirable to capture much of this diversity. A mouse has been estimated to have about 10^7

B and T cells. Thus, one needs the capability of representing repertoires of this size. In addition, one requires a method of computing the affinity of interactions between any of the 10^7 B or T cell receptors and a diversity of ligands. Here we will review methods used to represent B and T cell receptors as well as the antigens they interact with, and compare and contrast the advantages and disadvantages of the various approaches. The models fall into three categories, digit strings models in which the length of the string and the alphabet size determine the repertoire that can be represented, geometric models in which receptors and antigens are given shapes in one, two or three dimensions, and lastly random energy models in which receptors are thought of as proteins that fold and take on a shape such that the energy of interaction or affinity can be computed between a receptor and an antigen.

4.2 String Models

The hallmark of the immune system is specificity. In order to represent the antigen specific receptor on a B cell or a T cell string models have been used. For example, if a receptor is represented by a binary string of length 32 then 2^{32} or 4×10^9 different receptors can be represented. In addition to receptors, antigens, antibodies and MHC molecules have also been represented by strings. An advantage of this representation is that binding between molecules, e.g., between antibodies and antigens, each represented by a string can be converted into a string match problem, with the number of matches being monotonically related to the affinity.

The first use of a string representation was by [Farmer et al. 1986], who described the dynamics of interactions among antibodies and antigens by a set of differential equations, with the strength of the interactions determined by a string match score between the strings representing the various molecules. Binary strings were used and the strength of interaction was given by the number of complementary bits among sequences. In the simplest representation, and the one generally followed by later workers, the strings were of the same length and were aligned. The XOR operator was applied in order to efficiently evaluate when opposing bits were complementary (see Fig. 4.1). [Farmer et al. 1986] also raised the possibility of the sequences being of different sizes and the smaller sequence moved across the larger in all possible alignments. The largest number of complementary bits of all possible alignments is also another possible match rule. Rather than using complementary bits it is sometimes easier to say two strings match if they are identical and then use the number of mismatches, or the Hamming distance, as a metric between sequences; the closest sequences being the ones that have the highest affinity interactions.

[Farmer et al. 1986] also introduced the idea of "metadynamics". In immune system models the population dynamics of lymphocytes, antibodies and antigen are frequently followed by constructing a differential equation for the time rate of change of each entity. For instance, one might say an antigen, Ag, is eliminated if it is recognized by an antibody, with the elimination rate proportional to the affinity of the antibody-antigen interaction. If multiple antibodies are in the system then there is

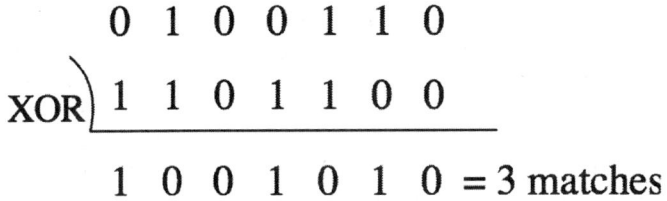

Fig. 4.1. The rule of complementary match for two binary strings. The XOR operator is used to determine the positions in which a 0 matches a 1.

one term for each antibody that binds the antigen. String matching determines which antibodies bind to the antigen and also give the strength of the match. The immune system is constantly evolving, new B cells are created in the bone marrow and enter the body. New antigens are encountered and hopefully eliminated. As these new entities are created or destroyed new differential equations need to be constructed to represent these entities, and the right hand side of these differential equations need to include the interactions between these newly formed entities, and molecules and cells already existing in the model immune system. The rules for generating these new equations are called metadynamics, and provide a way for automatically updating the model. Similar methods involving metadynamics are useful in models that aim to study the origin of life in which new molecules are being created from simpler ones and then interact with the existing system [Farmer *et al.* 1987]. The use of differential equation models in immunology is also described by Hone and Van Den Burg (Chapter 16) where the dynamics of establishing immune memory is discussed.

The binary string model was extended by Celada and Seiden so as to include peptides, MHC molecules and T cells. Because peptides bind in the "groove" of an MHC molecule, the strings representing the peptide and MHC molecule were each half as long as the T cell receptor string. The peptide and MHC strings were then combined into a single string representing the MHC-peptide complex and the combined string was matched against T cell strings to determine the match score. The model was then used to answer questions like optimal number of MHC types per individual [Celada & Seiden 1992].

String representations need not be restricted to binary strings. For example, [Percus *et al.* 1993] introduced strings with B letters, where B could be greater than 2. They had in mind representing the physical properties of amino acids, such as charge (positive or negative) and hydrophobicity, which would give rise to a 3 letter alphabet, i.e $B = 3$, where each letter would match only one other, e.g. positive matched negative, and hydrophobic matched hydrophobic. They also introduced a match rule in which the match score or affinity of the interaction was related to the size of the largest substring of continuous matches. The biological interpretation was that as two molecules interact they do not necessarily interact over their entire surfaces but only over a subdomain, which here would correspond to the largest contiguous match region.

Digit-string representations using large alphabet sizes, e.g. $m = 128$ or $m = 256$ proved useful in models of thymic selection [Detours & Perelson 1999, Detours et al. 1999, Detours & Perelson 2000, Chao et al. 2005]. Here T cells with receptors that match self peptides presented on MHC within a range of affinities survive, while T cells with affinities outside this range die. The idea is that T cells need to be able to recognize MHC and thus there is a minimal affinity needed to ensure their survival. This is called the positive selection threshold, K_P. However, if the affinity is too high, then the T cell could cause autoimmunity and hence it should be eliminated. This upper threshold is called the negative selection threshold, K_N. Only about 3% of T cells survive both positive and negative selection. Thus to model thymic selection one needs a model in which affinities change in very small increments. This can be done either by using very long bitstrings or by using digit strings over larger alphabets. Detours and Perelson in a series of papers exploited this representation to explain self MHC restriction and alloreactivity [Detours & Perelson 1999, Detours et al. 1999, Detours & Perelson 2000].

One difficulty in representing immunological phenomena is choosing the appropriate representation and match rule. Once a match rule is selected, parameters, such as string length and alphabet size, need to be calibrated to satisfy immunological criteria. [Smith et al. 1997] showed how the calibration could be done for models of antibody responses. For example, if one wants to represent a repertoire of 10^8 antibodies then one must choose a string length and alphabet size consistent with this requirement. About 1 in 10^5 B cells typically responds to an antigen. Thus, the match rule for determining when a B cell responds to a randomly encountered antigen needs to be set to be consistent with this criterion. Lastly, there is some scattered data, that suggests the degree of cross-reactivity between antibodies. Cross-reactivity is the phenomenon where an antibody raised against antigen x also reacts with antigen y. Such cross-reactivity only occurs if antigens x and y are closely related and such information can also inform the choice of string length and degree of match needed to stimulate a B cell into antibody production [Smith et al. 1997]. Once such calibration was done for antibody response, Smith et al. [Smith et al. 1999] were able to generate a realistic simulation of the antibody response to influenza vaccination, and then used the simulation to study the effects of repeated yearly vaccination. Interestingly, they found that depending on relative locations among the two vaccines and the epidemic strain of influenza in shape space, the two vaccines may interfere with one another and not produce optimal protection, during the second flu season.

4.3 Geometric Models

Another approach for modeling antibodies and antigens has been to represent the molecules as geometric shapes. For example, Segel and Perelson [Segal & Perelson 1988] introduced a one-dimensional shape space in which antibodies were represented as having binding sites with triangular grooves of depth x and antigens were represented as having epitopes that were triangles of height y (see Fig. 4.2). The base of the triangles were equal so that an epitope of height y exactly would fit into the binding site groove of an antibody of depth $x = y$. Mismatches, measured by $x - y$, would reduce affinity. A two-dimensional shape space could be constructed in a similar manner by also allowing the base of the triangles to be different. Antibodies and antigens have also been represented by a set of three dimensional units arranged in a variety of shapes [Weinand 1990]. The representation one uses depends on the application. In the work of Segel and Perelson [Segal & Perelson 1988] the idea was to develop a set of partial differential equations to represent the dynamics of the immune system. Changes in the B cell repertoire could then be visualized as densities in shape space. If the initial repertoire were uniform, then the density of B cells of shape x could be represented by a horizontal line, i.e. $B(x) = $ a constant. Due to interactions with antigen certain B cells would grow and others would die, so the $B(x)$ curve would now have peaks and valleys. This non-uniform distribution of B cells in the shape space would then provide a visually appealing representation of immune memory ("peaks"), and questions such as whether memory is localized within certain models could be addressed.

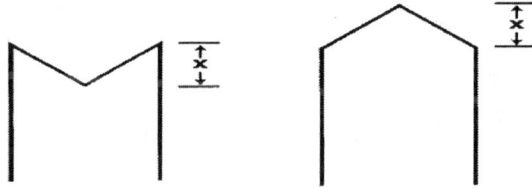

Fig. 4.2. Antibody shapes described by a real number x in a one-dimensional shape space.

An alternative method for modeling interactions among B cells is placing the B cells on a lattice rather than in continuous shape space [Weisbuch 1990]. One could then establish a rule of interaction among B cells, such as only nearest neighbor B cells interact. In general, one can define an interaction strength, J_{ij}, between B cell i and B cell j and utilize techniques from statistical mechanics to study the properties of lattices of interacting B cells [Perelson & Weisbuch 1997]. These methods were used for studying idiotypic networks [Perelson 1989]. Stepney (Chapter 12) reviews shape space models discussing various ways of mapping immune agents into geometrical space.

4.4 Random energy models

One problem in immunology that has attracted the attention of a number of modellers is affinity maturation. In this phenomena the average affinity of antibody for an antigen increases with time during an immune response. This increase of affinity is in part caused by mutational changes that occur in the genes that code for the variable region of the antibody. Various computational models have been developed to assign antibody affinity. Above we reviewed string match and geometric rules. Here we will look at other models that attempt to predict affinity and examine how affinity is predicted to change as the amino acid sequence of the antibody is varied. If one graphs affinity versus sequence, one would obtain a representation of an affinity or "energy landscape". It is known that some amino acid substitutions radically change affinity, whereas other substitutions do not change, or only slightly change the affinity [Berek & Ziegner 1993, Wysocki et al. 1986, Rudikoff et al. 1982, Panka et al. 1988, Roberts et al. 1987]. Given that the connection between affinity and amino acid sequence has not been established, random energy models have served as a starting point. Here the idea is that since we do not know the affinity of any given amino acid sequence, we assign it at random. Random energy theory has captured the essence of the correlated ruggedness of landscapes in a variety of physical systems, especially spin glasses and in protein folding [Derrida 1980, Bryngelson & Wolynes 1987, Shakhnovich 1993].

A representative random energy model is the NK model developed by Kauffman [Kauffman et al. 1988, Kauffman & Weinberger 1989]. In the NK model the fitness or affinity of a protein of length N is given by the sum of the fitnesses of each amino acid. The fitness of each amino acid in turn depends on the amino acid in that position and the amino acids in K other positions. When $K = 0$ the fitness of each amino acid is independent of all other amino acids. Hence at each position there is a "most fit" amino acid and the fitness landscape has a single peak corresponding to the antibody with the most fit amino acid at each position. As K increases, the ruggedness of the landscape increases from a single peaked landscape to a multi-peaked landscape. If one starts at a random position on the landscape and then moves on the landscape always taking steps uphill, i.e., evolving toward higher affinity, then one can compute the average number of steps to reach a local optimum [Kauffman et al. 1988]. This procedure was used by Kauffman and Weinberger [Kauffman & Weinberger 1989] to estimate the values K that corresponded to an affinity maturation process in which uphill steps corresponded to a mutation in an antibody V-region of a given length N. For $N = 112$ they estimated K as about 40. The contribution of each amino acid to the affinity is affected by around 40 others in the 112 amino acid long V region. Other random energy models of affinity maturation have been studied by [Macken & Perelson 1989, Perelson & Macken 1995].

A generalization of the NK model was developed by [Bogarad & Deem 1999, Deem & Lee 2003] in which more information about the structure of an antibody was used to estimate its affinity. Because antibodies can be viewed at the structural level as being constructed from domains, in the generalized NK model, an antibody is represented by an amino acid sequence of length of $N \times M$, consisting of M domains with N amino acids each. Motivated by the physics of chemical interactions, the affinity of the antibody was calculated by first determining the energy E of the interaction between antibody and antigen. The antigen, as in the NK model is not explicitly given, but is implicit in the energy calculation. The energy of the antibody-antigen complex is assumed to have three components, one due to molecular interactions within a domain or as Deem calls them subdomains, (U^{sd}), interactions between

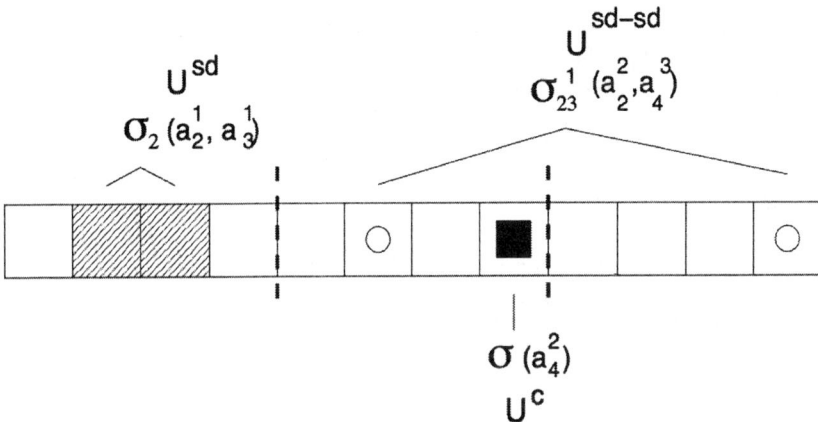

Fig. 4.3. Schematic diagram showing an example of subdomain interaction energy U^{sd}, subdomain-subdomain interaction energy U^{sd-sd}, and chemical binding energy U^c. Here a receptor is represented by a 12 letter sequence with M=3 subdomains (divided by dashed line) each of length N=4. Within a subdomain, interactions among all possible K=2 consecutive sites contribute to the subdomain energy, U^{sd}. For instance, the second and third "amino acids" in subdomain 1, (a_2^1, a_3^1) in shaded boxes contribute the interaction energy $\sigma_2(a_2^1, a_3^1)$ to U^{sd}. The type of the first subdomain, α_1 is 2 here. The second amino acid in subdomain 2 and the forth amino acid in subdomain 3, represented by the boxes with a circle, contribute to U^{sd-sd} as $\sigma_{23}(a_2^2, a_4^3)$. The amino acid in the box with filled square, a_4^2, contributes to the chemical binding energy, U^c.

subdomains (U^{sd-sd}), and direct binding energy with the antigen at the contact sites (U^c) as shown in Figure 4.3. Thus, the total energy is represented as

$$U = \sum_{i=1}^{M} U_{\alpha_i}^{sd} + \sum_{i>j=1}^{M} U_{ij}^{sd-sd} + \sum_{i=1}^{P} U_i^c, \tag{4.1}$$

where P is the number of antibody amino acids contributing directly to the binding.

The ith subdomain energy, $U_{\alpha_i}^{sd}$ of subdomain type α_i, is

$$U_{\alpha_i}^{sd} = \frac{1}{\sqrt{M(N - K + 1)}} \sum_{j=1}^{N-K+1} \sigma_{\alpha_i}(a_j^i, a_{j+1}^i, \cdots, a_{j+K-1}^i). \tag{4.2}$$

Here K is the number of interacting sites in a subdomain and the energy contribution by ith subdomain is the sum of the contributions from all the possible consecutive segments of neighboring K amino acid sequences within a subdomain. All subdomains belong to one of $L = 5$ different types, e.g. alpha helix, beta-sheet, etc. Although, it is not clear if this degree of generality is needed in a computational model we include it as this was done in the original model. The energy contribution by the segment of K amino acids, $\sigma_{\alpha_i}(a_j^i, a_{j+1}^i, \cdots, a_{j+K-1}^i)$, may differ depending on the subdomain type. Here a_j^i denotes the amino acid at jth position in subdomain i. Amino acids are classified into only five chemically distinct types (e.g., negative, positive, polar, hydrophobic, and other) and hence are represented with a 5 letter alphabet. For each segment of K amino acids, the energy contribution σ_{α_i} within subdomain type α_i is chosen from the normal distribution with mean 0 and variance 1, and stored in an array, so if the same sequence of amino acids reappears in the same subdomain type it is assigned the same energy.

The energy of interaction between subdomains i and j is given by

$$U_{ij}^{sd-sd} = \sqrt{\frac{2}{DM(M-1)}} \sum_{k=1}^{D} \sigma_{ij}^k (a_{j_1}^i, \cdots, a_{j_{K/2}}^i;$$
$$a_{j_{K/2+1}}^j, \cdots, a_{j_K}^j), \qquad (4.3)$$

where D is the number of pairs of interactions between subdomains. The energy contribution from the kth interaction among subdomain i and j, σ_{ij}^k is selected randomly depending on the amino acids in positions $\{j_1, \cdots, j_K\}$. The positions of interacting amino acids are selected at random for each interaction (k, i, j).

The ith chemical binding energy of the antibody to the antigen is given by

$$U_i^c = \frac{1}{\sqrt{P}} \sigma_i (a_{i1}^{i2}), \qquad (4.4)$$

where a_{i1}^{i2} denotes $i1$th amino acid in subdomain $i2$. The position of the ith contributing amino acid, $(i1, i2)$, and the unit-normal weight of the binding, σ_i are chosen at random. Once we fix an antigen interacting with an antibody, the antigen is represented with a set of parameters in the energy of interaction between subdomains and chemical binding energy. Hence the interaction pattern of amino acids in the antibody is uniquely determined by a specific antigen.

The affinity of an antibody to an antigen is a function of the total energy in Eq. (4.1),

$$K_i = \exp(a - bU_i), \qquad (4.5)$$

where a and b are constants chosen to give a realistic distribution of affinities [Deem & Lee 2003]. Note that the interaction strength U, and hence the binding affinity, depends not only on the sequence of the antibody but also on the antigen, since for a different antigen one would pick different contributions to U^{sd}, U^{sd-sd}, and U_i^c.

This generalized NK model is clearly very complex because it was designed to mimic many of the features of real antibody-antigen interactions. However, because the energy is computed by summing a large number of random numbers corresponding to energy contributions it tends to generate a Gaussian distribution of energies and hence a log-normal distribution of affinities. If one were willing to abandon the property that a particular affinity is associated with a given sequence, one could simply pick an affinity at random from a log-normal distribution. However, if one is interested in the effects of somatic hypermutation where antibody sequence is changed by point mutation, the generalized NK model has the advantage of keeping track of affinities of neighboring sequences. Thus this model has been used by Deem and colleagues to study affinity maturation of antibodies and a similar process of selection for higher affinity T cells [Deem & Lee 2003, Park & Deem 2004].

4.5 Affinity distribution by a point mutation

Because of the complexity of the generalized NK model it is of interest to compare the behavior of different models in predicting affinity when the antibody sequence is changed by point mutation. Experimentally, the effect of point mutation on the strength of binding has been systematically investigated by changing amino acids at specific points either in an antibody, T cell receptor, or antigen [Chen et al. 1992, Brown et al. 1996, Casson & Manser 1995, Churchill et al. 2000, Lee et al. 2000, Pantophlet et al. 2003].

4.5.1 Mutation and the string model

The simplest receptor model is one based on a string representation. Consider a receptor consisting of N amino acids, each amino acid classified into one of B different groups. Thus, the receptor is modeled by a string of length N using an alphabet size B. Assume each of the B amino acid types matches only one other type, e.g. positively charged matches negatively charged. Let the binding energy or match score between a receptor and an antigen be the number of complementary matches, E_0. For the given string with match score E_0, a single point mutation generates a receptor with an energy that is either E_0, $E_0 + 1$, or $E_0 - 1$. To generate E_0 a non-matching amino acid at a given position would be changed into a new amino acid at that position but one that still does not match the antigen. To generate $E_0 + 1$ a non-matching amino acid would need to be changed into a matching one. In a string of length N with E_0 matches there are $N - E_0$ non-matching positions. The probability of a mutation occurring in one of these non-matching positions is then $(N - E_0)/N$. Considering only mutations that change an amino acid of one type into another type, a single point mutation can change an amino acid to any of the other possible $B - 1$ amino acids. A change to a particular one will yield a match and a change to any of the remaining $B - 2$ will remain a mismatch. Lastly, to decrease the number of matches the mutation must occur in a matching position. This occurs with probability E_0/N. Summarizing, the probability of having each energy value after a single mutation is

$$P(E = E_0) = \frac{N - E_0}{N} \frac{B - 2}{B - 1}$$

$$P(E = E_0 + 1) = \frac{N - E_0}{N} \frac{1}{B - 1},$$

$$P(E = E_0 - 1) = \frac{E_0}{N}. \tag{4.6}$$

Figure 4.4a depicts the distributions of the new energies in a string of length $N = 20$ with an alphabet size $B = 3$, starting from $E_0 = 5$, 10, and 15. Note that if you mutate a string with a high number of matches, e.g., $E_0 = 15$, it is much more likely to decrease the match score than to increase it. Thus the probability of having 14 matches after one mutation is 0.75, while the probability of having 16 matches is 0.125. Conversely, if the original number of matches is low, then it is more likely that the match score will improve through mutation. For $E_0 = 5$, the probability of having 6 matches after mutation is 0.375, while the probability of having 4 matches is 0.25. Note that each distribution is asymmetric and that the probability of improving the match score is smaller and the variance of the distribution greater as the starting match score increases in value. More importantly, with this type of complementary match rule mutation can only change the match score by one unit and hence the fractional change is at most $1/N$. For large string length N such a rule thus gives very conservative changes in the match score under mutation. This does not seem very realistic for antibodies which can lose all binding or increase their affinity 10-fold under some circumstances [Brown *et al.* 1996].

A match rule that might more closely mimic this property of having large changes in match score as the potential outcome of a single point mutation is the consecutive match rule. The match score is given by the maximum number of continuous complementary matches, i.e. the length of the longest matching substring. The probability distribution of a new match score after one point mutation is shown in Fig. 4.4b. Computing this distribution in analytical form is beyond the scope of this paper, and we have used a Monte Carlo (MC) method to compute the distributions. With this match rule, a single point mutation generates diverse values of match scores by splitting matches, adding to the end of an existing run of matches, and by allowing two continuously matched segments separated by a single mismatch to join. Examining Fig. 4.4b one notices that when $E_0 = 15$ it is extremely unlikely to generate a match score of 16 since depending upon whether the run of 15 matches is in the middle of the sequence or at one end there are either one or two positions in which a mutation can increase the match score. At each of those positions the chance of the correct mutation is only $1/(B - 1)$. And the position next to the mutation point should be unmatched, an event that occurs with probability $(B - 1)/B$, assuming such a position exists, i.e. that the mutation site is not an end point. It is easy to see for $N = 20$ and $B = 3$ this probability is $2/6 \times 1/20 \times 1/3 + 2/6 \times 1/20 \times 1/3 + 2/6 \times 1/20 \times 1/2 + 2/6 \times 2/20 \times 1/3 = 0.031$. While this not surprising, it is somewhat surprising that when $E_0 = 5$ or $E_0 = 10$ that the probability of increasing the match score by mutation is still very small and that in all three cases it is much more likely to decrease the score than increase it. Similarly, when making mutations in highly evolved proteins such as antibodies it is much easier to lower affinity than to increase it. Whether this rule quantitatively captures the correct probability distribution for these events remains to be determined.

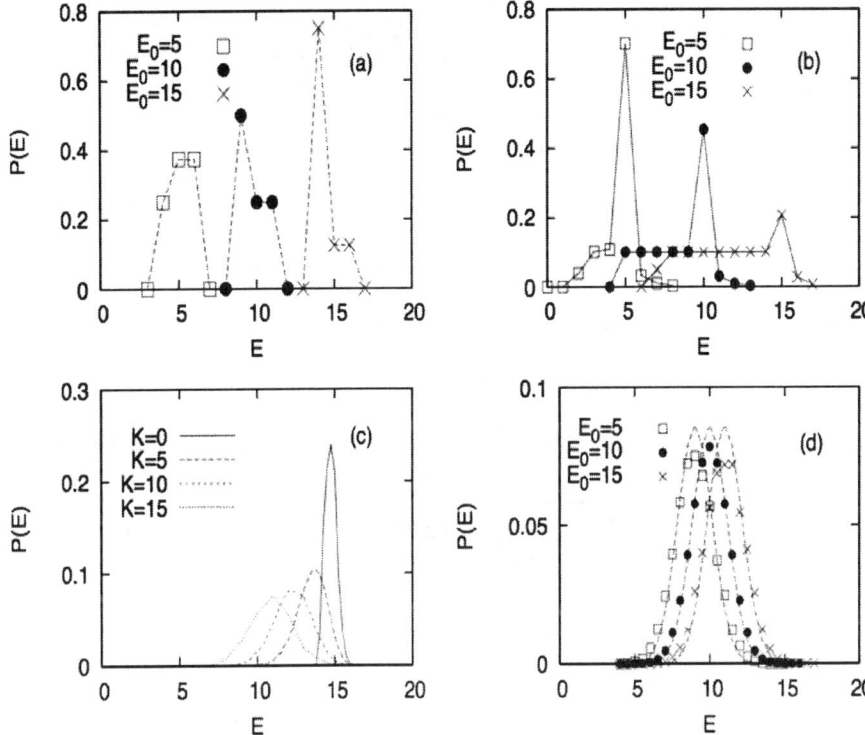

Fig. 4.4. The distribution of the new energy (E) or match score obtained by a point mutation of a given receptor with $E_0 = 5, 10, 15$ for the rule of complementary match (a), consecutive match rule (b) and NK model with $N = 20$ and $B = 3$ (c) and (d). In (a) and (d), points represent the results of MC simulations and dotted lines represent the analytical predictions in Eqs. (4.6) and (4.9). In (c), the distribution of the new energy $P(E)$ is shown for the NK model when the energy before the mutation, $E_0 = 15$, and the value of K is changed. For the MC simulations, we have done 10^{10} samplings averaging over random sequences. In (d), the distribution of the new energy with $K = 15$ is depicted for different values of the previous energy $E_0 = 5, 10$, and 15.

4.5.2 Mutation and the NK model

The other class of match rules involve the random energy models. For the standard NK model the energy of a receptor of length N is given by the sum of the energetic contributions of the amino acids at each position, i.e.

$$E = \sum_{\alpha=1}^{N} \sigma_\alpha(a_\alpha, a_\alpha^1, a_\alpha^2, \cdots, a_\alpha^K), \tag{4.7}$$

where each site energy σ_α is a random number uniformly distributed between 0 and 1, depending on the amino acid at site α, a_α, and the amino acids at K randomly chosen other sites, $a_\alpha^1, a_\alpha^2, \cdots, a_\alpha^K$. We estimate the probability distribution of the new energy by one point mutation using the central limit theorem. Since $(K + 1)$ sites contribute to the energy of each site, one point mutation alters $(K + 1)$ site energy values by $(K + 1)$ independent random samplings of the interval $[0, 1]$. In the limit of large K, the sum of the new site energies is normally distributed with mean $(K + 1)/2$ and variance $(K + 1)/12$. Before mutation the average energy per site is E_0/N and thus the average contribution of the $K + 1$ sites that change by mutation is $E_0(K + 1)/N$. The new energy contribution of these sites, x, replaces the old contribution. Hence a good approximation for the total energy, E, after a point mutation on a string with E_0, is given by

$$E = E_0 - E_0 \frac{(K + 1)}{N} + x, \tag{4.8}$$

with $P(x) = \sqrt{6/(\pi(K + 1))} \exp\{-6/(K+1)(x-(K+1)/2)^2\}$. Then the distribution of E is given by

$$P(E) = \sqrt{\frac{6}{\pi(K + 1)}} e^{-\frac{6}{(K+1)}(E - E_0 + E_0 \frac{(K+1)}{N} - \frac{(K+1)}{2})^2}. \tag{4.9}$$

As the value of K increases, a point mutation gives rise to a mean value of the energy, E, close to $(K + 1)/2$ independent of E_0.

The distribution of the new energies, $P(E)$, for a receptor with $E_0 = 15$ after one point mutation is shown in Fig. 4.4c for different values of K by MC simulations. The higher K, the less correlated are the energies before and after mutation. Thus, by changing the value of K, one can control the level of correlation between the energy values before and after a point mutation. Figure 4.4d presents the distribution for E for $E_0 = 5$, 10, and 15 for $N = 20$ and $K = 15$ obtained by MC simulations and by Eq. (4.9). With $K = 15$ and $N = 20$, the distribution of the new energy does not depend strongly on the energy before a mutation. We plot the distribution of E for $E_0 = 28$, 56, and 84 for $N = 112$ and $K = 40$, which are values estimated by Kauffman and Weinberger (1989) for affinity maturation [Fig. 4.5]. A single mutation shifts the distribution of new energy toward the mean energy of the NK model, $N/2$.

4.5.3 Mutation in the generalized NK model

For a generalized NK model, one is summing a large number of random energy contributions, each with mean 0 and variance 1. Hence the energy distribution will approximate a Gaussian distribution with the mean 0. Because the variance of each U^{sd}, U^{sd-sd}, U^c term is 1 and the variance of the random variable $X + Y + Z$, where X, Y, and Z are independent normal random variables with the variances σ_1, σ_2, and σ_3 is $\sigma_1^2 + \sigma_2^2 + \sigma_3^2$, the variance of the generalized NK model energy will be 3 regardless of the model parameters. Figure 4.6(a) shows that the energy distribution of random sequences in the generalized NK model is fit well by a Gaussian distribution with mean 0 and variance 3 for $K = 10$ and $K = 2$.

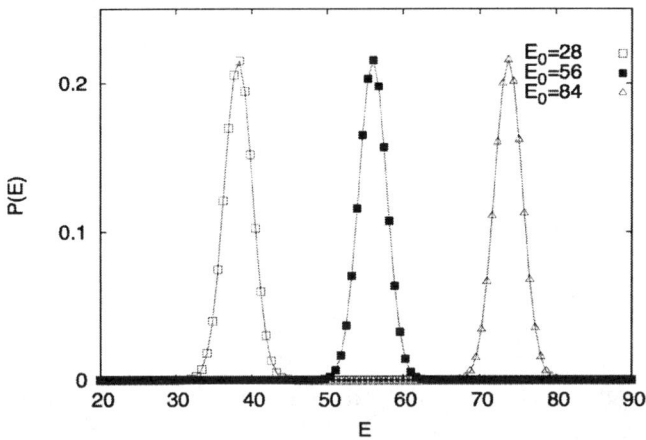

Fig. 4.5. The distribution of the new energies (E) by a point mutation of a given receptor with $E_0 = 28$, 56, 84 for NK model. Here we use $N = 112$ and $K = 40$ in the analytical prediction given by Eq. (4.9).

Let us now calculate for a generalized NK model the distribution of the new energies obtained after a point mutation in sequences with initial energy E_0. We first estimate the average number of changes in each energy term. For given N and K, the average number of new energy contributions to the subdomain energy, S^{sd}, is

$$
S^{sd} = \frac{1}{N} \sum_{i=1}^{K-1} i + \frac{1}{N} \sum_{i=K}^{N-K+1} K + \frac{1}{N} \sum_{i=N-K+2}^{N} (N - i + 1)
$$
$$
= \frac{K(N - K + 1)}{N}, \tag{4.10}
$$

since one point mutation changes K values of σ if the mutation site is located between positions K and $N - K + 1$, or changes i values of σ if the mutation site is located between positions $i = 1$ and $i = K - 1$, or changes $N - i + 1$ values of σ if the site is between $i = N - K + 2$ and N. One point mutation also changes $S^{sd-sd} = (M - 1) \, D \, K/(2N)$ values of σ within a subdomain interaction energy term, because $K/2$ random positions are selected out of N in each U^{sd-sd} term in Eq. (4.3). The average number of substitutions by one point mutation in U^c, S^c, is $P/(MN)$.

The average contribution of each σ term before a mutation, x, in a sequence with total energy E_0 in Eq. (4.1) is

$$
x = \frac{E_0}{\sqrt{M(N - K + 1)} + \sqrt{DM(M - 1)/2} + \sqrt{P}}. \tag{4.11}
$$

Since the average new contribution of each σ to the total energy is 0, the mean of the new total energy, $\bar{E}$ is

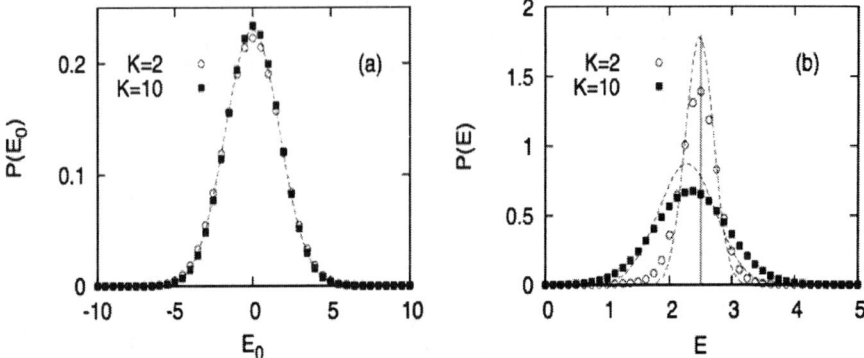

Fig. 4.6. (a) The distribution of the total energy (E) of random sequences in the generalized NK model with $M = 10$, $N = 10$, $D = 4$ and $P = 5$. The energy distributions from MC simulations with 10^8 samplings for $K = 2$ (circles) and $K = 10$ (squares) are well fitted by a Gaussian distribution with the mean 0 and the variance 3 (dashed line). (b) The distribution of the new energy, E, after a point mutation as a function of the energy $E_0 = 2.5$ (solid line) before the mutation. The circles (squares) represent the distribution of the new energy over 10^9 MC samplings for $K = 2$ ($K = 10$). The dashed lines present the analytical estimates of the distribution for $K = 2$ and $K = 10$ from Eq. (4.14).

$$\bar{E} = E_0 - \frac{x}{\sqrt{M(N - K + 1)}}S^{sd} - x\sqrt{\frac{2}{DM(M - 1)}}S^{sd-sd} - \frac{x}{\sqrt{P}}S^c$$

$$= E_0 - \frac{E_0\{K(N - K + 1)/N + (M - 1)DK/(2N) + P/(MN)\}}{\sqrt{M(N - K + 1)} + \sqrt{DM(M - 1)/2} + \sqrt{P}},$$

$$(4.12)$$

and the variance of the new energy, s^2, is

$$s^2 = \frac{1}{M(N - K + 1)}S^{sd} + \frac{2}{DM(M - 1)}S^{sd-sd} + \frac{1}{P}S^c$$

$$= \frac{2K + 1}{MN}.$$

$$(4.13)$$

The distribution of the new energy E is thus

$$P(E) = \sqrt{\frac{MN}{2\pi(2K + 1)}}e^{-\frac{MN}{2(2K+1)}(E - \bar{E})^2},$$

$$(4.14)$$

where $\bar{E}$ is given in Eq. (4.12).

Figure 4.6b shows the distribution of $P(E)$ for $E_0 = 2.5$ and $K = 10$ and $K = 2$ by MC simulations and the prediction from Eq. (4.14). As we increase the value of K, the mean of the new energy deviates from E_0 and the variance increases. In comparison with the standard NK model, we can control the mean and the variance of the new energy by one point mutation in more diverse ways. For example, the change in the mean of the new energy in the NK model always accompanies the change in the variance of the distribution of new energies, Eq. (4.9). However, in the generalized NK model one can change only the mean of the new energy by fixing the variance, as can be seen from Eqs. (4.12) and (4.13).

4.6 Connection between string models and shape space models

Experimentally, one can measure the correlation between amino acid sequence difference and antibody binding difference. This approach has been used to study the influence of evolution of viruses such as influenza A and HIV on the ability of antibodies to either bind or neutralize the virus [Richman et al. 2003, O'Connor 2002, Binley et al. 2004]. From a modeling prospective it is important to know if changes in amino acid sequence of a real protein, which we model say as changes in a string representation, translate in a simple manner into changes in antibody binding that can be predicted from simple models. Further, as measured by antibody binding one would expect that as antigen sequence changes antibody binding will be affected. Assuming this is the case, one in principle, can define the distance between two antigens by their difference in binding to a given antibody. We expect that the correlation between sequence and antigenic distances, identified through binding measures, may provide a connection between string models or shape space models and the real biology.

This type of approach has been developed by [Lapedes & Farber 2001, Smith et al. 2004] and can be used to describe the evolution of influenza virus. The antigenic properties of influenza viruses are commonly characterized using the hemagglutination inhibition (HI) assay. The major protein found on the surface influenza virus, hemagglutinin, can bind and aggregate red blood cells. When antibodies bind hemaglutinin they prevent red cell aggregation. The HI assay measures the titre of antibody needed to prevent agglutinination. In trying to determine the influenza strains to put into a vaccine, one obtains antibody raised against one flu strain and then tests its ability to inhibit agglutinination of other viral strains. If an antibody raised against one strain can prevent agglutination of the others then the hemaglutinin on the two strains should be antigenically similar. By doing multiple tests of this type, one can generate a matrix of titre values in which the i, j entry gives the titre of antibody raised against virus i needed to inhibit agglutination of virus j.

The HI matrix can then be analyzed by multidimensional scaling methods (MDS) [Shepard 1963, Shepard 1964]. In this method MDS is first used to construct the dimension of the shape space needed to represent this binding data. Here the idea is to represent the HI data as distances such that antibodies and antigens are close if they bind. First, the rank order of HI assay values are determined and ordinal MDS [Lapedes & Farber 2001] is used to match the distance order among points representing the virus strains and antibodies in shape space with the rank order of the HI assay values. The distances between points representing M antigens and N antisera are labeled as D_{i^1,j^1}, D_{i^2,j^2}, $\cdots$, $D_{i^{MN},j^{MN}}$ according to the rank order of HI data values, where the closest distance is D_{i^1,j^1} corresponding to the antigen antibody pair with the maximum HI value. Here distance is defined as a Euclidean distance among the coordinates of antigens and antisera in a shape space of dimension d. The goal of ordinal MDS is to find the coordinates of MN points in shape space that satisfy the distance order, $D_{i^1,j^1} < D_{i^2,j^2} < D_{i^3,j^3} \cdots < D_{i^{MN},j^{MN}}$ for the given rank order of HI values, $H_{i^1,j^1} > H_{i^2,j^2} > H_{i^3,j^3} \cdots > H_{i^{MN},j^{MN}}$. Lapedes and Farber [Lapedes & Farber 2001] used the following energy function, which is to be minimized to satisfy the distance order,

$$E = - \sum_{\alpha=1}^{\alpha=MN} \log(g(D_{i^{\alpha+1},j^{\alpha+1}} - D_{i^\alpha,j^\alpha})), \qquad (4.15)$$

where $g(x) = 0.5(1+\tanh(x))$. For the influenza data they were analyzing they found dimension 5 was the smallest dimension in which the distance order in shape space agrees with the rank order of HI values. Once the dimension of shape space was established, the distances D_{ij} were plotted against $\log_2(HI_{ij})$ and the relationship $D = C - \log_2(HI)$, where C is a constant, appeared to summarize the relationship between HI values and shape space distances.

The next procedure is to determine the relative coordinates of the points representing viruses and antisera using metric MDS for which the distance between the HI values itself is an input. From the relationship between the shape space distance and the HI assay value, $D = C - \log_2(HI)$, metric MDS uses the following energy function

$$E = \sum_{i=1}^{N} \sum_{i=1}^{M} (b_j - \log_2 HI_{ij} - D_{ij})^2, \qquad (4.16)$$

with b_j set to the $\log_2$ of the maximum HI value for antiserum j. D_{ij} is the shape space distance between antigen i and antiserum j. By minimizing the energy in Eq. (4.16), all the coordinates of antigens and antisera were identified for influenza A (H3N2) viruses from 1968 to 2003 [Smith $et\ al.$ 2004]. From the coordinates in shape space, the antigenic distances among viruses were then measured. Calculated antigenic distances were compared with the sequence distances, either amino acid substitution distance or maximum likelihood phylogenetic tree distance [Smith $et\ al.$ 2004].

Interestingly, antigenic distance among influenza viruses showed a linear relationship with sequence distance. Thus, at least in this case, shape space distance and genetic (sequence) distance were found to be well correlated. This may not be true for all viruses and in particular for HIV.

4.7 Summary

We have shown how various methods of representing B and T cell receptors, antibodies, and antigens have been used in computational models of the immune system. While these methods have provided insights into immunological phenomena there is, as yet, no consensus on the best or most accurate representation of recognition molecules in the immune system. Hopefully, future work will suggest new representations or provide more evidence about the relative merits of the different representations presented here.

Acknowledgments

This work was done under the auspices of the U.S. Department of Energy under contract W-7405-ENG-36 and supported by NIH grants AI28433 and RR06555 and the Human Frontiers Science Program grant RGP0010/2004.

Modelling Immunological Memory

Simon Garrett[1], Martin Robbins[1], Joanne Walker[1], William Wilson[2], and Uwe Aickelin[2]

[1] Computational Biology Group, Department of Computer Science, University of Wales, Aberystwyth, SY23 3PG. Wales, UK. {smg,mjr00}@aber.ac.uk
[2] School of Computer Science (ASAP), University of Nottingham, Nottingham, NG8 1BB. England, UK. {w.wilson,uwe.aickelin}@notts.ac.uk

Summary. Accurate immunological models offer the possibility of performing high-throughput experiments *in silico* that can predict, or at least suggest, *in vivo* phenomena. In this chapter, we compare various models of immunological memory. We first validate an experimental immunological simulator, developed by the authors, by simulating several theories of immunological memory with known results. We then use the same system to evaluate the predicted effects of a theory of immunological memory. The resulting model has not been explored before in artificial immune systems research, and we compare the simulated *in silico* output with *in vivo* measurements. Although the theory appears valid, we suggest that there are a common set of reasons why immunological memory models are a useful support tool; not conclusive in themselves.

5.1 Introduction

One of the fundamental features of the natural immune system (NIS) is its ability to maintain a memory of previous infections, so that in future it can respond more quickly to similar infections [Sawyer 1931]. The mechanisms for immunological memory are still poorly understood and, as a result, are usually highly simplified during the construction of artificial immune systems (AIS).

Although all AIS are *inspired* by the immune system, see Chapter 3 of this book, here we study more detailed immunological *models*. Immune system models will be required by theoretical immunologists if there is to be a significant increase in the generation of new ideas in the field because computational simulation is considerably faster than laboratory experiments. So far, however, this has not been practical because the granularity of the simulations has been far too large, and single systems are able to either generate high-level, global immune simulations, or detailed but partial simulations, but not both.

We differentiate between a model and a metaphor. In AIS there are several metaphors, such as clonal selection methods, negative selection methods, and network methods that provide computational tools for the AIS practitioner. These are not models. Models are an attempt to create an artificial system that displays the same behaviours as another (normally natural) system. Metaphors simply use the natural system as inspiration for an algorithmic device.

Here we focus on the creation and use of immunological models in immunology. There may be side-effect benefits from these models that inspire the discovery of new computational methods in AIS, but that is not our central aim here. We outline a system, still under development, that can provide fast, detailed immune simulations, and which is beginning to suggest *in vivo* effects with enough accuracy to be useful as an immunology support tool. We choose immunological memory as our application area. This chapter:

- Provides a survey of immunological memory, including well-known theories, and a new immunological memory theory that may be of interest to AIS practitioners.
- Provides a survey of existing immune simulation systems.
- Describes how we built and tested a simple set of immunological memory models, and then expanded this approach to a more advanced, generic simulator.
- Describes how we tested the validity of a new theory of immunological memory [Bernasconi *et al.* 2002]. First we used the advanced immune simulator to generate *in silico* results from the new theory. Then, since this theory was generated in response to *in vivo* results, we evaluated the reliability of that theory by comparing our *in silico* results with the *in vivo* results.

Our advanced simulator is fast, even when simulating 10^8 lymphocytes, the number present in a mouse. It also has the ability to simulate cytokine concentrations, which proved vital in simulating the work of [Bernasconi *et al.* 2002]. The simulator's speed and flexibility allows it to be applied to tasks that were previously impossible. Furthermore, our new simulator is not just a one-off immune simulation for a single task, rather it is designed from the ground-up as a reusable, flexible tool for research.

5.2 Background

5.2.1 Immune Memory

As with many aspects of immunology, our understanding of the processes underlying immunological memory is far from complete. As Zinkernagel et al say, in their seminal paper on viral immunological memory, *"Browsing through textbooks and authoritative texts quickly reveals that the definition of immunological memory is not straightforward."* [Zinkernagel *et al.* 1996]. Many of the questions they raised are still relevant almost ten years later. There are several theories, some of which appear mutually exclusive, and there is experimental evidence used to support almost all of these theories. Before examining the techniques for modelling theories of immunological memory, we need to discuss the theories themselves.

It is now widely accepted that hyper-sensitive *memory cells* exist, and research has been conducted in order to describe their attributes and behaviours, e.g. [McHeyzer-Williams & McHeyzer-Williams 2005]. Memory cells come in at least two varieties: memory B-cells and memory T-cells. These cells are formed during (or soon after) an immune response. Acute viral infections induce two types of long-term memory: humoral immunity, in which B-cells produce antibodies to tag cells infected by viruses, and cellular immunity, in which T-cells, activated by specific viral antigens, kill the virus-infected cells and also produce cytokines that prevent the growth of viruses and make cells resistant to viral infection[3].

It has been established that a memory of an infection is retained for several years or even decades [Sawyer 1931, Paul *et al.* 1951]. One way to measure the strength of this immune memory is by counting the population of specific memory cells. This figure tends to fall rapidly immediately after an infection, reaching a stable (but reproducing) level that is maintained over many years or decades, even in the absence of re-exposure to the antigen. The challenge facing immunologists is to discover how these cells are maintained.

Underlying these issues, it seems likely that some sort of homeostasis mechanism maintains a stable total population size of memory cells. Evidence suggests that the total number of memory cells in the body must remain roughly constant, and it has been shown that any increase rapidly returns to this resting concentration [Tanchot & Rocha 1995]; indeed, it is common sense that the number of cells could not increase indefinitely within the fixed volume of the immune system's host. One possible explanation for this is that memory cells (particularly T-cells) release cytokines that have an inhibitory effect on any enlarged antibody sub-population.

Overall, what differs in the theories of immunological memory is: (i) how memory cells are formed, and whether they are qualitatively different to other B- and T-cells, and (ii) how memory cells are maintained in the long term, so that the memory of the primary response is not lost by cell death.

Long-Lived Memory Cell Theory: Given that lymphocytes (both B- and T-cells) differentiate into 'memory cells', and that these memory cells are then highly responsive to the original antigenic trigger, the simplest way of implementing this in nature might be to invoke very long-lived memory cells. In this case, we would assume that there is no cell-division, the memory cells just live a very long time: moreover they *must* do so if they are to preserve immunity for many years. Is this

---

[3] from http://www.emory.edu/EMORY_REPORT/erarchive/2000/February/ er-february.21/2_21_00memory.html.

possible, since the majority of our cells have a life-span much shorter than that of the body as a whole, and so cells are continually dying, and being renewed?

Zinkernagel et al say that there's no convincing evidence for this type of pheno-type [Zinkernagel *et al.* 1996] and current opinion, such as McHeyzer-Williams and McHeyzer-Williams', agree [McHeyzer-Williams & McHeyzer-Williams 2005]. Fur-thermore, experimental evidence contradicts the long-lived memory cell theory. A series of experiments on mice showed that memory T-cells can continue to divide long after any primary response [Tough & Sprent 1994, Tough *et al.* 1996]. Since a stable population is maintained, this means that memory cells must also be dying at a similar rate, and are therefore not as long-lived as originally believed.

Furthermore, it has been known for decades [Sawyer 1931, Paul *et al.* 1951] that antibody produced in response to an antigen can persist at significant levels in serum for years after the initial infection has occurred. Antibodies cannot survive in the body for a particularly long length of time, so we can conclude that plasma cells are sustaining these concentrations (the primary source of antibody). The problem is that plasma cells, in mice, have been shown to have a life-span of just a few months [Slifka *et al.* 1998], and that they are *only produced by differentiating memory cells.* This evidence shatters the theory of long-lived memory B-cells, and draws us to the conclusion that memory B-cells – like their T-cell equivalents – are being continually cycled long after any infection has been dealt with.

Emergent Memory Theory: To address these issues, a Emergent Memory theory suggests that there are no special memory cells as such, rather the effector cells naturally evolve towards highly specific cells, and are preserved from apoptotic death via some sort of 'preservase' enzyme, such as telomerase [Weng *et al.* 1997]. Although it is unlikely that emergent memory is stable in itself [Wilson & Garrett 2004], the process would explain how memory cells are created: they are just specialised forms of effector cells.

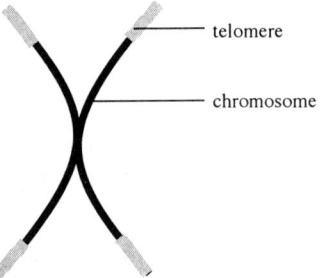

+ Telomeres protect the tips of the DNA in our cells -- including immune cells...
+ When they get too short, the cell cannot reproduce further.
+ Telomerase increases the length of the telomeres (adds TTAGGG x n)

Fig. 5.1. Telomeres protect the tips of our chromosomes, and allow cells to repro-duce successfully.

Each cell in our bodies can reproduce only a predefined number of times, as defined by the length of its *telomeres*. Telomeres are DNA sequences that 'cap' and protect the tips of our chromosomes, which are shorted each time the cell reproduces, indeed (Fig. 5.1), "*... each cycle of cell division results in a loss of 50 - 100 terminal nucleotides from the telomere end of each chromosome.*" [De Boer & Noest 1998]. What if the degree of telomere shortening were inversely proportional to the affinity between the cell's antibodies and antigen? In that case strongly matching immune cells would tend to survive longer than weakly matching ones.

This principle is not new in immunology – de Boer has suggested a model based on similar concepts [De Boer & Noest 1998]. Dutton, Bradley and Swain agree that the death rate is a vital component required in establishing robust memory. "*It stands to reason that activated cells must escape cell death if they are to go on to be memory. Thus, factors that promote the survival of otherwise death-susceptible T cells are candidates for memory factors.*" [Dutton et al. 1998].

Consider the impact of this hypothesis in the context of different types of immune cells. Grayson et al state that, "*... memory T-cells are more resistant to apoptosis than naïve cells ... Re-exposure of memory cells to Ag* [antigen] *through viral infection resulted in a more rapid expansion and diminished contraction compared with those of naïve cells.*" [Grayson et al. 2002]. This indicates that memory cells would have lower (but not zero) death rates, and higher proliferation rates, so the the cell population would naturally contract to long-lived (i.e. high-affinity) cells over time.

Telomerase may not be the only biological mechanism that can explain the evolution of immune cells into longer lived, higher affinity memory cells, an alternative explanation underpinning the longer life-span of memory cells is provided by Zanetti [Zanetti & Croft 2001]: the "*...selection of B-cells destined to become memory cells takes place in GCs* [germinal centres] *and is controlled by the expression of intracytoplasmic molecules (Bcl-2 and Bcl-x) which prevent a form of cell death ... together with the concomitant suppression of signals from cell surface proteins that lead to death.*" Although differing from the telomerase hypothesis, the implications would be the same: memory cells appear to reflect normal immune cells that have naturally evolved to develop a lower death rate, ensuring their survival over other cells such as effectors.

The problem with the Emergent Memory theory is that it is very cell-specific. How can a concentration of cytokines ensure a high affinity cell lives longer than a lower affinity cell in almost the same location?

Residual Antigen Theory: Several reports suggest that protein antigen can be retained in the lymph node (e.g. [Perelson & Weisbuch 1997]), suggesting that normal lymphocyte function cannot remove *all* traces of a particular class of antigen. This is a natural result of the immune system being focussed on particular locations in the body. Whilst most antigenic material will be cleared by the immune system, causing an immune response, some antigenic material will escape a localised immune response long enough to reproduce. The immune system then quickly establishes a steady state between immune response and antigenic population size, and the immune system's population is stimulated by the normal hypermutation response.

Therefore, it is possible that the immune system does not completely remove all antigenic material from the host, either because small concentrations of antigenic cells may remain long enough to reproduce, or because the immune system itself has retained some of the antigenic material in follicular dendritic cells (FDCs). These FDCs then slowly release the antigenic material into the host, to stimulate a low-level immune response. Zanetti et al say, "*The prevailing view is that maintenance of B cell memory ... is a function of the persistence of antigen on FDCs ... only a few hundred picograms of antigen are retained in the long term on FDCs, but these small amounts are sufficient to sustain durable and efficient memory response.*" [Zanetti & Croft 2001]. In either case, this would keep the immune system active enough to sustain memory cell populations. This idea has been supported by research suggesting that B-cell memory is particularly sensitive to residual antigen [Tew *et al.* 1990].

In recent years however, compelling evidence has been presented suggesting that the cycling of memory T-cells continues to occur without any of the specific antigen being present [Lau *et al.* 1994], which would mean that these cells must be responding to some other stimulus. Although some debate has occurred [Manz *et al.* 2002, Zinkernagel 2002], this view is now widely accepted by immunologists [Antia *et al.* 2005].

An additional objection stems from an evaluation of the performance of such a system. How could it be efficient, from an evolutionary point of view, to expend resources on what is essentially a rote learning approach to memory? We know, from Machine Learning, that rote learning is the least efficient method of storing learned information, and it does not allow for generalisation. Although, there *is* an element of generalisation inherent in the Residual Antigen theory due to the memories of previous infections overlapping with new infections, and providing a (weak) generalised response, it is questionable whether there is enough generalisation to make this an effective source of immune memory.

It may seem that antigen persistence is important for a model of immune memory, to ensure that the high affinity memory cells are sustained over long periods, but there is another, related possibility. Perhaps memory cells do not need stimulation by antigen; they simply proliferate periodically. Would this represent another evolutionary step for an immune cell in order for it to differentiate into a memory cell? Grayson et al identified the discrepancy between the long term behaviour of memory cells and naïve cells and state that, "*... memory cells undergo a slow homeostatic proliferation, while naïve cells undergo little or no proliferation.*" [Grayson *et al.* 2002] (our emphasis). If this is the case, do memory cells actually need persistence of the antigen to survive?

Even if re-exposure is not necessary, Antia et al conclude that "*... estimates for the half-life of immune memory suggest that persistent antigen or repeated exposure to antigen may not be required for the maintenance of immune memory in short-*

lived vertebrates; however, ... repeated exposure may play an additional role in the maintenance of memory of long-lived vertebrates." [Antia *et al.* 1998]. We choose to include antigen persistence in the model presented here.

Immune Network Theory: Network theory is based around the possibility that the immune system maintains and triggers memory by internal, not external stimulation. It suggests that immune cells, particularly lymphocytes, present regions of themselves that are antigenic to other immune cells. This causes cycles of stimulation and suppression, which, while begun by an external antigenic source, are continued and maintained even in their absence, and are thus a form of memory [Farmer *et al.* 1986]. A network of interactions between immune cells is widely believed to account for memory pool homeostasis [Zeng *et al.* 2005, Schluns & Lefrancois 2003], and certain immune cells are even able to form physically connected networks of tunneling nanotubules in vitro [Watkins & Salter 2005], but little evidence has been published recently in the major immunology journals for a strong role of the kind of co-stimulation described above.

Heterologous and Polyclonal Memory Theories: It has been observed that during an immune response, populations of memory T-cells unrelated to the antigen may also expand [Bernasconi *et al.* 2002, Tough *et al.* 1996], suggesting that perhaps serological memory could be heterologically maintained by a degree of *polyclonal stimulation* during all immune responses.

According to [Antia *et al.* 2005], two possible mechanisms have been suggested to explain these results - Bystander Stimulation and Cross-Reactive Stimulation:

(i) The Bystander Stimulation theory suggests that the antigen-specific T-cells produce a cytokine that stimulates all nearby (bystander) memory T-cells to divide. It has been suggested that bystander stimulation could be responsible for the continued cycling of memory B-cells, as well as for T-cells [Bernasconi *et al.* 2002]. The results of this high impact work showed that if memory B-cells are simultaneously exposed to an antigen that they are not specific to, and to the cytokine IL-15, they will undergo clonal expansion. This ability was shown to be unique to memory B-cells, and could not be repeated with their naïve equivalents.

(ii) the Cross-Reactive Stimulation theory is based on speculation that memory cells could be more sensitive to stimulation than naïve cells, and might therefore be stimulated by different antigens, perhaps even a self-antigen. In either case, it has been shown experimentally that memory T-cells specific to a particular antigen can be directly stimulated by a different, unrelated antigen [Selin *et al.* 1994].

Both of these theories suggest that once memory T-cells have been created, they can be stimulated during immune responses to unrelated antigen. The difference is that in one case the cells are directly stimulated by antigen, and in the other (polyclonal stimulation) they are stimulated by cytokines released by other, antigen-specific cells.

5.2.2 A Brief Survey of Immune Modelling

Mathematical Models: Mathematical models of immunological (sub)systems often use ordinary differential equations (ODE) or partial differential equations (PDE) to encapsulate their chosen dynamics (e.g. [Perelson 2002, Smith *et al.* 1999]). Perelson's HIV equations [Perelson 2002], and Smith's influenza dynamics [Smith *et al.* 1999], are illustrations of models of small parts of the immune system dynamics that have had significant benefits to human health, but which do not set out to model the immune system as a whole. In Chapter 4, we have already seen Perelson's detailed models of B cell and T cell receptors. When one considers the chemical complexity of amino acid binding it is not surprising that many balk at the idea of modelling the immune system at all. However, immunological simulations are possible because we observe gross-scale effects (such as primary/secondary responses) that are then modulated to a greater or lesser degree by small-scale processes, such as Perelson's discussion of B and T cell binding. Both are vital for truly accurate models, but larger scale models can be used successfully to explain gross-scale features of the immune system [Yates *et al.* 2001].

Immunological memory has also been modelled in a similar manner— the classic example being Farmer et al's work [Farmer *et al.* 1986] – but there are more recent attempts to model immunological memory too [Ahmed & Hashish 2003]. Although these models say a lot about certain details, they are not intended to be global models of immunological memory. For example, the important work of Antia et al on understanding $CD8^+$ T-cell memory [Antia *et al.* 2005] is based on a few, relatively simple equations. This is not to say that it is easy to generate such equations (it is not); rather, we are saying that the applicability of these equations is limited. Indeed, the difficulty in building and managing these equations is precisely the reason that a computational simulation approach is sometimes more appropriate.

Computational Models: Computational models are not as well established as mathematical models. Those that do exist are usually either population-based (entities that are tracked as they freely interact with each other), or cellular automata (entities that are tracked in a discrete grid-like structure, generally with local-only interactions [Wolfram 2002]). Nevertheless, computational models do have some advantages over mathematical models.

Firstly, it is possible to define, informally, the behaviour of a highly complex system, without formally defining it in terms of formal ODEs or PDEs—we can create a population of entities by mapping from objects in nature to objects in the computational simulation. Furthermore, many ODEs have no analytical solution and can *only* be solved by computational analysis, in software such as Matlab™ and Mathematica™.

Secondly, some forms of *in silico* experimentation may be difficult in mathematical models, and indeed in the immunology laboratory, such as tracking a single B-cell or antibody over its lifetime. It is possible, therefore, that computational immune simulators will provide the only means of investigating some immunological challenges.

In all computational simulations, we re-iterate the importance of the choice of binding mechanism, the type of cell-cell and cell-antigen interaction, (see [Garrett 2003] and Chapter 4 of this book), and we note that the few computational simulators that do exist are often underdeveloped and may not have been peer-reviewed by the academic community.

ImmSim: The work of Seiden et al, on ImmSim was the first real attempt to model the immune system as a whole, and it is still the only simulator to have been fairly widely peer reviewed [Kleinstein & Seiden 2000, Kleinstein *et al.* 2003]. It is similar in style to the work of Farmer et al [Farmer *et al.* 1986], but is a true simulation, not a set of ODEs[4].

Simmune: There are at least two "SIMMUNE" immunology simulators: Meier- Schellersheim's version [Meier-Schellersheim & Mack 1999], which was developed in the late-1990s, and Smith and Perelson's version. Of the two, Meier-Schellersheim is the more advanced, implemented as a full cellular automata with the ability to define almost any rules that the user desired, whereas Smith and Perelson's was a relatively simple, unpublished Lisp simulation.

Synthetic Immune System (SIS): Although SIS appears to be significantly faster and more powerful, it does much less. SIMMUNE can simulate large numbers of complex interactions, whereas SIS is designed only to investigate self-nonself relationships. SIS is a cellular automata; it can only be found on the web[5].

ImmunoSim: Ubaydli and Rashbass's Immunosim set out to provide researchers with an "Immunological sandbox" - it was a customizable modelling environment that simulated cell types, receptors, ligands, cascades, effects, and cell cycles, with experiments run *in silico*. A key requirement was that it should have a purely visual interface, with no programming necessary. It received the Fulton Roberts Immunology prize (twice) from Cambridge University but does not appear to be available as a publication.

Other systems: These simulations [Castiglione *et al.* 2003, Jacob *et al.* 2004] are smaller scale than that proposed here, but have still had benefits to chemotherapy and immunology, and/or highlight problems that need to be overcome. Others have emphasised the importance of the binding mechanism, the type of cell-cell and cell-antigen interaction chosen, and the multitude of other possibilities that should be considered [Garrett 2003].

5.3 Basic Simulations

Our work with a set of Basic Simulations set out to explore the gross-scale behaviour of some of the theories just studied, while keeping the models as simple as possible –

---

[4] ImmSim currently to be found at http://www.cs.princeton.edu/immsim/software.html
[5] at: http://www.cig.salk.edu/papers/SIS_manual_wp_M.pdf

here, the only the interactions simulated are those between antibodies and antigen. This begs the question, "how simple can an effective model be?" Assuming Occam's razor applies, our answer is, "as simple as possible, and no simpler." However, the models described in this section are deliberately *too* simple. This is partly because no one knows how complex a simulation must be before it can accurately reproduce *in vivo* results, partly because by starting as simple as possible we get a lower limit on the computational performance of simple models, and partly (more importantly) because it lets us explore the dynamics underlying simple immune simulations, so that later additional complications can be viewed as modulations of this basic model. Note that the lack of complexity should not be seen as an indication that the models described in this section are trivial. Although simple, great care was taken to ensure they were as realistic as possible, as we hope will become clear.

The Basic Simulations will also act as a primary validation for the *underlying mechanisms* of the more complex experiments. They do not validate any other aspect of the complex experiments. It is easier to verify and validate the performance of a simple model than a complex model; then if the complex and simple models share similar behaviour this partially validates the complex model. This raises another issue: how do we validate immunological models? If we apply standard Machine Learning methodology, where 'models' are 'hypotheses', then we should do some form of k-fold cross-validation to obtain a measure of the accuracy of the defined immunological hypotheses. But how do we do this when we have no well-established 'correct' data? To some extent, we can assume that if a model is able to *predict* what will be observed in nature, then the model is validated to some extent. Indeed, the ability to predict is one of the reasons for building models in the first place. We will return to this point later.

5.3.1 Basic Simulations: Methods and Materials

Each Basic Simulation was built from antibodies and antigen, and no models were allowed to directly create memory; memory had to evolve. This blurs the distinction between antibodies, B-cells and T-cells in order to explore the effects of immune cell/antibody proliferation in response to antigen. To indicate this blurring, we will call the simulated immune system elements 'reactive immune system elements', or RISEs. The RISEs were defined as being more likely to die as they got older; implemented by removing a RISE when $rnd().a > rnd().dr$, where rnd() is a uniform random number generator, a is the age of the RISE measured in generations from the current generation, and dr is a death rate integer, which was set to 30. A constant 50 RISEs were added each generation. This led to a stable population size, which returned to the stable level despite external perturbations. Fig. 5.2 demonstrates this effect: despite a large influx of new RISEs (the large peak) and a small culling of RISEs (the small trough), stability is maintained. The size of the peaks were also reversed, with the same result that the population size returned to a stable level – note also the differences in scale between Fig.s 2(a) and 2(b), which show that the size of perturbation is irrelevant. This implements a simple homeostatic population of RISEs.

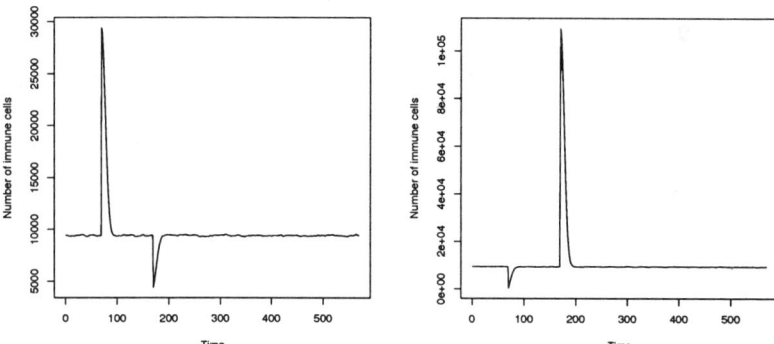

Fig. 5.2. The resting population of B-cells was in homeostasis. These graphs show the population stability that underlies all models that will follow. Any positive or negative change to the population size is quickly corrected, and stable population size is restored.

Antigen populations were 'injected' into the system as a whole, at a predefined times. The primary infection was always at generation 70, and the secondary infection was either at generation 120 ('smallGap' experiments), or generation 420 ('bigGap' experiments), to test the short- and long-term memory abilities of the population. An antigen was removed once is was bound to an RISE, and binding could only occur when the similarity between the RISE and antigen was within a distance of 100. The RISEs could take any value between zero and 10,000, and the antigen always had a randomly chosen value of 3.3, fixed at this value for all tests. In all cases we assume that the strongest affinity RISE will bind with the antigen. We implement this by a form of tournament selection, whereby the strongest matching RISE of ten randomly chosen RISEs is chosen to be the one that actually binds. Our more complex simulation, presented later in this chapter, uses simulated chemotaxis.

For each experiment, we measured the total number of RISEs, the total number of antigen, and the number of RISEs with affinity in the ranges, [0.01-0.1), [0.1-1), [1-10), [10-100), [100-1,000), [1,000-10,000) and [10,000-100,000). We recorded this information every generation for 600 generations.

5.3.2 Basic Simulations: Experiments and Tests

We performed the following experiments and tests.

Memory By External Stimulation These experiments tested the ability of the Basic Simulations to remember infections over a short and long period of time, assuming the only stimulation to be external, i.e. via antigenic interaction:

Control/None : On top of the homeostasis mechanism, we tested a standard implementation of clonal selection. This is activated by the presence of antigen, so that a good-matching RISE produces many clones, and a poor-matching RISE produced few clones. Furthermore, the good-matching clones are only slightly mutated from their parent cells, via a Gaussian centred on the parent, whereas the few poor-matching clones are often highly mutated, relative to their parents. This approximates Burnet's clonal selection theory [Burnet 1959] and acts as a control for these experiments. Since memory cells were not explicitly created, we would expect the RISE population to clear the antigen, and then forget the infection.

Emergent Memory : In the emergent memory tests, when a RISE was bound to antigen, the RISE's age was reduced in proportion to its affinity to the antigen, so that better-fitting RISEs tended to survive longer – this implemented the effects of 'preservase'. This should preserve the high matching RISEs to some extent, producing a form of memory.

Residual Antigen : Once the antigen population had been injected, a single antigen was then re-introduced into the simulation at random time intervals (on average, every three generations). Would this prevent the memory of the infection from being lost because this value is considerably smaller than dr? If so, under what conditions? It might be argued that this does not really represent residual antigen, as antigen are being reintroduced rather than maintained, however the purpose of this model is to show whether a small amount of stimulation can maintain memory, not to demonstrate mechanisms by which the antigen could be maintained, and thus in practical terms reintroduction performs the same role in our model as maintenance (keeping a small, stable population of antigen), with the advantage of allowing us to simplify the experiment.

Both Emergent Memory and Residual Antigen : Is there any benefit in implementing both the Emergent Memory and Residual Antigen theories?

Memory By Internal Stimulation These experiments tested the effects of adding internal stimulation to the Basic Simulations, so that one RISE could interact with another RISE, even in the absence of antigen. Although antibody-antibody interaction is not widely thought to be a form of memory in nature, it does occur, and is likely to have some function. These tests set out to suggest what that function might be. The graphs, described above, of affinity level distribution are of particular relevance to these experiments.

It is important to note that we do not use paratope-paratope binding here: i.e. we do not assume that a RISE/antibody's light chain will bind with the light chain of another RISE/antibody, for reasons outlined in [Garrett 2003] (e.g. the problems of positive feedback). Instead we shift ideal binding by 2,500 (in a circular range of 10,000) so that a RISE with value 1,000 would bind most strongly with another RISE of value 3,500. This means there would need to be a cycle of four RISEs if internal memory were to work. This implements paratope-epitope binding, although we note that there is still a functional relationship between the paratope's and the epitope's shape space, which is less than realistic, but was necessary to keep the simulation simple.

5.3.3 Basic Simulations: Results

Memory By External Stimulation The results are presented in Fig. 5.3 and Fig. 5.4. These are averages of ten runs. Remembering that the population is completely renewed on average every 30 generations, even the short gap experiments (left column of graphs; 50 generations between infections) should not have shown any memory of the previous infection.

In the "None" graphs (top line) we actually see a slight increase in response, but this is not statistically meaningful – there is no memory of previous infections. The statistics we used were the Wilcoxon Signed Rank Test, and the results are tabulated in Table 5.1. This test allowed us to decide when the difference in height between the primary and secondary responses was significant, and the ratio expresses the extent of that difference. This non-parametric test was chosen because it is likely that the secondary response is conditioned by the primary response, and that the data are *not* normally distributed.

The affinity graphs in Fig. 5.4 indicate there is an increase in RISEs that have affinities in the < 0.1, < 1.0 and < 10.0 ranges, but there is no memory between infections.

The "Emergent" results show a distinct secondary response in the short gap experiment, because the population members that were able to successfully bind were preserved beyond 30 generations; however, this effect is not enough to allow memory to persist over the big gap because the antibodies that were effective against the primary infection, tended to die over that time period. Nevertheless, the results indicate that memory can be preserved for at least 50 generations.

Now the affinity graphs show that high affinity RISEs are maintained between the infections that are separated by a small time gap, and how these types of RISE drain away over the longer time gap so that the simulated immune system needs to begin again to find a high affinity response to the antigen.

The "Residual Antigen" tests have a similar pattern in Fig. 5.3, with the population stimulated enough by the on-going, low-level antigen to promote a secondary response in the short gap experiment. In the big gap experiment, however, the effect is not statistically significant.

The affinity graphs show an elevated number of high- to mid-range affinity RISEs (in the < 1, < 10 and < 100 ranges) but indicates the very high affinity RISEs return to lower levels by 200 generations. This explains why the secondary response

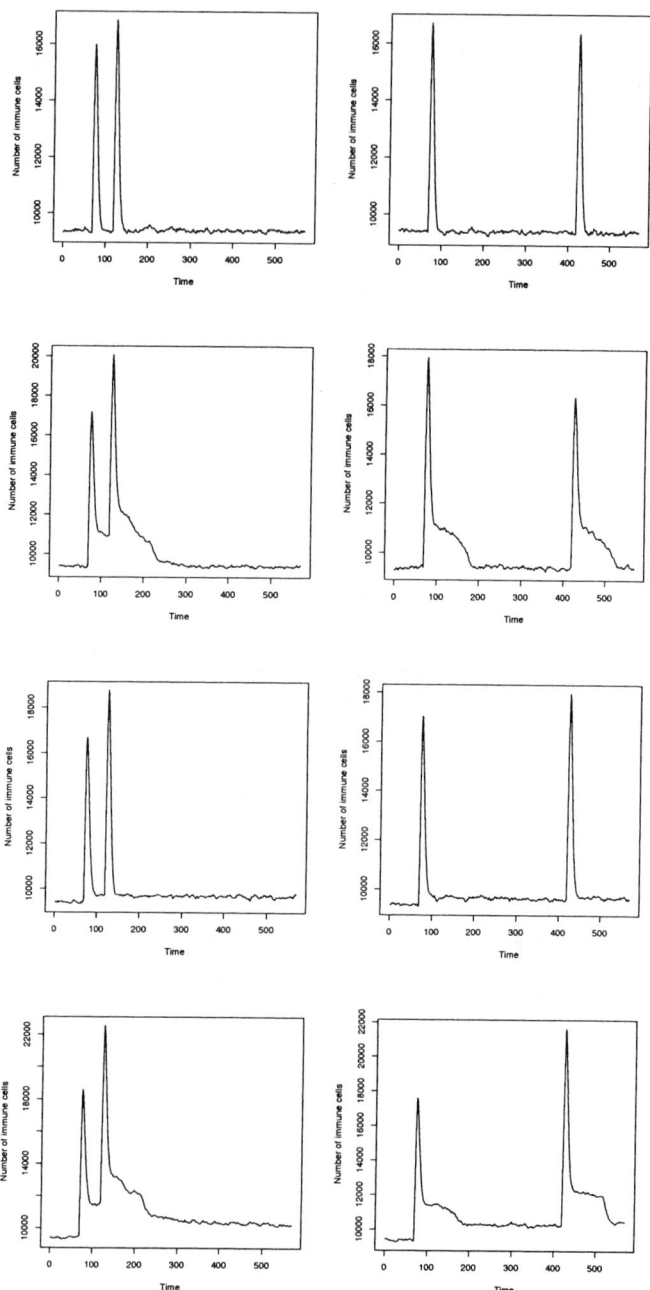

Fig. 5.3. Graphs of the theory simulations, "None", "Emergent", "Residual" and "Both" (top to bottom, in order) for a small time gap (50 generations, left column) and a longer time gap (350 generations, right column), averaged over 20 runs.

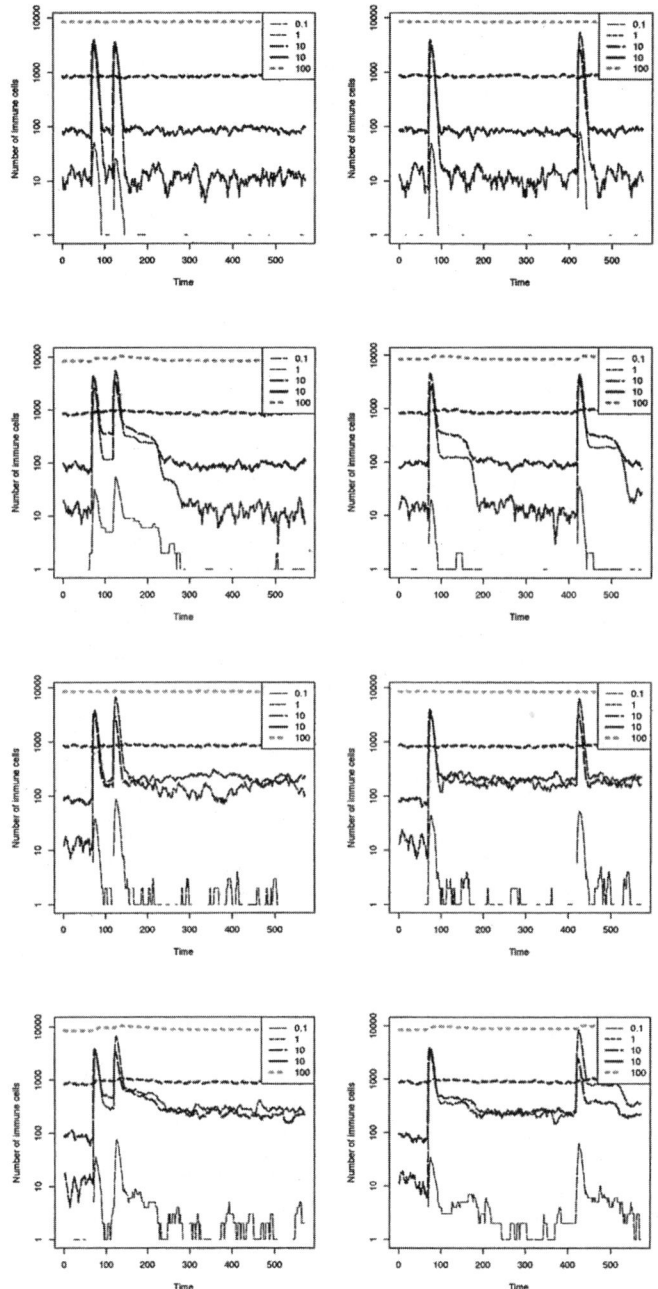

Fig. 5.4. Graphs of the theory simulations, "None", "Emergent", "Residual" and "Both" (top to bottom, in order) for a small time gap (50 generations, left column) and a longer time gap (350 generations, right column), averaged over 20 runs.

was not sufficient to be statistically meaningful when the antigenic injections were separated by a large gap.

With "Both" emergent and residual antigen implemented, the story is different. Now, we see strong secondary responses for both short and big gaps, although there is a slight sustained, global increase in RISE population after the first infection.

The affinity graphs also show that the high affinity < 0.1 RISEs never returned to the zero mark. This appears to have been crucial in maintaining a powerful secondary response, and corresponds to the existence of high affinity memory cells in nature.

Some may ask why the residual antigen phenomenon does not explain immune memory on its own. If the amount of residual antigen were high enough then surely the immune response would be enough to remember that infection? Indeed, this is true, but at the cost of a permanently raised antibody population level, which is not seen in nature. At the extreme, if the infection were to persist at the same high levels then it is obvious that the memory would not be lost, because the infection would be continuous and on-going, but this is also not a realistic state of affairs, except in pathological cases, such as in elderly patients who are infected with cytomegolovirus [Perelson 2002]. The level chosen is one that only very slightly raises the antibody population size: it is enough to maintain memory over a short period, but not in the longer term.

Furthermore, Residual Antigen does not explain why better matching cells tend to survive and worse matching cells tend to die off; nor does it explain how memory cells can naturally emerge as a result of immune cell evolution. As a result, both the apoptosis reduction (or telomerase memory maintainance mechanism), and the re-stimulation mechanism are required to evolve an effective immune response.

Memory By Internal Stimulation The results are presented in Fig. 5.5 and Fig. 5.6. For each run, each RISE attempted to bind to the RISE with the highest affinity out of ten randomly chosen RISEs. There does not appear to have been any memory effect; indeed, the opposite seems true – as soon as any subpopulation increased in size out of proportion to the population as a whole, the network effect reduced the size of that subpopulation. This made the levels in Fig. 5.6 more *stable* than the comparative graphs in Fig. 5.4.

We conclude that the memory effects of immune networks are limited — at least the types of network that we have implemented here. Since our aims in these basic experiments are to produce simple models of immunological interactions, we use non-symmetric, paratope-epitope binding, in which knowing that A binds B does not imply B binds A. In contrast, AIS network algorithms tend to use paratope-paratope binding because it is of interest from a computational point of view, even if it is less biologically tenable.

Experiment	p-Value	99%	Ratio
None Small Gap	0.240	No	0.948
None Big Gap	0.955	No	1.011
Emergent Small Gap	0.0000957	Yes	0.852
Emergent Big Gap	0.225	No	1.067
Residual Small Gap	0.00318	Yes	0.890
Residual Big Gap	0.332	No	0.945
Both Small Gap	0.0000957	Yes	0.822
Both Big Gap	0.0000957	Yes	0.813
Network Small Gap	0.765	No	0.999
Network Big Gap	0.896	No	1.001
All Small Gap	0.0000942	Yes	0.850
All Big Gap	0.000315	Yes	0.832

Table 5.1. Results of the Wilcoxon Signed Rank Test for difference between the size of the two peaks in each experiment. The p-values are shown to 3 significant figures and whether or not the difference can be regarded as significant at the 99% confidence level. The smaller the p-values the greater the degree of confidence that there is a difference between the primary and secondary responses, with 1.0 being zero confidence, and 0.0 being 100% confidence. The ratio gives the size and direction of the difference between the two peaks.

5.4 Experiments Using the Sentinel System

5.4.1 Method and Materials

The simulations that form the basis of this chapter were modelled using our software, 'Sentinel'. Sentinel is an agent-based complex system simulation platform for immunology and AIS research that currently exists as a prototype. Its design is based largely around the principles of cellular automata, with the environment divided into a discrete grid of locations. Entities within the simulation are free to move around in this environment, but are only able to respond to events that occur within closely neighbouring cells. 'Engines', such as those used in computer games for managing graphics, physics, etc., manage the physical and chemical interactions that occur within this environment.

The physics engine allows accurate simulation of the physical properties of agents, restricting their movements according to attributes such as size, mass or energy output. Whereas many simulations or differential equation models are exclusively based on cells that exhibit some form of Brownian motion, entities (cells) in Sentinel move according to the chemical stimuli they receive, their motor capabilities, and

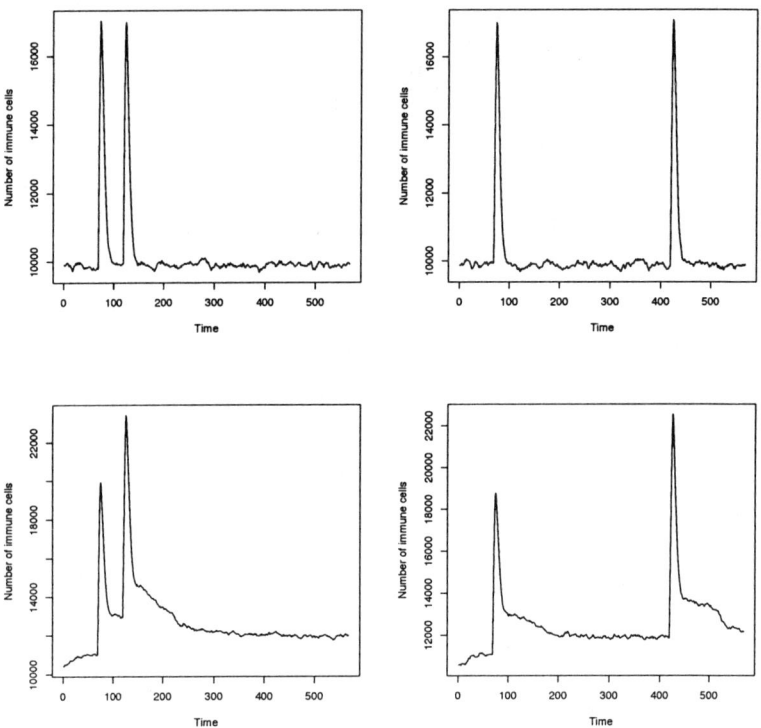

Fig. 5.5. Graphs of the theory simulations, "Network" and "All" (top to bottom) for a small time gap (50 generations, left column) and a longer time gap (350 generations, right column), averaged over 20 runs.

external forces acting upon them. The physics engine ensures that movement is as realistic as possible, and is a novel feature of our system.

A chemistry engine is responsible for managing chemical and biochemical reactions, and also the distribution of extra-cellular molecules throughout the environment. For example, if a cell releases a particular kind of cytokine at its location, the chemistry engine will cause that cytokine to gradually disperse across the environment (see Fig. 5.7, right, for an example map of densities) by diffusion. This feature is essential for the accurate simulation of cell movement by chemotaxis – the process by which immune cells move towards higher concentrations of chemotactic factors, i.e. chemicals that attract them. It also enables a cell to influence a larger expanse of its environment than would typically be allowed in a cellular automata.

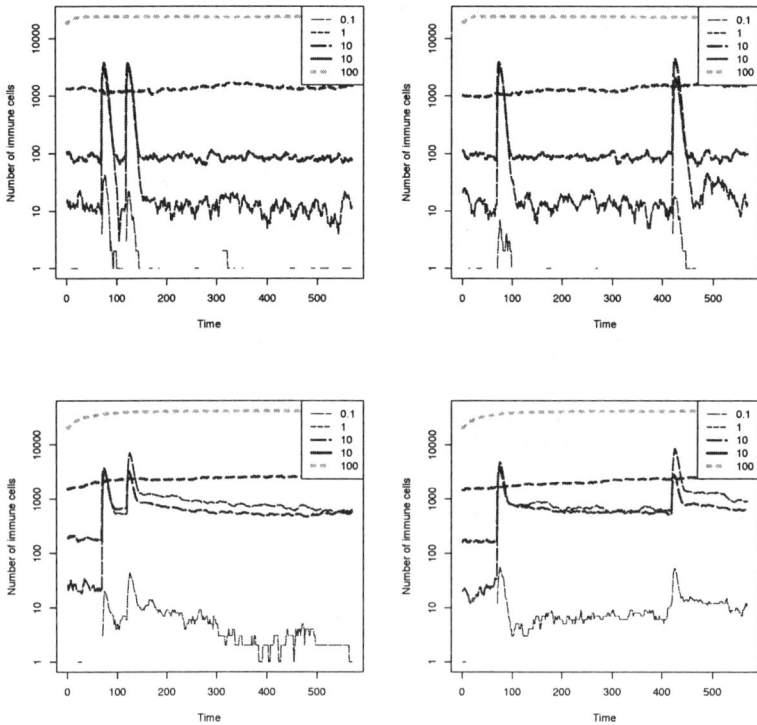

Fig. 5.6. Graphs of the theory simulations, "Network" and "All" (top to bottom) for a small time gap (50 generations, left column) and a longer time gap (350 generations, right column), averaged over 20 runs.

The implementation of chemotaxis is another novel feature of Sentinel. Cells *in vivo* are able to respond to various chemotactic molecules by detecting density gradients, and moving towards the highest or lowest density of that agent [Ramsay 1972]. The dispersal of chemotactic molecules in Sentinel is calculated by dispersing molecules from each location in the simulation to its neighbours over time. A cell in the simulation is able to access its eight neighbouring locations to find out the densities there, and retrieve the highest or lowest density of a particular molecule. It can then use this information to move accordingly.

Given a set of entities and chemicals (B-cells, antibodies, memory cells, cytokines, etc.), the influence of the physics and chemistry engines is defined by a number of *rules*. These rules define when an entity can interact with another cell, and the nature of that interaction; how one cell releases chemicals, or other entities, into its near environment, and any global features, such as blood flow that affect all entities and chemicals.

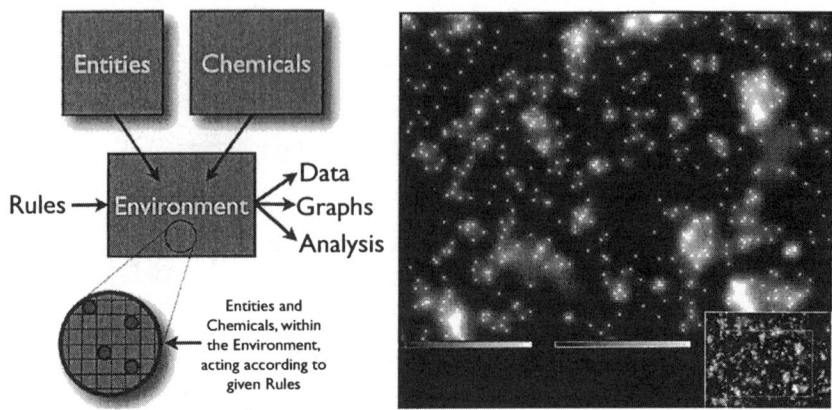

Fig. 5.7. (left) The structure of the Sentinel system. (right) Sentinel models the diffusion of chemicals to implement realistic chemotaxis and, crucially, to model the effects of cytokines (see text). The main figure shows the different concentrations of chemicals over a detailed view of the simulator's simulation environment. The inset shows the location of the detailed view in the whole space being modelled.

Having defined the simulation model, by choosing the entities, chemicals and rules, the simulator is run and information is output according to user-defined data-feeds. These data can then be viewed in the form of various graphs and samples, or streamed to log files for analysis, all within the Sentinel system. It seems likely that this simulator architecture will be useful in other areas too, such as biochemistry and abstract work in genetic and evolutionary computing.

The simulator is complemented by an Integrated Development Environment (IDE), that provides a set of powerful tools for the rapid development of new models. The drag-and-drop graphical interfaces allows the user to quickly choose sets of agents and establish the links between them, and to set up and connect areas of the environment, and describe the rules of physics that will operate within them. A code editor allows users to develop Java-based extensions to these basic models, with the assistance of automated code-generation tools, and a comprehensive Application Programmers Interface (API) that provides general-purpose functions for manipulating agents and the environment. In many respects, the system is somewhat similar in nature to platforms such as Robocode[6], but far more powerful.

Sentinel can simulate several million cells, hundreds of millions of antibodies, and their interactions, on a typical high-end desktop. Although this figure varies depending upon the complexity of the model, Sentinel appears to be the most powerful simulator currently available, especially in view of the complex interactions that it is simulating. Sentinel's ability to simulate diffusion is very important – cytokine

---

[6] See http://robocode.sourceforge.net

signalling between cells is a vital part of immunology. Indeed, one of the following experiments could not have been implemented without this ability.

5.4.2 Sentinel Experiments and Tests

Sentinel Validation Tests: Before using Sentinel to evaluate Bernasconi et al's theory, we validated its performance. Both the validation and the evaluation models ran with of the order of 10^8 B-cells. We recapitulated the "None", "Emergent" and "Residual" experiments, as in the previous section, but did not implement "Network" because it had little value for our goals here. By implementing the same tests as the Basic Simulations, we set out to show that Sentinel would work at least as well as the Basic Simulations. If the results are qualitatively the same then we will have demonstrated that Sentinel can reproduce previous results. Each of our simulations were run ten times, in order to ensure that the results were consistently reproduced.

Sentinel 'Theory Evaluation' Experiment: This experiment is designed to explore the veracity of Polyclonal Activation Memory, via simulation – something which has not been done before. We could not use our Basic Simulation tool because the experiment required implementation of cytokine gradients (of IL-15), and needed to be performed on a much larger scale to obtain meaningful results. Only Sentinel could meet these requirements.

The construction of Bernasconi et al's model is based on the theory described in [Bernasconi *et al.* 2002]. They suggested their theories as a result of *in vivo* experiments, and claim that the experimental results provide compelling evidence for bystander stimulation of memory B-cell populations. The comprehensive set of results published in [Bernasconi *et al.* 2002] will be tested against the data from our simulation, so our aim is to simulate the implications of Bernasconi et al's theory, and assess whether it could indeed be responsible for the *in vivo* results that they observed. Despite our validation efforts, the process described above is fairly limited and the process of parameterising any simulation is complex, therefore we can only safely look for qualitative similarities in between the results of Bernasconi et al and those produced by Sentinel.

5.4.3 Assumptions

In constructing these Sentinel models, a number of assumptions were made. These have been kept consistent through all the simulations conducted.

Repertoire: Sentinel's simulation repertoire included B-cells, antibodies, antigen, as well as a signalling chemical. It was more complex in terms of the entities used, and used many orders of magnitude more antibodies, than the RISEs in the Basic Simulations.

Longer-lived memory cells: Memory B-cells live longer than their naïve equivalents. In nature, a naïve B-cell tends to live for about 24 hours unless it receives stimulus, at which point it is "rescued, and may go on to live for a few months [Bernasconi *et al.* 2002] This is reflected in our models.

Antigen: Antigen does not reproduce or mutate during the simulation.

Simplified binding: As in the basic simulations, and in order to provide the best possible performance, a simplified binding mechanism was used. A strain of antigen is given a number between 0 and 20,000, which remains constant across the population. Every new B-cell is assigned a random number within that range, and the binding success is measured as the distance between the two numbers.

Clonal selection: In response to antigen, B-cells undergo clonal selection and hypermutation, as described by Burnet's 1959 theory. [Burnet 1959]. Cells that have been cloned retain the binding integer value (see previous bullet point) of their parent, mutated in inverse proportion to its binding strength.

Simplified Immune Repertoire: The simulation consists of B-cells, antibodies and antigen, plus one signalling cytokine. B-cell T-cell interaction is not simulated in these tests, but are planned (see Further Work). We needed to keep the model as similar to our previous system as possible (no plasma cells) to make the validation process as meaningful as possible.

5.4.4 Sentinel Results

Sentinel's Validation Results

The results in Fig. 5.8 show that Sentinel correctly produces a secondary response to a repeat infection of the same antigen, for both memory theories. Furthermore, Sentinel's results show that the Residual Antigen model maintained a considerably higher population of memory cells and antibodies – down to only about 10^6 antibodies before second infection, compared to of the order of 10^1 for the other memory models. This relates to the Basic Simulations that showed the residual antigen populations had more antibodies.

In both simulators, the models of the Emergent, 'preserveron' theories sustained good short-term memory, and in both simulators we observed the memories stored in this manner failing when the cells carrying them died. Unless we accept that the primary immune response produces memory cells that live for years, such models will always result in an immune memory that fades over time.

The model of the Residual Antigen theory sustained a stable level of memory cells in both simulators, and was able to produce a substantial secondary response regardless of the length of time between the first infection and subsequent re-infection. It appears to be a viable model of immune memory; however, the requirements to sustain such a system seem unlikely to be met in nature because the immune system would have to produce such material over a highly extended period. Indeed this point was debated several years ago [Matzinger 1994a].

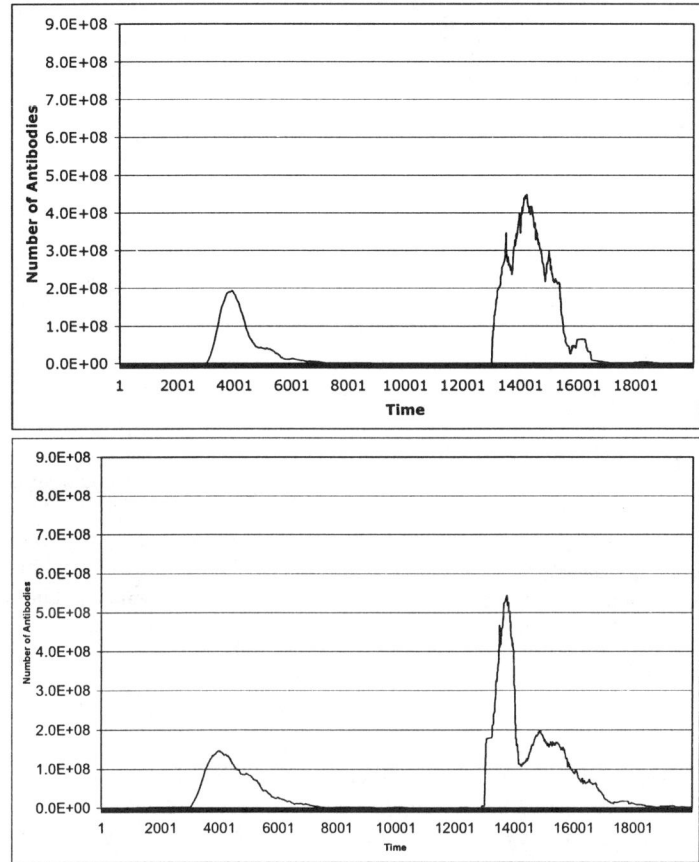

Fig. 5.8. Validation graphs for the number of cells over arbitrary time for: (i) the Emergent/'Preserveron' model, and (ii) the Residual Antigen model. Antigen A is injected at t=3000, and t=13,000.

Although there are some differences in the details, such as the more pronounced secondary peak in the secondary response, we consider the two simulators similar enough to proceed with the qualitative comparison of the *in vivo* and *in silico* results.

One advantage of Sentinel is that we can now distinguish between the secondary responses from the various theories: (i) the 'Preservon' model has a wide response, but it does not lead to as many antibodies being created, and the after-response is small, and (ii) the Residual Antigen model has a sharp, medium height secondary response, with a much extended, exponentially decreasing after-response.

Our previous experiments were too coarse-grained to provide results that had meaningful differences, and the curves they produced were an almost perfect exponential followed by a slower, almost perfect exponential decrease. Interestingly, there is a slight 'wobble' at the end of the exponential decrease, which is also consistent with the after response we see in the graphs of Fig. 5.8. These experiments show that we can reproduce the results of the Basic simulations, but with finer resolution.

Sentinel's 'Theory Evaluation' Results

Since we stated in the 'Experiments and Tests' subsection that we have not validated the finer-grained elements of Sentinel's results, we will compare the results, in a qualitative way. Fig. 5.9 shows two plots from Sentinel – each for different model parameters – and a presentation of the graph from [Bernasconi et al. 2002].

Note that the Anti-A plot, caused by re-injected Antigen A, in (top) and (middle) has a shallower peak than the plot of Anti-TT in the bottom plot. The parameter values for the (top) graph yield poor results, but in (middle) are better, assuming we use the section of graph from time index one to five. The need to find good parameters is discussed in the Further Work section below. Both parameter choices result in some features of the Bernasconi et al plot but the *relative* increases seem to indicate that here is some degree of match between the simulated (middle) and *in vivo* (bottom) results.

Although not perfectly confirmed, a simulation of Bernasconi et al's theory has been shown to be qualitatively reasonable, relative to the *in vivo* measurements. But what causes the quantitative differences? The disparities may be due to: (i) *incorrect modelling* of the Bernasconi et al theory; (ii) *lack of detail* in the model; (iii) *incorrect parameterisation* of that model, and/or (iv) a fundamentally *faulty theory* underlying the model. The next step is to isolate the cause of disparity. The first and last of these points can be addressed by opening a dialogue with Bernasconi's group, but points (ii) and (iii) will require significant further work, as described below.

In conclusion, the simulated theory of polyclonal activation produced interesting results, similar to those obtained by residual antigen theory, but without requiring a long-lived supply of antigen. The signalling provided by IL-15 seems to be essential for this phenomenon. It appears consistent with nature's efficient ways that the body would use the constant attack by antigen to strengthen itself, and we have demonstrated a polyclonal memory effect that is qualitatively similar to the experimental observations of [Bernasconi et al. 2002].

5.5 Further Work

The logical extension of our basic model of polyclonal memory is to create a more detailed B-cell/T-cell and APC model, and then to use that as the basis for a

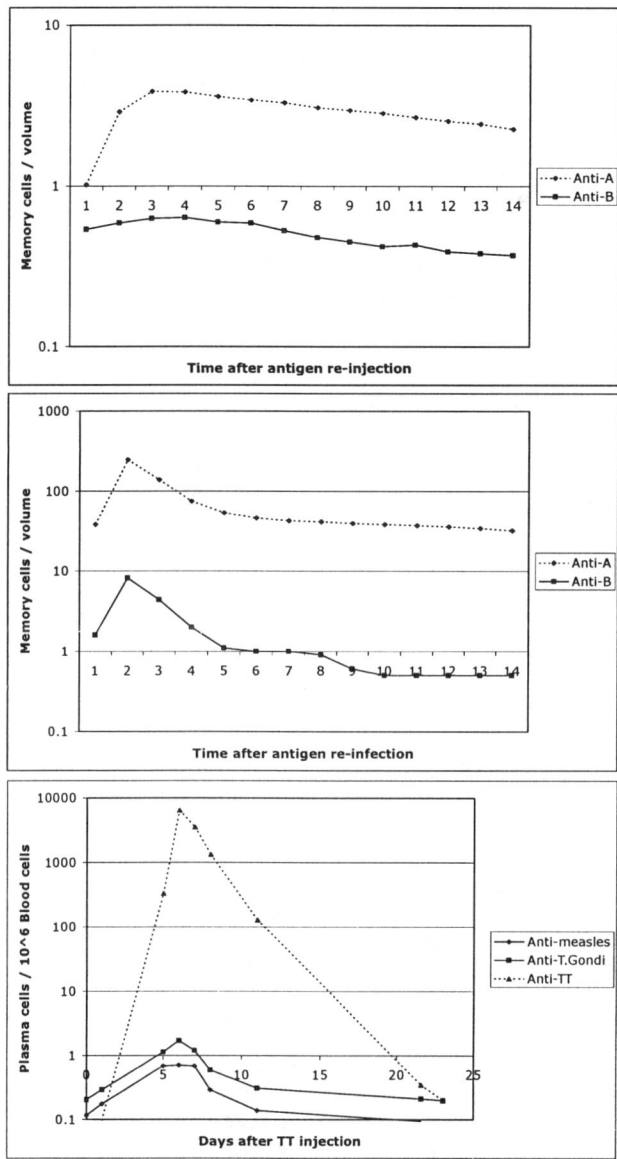

Fig. 5.9. (top and middle) Plots of the memory cell-levels per volume for two antigens, A and B, which are too dissimilar to directly cause a response in each other's memory cells. The immune system has already been exposed to both Antigen A and B; Antigen A is re-introduced at t=0. (top) and (middle) are for two different model parameterisations (see text). Both cases show an unexpected increase in the memory cells that are specific to the non-injected antigen. Since plasma cells levels are roughly linear, relative to memory cell levels, the *in silico* results are qualitatively consistent with the *in vivo* Bernasconi's results (bottom).

combined model attempting to simulate the latest theories of both B- and T-cell memory. Once such a model has been implemented, we can begin to explore questions specifically surrounding the relationship between B- and T-lymphocyte memory, and look at new rules for plasma cell and memory cell creation, death and homeostasis.

As mentioned above, the level of detail of a simulation should be as simple as possible, but a simulation that is too simple will not be as effective. This is a standard dilemma of machine learning hypothesis generation, and we intend to address this issue by means of automatic feedback. In other words, we will generate a *population* of simulations, and then evolve them to find the simplest, most effective candidate model.

The choice of parameters for any model is known to be a hard problem [Ljung 1999], but creation of the model is much harder [King *et al.* 2005]. We are examining several methods of assisted parameterisation of the models, so that a 'best-fit' can be found by Sentinel. This will allow the research to focus on the scientifically interesting model-building task, rather than the more mechanical parameterisation task, and will help to remove four of the possibilities for the differences in between the *in silico* and *in vivo* results in the previous section.

One of our long-term goals is to produce an integrated model of immunological memory that explains the experimental evidence used to support many of, if not all, the theories explored here. Such a model could be used to explore more detailed issues in immunological memory, such as the unusual effects of the SAP gene (which controls long-term memory, but has no effect on short-term memory) [Crotty *et al.* 2003]. Furthermore, a general theory of immunological memory would have implications for machine learning.

The applications described here are mostly related to immunology, and indeed that is the main focus of our work. Nonetheless, our Sentinel platform is likely to be useful in AIS endeavours in the future, in particular when it comes to understanding the dynamics of AIS algorithms that are based on complex systems of agents. In addition, simulating theories from immunology that have yet to be adapted by AIS researchers can provide assistance in determining the minimum set of features required in developing an abstract representation of an immune mechanism.

As Sentinel continues to develop, and becomes ever more sophisticated, we will be able to develop larger, more complex models than at present. It will be interesting to see if the increase in complexity is important, or whether there is a level of complexity that is sufficient for the majority of immunological research.

Capturing Degeneracy in the Immune System

Paolo Tieri[1,2], Gastone C. Castellani[2], Daniel Remondini[2], Silvana Valensin[1,2], Jonathan Loroni[1,2], Stefano Salvioli[1,2], and Claudio Franceschi[1,2]

[1] Department of Experimental Pathology, University of Bologna, 40126 Bologna, Italy
[2] C.I.G. "L. Galvani" Interdepartmental Centre, University of Bologna, 40126 Bologna, Italy. p.tieri@unibo.it

Summary. For host survival, the immune system (IS) is required to deliver high-level, specific and continuous performance, dealing with a very complex universe of stimuli and functions, as well as physical and resource constraints. From this perspective, the immune system needs an effective strategy to assure the requested operational functions, to survive and to evolve. The concept of degeneracy discussed in this chapter, is the ability of some immune receptors to bind many types of ligands and it would appear to be a fundamental characteristic for immune system functioning as well as a formidable weapon in the architecture of complex biological structures and systems. In this chapter, we will discuss how degeneracy acts as a strategy to optimize the necessary trade-off between the inescapable promiscuity of receptors and ligands, with the necessity to produce a specific response, and how the degeneracy principle acts to set up a memory of each immunological event, thus contributing to the fitness of the organism, and how degeneracy can be considered among the underlying causes for the evolution and robustness of the IS.

6.1 Introduction

In observing complex systems, one can notice how the capability to exploit structures and units in a *variable mode* often acts out as origin for their behavior and functioning. It appears that players within such systems have the possibility to differently interpret the input signals, and consequently to respond in a "ambiguous" manner, that give rise to very different paths and to emergent behavior: this is typical of a complex systems' performance. In the immune system (IS), signaling molecules, cells and organs form an intricate and highly reacting network, and show, at different levels, all the traits that characterize complexity such as non-linear outcomes, feedback loops, the significance of system's early history, the difficulty to define boundaries. Degeneracy, in this context, is the ability of an immune receptor to bind many different ligands. It appears to be another piece in the puzzle that is the chain of nested and interconnected subsystems which compose the whole IS. In this view, degeneracy can be seen as a further complication to the global picture, but at the same time its concept can help to explain the evolution of the IS, as well as many of the immune system operational capabilities.

6.2 The Evolution of the Immune System

The immune system is devoted to the neutralization of a variety of agents, including bacteria, viruses, and parasites, which can cause infectious diseases and threaten the survival of the organism (see Chapter 1 for an overview of the immune system).

Within this perspective, the main selective forces for the immune system has been the capability to counteract infections and in particular acute infections, that were likely the most common cause of death of our ancestors.

Immune responses are tightly connected to inflammation, i.e. the first and complex response which occurs at tissue and organ levels in order to get rid of infectious and damaging agents. Indeed, since the beginning, immune responses have always been involved in the responsiveness to other damaging agents besides infectious pathogens.

In fact, a large variety of data obtained from invertebrates has led to the concept that immune and stress responses, and inflammation, are part of an integrated and complex response, which is fundamental for survival in invertebrates. Additionally, it appears these processes were integrated with the newly emerging clonotypic immunity. Thus, the vertebrate evolving immune system had to face the problem to merge and harmonize a completely new cellular and organ apparatus within an old and stabilized system as that typical of invertebrates. Within this scenario, the first problem that such an immune system had to solve was to find an effective and economic way to recognise the enormous variety of molecular configurations present in the universe of potentially infectious agents.

In the last 15 years it has become clear that a limited number of peculiar types of receptors called Toll-like receptors (TLRs) are responsible for sensing the so-called "pathogen-associated molecular patterns" (PAMPs) [Schnare et al. 2001] and/or providing the "danger signal" as speculated by [Medzhitov & Janeway 2002] and [Matzinger 2002], respectively. This result is quite remarkable since it shows that a first degree of discrimination exists between infectious non-self and non-infectious self and is solved in a very economic way. Furthermore, recent data also suggests that the mechanisms of TLRs-based sensing relies on a synergistic interaction between diverse TLR types that cooperate with each other in dendritic cells (DCs) activation. In fact, each type of TLR is capable of binding a quite specific molecular pattern (e.g. TLR3 is triggered by double-stranded RNA, TLR4 by lypopolysaccharide –

LPS– etc.) and it seems clear that certain binary combinations of TLRs act in a synergistic way when simultaneously triggered. This results in the initiation of different pathways and drives different DC activation, each of which is followed by a peculiar immune cell response. Combinatorial TLR stimulation may in this way ensure tailored responses of specific-amplitude. For more information on the innate immune system and its interplay with the adaptive,see the chapter by Robins in this book (Chapter 1).

6.3 The Universally Degenerated Sensing of the Immune System

TLRs elucidate the concept of degeneracy which was originally proposed to describe the capability of a single immunoglobulin (Ig) to bind a variety of peptides (*one-to-many* rule). This concept was later applied to the T cell receptors (TCRs) and Major Histocompatibility Complex (MHC) molecules, i.e. the main sensors of the clonotypic IS.

As discussed in Chapter 7 by Andrews and Timmis, degeneracy seems to be an imperative weapon in biological systems, having to deal with a very complex universe of functions and physical and resource saving constraints. Thus, degeneracy could be considered the stratagem found to optimize the necessary trade off function. As an example, it has been argued that, assuming for a mouse a stringent *one-to-one* specificity of TCRs toward ligands, the weight of the T cells necessary to perform such a task would be 70 times higher of the entire weight of the mouse [Mason 1998]. The same considerations apply to MHC molecules, which perform the crucial steps of presenting epitopes within the groove of Class I and Class II molecules to CD8+ and CD4+ T cells, respectively. It is well known that the limited number of MHC variants have the capability to bind the entire set of short epitopes derived from the self repertoire as well as from the repertoire of foreign proteins of whatever origin [Mason 2001].

On this basis we can assume that all the sensor molecules of both innate and clonotypic immunity utilised by the immune system to recognize antigens are characterized by an intrinsic degeneracy which appears to be a structural feature of this category of molecules. The main message from a systemic point of view is that the old *one-to-one* rule originally hypothesised by Alric and Burnet, to quote only some of the major theoreticians, is far from being true[Burnet 1959]. On the contrary, it appears that the only way that the immune system has to recognise the full repertoire of molecules which threaten the body, and requires the triggering of an immune response in order to neutralise them and let the body survive, is to exploit the pervasive characteristic of biological systems that has been called degeneracy [Cohen *et al.* 2004].

6.3.1 Main Requirements of Degeneracy

The conclusion of the previous section is not unexpected, since degeneracy is apparently an ubiquitous property of biological systems at all levels of organization. According to [Edelman & Gally 2001], degeneracy is

"the ability of elements that are structurally different to perform the same function or yield the same output",

and its main characteristic [Edelman & Gally 2001] is that it implies

"structurally different elements, and may yield the same or different functions depending on the context in which it is expressed".

This concept is opposed to redundancy, in which the same function is performed by identical elements.

This exactly reflects the situation of the immune system and particularly of its adaptive branch, where each T and B cell mounts a molecularly unique receptor (TCR and BCR -or Ig-, respectively), characterised by large degeneracy despite their clonotypic distribution. In other words, these cells, in which apparently the immune system was able to achieve the maximum level of individual specificity, are capable of binding a large spectrum of epitopes and constitute, when considered with neurons as a whole, one of the most degenerated systems of the body. In the same way, dendritic cells, in which TLR are not clonotypically distributed but are concomitantly present on the same cells, not only exploit a degeneracy strategy but also a joint and cooperative action. Indeed, the capability of a single TLR to bind a variety of ligands, is joined with the above mentioned ability to integrate binding signals from different TLRs (to initiate diverse signalling pathways and tailored immune responses).

Thus, the immune system can be conceptualised as a system composed of a very large number of elements which fulfil requirements of the degeneracy. It is composed of components which create either structurally different receptors (T and B cells) which are able to bind very different epitopes, or degenerated receptors capable of a functional cooperativity between their signalling pathways (DCs). In all cases, the context becomes prominent as far as the outcome is concerned, i.e. the type and strength of the response. Context in this case is everything that can be sensed by immune cells. In DCs, the context is represented by the simultaneous presence and binding of different ligands to TLRs which in turn leads to a cooperative transcriptional regulation of target genes that act in concert to give rise to the immune response. In the case of T and B cells, the simultaneous binding of different ligands to co-receptors with inhibitory or activatory function constitutes the context where is situated the binding of possibly different epitopes (with different affinity) to the TCR or Ig [Mason 1998, Hiemstra et al. 2000].

6.3.2 Degeneracy and Evolution

The relationship between degeneracy and evolution is quite complex. [Edelman & Gally 2001] have suggested that

"degeneracy is not a property simply selected by evolution, but rather is a prerequisite for, and an inescapable product of, the process of natural selection itself".

Thus it is conceivable that starting from a low, inescapable level of degeneracy of simple forms of life, evolution progressively selected organisms where the degeneracy of biological networks was increased, thus allowing more complex performance [James & Tawfik 2003].

It is remarkable that this process utilised the same basic building blocks i.e. genes and proteins. Indeed the number of genes in animals like *Caenorhabditis elegans*, constituted by a little more than one thousand cells, and *Homo sapiens*, constituted by an astronomically higher number of cells, is of the same order of magnitude. As humans are definitively and undoubtedly more complex than *C. elegans*, it is possible to speculate that complexity is, rather than an increased number of elements, the result of an increasing number of interactions among them at all levels of biological organizations and networks, from proteins and genes, to cells and organs. Within this scenario, an initial degeneracy was the substrate and the target of natural selection for a progressively increased level of degeneracy and complexity. Thus, the more complex organisms and systems are, the higher their own level of degeneracy. This conclusion is quite counterintuitive to the traditional point of view, according to which degeneracy should be avoided in order to have solid and efficient machineries and structures.

Further studies in different model systems regarding different level of biological organization are thus needed in order to validate the hypothesis that an increased degeneracy is favoured by evolution. How can such a high level of degeneracy in the most complex animals and biological systems be reconsidered with their capability to perform specific tasks in a very efficient way? How is it possible to envisage degeneracy as the unavoidable background of the highest performances characteristic of immune system and nervous system? In immunological terms, how is it possible to explain the specificity of the immune response on this background of a degenerated interactions between ligands and receptors, and different cell types?

An increased level of degeneracy is often accompanied by a concomitant increase in complexity of spatial and temporal constraints. We can speculate that an increased sophistication in cellular compartmentalisation and anatomical topology represent the major constraints which allows the intrinsic degeneracy of the immune system and other systems like the NS to be effective and to decrease the number of ineffective interactions. As far as the immune system is concerned, the increase in degeneracy along with complexity in evolutionary terms must be envisaged as a process involving the anatomical remodelling as well as the emergence of new sophisticated anatomical sites and organs such as germinal centres and the thymus. The other strategy which allowed the emergence of specificity from this soup of degenerated interactions is the above mentioned capacity of immune cells such as DCs and T and B cells to integrate the variety of signals they are exposed to in their specific (anatomical) context.

The coevolution of receptors and ligands implies degeneracy, once given the presence of multiple "multi-purpose" structures and not of *one-to-one* matches: in such environment, evolution takes into account the multiplicity of players and the necessary degeneracy. In this sense, each player is necessary, and none is redundant but has its own essentiality, especially in a long term view, since each unit can contribute to an unforeseen function or can improve an existing process or metabolic pathway.

In general terms we can speculate that the force of natural selection has favoured the emergence of sets of genes, proteins, receptors with high degree of degeneracy, i.e. the emergence of pathways and functions which rely upon a variety of elements which can structurally slightly differ from each other. Hence, when necessary and despite performing non identical functions, these elements are capable of swapping and replacing each other. Natural selection probably shaped not a single element (*one-to-one* rule) but rather the entire set of interacting degenerated players. This means that degeneracy overlaps among context.

6.3.3 Degeneracy and Context

Degeneracy increases the connectivity capability of players in biological and immunological networks. Connectivity and the ability to communicate increase and stimulate the capacity of the integration of signals and their consequent interpretation. Recently it has been recognised that not only individual immune cells continuously integrate antigenic and other signals, but also that both individual cells and populations of cells respond to the rate of change in the level of stimulation, being capable of discriminating the magnitude of system perturbations. In other words both at the individual cell level and at cell population level the main systemic characteristic of the immune system is its capability to sense the rate of change in the level of stimulation (the temporal derivative of stimulation).

From this perspective the antigen, i.e. the stimulus capable of triggering an immune response can be defined as any given perturbation whose parameters vary at a rate above a certain threshold. We summarize that danger signals from this point of view are all those signals which perturb abruptly the context in which immune cells are immerged. Such a systemic point of view indicates that a proper definition of antigen cannot avoid a concomitant proper definition of the context and indicates that it could be more appropriate to define the "antigen" as a context-sensitive stimulus rather than a stimulus as such. In some extreme cases, the importance of the context can be so overwhelming that an immune response can be obtained in the absence of the antigen [Selin *et al.* 2004].

6.3.4 Degeneracy and Robustness

From an engineering point of view, it could be argued that for a system to be robust, any type of degeneracy as described here should be avoided. Typically, engineering solutions make use of the typical *one-to-one* (one structure-one function) rule and relies heavily on redundancy. In this context, redundancy means that whenever the critical persistence of a specific performance is required, there is a multiplication of identical structures, which perform the same function, to cover a failure of one of the structures. From the same view, when the optimisation of functions and structures is the primary goal, then the possibility that one structure could perform different concomitant functions is "disturbing" or not tractable at all (with classical engineering methodologies and approach).

On the contrary, optimisation in living organisms seems not to be tailored only for short-term objectives, but also for out-of-sight and unpredictable long-term goals. Many biological functions are based on flexible and degenerate interactions among elements, which slightly differ structurally in a way which can be quite subtle. Such a structural diversification results in large repertoires of scarcely dissimilar elements capable of a multitude of interactions dynamically varying in space and time as well as for their strength and duration. These characteristics are quite similar to those required by a robust system (a system able to maintain a feature in the face of perturbations), and thus degeneracy can be taken as a major ingredient of biological robustness, for the best adaptation to environments and contexts. In this case robustness and plasticity are both achieved by systems with high degrees of degeneracy. Moreover, degeneracy allows new combinations of interactions in different frameworks allowing adaptation over ontogenetic and phylogenetic times (at a somatic level and increasing fitness from generation to generation) [Wagner 2005, Jen 2001, Krakauer 2001, Carlson & Doyle 2001]. Indeed, this notion of degeneracy in Artificial Immune Systems is explored in greater depth by Andrews and Timmis (Chapter 7 of this book).

6.3.5 Integral degeneracy

There is experimental evidence of the importance of degeneracy in ruling the shape, the functioning and the evolvability of systems in the real biological world. At a cellular scale, in the integration of signals with different intensity from different sensors, at the cell population scale, in the behaviour of interacting immune cells, and a higher-level scale in the systemic interactions among major body structures resulting in a highly integrated immuno-neuro-endocrine system. In the T cell antigen recognition and response, the first level of TCR degeneracy is integrated with a second level of degeneracy given by the combination of the activation state of the APC presenting accessory signals. The activation state of an APC can significantly vary and relies on the identification of dsRNA, PAMPs or LPS by its set of non-clonal receptors (TLRs). Hence, antigen presentation by not fully activated APCs can result in T cell anergy or apoptosis, or instead, interaction between a completely mature APC and a pre-activated T cell can lead to a full-blown attack. Furthermore, it has been demonstrated that self-recognition by TCRs happens very often in standard physiological conditions and -in specific immunological spaces like secondary lymphoid tissues- lowers the activation thresholds for alien antigen responsivity [Kamradt & Volkmer-Engert 2004].

In the same way at the cellular interaction level, the multiplicity of immune mediators exchanged by immune cells leads to a network of interactions that strongly influences the reliability of the immune system and that significantly contribute to its robustness and adaptability. One of the studies on immune cells network topology [Tieri et al. 2005] shows how each cell can exploit a variety of mediators as signals in an information flux exchanged with another cell. Some of these mediators, each in their chemical and structural dissimilarity, can sometimes carry the same information signal: this is evidence of the result from an adaptation of different structures to similar functions.

6.4 A Model: Network Dynamics with Adaptive Degeneracy

The concept of degeneracy leads, in a natural way, to a class of network models where the selectivity/specificity properties can be relaxed from the maximum selectivity/specificity principle (*one-to-one* rule) to a more realistic one that allows the possibility to have various degrees of selectivity/specificity: in other words, a certain degree of degeneracy. The degree of degeneracy can be tuned by a "plasticity threshold" that depends in a non-linear way from the past history of the input-output environment that each network's element have experienced.

As discussed in Chapter 7 of this book, the immune system is a cognitive system, capable of recognition and action. The cognitive capability (learning and memory) in neural as well as immune system is influenced by a signalling system, the so called Kinase-Phosphatase (K-P) network. Among the various proteins involved in K-P network, Calmodulin-Dependent Protein Kinase II (CaMKII) and Calcineurin play a pivotal role during the memory induction in immune and neural system. Memory induction in neural system has been deeply investigated, and it has been divided into two elementary mechanisms, Long-term Potentiation (LTP) and Depotentiation (LTD), as postulated in Bienenstock-Cooper-Munro (BCM) theory of synaptic plasticity [Bienenstock et al. 1982]. These mechanisms are implemented through an energy-based learning rule [Bazzani et al. 2003], where the connections updating is obtained by minimization of a "risk" or "energy function" [Bazzani et al. 2003, Castellani et al. 1999]. This theory is based on the so-called "maximum selectivity principle", or in other words a *one-to-one* association between input and output without any degeneracy. We already established a mapping between neural and immune system mechanisms of memory induction [Remondini et al. 2003], both in maximum selectivity conditions and in case of degeneracy, that represents the possibility of recognizing more than one stimulus.

Within this framework, generic lymphocytes are the units of a network, and they communicate through mediators represented by the links between nodes (in analogy with synaptic connections in neurons). Each unit is specified by its inputs, outputs and an internal parameter: a history dependent threshold. This threshold depends on the time average of incoming and outgoing signals and maps the history of each lymphocyte to a plasticity function. The "plasticity" function Φ is responsible for the strengthening and weakening of communication efficiency between nodes (analogous to LTP and LTD). The definition of Φ is chosen according to statistical considerations and mathematical simplicity, but the general results are true for a wide class of changing sign functions [Castellani *et al.* 2005, Castellani *et al.* 2001]. The evolution equations are:

1) $\frac{du_{ij}}{dt} = \Phi_{ij}$ $i \neq j = 1, \ldots n$

2) $\Phi_{ij} = u_{ij} \cdot (u_{ij} - \Theta_{ij})$ $i \neq j = 1, \ldots n$

3) $\Theta_{ij} = \sum_{i \neq j \in \Omega_{ij}} u_{ij}^2$

where u_{ij} are the mediators concentration values and Ω_{ij} are appropriate subsets of input and output links. The choice of the Ω_{ij} subsets is crucial, since the development of different network structures critically depends on this choice. It is possible to show that according to the choice of Ω_{ij} we have various degree of degeneracy: lymphocytes can respond to a single antigen or to a number of antigens in relation to such value. This choice also determines the resulting network topology: the degree distribution (number of links of each node) can range from a quasi-random (not structured) to more fat-tailed distributions that reflect a hierarchical ordering of nodes. The possibility to have a non-one-to-one mapping between inputs (antigens or external signals) strongly increases the network capacity in terms of number of stable states [Castellani *et al.* 1998].

This is a first step towards more realistic immune models taking into account the internal structure [Tieri *et al.* 2005, Castellani *et al.* 2005] of cell communication and the role of signalling molecules in the induction of learning and memory and response to stressor signals. A possible algorithm inspired by this plasticity theory may be an agent-based model. Each agent communicates with a variable number of other agents, and adjusts its internal parameters (the Θ threshold and the number of its connections) on the basis of a threshold-based "weighting" of the incoming and outcoming signals (risk function minimization).

6.5 Conclusion

Evolution and degeneracy seem tightly linked, and furthermore, it is likely that degeneracy acts as a prerequisite for evolution. Indeed, the degeneracy principle itself, i.e. the flexibility in using the same structure for different tasks, contributes to the breakthrough of functions not foreseen *ab origine* in the system's abilities.

Experimental evidence now shows that immune system basic functioning relies on degeneracy of its receptors, since an exact *one-to-one* match principle would not be cost-effective and is unlikely to work at all. Degeneracy principle appears to pervade the system at many level of integration, from molecular to intercellular interaction, to main body systems communication. Further studies in this direction would allow us to elucidate how deep the degeneracy principle is nested in the system machinery. Degeneracy can also be considered as a form of robustness, since it assures continuous working conditions, with the unavoidable drawback of performance decrease.

To add a note on the terminology, the term "degeneracy", referred to the capacity of a single receptor to bind many different ligands, seems to come out from the misconstrued idea of absolute "fidelity bond" between receptor and ligand. Since this hypothesis appears now to be unrealistic, one can use the term "polygamy" instead of "degeneracy", taking into account the fact that "multiple mating" seems to be the natural condition of such a receptor.

It can be finally argued that the intrinsic nature of evolutionary success of such systems, and organisms, relies upon the reduction of system performance that should anyway remain secured to a minimal level, and on despecialization.

Alternative Inspiration For Artificial Immune Systems: Exploiting Cohen's Cognitive Immune Model

Paul Andrews[1] and Jon Timmis[1,2]

[1] Department of Computer Science, University of York, Heslington, York, UK.
[2] Department of Electronics, University of York, Heslington, York, UK.
  {psa,jtimmis}@cs.york.ac.uk

Summary. In this chapter, we highlight the idea that actively investigating alternative theories of the immune system can identify new and novel inspiration for artificial immune systems (AIS). It is clear that there is disagreement amongst immunologists concerning many mechanisms of the immune system, often providing an unclear picture for the engineer trying to exploit immune ideas. In spite of this, the abundance of immune theories can be beneficial to the AIS practioner if investigated in detail. Here we provide an example of this by describing Irun Cohen's cognitive immune model and showing how, from this, new ideas for AIS inspiraction can be identified and exploited.

7.1 Introduction

As discussed in Chapter 3, early artificial immune systems (AIS) were developed with an interdisciplinary slant. A notable example was [Bersini 1991] where they developed immune network models, and then applied those models to a control problem characterised by a discrete state vector in a state space. Other examples include the development of immune gene libraries leading to a bone marrow algorithm employed in AIS [Hightower *et al.* 1995], and the development of the negative selection algorithm with the first application to computer security [Forrest *et al.* 1994]. However, in more recent years, work on AIS has drifted away from the more biologically-appealing models and attention to biological detail, with a focus on more engineering-oriented approach.

This drift away from biological detail has led to systems that are examples of *reasoning by metaphor* [Stepney *et al.* 2004]. These include simple models of clonal selection, immune networks and negative selection algorithms as outlined in Chapter 3 in this book. For example, the clonal selection algorithm (CLONALG), whilst intuitively appealing, lacks any notion of interaction between B cells and T cells, MHC or cytokines. In addition, the large number of parameters associated with the algorithm, whilst well understood, make the algorithm less appealing from a computational perspective. aiNet, again, whilst somewhat affective, does not employ the immune network theory to a great extent. Only suppression between B-cells is employed, whereas in the immune network theory, there is suppression and stimulation between cells. More recent studies in [Hart & Ross 2004], point out that the main effect on immune network algorithms may well be the way in which interaction is defined. Through the development of a simple model Hart and Ross demonstrate the evolution of various immune network structures which are considerably affected by the choice of affinity measure between two B cells, which in turn effects how B cells interact with each other. With regards to negative selection, the simple random search strategy employed, combined with using a binary representation, makes the algorithm computational so expensive, that it is almost unusable in a real world setting [Stibor *et al.* 2004, Stibor *et al.* 2005a].

In order to attempt to draw back the eyes of the AIS practitioner, away from the *well known* immunology (at least to the AIS practitioner), in this chapter we first examine the disagreements amongst many immunologists trying to explain how the immune system works. Following on from this, we highlight one point of view, that of Irun Cohen, which dscribes the immune system as a cognitive system [Cohen 2000a], and investigate how this might be a useful exploration for the AIS practitioner. As with the chapter by Neal (Chapter 14 in this book), we argue that the simplicity that has been adopted thus far in the development of AIS, has potentially limited its scope. Neal discusses the role of interactions in the immune system, both internal (between innate and adaptive), and external forces. Here we focus on the role of *degeneracy* of immune receptors (highlighted by Neal, and by Tieri *et al.* in Chapter 6) in the context of Cohen's cognitive model, and suggest ways in which this may be exploited in future applications.

7.2 Immunological Arguments

By investigating the immunological research literature, it is clear that many immunologists fundamentally disagree on the immune mechanisms responsible for many of the key observed properties of the immune system. This disagreement gives rise to many different theories of immune function, yet any of these theories may be useful for inspiring AIS. In this section, an example is given on one such disagreement documented in an immunology journal, and it is shown that much of the disagreement arises from a clash of research methodologies being employed.

7.2.1 Immunology: The Science of Self–Non-Self Discrimination?

The formulation of the clonal selection theory gave to immunology an explicit notion of an immune self and a mechanism by which the adaptive immune system could discriminate between self and non-self molecules in the body. With the adoption of the clonal selection theory as the key principle of adaptive immunity, [Tauber 2000] notes that by the 1970s, the immune self had became the defining idea in immunology with the field itself being referred to as the science of self–non-self discrimination. Indeed, it is still a hotly debated topic, see Chapter 17 of this book for more detail regarding this matter. It appears evident, however, from articles appearing in volume 12 issue 3 of the journal *Seminars in Immunology* published in 2000 ([Anderson & Matzinger 2000b, Bretscher 2000, Cohen 2000a, Grossman & Paul 2000, Langman & Cohn 2000c, Medzhitov & Janeway 2000, Tauber 2000, Silverstein & Rose 2000]), that this view is not universely held. In this journal volume, many leading immunologists discuss their views on the nature and importance of self–non-self discrimination in the immune system. In their editorial introduction, [Langman & Cohn 2000a] state that this is the first time a large collection of competing immune theories have attempted to be published together.

On examination of the articles presented in the journal issue, the level to which many immunologists differ in their views becomes clear. This is summed up by [Langman & Cohn 2000b], who state in their editorial summary:

> "There is an obvious and dangerous potential for the immune system to kill its host; but it is equally obvious that the best minds in immunology are far from agreement on how the immune system manages to avoid this problem."

In his commentary on the models proposed by the other immunologists, [Tauber 2000] believes that they all fall to various degrees between the ideas of Burnet (clonal selection theory) and Jerne (immune network theory), and are thus a continuation of the arguments between these two points of view.

Of the immune models presented in the journal issue, Langman and Cohn's own minimal model of self–non-self discrimination [Langman & Cohn 2000c] is the one closest to the original ideas of Burnet, whereas [Cohen 2000a] is the closest to Jerne's. Langman and Cohn's model considers only the adaptive immune cells to be involved in the recognition of non-self antigen, and thus the initiation of the immune response, in the body. Conversely, Cohen's model removes the requirement for self–non-self discrimination in the immune system. In this model all immune cells recognise both self and non-self antigens and form an immune dialogue with the body's tissues in order to fulfil the role of body maintenance. The [Bretscher 2000] model is very similar to that of Langman and Cohn, but tries to reconcile the original ideas of Burnet with contemporary immunological observations. It does this by including the role of innate immune cells acting as antigen presenting cells (APCs) during the initiation of an immune response. The models presented by [Medzhitov & Janeway 2000] and

[Anderson & Matzinger 2000b] are essentially the infectious–non-self model and danger theory respectively. The model proposed in [Grossman & Paul 2000] is closest to that of Cohen's, in which lymphocytes react to the rate of change in excitatory signals. They do this by constantly tuning their activation thresholds, only responding when they can no longer adjust due to a rapid change in excitation. This response is dependent on many conditions that include antigen concentrations, activation levels of APCs and danger signals. In addition to these six models, other positions are presented, including that of [Silverstein & Rose 2000]. Although they do not present a model of the immune system, they state that the concept of self–non-self discrimination is a delusion, arguing that the immune system has evolved in complexity gradually over time. To cope with changing circumstances, control mechanisms have been evolved in parallel to amplify or dampen the immune response. Such an immune response thus becomes a balance between protection and damage dependant on the parameters (e.g. quantity and quality) of the immunogenic stimulus.

In [Anderson & Matzinger 2000a], the authors examine the relatedness of the six models just described (excluding the Silverstein and Rose commentary), based on the nature of the signals that turn the effector immune response on. The first signal is considered to be the antigen recognition by a lymphocyte receptor, and is required by all models. The model by Cohen and that of Grossman and Paul require many different signals on top of this for immune activation to occur, whereas the other models require just a critical second signal. These second signal models can be split into two groups dependent on where this signal comes from. Bretscher's model and Langman and Cohn's require the second signal to be provided by a T_H cell, whereas the model of Anderson and Matzinger and that of Medzhitov and Janeway state it comes from an APC. This classification highlights the uniformity of the differences in opinion regarding this key area of immune system function.

7.2.2 A Clash of Methodologies

Much of the difference of opinion regarding the immune models presented above, can be explained by the research methodology that has been employed in formulating the model. This is most apparent when comparing the minimal model of Langman and Cohn presented in [Langman & Cohn 2000c, Langman & Cohn 2002b], with the ideas of Cohen and his collaborator Efroni, who provide a commentary of the minimal model in [Efroni & Cohen 2002, Efroni & Cohen 2003]. To appreciate this commentary, the minimal model is described in more detail. Langman and Cohn state that for the immune system to counteract the attack of pathogen, it must generate and regulate new specificities during the lifetime of an individual, and that this requires a mechanism of self–non-self discrimination. The minimal model considers only the antigen specific lymphocytes of adaptive immunity, and assumes that during the embryonic development of an individual, maternal protection provides an environment containing only self antigen. Thus, during this early stage in life, a process can take place that produces a lymphocyte population capable of determining self from non-self for the duration of the individual's life. The minimal model proposes that B, T_C and T_H cells exist in one of 3 states, an initial state (i-state), an anticipatory state (a-state), and an effector state (e-state). All cells arise in the i-state and express no effector responses. Upon binding with an antigen, an i-state cell receives a signal (Signal 1) through its antigen receptor and becomes an a-state cell. To become an e-state cell, an a-state cell must receive a second signal (Signal 2) from a regulatory e-state T_H. If this Signal 2 is not received, then the a-state cell dies and tolerance to the antigen that provided the Signal 1 is achieved. During the embryonic development window, e-state T_H cells are assumed to be absent and so tolerance to self antigen is achieved. In order for this minimal model to logically hold, the occurrence of primer e-state T_H cells needs to be accounted for, as a small number of initial e-state T_H cells are required for the production of all other e-state cells. It is proposed that these primer e-state T_H cells can be slowly derived independently of antigen from i-state T_H cells early in life.

In their commentary, Efroni and Cohen criticise the reductionist logic used to try and show the minimal model as the only reasonable model of the immune system's functioning. They note that the reasoning used by Langman and Cohn neither matches the observed behaviour of the immune system, nor is an appropriate method for understanding its complexity. Instead, they advocate the use of complex systems research tools to understand how the immune system works, as in their view, the immune system is a complex system. Efroni and Cohen go on to concede that reductionism has provided immunology with much information regarding the specific agents of the immune system, but claim that it is not possible to deduce its functions by simply dismantling it. Instead, they believe that it is now time to acquire knowledge about the immune system from the bottom up, building models to examining how networks are created and properties emerge at the level of the system as whole. The differences between the research philosophy of Langman and Cohn and that of Efroni and Cohen are summed up by Cohn [Cohn 2003] in his response to the criticism of Efroni and Cohen. Here, Cohn equates complex systems research to computer modelling and believes that the major difference between the two positions is that they (Langman and Cohn) wish to put in place a framework for the functioning of the immune system before computer modelling, whereas Efroni and Cohen wish to build mathematical and computational models in order to discover such a framework. Cohn goes on to claim that:

> "Biological complexity is not a problem solved by building a mathematical, philosophical, or computer programming web around any random collection of observations."

This is in stark contrast to the Efroni and Cohen point of view, later summarised by [Cohen et al. 2004], who state:

> "Immunology needs precise mathematical modeling and computer simulation to help us understand the emergence of immune specificity from the collective co-response. The interactions are simply too complex to be grasped by intuition"

7.3 The Cognitive Immune System

In the previous section, Cohen's immune model [Cohen 2000a] was identified as an example of an alternative immune theory that viewed the immune system from a complex systems point of view. In [Cohen 2000b, Cohen 2001, Cohen et al. 2004], the author elaborates on these ideas, and presents a holistic immune system that is a complex, reactive and adaptive cognitive system. We provide here a description and explanation of many of these ideas that have been elicited from [Cohen 2000a, Cohen 2000b, Cohen 2001, Cohen et al. 2004].

7.3.1 The Role of the Immune System

The input to the immune system takes place at the molecular level, and constitutes the molecular shapes sensed when bonded to receptors on immune cells. The response of the immune system to this receptor input is the changing of immune cell states and activities, forming a complex reaction between the immune agents. These immune agents can have different effects on the body that include cell growth and replication, cell death, cell movements, cell differentiation and the modification of tissue support and supply systems. The term inflammation is given to this range of processes that the immune system has on the body. The output of the immune system can, therefore, be considered as inflammation. This inflammatory response can be of many different types and degrees depending on the receptor inputs of the immune agents.

The most popularly held purpose for the immune system is defence against pathogen, requiring the discrimination between self and non-self. In physiological terms, however, it was highlighted above that the output of the immune system is simply inflammation. The effect of this inflammation is to perform maintenance on the body keeping it fit for living, not the discrimination of self from non-self antigen. Cohen believes that the result of inflammation, and hence the role of the immune system, is to repair and maintain the body. As the removal of pathogen is beneficial to the health of the body, defence against pathogen can be seen as just a special case of body maintenance. In order to achieve body maintenance, the immune system must select and regulate the inflammatory response according to the current condition of the body. This condition is assessed by both the adaptive and innate immune agents, which are required to recognise both the presence of pathogen (non-self antigen) and the state of the body's own tissues (self antigen). The specificity of the immune response, therefore, is not just the discrimination of danger, or the distinction of self–non-self, but the diagnosis of varied situations, and the evocation of a suitable response. In summary, Cohen's maintenance role of the immune system requires it to provide three properties:

- *Recognition*: to determine what is right and wrong
- *Cognition*: to interpret the input signals, evaluate them, and make decisions
- *Action*: to carry out the decisions

The properties of immune recognition and cognition are covered in the following sections, whilst immune action is simply achieved via the effector functions of the various immune agents.

7.3.2 Immune Specificity

According to the clonal selection theory, immune specificity is a property of the somatically generated immune receptors of the T and B cells, which both initiates and regulates the immune response. Initiation is achieved via the binding between an antigen and a receptor that is specific to it. The response will then stop only when there is no antigen or receptor left for binding. Cohen, however, points out that immune receptors are intrinsically degenerate, i.e. they can bind more than one ligand. Immune specificity, therefore, cannot be purely dependent on molecular binding as no one receptor can be specific to a single antigen. Instead, affinity, the strength of binding between a receptor and its ligand, is a matter of degree. In Cohen's model, immune specificity requires diagnosing varied conditions in the body and producing a specific inflammatory response. This specificity emerges from the co-operation between immune agents, and does so despite receptor degeneracy and the fact that immune agents are also pleiotropic and functionally redundant. Pleiotropism refers to the fact that a single immune agent is able to produce more than one effect, for example the same cytokine is able to kill some cells whilst stimulate others. Functional redundancy concerns the ability of one class of immune agents to perform the same function as another, for example cell apoptosis can be induced by different immune cells. There are two processes provided by Cohen to explain the generation of immune specificity: co-respondence, and patterns of elements.

Co-respondence is a process whereby the agents of the immune system respond simultaneously to different aspects of its target, and to their own response. This results in a specific picture of an antigen emerging from immune agent co-operation via the following process. As previously noted, immune receptors provide the input to the immune system by recognising molecular shapes. There are three different types of immune receptor that recognise different aspects of antigen. These are the receptors of the T and B cells and the innate receptors of macrophages. The T cell receptors are restricted to recognising processed fragments of antigen peptides bound to a MHC molecule, whereas the B cell receptors (antibodies) recognise the conformation of a segment of antigen. The innate receptors of macrophages don't recognise antigen, but immune molecules. These molecules form a set of ancillary signals that describe the context in which lymphocytes are recognising antigen. These ancillary signals can be classified into three classes: the state of body tissues (some receptors detect molecules only expressed on damaged cells), the presence and effects of pathogen (some receptors are unique to infectious agents such as bacterial cell wall) and the states of activation of nearby lymphocytes (some receptors detect immune molecules produced by lymphocytes). In addition to interacting with their target object, the T cells, B cells and macrophages use immune molecules to communicate their response to each other, and other tissues of the body. This forms an immune dialogue comprised of an on-going exchange of chemical signals between the immune cells. Subject to this exchange of information, they update their own responses accordingly, be it to increase or decrease the vigour of their response.

The exchange of information between immune cells is also affected by the existence of networks of immune agents, such as cytokine and idiotypic networks, and by the processes of positive and negative feedback in these networks. For example, the cytokines can be split into two functionally similar groups, the T1 and T2 cytokines. T1 cytokines tend to activate destructive effects, whilst the T2 set are less destructive, often promoting healing. These sets of cytokines interact, with cytokines re-enforcing the production of other cytokines within their own set, whilst suppressing the activities and production of the cytokines in the other set. This produces both positive and negative feedback loops for the production of immune agents. Cytokines also express pleiotropism, with the effects of some cytokines being able switch between the roles of the T1 and T2 sets depending on the state of the cell receiving them. The process of co-respondence is illustrated by Fig. 7.1.

Patterns of elements help generate immune specificity as the specificity of a pattern can extend beyond that of the individual elements that make up the pattern. Immune patterns are a complex arrangement of populations of immune agents, which, through their individual activity, produce a specific pattern of activity. For example, a pattern can emerge toward a particular antigen from the overlapping reactions of a population of degenerate immune receptors. Even though each immune receptor is non-specific to its target, the result of all the receptor reactions together will be unique, and thus specific to that antigen. Patterns can also be built with the help of immune agent pleiotropism and functional redundancy. Here, the ability of different immune agents each being capable of responding to a situation with a number of different immune effects, allows more response options to be available than just having a single mapping between immune agent and its effect. Thus, specificity emerges through a co-operative pattern of degenerate, redundant and pleiotropic immune agents.

7.3.3 Cognitive Systems

Before describing the immune system in terms of a cognitive system, the definitions Cohen uses for 'cognition' and 'a system' are presented. In biology, cognition is often related to the workings of the brain, being equated to awareness or conscious thinking. This, however, rules out any entity other than the human brain. Instead, cognition can be defined as a particular way of dealing with the world, or adjusting to the environment. A system is defined as an arrangement of connected components that forms a coherent whole. Such a system transforms an input of energy and information into an output of different energy and information. Examples of systems in a biological organism include the respiratory, renal, nervous and immune systems. Of these, based on the definitions of cognition and a system, only the nervous and immune systems are considered to be cognitive. These systems differ from the others in the way they utilise internal images of their environment and the processes of self-organisation to make unconscious decisions. A description of how these unconscious

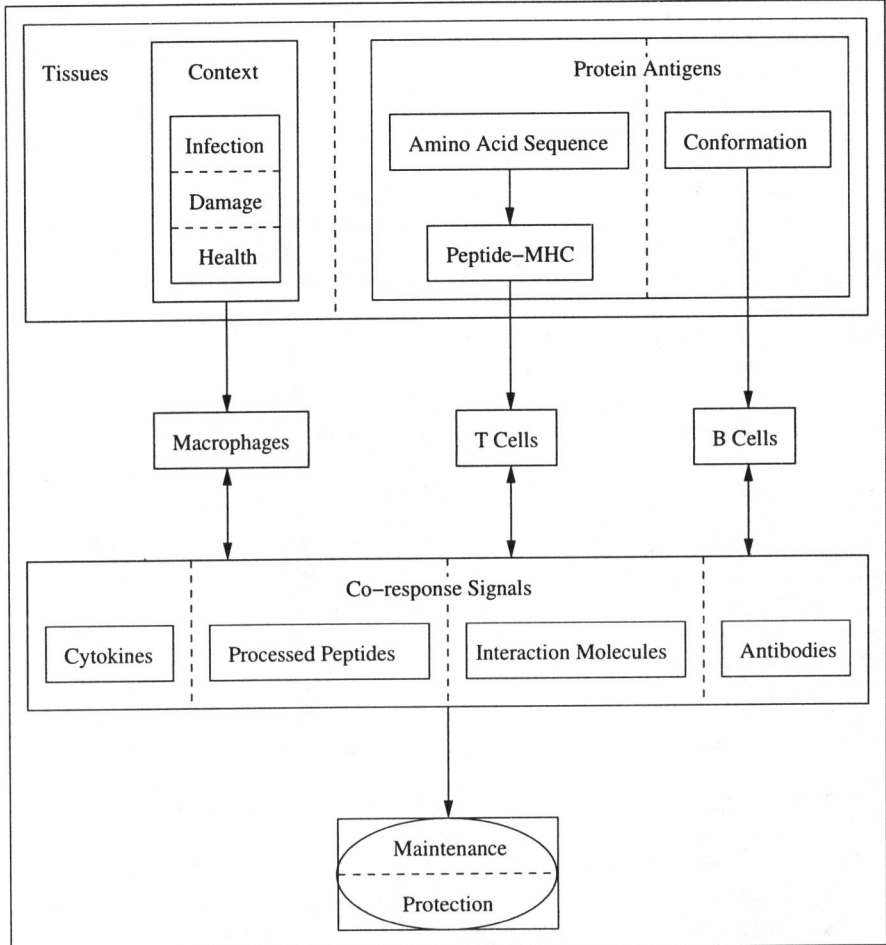

Fig. 7.1. Co-respondence, after [Cohen 2000b] page 161, with permission from Elsevier. Rectangles represent objects, ellipses within rectangles represent states of objects, arrows designate directions of relationships and items separated by broken lines can be combined to generate joint products.

decisions arise will follow an explanation of internal images and the processes of self-organisation.

Through their interaction and mutuality of information, entities are able to create abstract images of each other that exist in an information space. For example, the type of teeth possessed by a creature encodes a functional image of its diet. Such images can form a mirror image of what they represent, or provide re-enforcement through positive feedback. As a survival mechanism, cognitive systems build internal images that map the environment in which they exist. These images then help inform its host how to satisfy its needs. Cognitive systems such as the brain and immune system, contain two types of image: innate images inherited from parents, and acquired images that are built from life experience. The innate images can be split further into three categories: feature detectors and attention preferences, which regulate the input to the cognitive system, and motive forces, which influence the output of the system. Feature detectors filter the system's input providing it with information vital for survival, whilst ignoring other input. Attention preferences act to direct the cognitive system to obtain particular types of information from the environment, thus acting as feedback loops and enabling the selection of input. Motive forces cause the cognitive system to act, influencing the behaviour of the system. Examples of motive forces in the brain include emotions and feeling.

Self-organisation in cognitive systems is typified by the processes of learning and acquisition of memory. This involves the attainment of new behaviours and abilities from experience, which are then stored for later reference. To achieve this requires the unprogrammed and progressive creation of information within the system. This requires not only the generation of new information, but the retention of old information, resulting in a net increase of information. A formal theory of self-organisation in biological systems has been developed by Atlan and is summarised in the context of the immune system by [Atlan & Cohen 1998]. This highlights two conditions needed in a system for self-organisation to occur: redundancy of information and unpredictability (i.e. noise). To generate new information, copies of old redundant information are perturbed by the addition of noise, thus changing the old into new information. The old information is kept, as only a copy of it is changed, and so resulting in the net gain of information. Biological cognitive systems are seen to self-organise on two levels: the species level, and the individual level. Species self-organisation refers to the way a species learns to survive in its environmental niche. This information is passed on though genetics from one generation to the next, and forms the basis for individual learning by providing the innate images. Individual self-organisation is more specific, being shaped by the information generated through the individual's experience. Individual self-organisation, therefore, extends the innate images (e.g. motive forces) by incorporating new information into them to create adaptive images.

To make a decision is to choose from a number of available options. This is made possible in a deterministic system if two conditions are satisfied. Firstly, it is clear that to make a decision, the system must have options available to it. This means being able to relate to a set of inputs in different ways. Secondly, the system must possess an internal history detailing its experience. This internal history, constructed from the type of images described above, is used to influence the impact of the external inputs to the system. The output generated by a decision is deterministic, but the elements of the internal history are so many, and so complex, that the choice made appears to be unpredictable. This decision making process can thus be considered a process of associating a particular circumstance encountered in the external environment, with a class of feeling present within the system in the form of internal images. To put it another way, the decision emerges from the match of the environmental circumstances with the internal motive. A cognitive system that makes such decisions without consciousness can be summed up by the following cyclical process. The cognitive system impacts its environment by making choices. These choices emerge from the internal images of the system that have been shaped by the processes of self-organisation. This self-organisation is in turn driven by the system's environment. This type of cognitive system is both a concurrent and reactive system.

7.3.4 Immune Cognition

According to Cohen, the immune system achieves its role of body maintenance through a cognitive strategy, utilising the processes described in the previous section. Here, a cognitive system was described as using internal images and the forces of self-organisation to make deterministic decisions. It is the outcome of these decisions that are proposed to perform body maintenance. Both the adaptive and innate arms of the immune system are seen to self-organise via the creation of information from noise and redundancy. Self-organisation of the adaptive arm is seen to occur in the construction of the T cell and B cell antigen receptor repertoires. This receptor repertoire is produced using redundancy provided by the proliferation of cellular clones, and noise provided by the random genetic mutations that produce the variable regions of the clone's antigen receptor. These clones are then subjected to selection (T cell maturation in the thymus and the affinity maturation of B cells) to produce the immune receptor repertoire. Self-organisation of the innate arm of the immune system is seen to occur by the generation of the actual response repertoire of the immune system via the fine tuning of the set of innate immune responses. This determines the types of response that will be connected to the signals perceived by the lymphocyte receptor repertoire. The process of self-organisation to produce this actual response repertoire also involves the generation of a particular cytokine profile from their redundant and pleiotropic nature.

Cohen states that the immune system utilises both innate and adaptive images. In the process of co-respondence, the immune cells can be seen to use innate feature detectors and attention preferences to filter out the information they require from their target. Innate motive forces are said to exist in the form of the innate response repertoire that undergoes the self-organisation process described above to produce the actual immune response repertoire. This resulting response repertoire can thus be viewed as an adaptive image, providing the immune system with an internal map of its possible immune responses. Other adaptive images are generated by the immune system, which can be categorised into concrete, abstract and distributed images. Concrete images are those based on physical contact points such as antigen receptors forming an image of the antigens it binds. Abstract images are not constrained to physical space, and are formed out of the mutuality of information. For example, an immune reaction made up of cells and cytokines maps to the antigen and other signals that elicited the response. Distributed images are expressed as the patterns of immune agents distributed around the body, for example the entire T cell repertoire is an image of the T cell selections that have occurred in the body.

Cognitive choice, as highlighted above, is deterministic and emerges from the exercising of options under the influence of an internal history. The decision that the immune system must make is the choice of an inflammatory response to the perceived input. Decision making in a cognitive system was likened to the association between a particular in the environment, with a class of action. In the immune system, therefore, the decision making process is the association between the current immune environment perceived through the receptor repertoire (images of perceptions), with a suitable inflammatory response from the response repertoire (images of responses). This decision is achieved through the process of co-respondence described above.

Memory is also a property of cognitive systems and is formed via learning from past experiences. In the immune system, memory is expressed through the differences that are seen to arise between a primary and secondary infection from a particular pathogen. During the first infection, the immune system must expand its T and B cell repertoire to recognise the pathogenic antigen, and memory T and B cells are selected that express the innate effector response needed to kill the pathogen. During the second infection, the immune system can draw on its experience from the first infection to rid itself of the pathogen immediately. Thus, the memory T and B cells no longer need the full string of signals needed to produce the effector response in the first infection. Immune memory can, therefore, be considered as the replacement of the context of an infection. This memory is never in a final state, and continues to evolve during the lifetime of the individual.

7.4 Alternative Inspiration for AIS

From Cohen's immune model, a number of properties can be identified that could be useful computationally in an AIS. It is noted that some of these ideas are present in Jerne's immune theory and so have already been used as inspiration for AIS (see Chapter 3 in this book for more details). However, these properties are still highlighted here as they form key parts of Cohen's model, and the way in which they are integrated into Cohen's model may highlight alternative methods of inspiration. There are two levels of scale that can be identified for AIS inspiration from Cohen's model, the high level ideas and paradigms that describe the functioning and behaviour of the immune system, and the lower level processes that are proposed to achieve the described functions. The high level ideas include:

- *Cognition*: The immune system is a cognitive system that can make unconscious decisions dependant on the information presented to it
- *Images*: Adaptive and innate images provide a history of past immune encounters.
- *Self-Organisation*: Learning and memory arise through self-organisation.
- *Maintenance*: The role that Cohen sees the immune system as fulfilling, rather than the discrimination of self from non-self.
- *Co-operation*: The immune response is a collaborative effort between the innate and adaptive immune agents.
- *Emergent Behaviours*: The observed immune responses and properties, such as immune specificity, emerge from the functioning of immune agents rather than a one-to-one mapping between a receptor and antigen.

Examples of the low level processes include:

- *Multiple Immune Agents*: Co-respondence involves the interactions of different agents such as macrophages, T and B cells.
- *Signalling Networks*: Immune agents communicate using an immune dialogue of signalling molecules.
- *Feedback*: Positive and negative feedback help to co-ordinate the immune response.
- *Degeneracy*: The degeneracy of antigen receptors provides a many-to-one relationship between the receptors and specificity of recognition (see Chapter 6 in this book for a further discussion on this topic).
- *Pleiotropia and Redundancy*: The pleiotropic and redundant nature of immune agents are also important in providing specificity of response.

By highlighting these properties, it is clear that the investigation of an alternative immune theory can broadened the scope for AIS inspiration. It is also possible to identify some application areas to which these properties might be applied. For example, the notion of the immune system as a concurrent and reactive maintenance system, lends itself well to application domains that operate in dynamic environments, such as embedded systems and robotics. For other properties such as degeneracy, however, it is not entirely clear how they can be beneficial to an AIS, and so it can be useful to investigate these properties in more detail. This is carried out for degeneracy in section 7.5, identifying it as an important biological property and how it might be a useful algorithmic recogntion property. First, however, we provide a general discussion on how the engineer might best exploit immune properties to build AIS.

7.4.1 Exploiting the Inspiration

As the field of AIS has matured, a better understanding of how they work and how they should be designed is currently being sought [de Castro & Timmis 2002a, Freitas & Timmis 2003, Hart & Ross 2004, Stepney *et al.* 2004]. Common to all of these works, is the requirement on the designer to consider both the biology on which the AIS is based, and the AIS application area. By doing this, it is hoped that the designer will take a more principled approach to algorithm design, leading to a better suited and performing algorithm. As highlighted in Chapter 3, [de Castro & Timmis 2002a] introduced a layered framework for engineering AIS that identifies three basic system design elements: component representations, affinity measures and immune algorithms. [Freitas & Timmis 2003] highlight the need to consider the problem for which the AIS is being designed, thus advocate a problem-oriented approach to AIS design. They review AIS for the application of classification, and show how the selection of an algorithm, representation scheme and evaluation function in AIS can all bias the results. [Hart & Ross 2004] argue that the set of matching mechanisms (affinity measure) of de Castro and Timmis' layered framework makes AIS distinct from other bio-inspired algorithms, thus suggest this should be chosen with care.

One of the main problems involved in designing bio-inspired algorithms, is deciding which aspects of the biology are necessary to generate the required behaviour, and which aspects are surplus to requirements. To help tackle this and enable the development of bio-inspired algorithms in a more principled way, [Stepney et al. 2004] have suggested a conceptual framework for designing these algorithms. This framework promotes the use of an interdisciplinary approach to developing and analysing these algorithms. It encompasses a number of modelling stages, the first of which utilises biological observations and experiments to provide a partial view of the biological system from which inspiration is being taken. This view is used to build abstract models of the biology, which are then open to validation. Frameworks can then be built and validated from these models to provide the principles for designing and analysing the required bio-inspired algorithms. Using such a framework aims to stop the designer from making naive assumptions about the biological processes that are providing the inspiration, and thus preventing the development of algorithms that are just a weak analogy of the process on which they are based.

We suggest that when taking inspiration from an alternative immune ideas, the designer should follow the suggestions just outline. In particular, the adoption of the conceptual framework is deemed especially beneficial. For instance, the immune ideas may not be fully formed or well understood, so it is possible that unexpected or unexplained behaviours will arise from the properties being modelled. Following the ideas of the conceptual framework should help capture such occurrences. Additionally, the immune models built by the AIS practitioner to investigate the properties of immune theories may be of help to immunologists. By following a principled design methodology, these models should be able to provide experimental evidence for, or against, the assertions made by immunologists in their theories. This should provide useful insights and help to develop these theories further.

The process of following the conceptual framework approach requires the building of models of the immune processes. Building these models is an aid to understanding how these processes work, and has been carried out by theoretical immunologists and AIS practitioners for many years. The immune models used in these fields naturally fall into two main classes: mathematical models and computational models. The majority of the mathematical models consist of differential equations that model the population dynamics of interacting immune agents. An overview of many of these techniques is provided by Perelson [Perelson & Weisbuch 1997]. Successful computational models that have been used for immune process modelling include cellular automata [Kleinstein & Seiden 2000], Boolean networks [Weisbuch & Atlan 1988] and UML statecharts [Efroni et al. 2003]. When choosing modelling techniques for immune processes, it is important that they are suited to the nature of the processes being modelled. For example, if the immune process is an emergent behaviour, then the designer needs to choose a modelling technique that allows emergence to occur. Such modelling tools are used extensively by the Artificial Life (ALife) community. Examples of these ALife modelling techniques include recursive developmental systems, such as cellular automata and L-systems, evolutionary systems, such as genetic algorithms, (multi-)agent based systems, such as swarms, and networks of automata, such as Boolean networks. In addition to these tools, techniques that have been utilised in other biological modelling areas may be appropriate to immune modelling. For example the work by Johnson et al. [Johnson et al. 2004] utilises object-oriented methods to model intra-cellular processes, and a similar approach could be used successfully in immune modelling. An effective example of a modelling technique taken from a different area of modelling is given by Chao et al. [Chao *et al.* 2004], where a stochastic age-structured model, often used in ecology, has been applied to modelling immune cell populations and transitions.

7.5 Degeneracy: An Alternative Inspiration Example

It was previously identified that the lymphocyte receptors of immune agents are degenerate, and that this may provide possible inspiration for AIS. Degeneracy, however, is not a unique property of just the immune system in the biological world. According to [Edelman & Gally 2001], degeneracy, which they define as:

> "the ability of elements that are structurally different to perform the same function or yield the same output"

is a ubiquitous biological property present at most levels of biological organisation. Considering this, the impact of degeneracy on biology as a whole as well as the immune system are investigated in this section. This then leads to the identification of how degeneracy might be exploited for the benefit of AIS.

7.5.1 Degeneracy in Biological Systems

[Edelman & Gally 2001] state that degeneracy appears at each of the genetic, cellular, system and population levels of biology. To highlight this, they provide a long list of examples of degeneracy in biology, which includes:

- *Genetic code*: different sequences can encode a polypeptide
- *Protein folding*: different polypeptides can fold to become structurally and functionally equivalent
- *Intra-cellular signalling*: parallel pathways of e.g. hormones transmit degenerate signals
- *Connectivity in neural networks*: connections and dynamics are degenerate
- *Body movements*: different patterns of muscle contractions can produce equivalent movements
- *Inter-animal communication*: there are many ways to transmit the same message, e.g. via language

Edelman and Gally go on to argue that the omnipresence of degeneracy in biology is a result of it being conserved and favoured by natural selection. As natural selection can only operate on populations of genetically dissimilar organisms, many different overlapping genes and gene networks will tend to contribute to a phenotypic feature that is undergoing selection. Degenerate systems will thus be maintained as the selection process cannot assign the responsibility of this feature to any particular gene or network.

Edelman and Gally also note that degeneracy in biological systems is typically accompanied with complexity, and suggest that degeneracy plays a key role in complex systems. The definition of a complex system used by [Edelman & Gally 2001] is:

> "one in which smaller parts are functionally segregated or differentiated across a diversity of functions but also as one that shows increasing degrees of integration when more and more of its parts interact"

This relationship between complexity and degeneracy is based on earlier work reported by [Tononi et al. 1999]. Here, information theoretical concepts are used to develop functional measures of degeneracy and redundancy in a neural network with respect to a set of outputs. These measures are, however, considered to be applicable to any biological network or complex system. The definition of degeneracy used by Tononi et al. is the same as that provided above by Edelman and Gally, and redundancy is defined as occurring when the same function is carried out by identical elements. Experiments using the neural network model and the degeneracy and redundancy measures showed that degeneracy is low in systems where the individual elements can affect the output independently. Degeneracy is high, however, in systems where different elements can at the same time affect the output in similar ways and have independent effects. Additionally, using a complexity metric defined as the measure of average mutual information between subsets of the system, it was shown that systems with high degeneracy also expressed high complexity. Degenerate elements were also observed to produce different outputs in different contexts, thus making degenerate systems extremely adaptable to changes in their environment. Lastly, the relationship between degeneracy and redundancy was examined, showing that degenerate systems must express a degree of functional redundancy, whereas a fully redundant system is not necessarily degenerate.

7.5.2 Degeneracy in the Immune System

Degeneracy in the immune system is typified by the degenerate nature of lymphocyte receptors described above in section 7.3.2. The discussion here will focus on the repercussions that receptor degeneracy has for the functioning of the immune system. [Parnes 2004] provides a representative chronology of the usages of the concept of degeneracy in immunology over the last 35 years. However, this chronology shows no rigorous definition of the term, so for this discussion, the definition from [Cohen et al. 2004] is adopted, which describes antigen receptor degeneracy as the:

> "capacity of any single antigen receptor to bind and respond to (recognize) many different ligands"

It can be seen that this definition holds when compared to the general definition of degeneracy in biological systems by Edelman and Gally that is stated above.

[Parnes 2004] argues that in the light of receptor degeneracy the challenge for immunology is to retain a meaningful explanation of immune activity, and that when taken seriously, the implication of degeneracy produces an alternate view of the immune system that does not fit with existing ideas. This can been seen from the description of Cohen's cognitive model given in section 7.3. Indeed, as [Parnes 2004] points out, Cohen's model attempts to cover the gap between the immunologist's idea of an immune response and what is known about the properties of adaptive immune components, such as the the inherent degeneracy of antigen receptors.

[Cohen *et al.* 2004] reports that the degeneracy of antigen receptors leads to two consequences for immune receptor recognition, both of which have been proven experimentally (see [Cohen *et al.* 2004] for appropriate references):

- *Poly-Clonality*: A single antigen epitope can activate different lymphocyte clones
- *Poly-Recognition*: A single lymphocyte clone can recognise different antigen epitopes

It is noted, however, that most immune responses do not express extreme poly-clonality as mechanisms such as clonal competition must exist to restrict it. Clonal competition should favour those lymphocyte clones whose receptors have the greatest affinity to an antigen epitope, thus they should proliferate over other clones. However, the existence of poly-recognition causes more problems for the traditional clonal selection theory view that relies on the strict specificity of lymphocyte clones. Thus, the degeneracy of antigen receptors above all provides the biggest challenge to the validity of the clonal selection theory. Cohen goes on to provide a description of colour vision as an example of the power of receptor degeneracy. The human eye possesses millions of colour receptors called cones, of which there are only three types (red, green and blue). These receptors are degenerate, each responding to broad range of light wavelengths, which overlap between the different cone types. The human brain, however, is able to perceive thousands of specific different colours, thus colour specificity is not encoded by the cones, but achieved via subsequent neuronal firings. Likewise, Cohen envisages immune specificity to be encoded in the patterns of degenerate, co-responding lymphocytes and their allied cells, not in the initial clonal activation of lymphocytes.

Other than Cohen's cognitive model of the immune system, other models exist that try and incorporate receptor degeneracy into the functioning of the immune system. Most notabe from a computational view point, is that of [Leng & Bentwich 2002], who present the immune system as a fuzzy system in order to compensate for the defects in the two-valued self–non-self classification. This model holds on to all the basic tenets of the clonal selection theory, but replaces the antigen-receptor selection and binding mechanism with a fuzzy recognition process. This process uses stimulation thresholds of a set of fuzzy lymphocytes functioning as a statistical clone for the activation of an immune response [Parnes 2004]. Thus, Leng and Bentwich conclude that in a fuzzy immune system the response to self or non-self antigen may include many effector functions, not just the response or no response of the original clonal selection theory. This fuzzy immune system view is shared by others such as [Sercarz & Maverakis 2004], who advocate the use of fuzzy logic models to help understand receptor degeneracy and its implication to the functioning of the immune system.

7.5.3 Degeneracy and AIS

The description of degeneracy just presented pitches it as an important, advantageous and powerful property at all levels of biological organisation including the immune system. At present there are no instances within the AIS literature where degenerate detectors have been directly addressed, although degeneracy is an issue that is both being discussed [Cohen et al. 2004, Parnes 2004, Sercarz & Maverakis 2004] and modelled (see Chapter 6 in this book) by immunologists. Fuzzy logic, however, has been used in association with AIS by a number of authors. Work by [Alves et al. 2004], use an AIS to induce a set of fuzzy classification rules for the purposes of data mining. These discovered rules are of the form "IF (fuzzy conditions) THEN (class)", but fuzzy logic is not used as part of the AIS itself. [Gomez et al. 2003] use a real-valued negative selection algorithm to generate a set of fuzzy detector rules given a set of self samples. These detectors are then used to determine whether a new sample is self or non-self. Thus again, this approach does not incorporate fuzzy logic into the actual AIS. The work by [Nasaroui et al. 2002], however, does use fuzzy logic within the actual antigen-antibody matching mechanism of an AIS used to mine web data. Their AIS is based on an immune network model by [Knight & Timmis 2001] called AINE. This algorithm uses the idea of an artificial recognition ball (ARB) to represent an n-dimensional data item that uses a threshold measure to match against antigen and other ARBs in the network. [Nasaroui et al. 2002] adapt the ARBs to represent fuzzy sets of data items instead of a single data item, removing the need for a crisp thresholding measure. The main motivation behind introducing this fuzzy mechanism into the AIS appears to be to deal with the weaknesses of previous AIS approaches, and is thus driven from an application viewpoint, not from specific immunological inspiration.

It is clear that incorporating degenerate detectors into AIS will affect the dynamics of the AIS algorithm. Instead of recognition being the responsibility of a single detector, recognition will emerge from the collective response of a set of detectors. The assumed benefit of an AIS with degenerate detectors will be to provide greater scalability and generalisation over existing classifier AIS. Greater scalability can be achieved as the capacity to discriminate patterns collectively by a set of degenerate detectors should be greater than by single detectors. Thus, as the number of patterns to be recognised increases, the number of detectors needed in an AIS with degenerate recognition should be less than that of existing AIS. Better generalisation ability to recognise unseen patterns could be achieved as similar patterns should produce a similar pattern of response from the set of detectors. In [Andrews & Timmis 2006], we have begun initial investigations into degenerate detectors for AIS. Here, we follow the conceptual framework approach and build an abstract computational model in order to understand the properties of degenerate detectors free of any application bias. The model is based on the activation of T_H cells in the lymph node, the sites in the body where the adaptive immune response to foreign antigen in the lymph are activated. Contained within the model are APC, antigen and T_H cell agents that move and interact in a 2-dimensional cellular space. The T_H cell agent receptors are assumed to be degenerate and their response to different antigen agents is measured. This model should help ascertain whether the perceived benefits of degenerate detectors in AIS described above are obtainable.

7.6 Conclusions

Since Jerne presented his novel idiotypic network theory, immunologists have argued over the meaning, nature and importance of self–nonself discrimination in the immune system, and these arguments are still abound today [Langman & Cohn 2000c, Anderson & Matzinger 2000b, Cohen 2000a, Tauber 2000, Langman & Cohn 2002b, Efroni & Cohen 2002, Cohn 2003, Efroni & Cohen 2003]. So where does this leave the engineer when deciding what aspects of immunological theory to take inspiration from? We suggest that in order for the AIS practitioner to make a more informed choice, he keeps an open mind on the type of immunological ideas he can take inspiration from. By investigating alternative theories of the workings of the immune system, different immune properties may become apparent that could be used as inspiration for AIS in new application domains. Alternatively, inspiration may be taken from ideas in these theories that explain the emergence of immune properties by different mechanisms. A good AIS doesn't always have to be based the theories that are currently most popular amongst immunologists.

In this chapter we have highlighted the disagreements and methodology clashes amongst a number of immunologists concerning how to understand the functioning of the immune system. Based on this, we have investigated the ideas of Irun Cohen who views the immune system as a cognitive system, to be understood through complex systems modelling approaches. From Cohen's ideas, we have identified a number of computationally appealing properties as possible areas of inspiration for novel AIS. From these it may be possible to highlight application areas to which they might be applied, although further investigations should reveal more insight. Faced with an idea for AIS inspiration, we suggest that following a suitable methodology such as the conceptual framework approach is advantageous to try and extract the key biological properties free of any application bias. In the final section, we have provided an example of investigating an alternative inspiration property for AIS. This property, degeneracy, is investigated in detail to reveal possible generalisation and scalability benefits for an AIS utilising degenerate detectors.

8

Empirical, AI, and QSAR Approaches to Peptide-MHC Binding Prediction

Channa K Hattotuwagama, Pingping Guan, Matthew Davies, Debra J Taylor, Valerie Walshe, Shelley L Hemsley, Christopher Toseland, Irini A Doytchinova, Persephone Borrow, and Darren R Flower

The Jenner Institute, University of Oxford, Compton, Berkshire, RG20 7NN, UK. darren.flower@jenner.ac.uk

Summary. Immunoinformatics is facilitating important change within immunology and is helping it to engage more completely with the dynamic post-genomic revolution sweeping through bioscience. Historically, predicting the specificity of peptide Major Histocompatibility Complex (MHC) interactions has been the major contribution made by bioinformatics disciplines to research in immunology and the vaccinology. This will be the focus of the current chapter. Initially, we will review some background to this problem, such as the thermodynamics of peptide binding and the known constraints on peptide selectivity by the MHC. We will then review artificial intelligence and machine learning approaches to the prediction problem. Finally, we will outline our own contribution to this field: the application of QSAR techniques to the prediction of peptide-MHC binding.

8.1 Introduction

Immunoinformatics - a newly emergent sub-discipline of bioinformatics, which addresses informatic problems within immunology - uses computational methodology to attack the critical immunological problem of epitope prediction. As high throughput biology begins to reveal the genomes of pathogenic bacteria, viruses, and parasites, accurate and reliable predictions will become increasingly important tools for the discovery of novel vaccines, reagents, and diagnostics. In this chapter we rehearse some of the key issues adumbrated in Chapter 2, *Immunoinformatics and Computational Vaccinology: a brief introduction,* which should be consulted for background information by the interested reader. The accurate prediction of T cell epitopes remains problematic and confounding. Current methods for the *in silico* identification of T cell epitopes rely on the accurate prediction of peptide-MHC affinity. As such, predictive computational models of peptide-Major Histocompatibility Complex (MHC) binding affinity are becoming a vital, if underappreciated and thus underused, component of modern day computational immunovaccinology. Historically, such approaches have been built around semi-qualitative, classification methods, but these are now giving way to quantitative regression techniques.

The products of the MHC play a fundamental role in regulating immune responses. T cells recognise antigen as peptide fragments complexed with MHC molecules, a process requiring antigen degradation through complex proteolytic digestion prior to complexation. The primary, but not the sole, role of MHC proteins within the adaptive immune system is thus to bind peptides and "present" these at the cell surface for inspection by T cell antigen receptors (TCRs). In order to better understand the sequence dependence of peptide binding to MHCs (Major Histocompatibility Complex), and thus identify immunogenic epitopes, the science of immunology has sought to explore the amino acid preferences exhibited by different MHC alleles.

Within the human population, for example, there are an enormous number of different genes coding for MHC proteins, each exhibiting a different sequence selectivity for peptide-binding. T cell receptors, in their turn, also exhibit different affinities for peptide-MHC complexes (pMHCs). The combination of MHC and TCR selectivies thus determines the power of peptide recognition in the immune system and thus the recognition of foreign proteins and pathogens. However, typical affinities exhibited by peptides for MHC are, on average, several orders of magnitude greater than those of TCR for pMHC.

A simple way of looking at the phenomenon of peptide-MHC interaction is to say that each MHC allele, be it class I or class II, binds peptides with particular sequence patterns. A more accurate description would be to say that MHCs bind peptides with an equilibrium binding constant which is dependant on the nature of the sequence of the bound peptide. Peptide-MHC interactions obey the same underlying physical laws, the same fundamnetal molecular mechanisms as do any and all other forms of biomolecular association event. The driving forces behind this binding are precisely the same as those driving the binding of drug molecules to their receptors or inhibitors to their target enzymes.

Peptide binding to MHCs is, arguably, the most important, and is certainly the best studied, aspect of the epitope recognition process. Once a peptide has bound to a MHC, and the resulting complex displayed on the cell surface, it is, as we have said, available for T cell-mediated surveillance by the immune system. It is generally accepted that a pMHC complex will only be recognized by a TCR if the epitope binds with an affinity greater pIC50 > 6.3, or a half-life > 5 minutes, or some similar figure for other binding measures. Some peptides binding at these affinities will become immunodominant epitopes, others will be weaker epitopes, and still others will show no T cell activity. This is partly a function of the T cell repertoire: the full and exact enumeration of the repertoire is something usually not known, at the molecular level, for an individual organism. Its structure, and the relative distribution of different specificities within it, is one of a prodigious number of alternatives and arises as a consequence of intrinsic factors, such as the organism's genome, as well as extrinsic factors, such as the infection history and environmental exposure to microbes experienced by that individual.

The selectivity for peptides exhibited by an MHC molecule is, by contrast, determined solely by its molecular structure. It undergoes no form of somatic hyper mutation or affinity maturation or thymic selection. It is, as part of a restricted set of alleles, an inherited characteristic of individual organisms. However, in man, as in most other species, the MHC is both polygenic (there are several MHC class I and MHC class II genes) and polymorphic (there are multiple variants, or alleles, of each gene). Each class of MHC is represented by several loci: HLA-A, HLA-B and HLA-C for class I and HLA-DR, HLA-DQ and HLA-DP for class II. All MHC loci are co-dominant: both maternally and paternally inherited sets of alleles are expressed. The linked set of MHC alleles found on one chromosome is called a haplotype. MHCs exhibit extreme polymorphism: within the human population there are, at each genetic locus, a great number of alleles. The October 2005 release of the IMGT/HLA sequence database, for example, contains 2,280 allele sequences, of which over 1352 are class I (414 HLA-A, 728 HLA-B, 210 HLA-C) and over 749 are class II (506 HLA-DR, 100 HLA-DQ, 143 HLA-DP) MHC molecules. This number is constantly increasing. Many of these alleles exist at a significant frequency ($> 1\%$) within the human population. MHC alleles may differ by as many as 30 amino acid substitutions. Such a level of polymorphism implies a selective pressure to create and maintain it; see Chapter 9, by Borghans *et al*. Pairs of MHC proteins will have either very similar or very different selectivies, leading to the possibility of grouping MHCs into so-called super-types, which is described in more detail elsewhere in this book; see Chapter 10, by Guan *et al*.

There is also some evidence suggesting that the probability of a particular peptide being a T cell epitope is proportional to the MHC binding affinity of that peptide. The still unsolved trick is to establish which will, or will not, be recognized by the TCR. Generally, the approach taken has been to reduce the number of epitopes to a small value using prediction. These peptides are then tested as potential epitopes in one of a great variety of assays based on different measures of T cell activation, such as T cell killing or thymidine incorporation, *inter alia*. We shall assume, as others do, that the prediction of MHC binding is the most discriminating step in the presentation-recognition pathway and thus progress to explore methods able to predict such binding. Firstly, however, we shall examine the nature of peptide MHC binding as a physical event and how its strength can be measured.

8.2 The Underlying Molecular Phenomenology of MHC Binding

Many ways exist to measure the binding of molecules. These include equilibrium constants, which include association (K_a) and dissociation constants (K_d), as well radiolabeled and fluorescent IC_{50} values that approximate equilibrium binding constants under certain conditions. Other types of measurement for peptide-MHC binding include BL_{50} values (also known as SC_{50}, EC_{50} and C_{50} values), as calculated in a peptide binding stabilization assay, T_m values (the temperature at which 50% of MHC protein is denatured), and β_2-microglobulin dissociation half-life. This final value, while strictly a kinetic measurement, is commonly believed to correlate well with binding affinity. These different measures form a hierarchy, with equilibrium constants, when calculated correctly, being the most reliable and accurate.

Peptide binding to MHC molecules can be quantified as one would quantify any other biomolecular receptor-ligand interaction:

$$R + L \longleftrightarrow RL \ [1]$$

Where R is the receptor, or MHC in this case; L the ligand, or peptide in our case; and RL is the receptor-ligand, or peptide-MHC, complex. Such interactions obey the law of mass action: the rate of reaction is proportional to the concentration of reactants. Thus, the forward reaction rate is proportional to [L] [R]. The rate of the reverse reaction is proportional to [RL], since there is no other species involved in the dissociation. At equilibrium, the rate of the forward reaction is equal to the rate of the reverse reactions, and so (using k_1 and k_{-1} as the respective proportionality constants):

$$k_1[R][L] = k_{-1}[RL] \ [2]$$

Rearranging:

$$\frac{[R][L]}{[RL]} = \frac{k_{-1}}{k_1} = K_D = K_A^{-1} \ [3]$$

Where K_A is the equilibrium association constant and K_D is the equilibrium dissociation constant, which also represents the concentration of ligand which occupies 50% of the receptor population at equilibrium.

The free energy of binding is related directly to the equilibrium constant:

$$\Delta G_{bind} = -RT \ln (K_D) \ [4]$$

Where ΔG_{bind} is the Gibbs free energy of binding, R is the gas constant, and T is the absolute temperature. The free energy (ΔG) is a product of enthalpy (ΔH) and entropy (ΔS) components related by the Gibbs-Helmhotz equation:

$$\Delta G = \Delta H - T\Delta S \ [5]$$

In the absence of non-linear effects, the enthalpy and entropy term can be obtained using the van't Hoff relation:

$$\ln K_D = \frac{\Delta H}{RT} - \frac{\Delta S}{R} \quad , \quad \frac{d(\ln K_D)}{dT} = \frac{\Delta H}{RT^2} \ [6]$$

The potential usefulness of this obvious: plotting $\ln(K_D)$ vs $1/T$ should describe a straight line with a slope equal to $(\Delta H/R)$ and an intercept on the y axis of $(\Delta S/R)$. Van't Hoff plots only identify that part of the binding enthalpy related to the observed measurement signal. So, only for a direct transformation from a defined initial state to a final state will the enthalpy be equivalent to ΔH_{bind}, as obtained, say, by ITC. No intermediate states are allowed nor should other steps be involved. ΔG typically has only moderate temperature dependence within biological systems, thus an accurate estimate of enthalpy and entropy is not usually possible using van't Hoff plots.

Energy is a property that invests a system with the ability to produce heat or perform work. Enthalpy is defined formally by:

$$H = U + PV \quad [7]$$

Where V is volume, P is pressure, and U is the total internal energy of a system. When pressure is constant, ΔH is the heat change absorbed from its surroundings by a system. For a molecular system, it is typically a function of the system's kinetic and potential energies. Entropy is often described as a measure of *disorder* within a system. However, increasing entropy can be better described as the partitioning of energy into an increasing number of explicit microstates. Within complex multicomponent systems such as these, it is often difficult to properly decompose entropies and enthalpies into clearly separable molecular contributions.

Favourable enthalpic contributions to the free energy can include complementary electrostatic interactions, such as salt bridges, hydrogen bonds, dipole-dipole interactions, and interactions with metal ions; and van der waals interactions between ligand and receptor atoms. Entropic contributions can include global properties of the system, such as the loss of three rotational and three vibrational degrees of freedom on binding, and local properties, such as conformational effects including ‿he loss of internal flexibility in both protein and ligand. Unfavourable entropic contributions from the increased rigidity of backbone and side-chain residues on ligand binding within the binding pocket are, in part, offset by favourable increases in conformational freedom at nearby residues. Strictly, all protein-ligand binding also involves multiple interactions with the solvent, typically a weakly ionic aqueous solution. These solvent interactions lead to so-called solvation, desolvation, and hydrophobic effects, each with both an enthalpic and an entropic component.

Experimentally, the measurement of equilibrium dissociation constants has most often been addressed using radioligand binding assays. There are many other ways to determine equilibrium constants more exactly, such as BIAcore and isothermal titration calorimetry. However, most have yet to be used to study peptide MHC interactions. Saturation analysis measures equilibrium binding at various radioligand concentrations to determine affinity (K_D), while competitive binding experiments measure binding at single concentrations of labeled ligand in the presence of varying concentrations of unlabeled ligand. Competition experiments can be either homologous (where the labeled and unlabeled peptides are the same) or, more commonly, heterologous (where labeled and unlabeled peptides are different) inhibition assays.

IC_{50} values, obtained from a competitive radioligand or fluorescence binding assay, are the most frequently reported affinity measures. The value given is the concentration required for 50% inhibition of a standard labeled peptide by the test peptide. Therefore nominal binding affinity is inversely proportional to the IC_{50} value. Values obtained from radioligand or fluorescence methods may be significantly different. IC_{50} values for a peptide may vary between experiments depending on the intrinsic affinity and concentration of the standard radiolabeled reference peptide, as well as the intrinsic affinity of the test peptide.

The K_D of the test peptide can be obtained from the IC_{50} value using the relationship derived by Cheng and Prussoff:

$$K_D^i = \frac{IC_{50}}{\left(1 + \frac{[L_{tot}^S]}{K_D^S}\right)} \quad [8]$$

Where K_D^i is the dissociation constant for the inhibitor or test peptide, K_D^S is the dissociation constant for the standard radiolabeled peptide, $[L_{tot}^s]$ is the total concentration of the radiolabel. This relation holds at the midpoint of the inhibition curve under two principal constraints: the total amount of radiolabel is much greater than the concentration of bound radiolabel and that the concentration of bound test peptide is much less than the IC_{50}. This relation, although an approximation, holds well under typical assay conditions. In practice, the variation in IC_{50} is often sufficiently small that values can be compared between experiments.

BL_{50} values are also obtained from a peptide binding assay. They are the half maximal binding levels calculated from mean fluorescence intensities (M.F.I.) of MHC expression by RMA-S or T2 cells. Cells are incubated with the test peptide and then labeled with a fluorescent monoclonal antibody. The nominal binding strength is again inversely proportional to the BL_{50} value. These assays are often termed stabilization assays, as it is presumed that cell surface MHCs are only stable when they have bound peptide. Given that peptides are typically administered extracellularly, there remain questions about the precise molecular mechanism of peptide induced MHC stabilization. Moreover, the measured BL_{50} values also represent an approximate overall value from a complex multi-component equilibrium. The interaction between peptide and MHC, as reflected in complex stability, is measured by binding to it either an allele- or class I-specific antibody, which is then bound by a fluorescently labeled antibody specific for the first antibody. The resulting complex is then assayed spectrophotometrically using flow cytometry or an equivalent technique.

The half-life for radioisotope labeled β_2-microglobulin dissociation from an MHC class I complex, as measured at $37\,^\circ$C is a commonly reported alternative binding measure. This is a kinetic measurement rather than a thermodynamic one, although it is often assumed that the greater the half-life the stronger the peptide-MHC complex. The half-life ($t_{1/2}$) equals:

$$t_{1/2} = \frac{\ln 2}{k_{-1}} \sim \frac{0.693}{k_{off}} \quad [9]$$

Here the $t_{1/2}$ corresponds to the dissociation of the MHC-β_2 micro-globulin complex rather than the kinetics of the protein-ligand interaction. One would anticipate that the peptide dissociation would be related to the overall dissociation of the complex, but quite what this relationship is, has not been characterized.

We may wish to ask the question: which of these measures is best? Unfortunately, there is no simple answer. The *prima facie* response might be "equilibrium constant", but in what context? K_d and IC_{50} values are probably the most accurately measured constants, as they are usually assayed using soluble protein. However, as MHCs are membrane bound in their functional context, a value, such as a BL_{50}, might be more relevant to processes *in vivo*. However, BL_{50} values are typically measured using a cascade of antibodies. The multiple-equilibrium that results may obscure salient experimental details. Other binding measures are also sometimes reported in the context of MHC-peptide interaction. As yet, no clear consensus has emerged on the most appropriate type of affinity measurement or assay strategy. It remains a prime concern to establish a correlation between these different binding measures, so that information-rich measures can work synergistically with those that are facile to perform. Moreover, it reinforces the need to establish effective predictive methodology that can substitute effectively for experimental assays.

8.3 The Long and the Short of MHC-peptide Interaction

Peptide binding to MHCs and TCR binding to peptide-MHC complexes are the key biomolecular interactions underlying the recognition process which enables the immune system to discriminate properly between proteins of benign origin and those produced by pathogens. MHC-mediated recognition of pathogens thus enables the immune system to do its work; the importance of these two processes can not be underestimated: its success is often what divides life from death. The study of immunology and vaccinology remains distinctly empirical in nature, yet such peptide binding phenomena are as amendable to rigorous biophysical characterisation as any other biochemical reaction.

MHC proteins are grouped into two classes on the basis of their chemical structure and biological properties. The two types of MHC protein have related secondary and tertiary structure but with important functional differences. Class I molecules are composed of a heavy chain complexed to $\beta2$-microglobulin, while class II molecules consist of two chains (α and β) of similar size. MHC class I molecules present, in the main, cytosolic and endogenous peptides, although exogenous antigens can also be presented by MHC class I molecules via mechanisms of cross-presentation. Following the receptor-mediated endocytosis of exogenous antigens by so-called antigen-presenting cells, such as macrophages and dendritic cells (DCs), antigenic proteins pass into endosomes, from where they pass, in turn, to late endosomes and then lysosomes, where proteolytically fragmented peptides are bound by MHC class II. Other peptides either escape from or are degraded within the endosome and pass into the cytoplasm where they enter the familiar compartmentalised class I pathway comprising the proteasome, TAP and MHC. Class I MHC is expressed on almost all cells in the body and presents a complex population of peptides on the cell surface which is dominated by a combination of self and viral peptides.

MHCs bind peptides, which are themselves derived through the degradation of proteins. The proteolytic pathway by which peptides become available to MHCs remains complicated and poorly understood. However, MHCs also bind a variety of other molecules other than peptides whose sequences are composed of the 20 commonly occurring amino acids. Peptides bearing diverse post-translational modifications (PTMs) can also form pMHCs and be recognized by TCRs. Such PTMs include phosphorylation, glycosylation, and lipidation [Kastrup et al. 2000, Zarling et al. 2001]. Additionally, MHCs will also bind other molecules including chemically modified peptides, synthetically derived peptide mimetics, and even small molecule drug-like compounds [Pichler 2002]. It is now well known that many drug-like molecules exhibit pathological effects through binding to MHCs. Indeed, small molecule binding has functional implications in behaviour-modifying odour recognition. The modern pharmaceutical industry does not currently favour peptides as drug candidates, but it has been proposed recently that a MHC-drug-TCR complex may be a suitable approach for developing immunotherapeutic inhibitors of T cell mediated processes.

Both classes of MHC molecule have similar 3-D structures. The MHC peptide-binding site consists of a β-sheet, forming the base, flanked by two α-helices, which together form a narrow cleft or groove accommodating bound peptides. The principal difference between the two classes are the dimensions of the peptide-binding groove: in class I it is closed at both ends while both ends of the MHC class II binding site are open. This has important implications for the length dependence of MHC peptide selectivity. Peptides bound by class I MHCs are perceived to be typically relatively short while peptides bound by class II MHCs are, by contrast, essentially unconstrained in terms of length. This creates particular problems of its own, which we shall examine in more detail below.

It has long been thought that Class I preferentially bind peptides with sequences which are 8-11 amino acids in length. This view is now beginning to be challenged. For example, there is increasing experimental evidence that a wide variety of other peptide lengths also form stable, high affinity complexes with class I MHC class I, including both very short peptides (4-5-mers) and very long peptides (up to 18-mers). This is in addition to the diverse array of post-translationally modified or synthetically treated peptides, all of which give rise to cytotoxic T cells (CTL). This view arose partly as a result of crystal structures showing that the length of MHC binding site most obviously accommodates peptides of 8-10 amino acids, and partly from elution, and other biochemical, studies, which also point to a similar restricted spectrum of lengths.

There is some experimental data to support the binding of short peptide sequences ("Shortmers") to MHCs. Mouse CTLs recognize MUC1 epitopes bound to H-2Kb: MUC1-9 (SAPDTRPAP), MUC1-8 (SAPDTRPA), as well as truncated versions: SAPDTRP (7-mer), SAPDTR (6mer), and SAPDT (5-mer) [Apostolopoulos et al.

2001]. Other possible shortmers which bind class I MHC molecules include 3-mers (QNH), 4-mers (QNHR, ALDL, PFDL) and 5-mers (RALDL, HFMPT) [Gillanders *et al.* 1997, Reddehase *et al.* 1989]. Since such observations lie outside normal expectation, and are thus open to question, how such phenomena manifest themselves are of not inconsiderable interest.

Evidence supporting the binding of long peptides sequences to MHC ("Longmers") is considerably more abundant and much, much stronger.Indeed, it is so abundant and so strong that we will eschew parenthetical results and concentrate instead on recent unequivocal crystal data. Crystallographic studies of MHC class I molecules revealed similarly bound conformations of MHC ligands, despite significant differences in primary structure. In general, peptides bind to MHC with their N- and C-termini fixed at either end, and with the central of the peptide bulging out slightly. Residues lining the MHC binding groove interact with the peptide, and thus define the specificity of the HLA-peptide interaction. Peptides of length 8 or 9 evince a more or less invariant conformation, while peptides of length 10 tend to "bulge" in the central region of the peptide, with the conformation of the bulged region seen in 10mer complexes varying considerably between structures. In all 3 cases, N- and C-terminal interactions remain essentially unchanged. Such trends are exacerbated as the length of bound peptides increases. [Probst-Kepper *et al.* 2004] report the 1.5Å crystal structure of the human MHC HLA-B*3501 in complex with a 14mer peptide (LPAVVGLSPGEQEY). This 14-mer originates from an alternate open reading frame (ORF) of macrophage colony-stimulating factor, which is expressed in several tumour lines. Elution studies of tumors demonstrated that the peptide is presented naturally and is recognized by CTL [Probst-Kepper *et al.* 2001]. The crystal structure showed that both the N and C termini of the peptide are embedded in the A and F pockets of the MHC peptide binding site, in a similar way to that of an 8mer or a 9mer. The centre of the 14mer peptide bulges, in a conformationally flexible manner, from the binding groove in an unpredictable way. [Speir *et al.* 2001] had reported earlier the 2.55Å crystal structure of a complex between the rat MHC allele RT1-A^a and a 13mer peptide (ILFPSSERLISNR), a natural ligand eluted from the cell surface, which was derived from a mitochondrial ATPase. More recently, [Tynan *et al.* 2005b, Tynan *et al.* 2005a] report the 1.5Å crystal structure of a 13-mer viral epitope (LPEPLPQGQLTAY) bound to the human MHC HLA-B*3508. These accounts, which detail crystal structures of "bulged" Longmers, are in contrast to earlier reports which describe binding to class I MHC via a C-terminal protrusion mechanism: a 10-mer peptide bound to HLA-A*0201 where the additional glycine residue extended out from the C-terminus.

These recent reports are in turn supported by observations at the functional level of CTL recognition of Longmers. Reports of human 10mer and 11mer CD8 epitopes restricted by class I MHC alleles are commonplace, as are 10mer peptides binding to mouse class I. SYFPEITHI is awash with class I peptides that are up to 15 amino acids in length in humans and up to 16 amino acids in mice. [Jiang *et al.* 2002] report convincing evidence of a H2-K^d-restricted CTL response to a 15mer epitope (ELQLLMQSTPPTNNR) from the F protein of respiratory syncytial virus. A 12-mer peptide (LLLDVPTAAVQA) is known to be processed naturally and presented in a HLA-A*0201 restricted manner [Chen *et al.* 1994]. 12mer C-terminal variants of VSV8 have been shown to bind H-2Kb (RGYVYQGLKSGN) [Horig *et al.* 1999].

Thus, the apparent potency of responses to such Longmers is indicative that peptides longer than the norm do not represent a major structural barrier to recognition by TCRs. Clearly, then, the evidence, as adumbrated above, suggests strongly that the natural repertoire of class I MHC presented peptides is far broader than has generally been supposed.

Although it is increasingly clear that the apparent repertoire of peptide lengths able to induce class I restricted CTL responses has widened considerably in recent years, most information on class I MHC-peptide binding remains focussed on 9mers and HLA-A*0201; there is far more data available for this allele than for any other, and by a not inconsiderable margin. It was long ago realised that logistic restrictions make exhaustive testing of overlapping peptides unrealistic, making computational prediction an attractive option. The first attempts to computerize the identification of MHC binding peptides led to the development of motifs characterizing the peptide specificity of different MHC alleles. Such motifs - a concept with wide popularity amongst immunologists - characterize a short peptide in terms of dominant anchor positions with a strong preference for certain amino acids. Probably the first real attempt to analyze MHC binding in terms of specific allele-dependant sequence motifs was undertaken by [Sette et al. 1989]. They defined motifs for the mouse alleles I-Ad and I-Ed after measuring affinity for a large set of synthetic peptides originating from eukaryotic and prokaryotic organisms, as well as viruses; in addition they also assayed a set of overlapping peptides encompassing the entire staphylococcal nuclease molecule. [Sette et al. 1994b] quote prediction rates at the 75% level for these two alleles. A large number of succeeding papers, both from this group and others, have extended this approach to include many other class I MHC alleles.

Motifs are usually expressed in terms of anchor residues: the presence of certain amino acids at particular positions that are thought to be essential for binding. For example, human Class I allele HLA-A*0201 has anchor residues at peptide positions P2 and P9 for a nine amino acid peptide. At P2, acceptable amino acids would be Leucine and Methionine, and at the P9 anchor position would be amino acids Valine and Leucine. Primary anchor residues, although generally deemed to be necessary, are not sufficient for peptide binding, and secondary anchors, residues that are favourable, but not essential, for binding may also be required; other positions show positional preferences for particular amino acids. Moreover, the presence of certain residues at specific positions of a peptide can have a negative effect on binding [Smith et al. 1996b, Southwood et al. 1998, Amaro et al. 1995]. It is possible to rationalise apparent preferences expressed within motifs in terms of physico-chemical complementarity of peptide structure with the structure of the MHC. Position P1 corresponds to pocket A of the cleft of the peptide-binding site on HLA-A*0201 [Saper et al. 1991]. Anchor residues at position 2 and at the C-terminus (position 9) are seen to be of primary importance for binding, where pocket B interacts with the side chain of the residue at position 2. The structure of pocket A is mainly polar residues and consists of a network of hydrogen bonding residues. A hydrophobic ridge cuts through the binding cleft forcing the peptide to arch between position 5 and the carboxyl-terminal residue (position 9) which are anchored into the D and F pockets in the floor of the cleft [Fremont et al. 1998]. Taking another example: human class I allele HLA-B*3501 has anchor residues at position P2 (Pro) and P9 (hydrophobic or aromatic residues, such as Phe, Met, Leu, Ile and especially Tyr).

Class I mouse MHC alleles also bind diverse peptides, typically 8-10 amino acids in length, with each allele exhibiting definite peptide specificity. From our work, [Doytchinova & Flower 2002b, Doytchinova & Flower 2002a, Doytchinova et al. 2002, Guan et al. 2003b, Guan et al. 2003a, Hattotuwagama et al. 2004, Doytchinova & Flower 2003] and previous peptide binding experiments, it is clear that the molecule binds short peptides, which are most often nonamers [Bjorkman et al. 1987]. The crystal structure of several mouse class I molecules has revealed that the peptide binding cleft is also closed at both ends, that the length of the cleft is similar for all class I molecules, [Fremont et al. 1998, Zhang et al. 1992, Young et al. 1994, Smith et al. 1996b, Smith et al. 1996a] and that the carboxyl-terminal peptide position is an anchor residue deeply buried in the F pocket. Analysis of the structure and binding results of the H2-K^b and H2-K^b octameric complex reveals that there is a strong preference for an aromatic and hydrophobic residues Tyr and Phe (H2-K^b) and Leu (H2-K^b) at positions 3 and 5 and for a strong hydrophobic residue Val (H2-K^b) and Ile, Val and Phe (H2-K^b) at position 8, which is in accordance to the studies of Falk et al., (1991). In H2-K^b the B pocket is large enough to accommodate a bulky Ile residue at position 2, which is in accordance with the crystal structure of the antigenic peptide from the ovalbumin complex OVA-8 (SIINFEKL). The anchor carboxyl-terminal (position 8) prefers hydrophobic residues, which fall into pocket F.

The motif method is admirably simple and it is straightforward to implement either by eye or, more systematically, using a computer to scan through protein sequences. However, there are many problems with the motif approach. The most significant of which is that the method is, essentially, deterministic: a peptide either binds or does not bind. Inspection of sequence patterns from epitopes and non-epitopes shows the inadequacies of the motif approach. For example, Table 8.1 gives a list of example peptides for human MHC HLA-A*0201 which are both affine and motif-negative, that is they possess non-canonical sets of amino acids lying outside established definitions for A2. There are many more examples of motif-positive peptides which do not bind. Even a brief reading of the immunological literature shows that matches to motifs produce many false positives, and are, in all probability, producing an equal number of false negatives, though peptides predicted to be non-binders are seldom screened. The deterministic dependence on the presence of anchor residues is, arguably, the most vexatious feature of motifs. Proponents of motifs characterise one or more anchor positions as being all important. Anchors are, as we have said, residues upon which binding is deemed to depend. This argument, based on observations of characteristic sequence patterns evident in pool sequencing and the sequences of individual epitopes, holds that only peptides matching a sparse pattern of residues can bind. This clearly makes little sense and is utterly incompatible with everything we know about molecular binding events. The whole of the biophysical and chemical literatures argues against this view. However, rather than review thousands of papers, we might, in the present context, examine evidence within the immunological literature. We have already seen that motifs are not consistent with the sequences of real epitopes and MHC binders. Moreover, there have been several analyses which have addressed this issue directly. [Stryhn et al. 1996] generated all systematic single mutants of the 8mer peptide FESTGNLI and observed that mutations at positions 2 and 8 resulted in the greatest changes in affinity, but that, contrary to expectation, few substitutions abolished activity. Thus it is possible to lose canonical anchors without eradicating binding. Likewise, [Doytchinova *et al.* 2004c] made all 40 mutations at positions 2 and 9 of a super-binding 9mer peptide that bound to HLA-A*0201. Again, only a few residues caused a complete abrogation of affinity and about half of these peptides were still sufficiently affine to be potential epitopes, with affinities beyond the well known $pIC_{50} > 6.3$ threshold [Sette *et al.* 1994b].This study demonstrates that it is possible to manipulate A2 binding affinity in a rational way, raising resulting peptides to affinity levels two orders of magnitude greater than previously reported. The study also showed that optimised non-anchor residues can more than compensate for non-optimal substitutions at anchor positions. Clearly, the whole of a peptide contributes to binding, albeit weighted differently at different positions. This is made very clear by data-driven modelling, which we shall describe in greater detail below. For class I, and probably also for class II, it is quite straightforward to generate high affinity peptides, and affine peptides without non-canonical anchors, at least for well characterised alleles, with extra affinity arising from favourable interactions made by non-anchor residues.

There is some evidence suggesting that as the MHC binding affinity of a peptide rises, the greater the probability that it will be a T cell epitope. The prediction of MHC binding is both the best understood and the most discriminating step in the presentation-recognition pathway. A pragmatic solution to the as yet unsolved problem of what will be recognised by the TCR, and thus activate the T cell, is to greatly reduce the number of possible epitopes using MHC binding prediction, and then test the remaining candidates using some measure of T cell activation, such as T cell killing or thymidine incorporation.

8.4 A Class Apart

As mentioned above, class II MHC molecules are a non-covalent heterodimer. Peptides binding to class II MHC molecules are usually 10-25 residues long, with peptide lengths of 13-16 amino acids being the most frequently observed [Rognan *et al.* 1999, Hunt *et al.* 1992, Chicz *et al.* 1992, Chicz *et al.* 1993]. From X-ray crystallographic data of MHC class II and TCR-peptide-MHC class II complexes, [Dessen *et al.* 1997, Hennecke & Wiley 2002] it is clear that nine amino acids are bound in an extended conformation within the class II binding site. They are not anchored at their amino and carboxyl termini, but stretch along the binding groove, with residues accommodated by binding pockets along the cleft. However, because the two ends of the peptide binding site are both open in class II MHC molecules, peptides are, as we have already noted, usually much longer than the minimal 9 amino acids required for binding. One of the most challenging aspects of the class II prediction problem is the initial identification of the *core* nonameric binding subsequence within the much longer peptides, which were typically generated by the quasi stochastic process of Cathepsin-mediated proteolysis of endocytosed extracellular protein. Various methods have been proposed for this purpose. Implicit in most of these efforts are attempts to identify some kind of semi-invariant sequence pattern. Such attempts have included the use of conventional sequence alignment programs and hidden Markov models, amongst others. Other methods, such as [Mallios

2001, Doytchinova & Flower 2003], have used an iterative approach to identifying a binding model. No extant method is wholly successful and this remains a continuing challenge.

Inspired by the relative success of investigations into class I motifs, previous interpretations of experimental data, extant within the literature, have suggested that class II peptides also have a small number of anchor residues upon which binding depends. These anchors are residues of an appropriate type, which must sit at particular spacings along the peptide in order for allele-restricted binding to occur; residues at other peptide positions are less constrained. It is seems clear from experimental studies of T-cell epitope analogue binding and data from X-ray crystallography, that peptides bind to MHC molecules through the interaction of side chains of certain peptide residues with pockets situated in the MHC class II peptide-binding site: these side-chains extend into discreet pockets within the binding groove [Hennecke & Wiley 2002, Fremont et al. 1998, Corper et al. 2000]. Peptide side-chains form favourable (polar, hydrophobic or steric) interactions with MHC side-chains within these pockets [Corper et al. 2000]; the most critical determinant of binding, other than the presence of appropriate types of side chain, is their relative spacing. The side chain at peptide position P1 binds into a deep pocket while four shallow pockets bind side chains at peptide positions 4, 6, 7 and 9. The side chains at positions 2, 3, 5 and 8 point towards the T cell receptor.

Results from data-driven modelling are consistent with an emerging view of MHC peptide interactions: motifs are an inadequate representation of the underlying process of binding. This is likely to characterise class II binding as much as it obviously does for class I. For example, [Liu et al. 2002] showed that for I-A^b it was possible for a peptide bearing alanines to bind to its four main pockets - which correspond to positions P1, P4, P6 and P9 and which usually bind larger peptide side-chains - with compensatory interactions, made by residues at other positions, maintaining the overall affinity. Both structure-driven and data-driven modelling of class II binding is strongly suggestive that the relative contributions, of particular residues, to binding are spread more evenly through the peptide than is generally supposed, rather than being concentrated solely in a few so-called anchor positions.

Complicated as this view of class II binding may appear, the situation may be even more complex. For example, it has been suggested that different MHC class II molecules may bind the same peptide in multiple binding registers, whereby the peptide is displaced longitudinally within the binding groove with side chains being bound by different pockets [Li et al. 2000, McFarland et al. 1999, Vidal et al. 2000]. Two main alternative scenarios have been proposed hitherto [Bankovich et al. 2004]: binding of the same peptide in different registers by the same or by different alleles. The more common second alternative is well demonstrated [Li et al. 2000, Vidal et al. 2000] and results from minor polymorphic differences in the amino acid residue composition of the binding groove. In the DRB$_5$ complex, the large P1 pocket accommodates Phe from the peptide and Ile occupies the shallow pocket at P4. However, in the DRB$_1$ allele, the small pocket at P1 is occupied by Val shifting the peptide to the right, while Phe occupies a deeper pocket at P4. This also causes certain peptide side-chains, which are orientated toward the TCR, to change [Li *et al.* 2000]. Unequivocal evidence supporting the former alternative is somewhat scarce: there are few, if any, proper examples of exactly the *same* peptide binding in different registers to exactly the *same* MHC molecule.

Logistical difficulties inherent in deciding upon the binding register and the weak motif-dependence are not the only problems which confound the study of class II peptide-MHC binding. A further important issue is the influence of *flanking* residues on affinity and recognition: [Arnold *et al.* 2002] identified residues at +2 or -2, relative to the core nonamer, as important for effective recognition by T-cells. [Godkin *et al.* 2001] also looked at how flanking regions influence immunogenicity, noting that active class II peptides show residue enrichment outside of the central core regions, but conclude that, for HAL-DR2 at least, such phenomena result from conferred preferences for both cellular processing and MHC-mediated T cell activation. Further studies showed similar patterns in another nine HLA alleles, where C-terminal basic residues were as highly conserved as previously identified N-terminal prolines. This, and other issues - such as do certain class II epitopes bulge as class I peptides do or can a class II ligand bind so that the full 9 positions in the class II groove are not used? - makes it clear that class II peptide binding is potentially an order of magnitude more difficult a problem to solve.

8.5 Empirical and Artificial Intelligence Approaches

In order to quantify adequately the affinities of different MHCs for antigenic peptides, many different methods have been developed. It is possible to group these methods together thematically, based on the kind of underlying techniques they employ, and we shall endeavour to review them in this fashion below. As a preliminary, it is perhaps appropriate to mention some of the underlying issues involved. A widely used conceptual simplification, often used to help combine this bewildering set of binding measures, is to reclassify peptides as either Non-binders or as High-binders, Medium binders, and Low binders. For example, the schema used by [Brusic *et al.* 1998] classifies binders using these criteria: Non-Binders > 10uM, 10uM > Low Binders > 100nM, 100nM > Medium Binders > 1 nM, High Binders < 1nM. Such broad schemes also allow for the inherent inaccuracy in MHC binding measurements.

While useful in themselves, binding motifs are, as we have said, very simplistic. They are not quantitative and their over-reliance on anchor positions can lead to unacceptable levels of false positives and false negatives. Alternative approaches abound. The different types have, as one might expect, different strengths and different weaknesses. The strategy adopted by many workers is to use data from binding experiments to generate matrices able to predict MHC binding. For want of a better term, we refer to these approaches as experimental matrix methods, as most such methods use their own measured data and relatively uncomplicated statistical treatments to produce their predictive models. For example, [Rothbard *et al.* 1994], developed a method to predict the strength of binding to human Class II allele HLA DRB1*0401. They assumed, as do many other workers, that peptides of the same length bind similarly and that the contribution made by each side chain is independent and can be treated as a simple sum of residue interactions. Within the context of an otherwise polyalanine backbone, the contributions made by the central 11 positions, of a 13 amino acid peptide, were quantified by measuring the effects of changes in amino acid identity at each position within the peptide.

An alternative strategy is the use of positional scanning peptide libraries (PSPLs) to generate such matrices. A number of such studies have aimed to investigate MHC-peptide interaction or to evaluate how variations in peptide sequence alter TCR recognition and T cell activation. One of the most recent of these is also one of the most promising: [Udaka et al. 2001] have used PSPLs to investigate the influence of positional sequence variation on binding to the mouse Class I alleles Kb, Db, and Ld. From their analysis a program that could score MHC-peptide interaction was developed and used to predict the experimental binding of an independent test set. Their results showed a linear correlation but with substantial deviation. About 80% of peptides could be predicted within a log unit.

There are many other papers developing methods of this ilk. Though valuable contributions, it is clear that that they betray a series of important limitations. Firstly, they do not, in general, constitute a systematic approach to solving the MHC-peptide binding problem. Rather, they are a set of different - essentially individual, independent, and inconsistent - solutions to the same, or nearly the same, problem. The measures of binding are different, the degree of quantitation is different for different methods and they also lack subsequent applications corroborating their predictive power.

A further step forward from motifs came with the work of [Parker et al. 1994]. This method, which is based on regression analysis, gives quantitative predictions in terms of half-lives for the dissociation of β_2-microglobulin from the MHC complex. Moreover, apart from its intrinsic utility, one of the other important contributions of this approach is that it was the first to be made available on-line (http://bimas.dcrt.nih.gov/molbio/hla_bind/).

[de Groot et al. 2001] have developed several computer programs, principally EpiMer and EpiMatrix, and have used them in various practical applications, with a particular focus on HIV. EpiMatrix and EpiMer are pattern-matching algorithms that attempt to identify putative MHC-restricted T cell epitopes as a preliminary to constructing multi-epitope vaccines. These algorithms are themselves based on matrix representations of positional amino acid preferences within MHC-bound peptides. The general utility of these methods has been limited by the commercial exploitation of the EpiMatrix and EpiMer technology. Hammer and co-workers have developed an alternative computational strategy called TEPITOPE [Sturniolo et al. 1999]. Although the program can provide allele specific predictions, its main focus is on the identification of promiscuous Class II binding peptides. One of the most interesting aspects of Hammer's work has been the development of so-called virtual matrices, which, in principal, provides an elegant solution to the problem of predicting binding preferences for alleles for which we do not have extant binding data. Within the three-dimensional structure of MHC molecules, binding site pockets are shaped by clusters of polymorphic residues and thus have distinct characteristics in different alleles. Each pocket can be characterized by "pocket profiles," a representation of all possible amino acid interactions within that pocket. A simplifying assumption is that pocket profiles are, essentially, independent of the rest of the binding site. A small database of profiles was sufficient to generate, in a combinatorial fashion, a large number of matrices representing the peptide specificity of different alleles. This concept has wide applicability and underlies, for example, attempts to use fold prediction methods to identify peptide selectivity.

A number of groups have used artificial intelligence techniques, such as artificial neural networks (ANNs) and hidden Markov models (HMMs), to undertake the prediction of peptide-MHC affinity. ANNs and HMMs, are, for slightly different applications, the particular favourites when bioinformaticians look for tools to build predictive models. However, the development of ANNs is often complicated by several adjustable factors whose optimal values are seldom known initially. These can include, *inter alia*, the initial distribution of weights between neurons, the number of hidden neurons, the gradient of the neuron activation function, and the training tolerance. Other than chance effects, neural networks have, in their application, suffered from three kinds of limiting factor: over-fitting, overtraining (or memorization), and interpretation. As new, more sophisticated neural network methods have been developed, and basic statistics applied to their use, over-fitting and overtraining have been largely overcome. Interpretation, however, remains an intractable problem.

Notwithstanding these potential problems, many workers have adopted an ANN-based strategy. [Bisset & Fierz 1993] were amongst the first to use ANN in this context, when they trained an ANN to relate binding to the Class II allele HLA-DR1 to peptide structure. [Adams & Koziol 1995] used ANN to predict peptide binding to HLA-A*0201. They took a dataset of 552 nonamers and 486 decamers and generated a predictive hit rate of 0.78 for classifying peptides into two classes, one showing good or intermediate binding and another demonstrating weak or non-binding. [Gulukota *et al.* 1997, Gulukota & DeLisi 2001] developed two complementary methods for predicting binding of 463 9mer peptides to HLA-A*0201. One method used an ANN and the other used statistical parameter estimation. They found the ANN was better than motif methods for rejecting false positives, while their other alternative method was superior for eliminating false negatives. [Milik *et al.* 1998] used ANN to predict binding to the mouse Class I molecule Kb based on a training set of binding and nonbinding peptides derived from a phage display library. While it was easy to identify strongly affine peptides with a number of different methods, they found that ANNs predicted medium binding peptides better than simple statistical approaches.

[Mamitsuka 1998] has applied supervised learning to the problem of predicting MHC binding using an HMM as his inference engine. In a cross-validated test, the discrimination exhibited by his supervised learning method is usually approximately 2-15% better than other methods, including back propagation neural networks. Interestingly, his HMM model allowed the straightforward identification of new, non-natural peptide sequences that have a high probability of binding.

8.6 QSAR approaches

QSAR analysis, as a predictive tool of wide applicability, is one of the main cornerstones of modern cheminformatics and increasingly, bioinformatics. Quantitative Structure-Activity Relationship (QSAR) methods are now establishing themselves as immunoinformatic techniques useful in the quantitative prediction of peptide-protein affinity. QSAR methods have long proved themselves to be powerful tools for the prediction and rationalization of structure-property relationships within physical science. The fundamental objective of QSAR is to take a set of molecular structures (peptides in this case), for which a biological response has been measured (an IC_{50} or other affinity measure), and, using some statistical method (Partial Least Squares or other robust multivariate method), to relate this measured activity to some description of their structure. We now review how we have deployed QSAR techniques in immunological research. Our QSAR-inspired approaches to the development of immunoinformatic methods - the Additive method and its extension, ISC-PLS, and the CoMSIA method - were generated for 28 human and mouse class I and class II alleles. Relevant statistical parametersare given in Tables 3-5.

Cross Validation: QSAR style

Cross-Validation (CV) is a reliable technique for testing the predictivity of models. With QSAR analysis in general and PLS methods in particular, CV is a standard approach to validation. CV works by dividing the data set into a set of groups, developing several parallel models from the reduced data with one or more of the groups excluded, and then predicting the activities of the excluded peptides. When the number of excluded groups is the same as the number in the set, the technique is called Leave-One-Out Cross-Validation (LOO-CV). The predictive power of the model is assessed using the following parameters: cross-validated coefficient (q^2) and the Standard Error of Prediction (SEP), which are defined in equations 11 and 12.

$$q^2 = 1.0 - \frac{\sum\limits_{i=1} (pIC_{50(exp)} - pIC_{50(pred)})^2}{\sum\limits_{i=1} (pIC_{50(exp)} - pIC_{50(mean)})^2} \text{ Or simplified to } q^2 = 1.0 - \frac{PRESS}{SSQ} \quad [10]$$

Where $pIC_{50(pred)}$ is a predicted value and $pIC_{50(exp)}$ is an actual or experimental value. The summations are over the same set of pIC$_{50}$ values. PRESS is the PRedictive Error Sum of Squares and SSQ is the Sum of Squares of pIC$_{50(exp)}$ corrected for the mean.

$$SEP = \sqrt{\frac{PRESS}{p-1}} \quad [11]$$

Where p is the number of the peptides omitted from the data set. The optimal number of components (NC) resulting from the LOO-CV is then used in the non-cross validated model which was assessed using standard MLR validation terms, explained by variance r^2 and Standard Error of Estimate (SEE) which are defined in equations 13 to 14.

[12]

$$r^2 = \frac{\sum_{i=1}^{n}(Y_{obs} - Y_{pred})^2}{\sum_{i=1}^{n}(Y_{obs} - \bar{Y})^2}$$

where Y_{pred} is the predicted, Y_{obs} the observed, and $\bar{Y}$ is the average dependent variable, in this case IC_{50} values.

$$SEE = \sqrt{\frac{PRESS}{n-c-1}} \quad [13]$$

Where n is the number of peptides and c is the number of components. In the present case, a component in PLS is an independent trend relating measured biological activity to the underlying pattern of amino acids within a set of peptide sequences. Increasing the number of components improves the fit between target and explanatory properties; the optimal number of components corresponds to the best q^2. Both SEP and SEE are standard errors of prediction and assess the distribution of errors between the observed and predicted values in the regression models.

8.6.1 2D QSAR techniques: Additive and ICS-PLS methods

We have recently developed an immunoinformatic technique for the prediction of peptide-MHC affinities, known as the Additive Method, a 2D-QSAR technique which is based on the Free-Wilson principle, [Kubinyi & Kehrhahn 1976] whereby the presence or absence of groups is correlated with biological activity. For a peptide, the binding affinity is thus represented as the sum of amino acid contributions at each position. We have extended the classical Free-Wilson model with terms, which account for interactions between amino acids side chains. The Additive method uses Partial Least Squares (PLS), an extension of Multiple Linear Regression (MLR) (Wold, 1995) to identify the sequence dependence of peptide binding specificity for various class I MHC alleles with known binding affinities. The binding affinities were originally assessed by a competition assay based on the inhibition of binding of the radiolabeled standard peptide to detergent-solubilised MHC molecule [Ruppert *et al.* 1993, Sette *et al.* 1994a] and are usually expressed as IC50 values. Extracted IC_{50} values were first converted to $\log[1/IC_{50}]$ values (or $-\log_{10}[IC_{50}]$ or pIC_{50}) and used as the dependent variables in a QSAR regression. pIC_{50} can be related to changes in the free energy of binding: $\Delta G_{bind} \quad \alpha -RT \ln IC_{50}$. The values were predicted from a combination of the contributions (p) of individual amino acids at each position of the peptide and used as the dependent variables in a QSAR analysis. Using literature data, we applied the Additive Method to peptides binding to several human class I alleles [Doytchinova *et al.* 2002, Guan *et al.* 2003b, Hattotuwagama *et al.* 2004, Doytchinova & Flower 2003]. Peptide sequences and their binding affinities were obtained from the AntiJen database, a development of Jen-Pep [URL: http://www.jenner.ac.uk/AntiJen] [Blythe *et al.* 2002, McSparron *et al.* 2003]. Compilations of quantitative affinity measures for peptides binding to class I and class II MHCs were carried out with known binding affinities (IC_{50}).

A program was developed to transform the nine amino acid peptide sequences into a matrix of 1 and 0. A term is equal to 1 when a certain amino acid at a certain position or a certain interaction between two side chains exists and 0 when they are absent. For example, 180 columns account for the amino acids contributions (20aa x 9 positions); 3200 columns account for the adjacent side chains, or 1-2 interactions (20 x 20 x 8); and 2800 columns account for every second side chain, or 1-3 interactions (20 x 20 x 7). As these two models were roughly equivalent in terms of statistical quality, we applied the principle of Occam's razor and sought the simplest explanation, choosing the amino acids only model, which will be discussed in this study. The matrix was assessed using PLS. The method works by producing an equation or QSAR, which relates one or more dependent variables to the values of descriptors and uses them as predictors of the dependent variables (or biological activity), [Wold 1995]. The IC_{50} values (the dependent variable y) were represented as negative logarithms (pIC_{50}). The predictive ability of the model was validated using "Leave-One-Out" Cross-Validation (LOO-CV) method.

The generated models (n=30-335) as shown in Table 8.8 has an acceptable level of predictivity: Leave-One-Out Cross-Validation (LOO-CV) statistical terms - SEP and q^2 - ranged between 0.522-1.005 and 0.317-0.652 respectively. The non-cross validated statistical terms NC, SEE and r^2 ranged between 2-9, 0.085-0.456 and 0.731-0.997 respectively. An extended motif, as defined by the class I models is summarised in Table 8.8 showing anchor and non-anchor residues related with strong and weak binding residues. For simplicity, the quantitative contributions of amino acids at each position for the class I mouse alleles are shown in Figure 8.8.

The 2D-QSAR additive method has been applied to the peptide binding specificities of the A3 superfamily human class I alleles: A*1101, A*0301, A*3101 and A*6801. Sequence analysis showed that only 11 of the residues inside the binding pockets are polymorphic. A good, if incomplete, consensus was found in the preferences at the primary anchor positions 2 and 9. Thr and short hydrophobic residues such as Ala and Ile were favoured at P2 and nearly all the peptides bound to A3 alleles had positively charged residues Arg or Lys at the C-terminus. The amino acids involved in peptide binding are similar in HLA-A2 and the A3 family. Pocket B interacts with the side chain of the residue at position 2, which was one of the anchor positions in nearly all the MHC class I alleles. Most of the amino acids in pocket B are conserved in the A2 and A3 families; both families accept hydrophobic residues. The amino acid at sequence position 9 of the MHC protein is important in peptide binding in the two families. Alleles with small to medium sized residues (Phe9 or Thr9) were able to accept residues with long side chains such as Leu, such as A*3101, A*0301 and A*0201. On the other hand, only small residues such as Ala and Val, could bind to A*6801, A*1101 and A*0206, all of which had the larger residue Tyr9. The five residues that directly interacted with the peptide in the F pocket are identical in both the A3 family and HLA-B27 (Leu81, Asp116, Tyr123, Thr143 and Trp147). Arg and Lys bound to pocket F and interacted with negatively charged residues Asp116 or Asp77 in both the A3 family and HLA-B27. B27 had been shown to accept hydrophobic residues such as Leu, Ala and Tyr because of their interaction with Leu81, Tyr123, Thr143 and Trp147 in the binding pocket [Jardetzky et al. 1991]. In the present study, the specificity at position 9 was restricted to Arg and Lys only; both Ala and Tyr had deleterious effects on peptide binding. This suggests a possible difference in the conformation of the binding pocket in spite of sequence similarity. Also, this may be the result of a change in conformation after the binding of other amino acids in the peptide. A peptide-binding motif for the HLA-A3 superfamily has been defined previously by [Sidney et al. 1996b] and [Rammensee et al. 1995]. Some useful similarities can be found on comparing the present motif with those defined by the above two groups. The amino acid preferences for the primary anchor residues are similar. All the motifs show preference for Arg and Lys at position 9 and have a preference for various hydrophobic residues at position 2, such as Ile and Thr. The preferences for secondary anchor residue positions 3 and 7 in the three motifs are hydrophobic amino acids such as Phe.

The amino acid contributions to the affinity of peptides binding to the A2 family: A*0201, A*0202, A*0203, A*0206 and A*6802 alleles using the Additive-PLS Method has also been analysed quantitatively. Certain discrepancies between A*6802 and A*02 molecules concerning the amino acid preferences at P1–P9 were seen in the present study. These discrepancies throw doubt on whether the A*6802 allele belongs to the A2-supertype. The sequence comparison showed that there are only one or two differences in the residues forming the 6 pockets of A*0201, A*0202, A*0203 and A*0206 molecules. The number of these differences between A*6802 and A*02 molecules is seven residues. Five of them concern pockets A, B and C and are so substantial that they alter the amino acid preferences at the primary anchor P2 and the secondary anchors P1 and P6. The preferred Val and Thr for P2 brings the A*6802 allele closer to the A3-supertype [Sidney et al. 1996b] rather than to the A2-one. But the A3 supermotif requires positively charged residues, such as Arg and Lys, at the C-terminus, [Sidney et al. 1996b] which is not true in the case of the A*6802 allele. Obviously, A*6802 is an intermediate allele standing between A2 and A3 supertypes: in anchor position 2 it is closer to A3 and in anchor position 9 it is nearer to A2. Residues identified as preferred for two or more A*02 molecules, without being deleterious for any molecule, are considered as preferred. Residues identified as deleterious for two or more molecules are considered as deleterious in the common motif. The expansion concerns all positions and especially the anchor P2.

The Additive-PLS results for the mouse alleles are in good agreement with previous studies of the preferred primary anchor positions: 5 and 9 (nonamers); 2, 5 and 8 (octamers – H2-K^b and H2-K^b, respectively). All three models also agree with previous analyses of the preferred residue type at the anchor positions. For H2-D^b: Asn at position 5 and Leu at position 9; for H2-K^b: Phe at position 5 and Val at position 8; and for H2-K^b: Glu, Pro, Gly (best three favoured residues) at position 2 and Ile, Val, Phe (best three favoured residues) at position 8. The nonameric and octameric alleles show both similarities and differences in amino acids preferred at various binding positions. Preferences for primary anchors show certain similarities: all models exhibit some preference for small amino acids (H2-D^b (Asn), H2-K^b (Val) and H2-K^b (Pro, Ala)), while C-terminal amino acids are strongly hydrophobic: H2-D^b (Leu), H2-K^b (Val) and H2-K^b (Ile, Val). The most noticeable difference between the nonameric and octameric alleles is at position 5, where H2-D^b exhibits a preference for polar Asn, while H2-K^b shows a preference for Phe (aromatic hydrophobic residue) and H2-K^b for Pro (small amino acid residue).

8.6.2 Iterative Self-Consistent (ISC) Algorithm – Class II Alleles

We have examined a recently developed bioinformatics method: the Iterative Self-Consistent (ISC) Partial Least Squares (PLS)-based Additive Method, which was applied to the prediction of class II Major Histocompatibility Complex (MHC)-peptide binding affinity. We have shown previously that ISC is a reliable, quantitative method for binding affinity prediction ([Doytchinova & Flower 2003, Doytchinova et al. 2002, Guan et al. 2003b], developing a series of quantitative, systematic models, based on literature IC_{50} values. For each set of class II alleles, peptide lengths of 10 to 25 were obtained from the AntiJen database. We now address binding to class II human and mouse alleles for peptides of up to 25 amino acids in length. The ISC additive method assumes that the binding affinity of a large peptide is principally derived from the interaction, with an MHC molecule, of a continuous subsequence of amino acids within it. The ISC is able to factor out the contribution of individual amino acids within the subsequence, which is initially identified in an iterative manner.

The method works by generating a set of nonameric sub-sequences extracted from the parent peptide. Values for pIC_{50} corresponding to this set of peptides were predicted using PLS and compared to the experimental pIC_{50} value for each parent peptide. The best predicted nonamer were selected for each peptide i.e. those with the lowest residual between the experimental and predicted pIC_{50}. LOO-CV was then employed to extract the optimal number of components, which was then used to generate the non-cross-validated model. Each new model is built from the chosen set of optimally scored nonamers. By comparing the new set of peptide sequences with the old set and if the new set is different, the next iteration is begun. The process is repeated until the set of extracted nonameric peptide sequences identified by the procedure have converged. The resulting coefficients of the final non-cross validated model describe the quantitative contributions of each amino acid at each of the nine positions. An example coefficient matrix for the I-A^b allele is shown in Table 8.8.

The generated models (n=44-185) as shown in Table 8.8 has an acceptable level of predictivity: Leave-One-Out Cross-Validation (LOO-CV) statistical terms - SEP and q^2 - ranged between 0.418-0.816 and 0.649-0.925 respectively. The non-cross validated statistical terms NC, SEE and r^2 ranged between 4-8, 0.051-0.180 and 0.967-0.999 respectively. Convergence was ranged between the 4th and 17th iteration. An extended motif, as defined by the class I models is summarised in Table 8.8 showing anchor and non-anchor residues related with strong and weak binding residues.

Our iterative method is different to the manual identification of anchor-based motifs by visual inspection. Such methods are intrinsically tendentious, arbitrary, subjective, and potentially inaccurate. Our method, which is, however, by no means perfect, is, by contrast, an objective and unsupervised approach. It is dependent, however, on the quantity and degeneracy of the data itself, and also upon its quality. The ISC algorithm described above combines an iterative approach to selecting the best predicted binders with PLS, a robust multivariate statistical tool for model generation. The ISC method is universal in that it can be used for any peptide-protein binding interaction where the peptide length is unrestricted but the binding is limited to a fixed, if unknown, part of the peptide. Implementation is relatively uncomplicated, the method executes quickly, and its interpretation is straightforward.

8.7 3D QSAR methods

8.7.1 Comparative Molecular Similarity Index Analysis (CoMSIA)

Three-dimensional QSARs are a technique of incalculable value in identifying correlations between ligand structure and binding affinity. This value is often enhanced greatly when analysed in the context of high-resolution ligand-receptor structures. In such cases, enthalpic changes – van der Waals and electrostatic interactions - and entropic changes – conformational and solvent mediated interactions - in ligand binding can be compared with structural changes in both ligand and macromolecule, providing insight into the binding mechanism [Klebe *et al.* 1994, Klebe & Abraham 1999]. Although there are many molecular descriptors that account for free energy changes, 3D-QSAR techniques which use multivariate statistics to relate molecular descriptors in the space around to binding affinities, have become pre-eminent because of their robustness and interpretability [Bohm *et al.* 1999]. In the case of CoMSIA (Comparative Molecular Similarity Index Analysis), a Gaussian-type functional form is used so that no arbitrary definition of cut-off threshold is required and interactions can be calculated at all grid points. The obtained values are evaluated using PLS. CoMSIA allows each physicochemical descriptor to be visualised in 3D using a map, which donates binding positions that are either "favoured" or "disfavoured".

Recently, CoMSIA has been used to produce predictive models for peptide binding to human MHCs: HLA-A*0201 [Doytchinova & Flower 2002b] and the HLA-A2 and HLA-A3 supertypes [Doytchinova & Flower 2002a, Guan *et al.* 2003a]. In this study, we show how CoMSIA has been applied to some of these class I MHC alleles. These models were used to evaluate both physicochemical requirements for binding, and to explore and define preferred amino acids within each pocket. The explanatory power of such a 3D-QSAR method is considerable, not only in its direct prediction accuracy but also in its ability to map advantageous and disadvantageous interaction potentials onto the structures of the peptides being studied. The data is highly complementary to the detailed information obtained from crystal structures of individual peptide-MHC complexes.

The 12 alleles, all peptides were built and aligned in three dimensions. Wherever possible an X-ray crystallographic structure for the nonameric/octameric peptide binding to the various class I alleles was chosen as a starting conformation. Using the crystallographic peptide as a template, all the studied peptides were built, and then subjected to an initial geometry optimisation using the Tripos molecular force field and charges derived using the MOPAC AM1 Hamiltonian semi-empirical method [Dewar *et al.* 1985]. Molecular alignment was based on the backbone atoms of the peptides, which was defined as an aggregate during optimisation. The peptides were placed within individual 3D grids. The final settings for the three models are shown in Table 8.8. The generated models (n=30-236) as shown in Table 8.8 has an acceptable level of predictivity: Leave-One-Out Cross-Validation (LOO-CV) statistical terms - SEP and q^2 - ranged between 0.443-0.889 and 0.385-0.700 respectively. The non-cross validated statistical terms NC, SEE and r^2 ranged between 4-12, 0.071-0.411 and 0.867-0.991 respectively.

To generate CoMSIA coefficient contour maps for each allele, which describes the relationship between the binding affinity and each physicochemical descriptor, three non-cross validated "all fields" models were created based on the five physicochemical descriptors (Steric, Electrostatic, Hydrophobic, Hydrogen Bond Donor and Acceptor). Table 8.8 shows a summary of the position specificities between the physicochemical descriptors and peptides positions for the A2 supermotif and class I mouse alleles. CoMSIA analysis for each allele was carried out using Partial Least Squares (PLS) [Young 2001] and models were then validated via the Leave-One-Out Cross-Validation method as previously described.

The motif of HLA-A3 superfamily includes main anchor positions 2 and 9 [Zhang *et al.* 1993]. Peptides bound to members of the A3 family usually had a positively charged residue—Arginine or Lysine—at the C terminus, and a variety of hydrophobic residues at position 2. It was found that steric bulk was favoured at position 2 for A*0301 and A*3101 but disfavoured in A*1101 and A*6801 models. The study of crystal structures of MHC molecules showed that the residue at peptide position 2 bound in pocket B [Saper *et al.* 1991, Madden *et al.* 1991]. There are different residues lining pocket B in the different MHC-A3 molecules: Tyr9 in A*1101 and A*6801, Phe9 in A*0301 and Thr9 A*3101 [Schonbach *et al.* 2000]. This means more space in pocket B for A*0301 and A*3101, allowing them to accommodate larger side chains. Electrostatic potential, hydrophobicity and hydrogen bond acceptance maps were very varied at this position. This was in good agreement with the broad spectrum of amino acids observed at this position, from the bulky hydrophobic Leu to the small polar Thr. The most important property for the amino acid at position 9 was hydrogen-bond donor ability. It was favoured by A*6801 and A*3101, and was disfavoured by A*1101. For A*0301 were found areas of favoured and disfavoured hydrogen bond donor groups at this position. In some cases, the change of Lys to the larger residue Arg could affect the expression of the molecule [Zhang *et al.* 1993]. Results from the present study suggested the interaction between the residue at peptide position 9 and the MHC molecule may play an important role. The side chain of larger basic residue Arg could extend to the bottom of the pocket F of A*6801 and A*3101, forming complex stabilising hydrogen bonds with residues at the bottom of the pocket.

Among the secondary anchors, positions 1, 3, 5, 6 and 7 were of great importance. The common favoured property for position 1 was hydrogen-bond donor/acceptor ability. Hydrogen-bond donor groups with negative electrostatic potential were preferred at position 3 for three of the alleles. [Sidney *et al.* 1996b] found that peptides with an aromatic residue, like Tyr, Phe and Trp, had a 31-fold increase in binding affinity to A*0301. Bulky side chains with negative electrostatic potential were preferred at position 5. Hydrogen-bond donors and acceptors were disfavoured here. Hydrophilic amino acids capable of forming hydrogen bonds were well accommodated at position 6. The only common favoured property for position 7 was hydrophobicity. Positions 4 and 8 face the T-cell receptor, [Silver *et al.* 1992], but can still contribute to the affinity. Hydrogen-bond donor ability was important for position 4. Steric bulk and negative electrostatic potential were favoured at position 8.

Looking at the CoMSIA results for the mouse alleles, we see that with the H2-D^ballele, steric bulk is favoured with the side chains of positions 3 and 6 falling into pockets D and C respectively. For the electrostatic potential field, the alkyl side chain of position 1 falls into pocket A which consists of Valine and Serine residues [Saper *et al.* 1991]. At position 2, where the side chain falls into pocket B, electrostatic potential interaction is favoured [Saper *et al.* 1991]. In the remaining positions there are no favourable electrostatic potential interactions. There is a strongly favoured hydrophobic interaction at position 8 where the side chain is solvent exposed and contacts the T cell. The major favoured interactions of the hydrogen bond donor fields are found at position 1 and across the peptide back bone between position 3 and 4. The hydrogen bond acceptor map shows position 2 to be favoured and, to a lesser extent, at positions 5 and 7.

For H2-K^ballele, steric bulk is favoured at positions 1, 3, 4 and 5. The side chain at position 1 makes a weak electrostatic interaction; while at position 2 the electrostatic potential map indicates that aromatic-type residues, such as Tyr or Phe, are well tolerated. This is in good agreement with experimental data [Ruppert *et al.* 1993, Parker *et al.* 1994]. There is no major interaction between side chains at position 3 and pocket D indicated by our model, and in the remaining positions there are no clear favourable electrostatic interactions. The hydrophobic interaction field identifies a favourable interaction at positions 3 and 5. Pocket D is a hydrophobic cavity and amino acids such as Tyr and Ile are well-tolerated here which would significantly deepen the depth and volume of Pocket D [Fremont *et al.* 1998]. The major favoured interactions of the hydrogen bond donor fields are found at positions 1, 3 and 4 (pockets A, D and the "flag" pocket, respectively), [Saper *et al.* 1991] with a major disfavoured interaction found at position 6 (pocket C). The hydrogen bond acceptor map has favoured interactions at positions 1 and 4, pocket A and the "flag" pocket respectively, but major disfavoured interactions between the side chain positions 3 and 5.

For the H2-K^b allele, steric bulk field is favoured at positions 1, 7 and 8. There is no favourable electrostatic interaction at position 1, while at position 2 electrostatic potential is favoured. Position 3 falls into pocket D but makes little interaction with the H2-K^b allele. In the remaining positions there seems to be no discernibly favoured interactions. Hydrophobic interaction shows a major disfavoured interaction at position 2 covering the whole side chain. The only favoured interaction in the hydrogen bond donor map in the H2-K^b allele lies between positions 7 and 8. The main disfavoured interaction is found at position 2. Within the hydrogen bond acceptor map, there is a strong disfavoured interaction between the side chains at positions 2 and 3.

8.8 Discussion

In this chapter, we have attempted to do two things: to explore the complex and perplexing nature of peptide-MHC binding and also to explore our faltering steps to predict it. We say faltering, because present prediction methods, based on available data, are still some way from addressing all aspects of the problem. Why is this? One of the principal reasons concerns the nature of the data we are trying to predict. Bias within it places extreme strictures upon the generality and interpretability of any models derived from it. In general, for MHC-peptide binding experiments, the sequences of peptides studied are indeed very biased in terms of amino acid composition, often favouring hydrophobic sequences. This arises, in part, from a process of pre-selection processes that typically results in self-reinforcement. Binding motifs are often used to reduce the experimental burden of epitope discovery. Very sparse sequence patterns are matched and the corresponding subset of peptides tested, with an enormous reduction in sequence diversity. Nonetheless, the peptide sets which are analysed by immunoinformaticans are still much larger than those typical in the pharmaceutical literature. The peptides themselves are physically large in themselves, and their physical properties are extreme. They can be multiply charged, zwitterionic, and/or exhibit a huge range in hydrophobicity. Affinity data itself is often of an inherently inferior quality: multiple measurements of the same peptide may vary by several orders of magnitude, some values are clearly wrong, a mix of different standard peptides are used in radioligand competition assays, experiments are conducted at different temperatures and over different concentration ranges. We are also performing a *meta-analysis*: almost certainly forcing many distinct binding modes into a single QSAR model.

Compared to such caveats concerning the quality of data, concerns about parameterization seem pale by comparison. Nonetheless, one may also feel justifiable unease about aspects of the methodology used. No method is perfect, nor, in the present context, are they ever likely to be. One principal criticism of most statistical and artificial intelligence methodology is the chance of over-fitting models. Usually the data block has too few degrees of freedom for a completely robust analysis to be undertaken. Many terms will be poorly populated, with only a handful of observations, inflating the associated errors and reducing the associated reliability of prediction. Moreover, it is always possible to over-emphasize the usefulness of cross-validation and q^2 as measures of performance [Golbraikh & Tropsha 2002]: high values of Leave-One-Out q^2 are a necessary, but not a sufficient, condition for a model to possess high predictivity. Likewise, high values of sensitivity or specificity or area-under-the-curve from ROC analysis are useful but flawed measurements of performance. External test sets and randomization of training data are also important criteria for assessing model quality. There is no single value or criteria that can give an adequate appraisal of methodological perfection. Combinations of all these criteria are required. However, most groups are certainly aware of most, if not all, the inherent dangers within immunoinformatics, and most actively seek to minimize them.

Ultimately, however, the greatest limiting factor is imposed by the data itself. This is, as we have said time and again before, undoubtedly the overriding issue. With a properly designed training set most issues would be resolved, but such sets are not subsets of available data. Rather they require new data to be produced and that necessitates close, if often elusive, synergistic interactions between theoreticians and experimentalists. No data-driven method can go beyond the training data: all methods are better at interpolating than extrapolating. It is only by having excellent quality data in abundance that we can hope for general and excellent models. Eventually, as more sophisticated methods, such as molecular dynamics (MD), become habitually used tools, we will, in concert with measured data, develop considerably more accurate and predictive models. MD escapes from the innate limitations imposed by data through its ability to offering us the chance of true *de novo* prediction of binding affinities and many other thermodynamic properties.

More sophisticated approaches will allow us to not only predict affinity but, through simulation, to drive forward experimental studies in a similar fashion to the way theoretical physics drives forward experimental physics. There are many aspects of the biophysics of peptide binding to MHCs which is only poorly understood. For example, as affinity rises, the phenomenon of enthalpy-entropy compensation becomes important. Where multiple weak non-covalent interactions hold a molecular complex together, the enthalpy of all of the individual intermolecular bonding interactions is reduced by extensive intermolecular motion. Additional interaction sites will generate a complex which is more strongly bound. This results in part from the dampening of intermolecular motion, with all individual interactions becoming more favourable. This was observed in the case of our A2 superbinders, where our additive models greatly under-predicted the actual affinity values we subsequently measured. To analyse such phenomena requires more subtle and sensitive techniques than are provided by radioligand competition assays. Currently, the best single methodology for obtaining relevant thermodynamic properties of binding reactions is undoubtedly ITC, which is rapidly becoming the method of choice for such studies. No method however readily addresses the joint goals of effectively mimicking the *in vivo* membrane-bound nature of the interaction and the need for accuracy. In an ideal world we would look at a variety of "internally rich" data from ITC, volumetric analysis, and fluorescence spectroscopy. To do this on an appropriate scale would however be prohibitively time consuming and expensive. Where one might conceive of doing this for one allele, there are dozens of frequent alleles within the human population. To pursue internally-rich assays for all *interesting* alleles is clearly beyond the scope of existing methodology.

Thus, many problems remain in attempting to predict peptide-MHC binding. This at least is clear. Some theoretical, some experimental. However, extant techniques do work. We have recently compared the performance of web-server implementations of various published methods and find the best of these to perform well as epitope identification engines (Guan et al., unpublished), at least for those alleles where binding data is reasonably plentiful. Thus the future for such approaches seems bright, assuming that reasonable quantities of data are available. Where data is scant we are less sanguine in our outlook. Thus, finally, two needs are clear. Firstly, for vaccinologists to use methods for those alleles where methods are seen to work well, and secondly for experimentalists to undertake the necessary assays for alleles where methods require improvement.

EPITOPE	IC$_{50}$ (Nm)	CATERGORY	SWISS DB REF.	REFERECNCE
KYADKIYSI	15	SELF PEPTIDE	FOH1_HUMAN	Horiguchi et al., 2001
YTAFTIPSI	26.3	VIRAL	POL_HV1OY	Altfeld et al., 2001
NYARTEDFF	67	SELF PEPTIDE	FOH1_HUMAN	Horiguchi et al., 2001
SIISAVVGI	69.4	CANCER	ERB2_HUMAN	Rongcun et al., 1999
AIHNVVHAI	166	ALLERGEN	GDA1_WHEAT	Gianfrani et al., 2003
KAACWWAGI	277.7	VIRAL	POL_HV1OY	Altfeld et al., 2001
QIIGYVIGT	312.5	CANCER	CCEM_HUMAN	Keogh et al., 2001
LYSDPADYF	406	SELF PEPTIDE	FOH1_HUMAN	Horiguchi et al., 2001
KCIDFYSRI	417	VIRAL	VE6_HPV18	Rudolf et al., 2001
LVGPTPVNI	454.5	VIRAL	POL_HV1OY	Altfeld et al., 2001
DIEITCVYC	649	VIRAL	VE6_HPV18	Rudolf et al., 2001
LQTTIHDII	1100	VIRAL	VE6_HPV16	Kast et al., 1994
VIHAFQYVI	1220	SELF PEPTIDE	MYPR_HUMAN	Dressel et al., 1997
SAICSVVRR	1429	VIRAL	DPOL_HPBVY	Bertoni et al., 1998
KTWGQYWQA	1600	CANCER	PM17_HUMAN	Bakker et al., 1997
LQTTIHDII	3157	VIRAL	VE6_HPV16	van der Berg et al., 1995
LQTTIHDII	3158	VIRAL	VE6_HPV16	van der Burg et al., 1996
TTAEEAAGI	4167	CANCER	MAR1_HUMAN	Rivoltini et al., 1995
RVIEASFPA	5440	CANCER	CAH9_HUMAN	Vissers et al., 1999
RTFYDPEPI	15800	CANCER	MAPE_HUMAN	Kessler et al., 2001
MVQAWPFTC	18500	CANCER	MAPE_HUMAN	Kessler et al., 2001
LQTTIHDII	80000	VIRAL	VE6_HPV16	Ressing et al., 1999

Table 8.1. Non-Canonical Motif-negative HLA-A*0201 binders. Assuming a cut-off of 500nM for potential immunogenicity, about half this list are putative epitopes. Following this definition, epitopes are shaded white and non-epitopes in grey.

	P1	P2	P3	P4	P5	P6	P7	P8	P9
A	-0.016	-0.008	0.265	-0.115	0.066	-0.442	0.050	0.447	-0.034
C	**0.000**	0.083	0.037	-0.051	0.090	0.050	0.216	0.079	-0.139
D	-0.065	**0.000**	-0.067	**0.000**	**0.000**	0.107	-0.077	-0.041	-0.203
E	-0.028	-0.129	**0.000**	**0.000**	**0.000**	**0.000**	**0.000**	-0.048	**0.000**
F	**0.000**	**0.000**	**0.000**	-0.283	**0.000**	**0.000**	**0.000**	**0.000**	**0.000**
G	-0.286	-0.039	0.050	-0.011	**0.000**	**0.000**	-0.003	**0.000**	-0.067
H	-0.003	-0.013	**0.000**	**0.000**	**0.000**	**0.000**	**0.000**	0.213	**0.000**
I	-0.043	0.090	-0.364	-0.090	**0.000**	-0.244	-0.351	**0.000**	-0.069
K	0.094	**0.000**	**0.000**	**0.000**	-0.069	**0.000**	**0.000**	**0.000**	**0.000**
L	**0.000**	-0.215	-0.110	0.094	-0.162	**0.000**	-0.003	-0.242	0.066
M	0.008	-0.067	**0.000**	0.258	0.223	0.154	0.017	-0.027	0.082
N	**0.000**	0.298	**0.000**	0.042	-0.003	-0.069	0.064	-0.097	-0.455
P	0.100	**0.000**	0.032	0.090	0.030	0.201	0.080	**0.000**	0.280
Q	-0.013	**0.000**	-0.235	**0.000**	**0.000**	**0.000**	**0.000**	-0.067	-0.051
R	0.164	-0.286	0.066	0.122	-0.233	0.120	0.213	-0.229	0.216
S	-0.051	0.090	0.161	0.036	-0.078	0.041	-0.125	**0.000**	0.213
T	0.054	0.151	0.079	-0.060	0.233	**0.000**	-0.079	0.012	0.161
V	-0.069	-0.048	**0.000**	0.064	**0.000**	**0.000**	**0.000**	**0.000**	**0.000**
W	**0.000**	**0.000**	-0.029	**0.000**	-0.097	-0.003	**0.000**	**0.000**	**0.000**
Y	0.155	0.092	0.116	-0.097	**0.000**	0.085	**0.000**	**0.000**	**0.000**

Table 8.2. Additive model for the binding affinity prediction to the I-A^b allele. *
constant = 6.044 and ** 0.000 represents position where amino acids are absent
within matrix

	Epitope	n [a]	LOO-CV		Non-Cross Validation		
			SEP [b]	q^{2c}	NC [d]	SEE [e]	r^2
Human	A*0101	95	0.907	0.420	4	0.146	0.997
	A*0201	335	0.694	0.377	6	0.456	0.731
	A*0202	69	0.606	0.317	9	0.193	0.943
	A*0203	62	0.841	0.327	6	0.197	0.963
	A*0206	57	0.576	0.475	6	0.085	0.989
	A*0301	72	0.680	0.436	6	0.181	0.959
	A*1101	62	0.572	0.458	2	0.321	0.829
	A*3101	30	0.710	0.482	3	0.325	0.892
	A*6801	38	0.594	0.531	4	0.175	0.959
	A*6802	46	0.647	0.500	7	0.119	0.983
	B*0702	78	0.707	0.488	6	0.150	0.977
	B*2705	89	0.522	0.434	6	0.089	0.984
	B*3501	52	0.710	0.435	6	0.118	0.984
	B*5301	63	0.868	0.508	6	0.154	0.985
	B*5401	74	1.005	0.458	6	0.288	0.956
	Cw*0102	57	0.722	0.652	5	0.180	0.978
Mouse	H2-K^b	154	0.565	0.456	6	0.198	0.933
	H2-K^b	62	0.894	0.454	6	0.128	0.989
	H2-D^b	65	0.837	0.493	5	0.268	0.948

Table 8.3. Class I Additive-PLS Method results. [a] Number of epitopes. [b] Standard Error of Prediction. [c] Obtained after Leave-One-Out Cross-Validation. [d] Number of components.[e] Standard Error of Estimate

| | Epitope | n [a] | Iterations | LOO-CV | | Non-Cross Validation | | |
				SEP [b]	q[2c]	NC [d]	SEE [e]	r[2]
Human	DRB1*0101	90	13	0.567	0.808	8	0.075	0.994
	DRB1*0401	185	7	0.701	0.716	4	0.174	0.967
	DRB1*0701	84	11	0.562	0.649	7	0.051	0.999
Mouse	I-A[b]	44	7	0.459	0.850	6	0.089	0.994
	I-A[d]	145	14	0.534	0.898	6	0.136	0.993
	I-A[b]	55	4	0.816	0.790	6	0.180	0.990
	I-A[s]	81	17	0.588	0.783	6	0.177	0.980
	I-E[d]	69	8	0.557	0.732	6	0.096	0.992
	I-E[b]	52	8	0.418	0.925	6	0.106	0.995

Table 8.4. Class II ISC-Additive Method results

	Epitope	n [a]	LOO-CV		Non-Cross Validation			Size	Spacing	Steps
			SEP [b]	q^{2c}	NC [d]	SEE [e]	r^2			
Human	A*0201	236	0.443	0.683	7	0.260	0.891	22x15x15	2.0	0.5
	A*0202	63	0.509	0.534	8	0.190	0.935	22x15x15	3.0	0.5
	A*0203	60	0.595	0.621	6	0.179	0.966	22x15x15	3.0	0.5
	A*0206	54	0.505	0.523	12	0.071	0.991	22x15x15	2.0	0.5
	A*0301	69	0.629	0.486	6	0.177	0.959	22x15x15	2.0	0.5
	A*1101	59	0.588	0.496	8	0.141	0.972	22x15x15	2.0	0.5
	A*3101	30	0.551	0.700	4	0.282	0.921	22x15x15	1.5	0.5
	A*6801	39	0.674	0.430	5	0.119	0.950	22x15x15	2.0	0.5
	A*6802	45	0.652	0.385	4	0.197	0.944	22x15x15	2.0	0.5
Mouse	H2-K[b]	154	0.525	0.611	6	0.248	0.913	18x13x12	2.0	0.5
	H2-K[b]	62	0.889	0.490	6	0.244	0.962	19x13x11	2.0	0.5
	H2-D[b]	65	0.783	0.518	4	0.411	0.867	18x14x11	2.0	0.5

Table 8.5. Class I Additive-PLS method results for CoMSIA method

Favoured binding

	Y	T	D	H	Y	P	N	V	A,Y
A*0101									
A*0201	F,Y	L,M	F,L,M,W,Y	T	F,Y	I	H,P	P,Q	V
A*0202	K	L	V	M	K	M,P	F,N	F,Y	L,V
A*0203	D,K,W,L,M	L,M	A,N,S,V	P,T	L,R,T,I	Q,V		T	L
A*0206	A,K	I	L	L	L	F	F	T	V
A*0301	G	I,T	F	R,T	Y	G	I,F	M	K
A*1101	S	V	M	N	V	S	F	L	K
A*3101	M	L	G	P	R	R	P	E	R
A*6801	Q	A	F	F	M	L	A	L	R
A*6802	*F*	*V*	*I,M*	*C,G*	*P*	*I*	*P,V*	*R*	*V*
B*0702	S	V	A,R,V	G,H,K,S	K	T	I,L	A,C,P	L,M
B*2705	*R*	*R*	*I*	*M*	*R*	*S*	*W*	*G*	*K*
B*3501	*F*	*P*	*I*	*H*	*T*	*F*	*L*	A,K	M
B*5301	M,F	P	W	P	R,F	Q,L,F	S	S	I,F,W
B*5401	M,F,Y	P	I,F,S,W	E,H,L,R	R	L,F,P,Y	S,Y	S,Y	A,I,T,V
Cw*0102	*C*	*M*	*A*	*T*	*A,G*	*F*	*A*	*T*	*L*
H2-D^b	A,Q,F	Q,S	I,L,P	C,T	*N*	*Q*	D,E	*Y*	*L*
H2-K^b	K,S	G,I,S,Y,R,F,Y	A	A	F	A,G,Q,G,P,V			
H2-K^b	*F*	A,D,E,G,L,P,S,T,V	L	P	P	F	R,L,F,A,N,I,M,F,S,T,W,V		

Table 8.6. Class I Additive-PLS Method: non-anchor residues related with strong and weak binding for amino acids only.[1] A cut-off value of >+/- 0.3 is applied to favoured binding amino acids. Where no amino acid residue lies outside the cut-off limit (>+/- 0.3), the next best residue is chosen (as shown in italics).

	P1	P2	P3	P4	P5	P6	P7	P8	P9
A*0101	F	N	K	G	D	K	A	E	D
A*0201	T	T	C, E, H	S, A, F, I	R	R	G, N, Q	D, I	S
A*0202	G, I	A, V	C	N	N	Q, S	S, T	D, E	A, R
A*0203	D, L, R	T, V	D, F, M	A, D, L, S, W, N		D	F, L, T	D, L, Q	A
A*0206	G	M	N	E	P, Y	P	V	R	A
A*0301	L	N	L	E	S	K	H	E	A, Q
A*1101	L	L	A	C	R	R	L	A	Y
A*3101	Q	A	F	T	S	F	A	R	K
A*6801	A	N	G	V	V	G	R	S	Y
A*6802	K	M	S, T, V	S	V	G	F, Q	D	L
B*0702	G	–	F, P, S, Y	D, E, V	V	K	Q	N, G, T, H	H, W
B*2705	Q	P	E	T	G	D	E	S	E
B*3501	T	A	Q, K	S	E	V	Q, S	T	T, W
B*5301	T, V	A	K	L, K, F	G, V, C	C	Q, E	P	A, V
B*5401	T, V	A	E, K, P	Q, G, P, S, T	D, E, G, P	D, E, G, R, V	H, I, S	N, D, F	C, L, F, Y
Cw*0102	S	L	E	L	C, L	D	K	E, G	I
H2-D[b]	L, V	E	G, S	A, I, F	G	P	G, I, M, V	C, G	Y
H2-K[b]	D, L, Y	D, Q, L	E, H, S	G, I	L, S	F, S, V, I, F, Y	D, K	M	
H2-K[b]	A	N, H, I, K, F, W, Y	K, S	S	K	G		R, D, Q, G, H, K, P, Y	

(Left-margin vertical label spanning lower rows: **Disfavoured binding**)

Table 8.7. Class I Additive-PLS Method: non-anchor residues related with strong and weak binding for amino acids only.[1] A cut-off value of > +/– 0.3 is applied to disfavoured binding amino acids. Where no amino acid residue lies outside the cut-off limit (> +/– 0.3), the next best residue is chosen (as shown in italics).

Favoured binding [1]

	P1	P2	P3	P4	P5	P6	P7	P8	P9
DRB1*0101	Y, F	I, Y	T	P	L	A	N	P	S
DRB1*0401	V, W	V, Y	L	A	A	S	P, V	L	N
DRB1*0701	Y	P	V	L	T	A	V	N	S, V
I-A^b	P	K	A	L	L, T	M	R	A	M
I-A^d	C, T, W	A, G, M, *T*		C, M, S, W	Q, L, V		A, I, Y	G, I, L	A, C, F, T
I-A^b	*T*	*T*	*G*	*C*	*G, S*	*F*	*E, Y*	*Q*	*C*
I-A^s	*F*	*C*	*N, L*	*L, F*	*A*	*I*	*I*	*G*	*H*
I-E^d	*M*	*Q*	*W*	*W*	*S*	*A*	*W*	*R*	*C*
I-E^b	*I*	*A*	*Y*	*R*	*R*	*Q*	*L*	*T*	*A*

Disfavoured binding [1]

	P1	P2	P3	P4	P5	P6	P7	P8	P9
DRB1*0101	A, V	K	A, G	K	H	Y, N	*E*	T	G
DRB1*0401	L	*F, W*	D	L	S	A	A	R	L
DRB1*0701	S	*N*	Y	S	A	L, N	T	S	N
I-A^b	Q	P	G	C	P	A	G	I	K
I-A^d	N, E, L	N	N, V, G, T		N, I	I	H, S	H, Y	E, I, K, S
I-A^b	N	*H*	N, H	H	A	A	A	S	A
I-A^s	K	*A, S*	*R*	*H*	G	*K*	G	*P*	R
I-E^d	I	*I*	*H*	*I*	L	*C*	L	*Y*	I
I-E^b	R	*R*	*L*	G	A, Y	*I*	E	K	E

Table 8.8. Class II ISC-Additive Method: non-anchor residues related with strong and weak binding for amino acids only. [1] A cut-off value of >+/- 0.4 is applied to favoured and disfavoured binding amino acids. Where no amino acid residue lies outside the cut-off limit (>+/- 0.4), the next best residue is chosen (as shown in italics).

Position	A2 Supermotif (Class I HLA-A*0201, A*0202, A*0203, A*0206 and A*6802)					Class I Mouse ($H2-D^b H2-K^b, H2-K^b$)				
P1 Side chain falls into pocket A		+ Aromatic amino acids preferred.	+			+	-	-	-	+
P2 Side chain falls into pocket B	+		+			-	-	-	+	+
P3 Side chain falls into pocket D						-	-	+	+	+
P4 Side chain is solvent exposed & can contact T cell	-	- Aliphatic amino acids preferred.		+				-	+	+
P5		- Aliphatic amino acids preferred.							+	+
P6 Side chain falls into pocket C	+		+			-		+		
P7 Side chain falls into pocket E	-	- Aliphatic amino acids preferred.						+		-
P8 Side chain is solvent exposed & can contact T cell	-	- Aliphatic amino acids preferred.	-	+	+	+	+	+	+	+
P9 Side chain falls into pocket F	+		+							

Table 8.9. Summary of CoMSIA position specificities for the A2 supermotif [17] (Class I HLA-A*0201, A*0202, A*0203, A*0206 and A*6802) and Class I Mouse H2-K^b, H2-K^b, H2-D^b. Key: S - Steric Bulk, E Eletron Density, H hydrophobicity, D H bond donor, A H bond acceptor. + favoured, - disfavoured

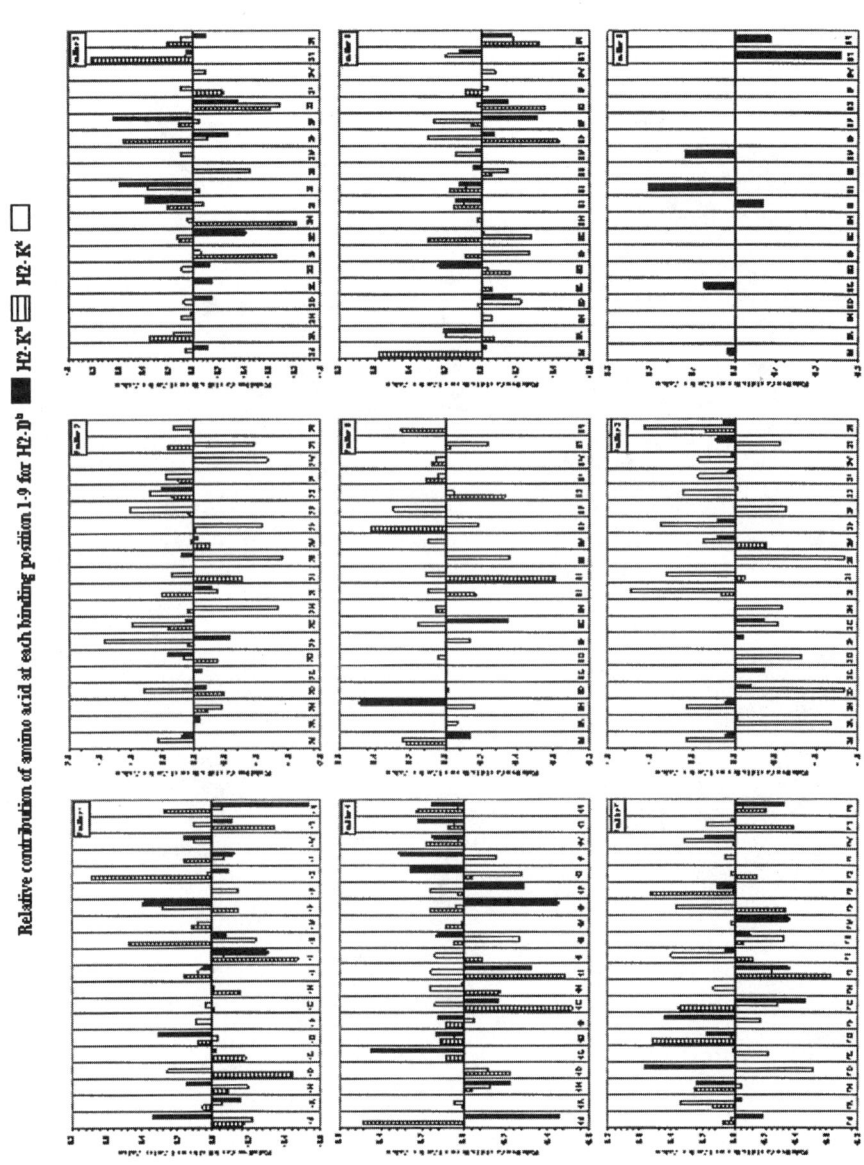

Figure 1 Relative contributions of position-wise amino acids at each binding positions 1 to 9 for the H2-D^b, H2-K^b and H2-K^d alleles. The contribution made by different individual amino acids at each position of the 9mer H2-D^b, H2-K^b and H2-K^d binding peptide. The contribution is equivalent to a position-wise amino acid regression coefficient obtained by PLS regression (as described in the text).

Relative contribution of amino acid at each binding position 1-9 for H2-D^b ■ H2-K^b ▦ H2-K^d □

MHC diversity in Individuals and Populations

José A. M. Borghans[1,2], Can Keşmir[1,3], and Rob J. De Boer[1]

[1] Theoretical Biology, Utrecht University Padualaan 8, 3584 CH Utrecht, The Netherlands
[2] Current Address: Dpt. Immunology, UMCU, Lundlaan 6, 3584 EA Utrecht, The Netherlands
[3] Center for Biological Sequence Analysis, BioCentrum-DTU, Technical University of Denmark.

Summary. The genes encoding the major histocompatibility (MHC) molecules are among the most polymorphic genes known in vertebrates. Since MHC molecules play an important role in the induction of immune responses, this polymorphism is probably due to selection for increased protection of hosts against pathogens. In contrast to the large population diversity of MHC molecules, each individual expresses only a limited number of different MHC molecules. This is widely believed to represent a trade-off between maximizing the detection of foreign antigens, and minimizing the loss of T cell clones during self tolerance induction in the thymus.

Here we review theoretical models and bioinformatic analyse that we have developed to study the diversity of MHC molecules, both at the individual and at the population level. We have found that thymic selection does not limit the individual MHC diversity. Expression of extra MHC types decreases the number of clones surviving negative selection, but increases the number of positively selected clones. The net effect is that the number of clones in the functional T cell repertoire would increase if the MHC diversity within an individual were to exceed its normal value.

It has been proposed that the large population diversity of the MHC is due to selection favoring MHC heterozygosity. Since MHC heterozygous individuals can present more peptides to the immune system, they are better protected against infections than MHC homozygous individuals. Using a population genetics model, we found however that this heterozygote advantage is insufficient to explain the large degree of MHC polymorphism found in nature. Only if all MHC alleles in the population were to confer unrealistically similar fitness contributions to their hosts, could heterozygote advantage account for an MHC polymorphism of more than ten alleles. By predicting the immunodominant peptides from various common viruses we found that different MHC alleles are expected to provide quite different levels of protection. Thus, additional selection pressures seem to be involved. Using a computer simulation model we found that frequency-dependent selection by host–

pathogen coevolution provides such an additional selection pressure that can account for realistic degrees of polymorphism of the MHC. The polymorphism of the MHC thus seems a result of host–pathogen coevolution, giving rise to a large population diversity despite the limited degree of MHC diversity within individuals.

9.1 Introduction

MHC molecules play a central role in the induction of cellular immune responses. The proteins of infected cells are degraded intracellularly and presented by MHC molecules on the surface of the cell. When T lymphocytes recognize the resulting MHC–peptide complexes, they can mount an immune response against the infected cells, see Chapter 1 and Chapter 8. The MHC is the textbook example of genetic polymorphism: for some MHC loci, more than a hundred different alleles have been identified [Parham & Ohta 1996, Vogel *et al.* 1999]. This polymorphism is most likely due to Darwinian selection for diversity in MHC–peptide binding. Indeed, the ratio of non-synonymous versus synonymous substitutions within MHC–peptide binding regions is much higher than in other genes or other regions of the MHC [Hughes & Nei 1988, Hughes & Nei 1989, Parham *et al.* 1989a, Parham *et al.* 1989b]. MHC polymorphism creates individual differences in immunity against pathogens [Barouch *et al.* 1995]. Thanks to this variation, a pathogen that manages to evade presentation in one particular host, may not be able to evade presentation in another host with different MHC molecules.

The diversity of MHC molecules within an individual is only a tiny fraction of the MHC population diversity. Each human being, for example, expresses only three classical MHC class I genes (HLA A, B, and C), and three classical MHC class II gene pairs (coding for the α and β chains of HLA DP, DQ, and DR). A fully heterozygous individual therefore expresses maximally six different class I MHC molecules and twelve different class II MHC molecules (due to trans-association of the α and β chains within HLA DP, DQ, and DR) [Paul 1999]. Since MHC genes are codominantly expressed, MHC heterozygous individuals can present a larger variety of peptides to the immune system than homozygous individuals, and are thereby better protected against infections [Penn *et al.* 2002]. Evidence for this MHC heterozygote advantage has been found in HIV [Carrington *et al.* 1999], HTLV-1 [Jeffery *et al.* 2000], and LCMV infection [Weidt *et al.* 1995]. Mate choice experiments also suggest that females tend to increase the degree of MHC heterozygosity of their offspring by choosing mates with MHC alleles that differ from their own MHC alleles [Potts *et al.* 1991, Wedekind *et al.* 1995, Ober *et al.* 1997, Reusch *et al.* 2001].

Prompted by the insight that expression of different MHC molecules provides a selective advantage, we used theoretical models and bioinformatic analyse to study the following three questions:

1. Why does each individual express only a limited number of MHC genes?

2. Can selection for MHC heterozygosity readily explain the extreme population diversity of the MHC?

3. Could the large degree of MHC polymorphism be due to host–pathogen coevolution?

The mathematical modeling of the immune system in this Chapter differs from the ordinary differential equation (ODE) models presented in Chapter 4 and Chapter 13 because both Chapters describe the kinetics of immune responses within one individual host. The models presented here have a dynamics on an evolutionary time scale. The first model considers one host, and provides a simple probabilistic expression for the likelihood that a T cell survives selection in the thymus. The other models consider evolution in a population of hosts by either a population genetics model, or by a computer simulation model in which the fitness of individual hosts and pathogens is computed.

9.2 MHC diversity within the individual

Since individual MHC diversity increases the presentation of pathogens to the immune system, one may wonder why the number of MHC genes is not much higher than it is. The argument that is mostly invoked is that more MHC diversity within the individual would lead to T cell repertoire depletion during self tolerance induction [Matzinger et al. 1984, Vidović & Matzinger 1988, Parham et al. 1989a, Nowak et al. 1992, De Boer & Perelson 1993, Takahata 1995, Janeway & Travers 1997, Cohn 1985, Celada & Seiden 1992, Lawlor et al. 1990]. Since all T lymphocytes recognizing self peptide–MHC complexes with too high avidity have to be tolerated to avoid autoimmune reactions [Nossal 1994], excessive MHC diversity within an individual would hamper the immune system [Matzinger et al. 1984, Vidović & Matzinger 1988, Parham et al. 1989a, Nowak et al. 1992, De Boer & Perelson 1993, Takahata 1995, Janeway & Travers 1997, Cohn 1985, Celada & Seiden 1992, Lawlor et al. 1990]. This argument is incomplete, however, because more MHC diversity could also increase the number of clones in the T cell repertoire through positive selection. In order to be rescued in the thymus, lymphocytes need to recognize MHC–self peptide complexes with sufficient avidity [Von Boehmer 1994, Fink & Bevan 1995]. A high MHC diversity thus increases both the number of lymphocyte clones that are positively selected and the number of clones that are negatively selected. We have calculated the net effect of these two opposing processes using a simple mathematical model [Borghans et al. 2003].

Consider an individual with M different MHC molecules and an initial T lymphocyte repertoire consisting of R_0 different clones. Let p and n denote the (unconditional) chances that a clone is positively selected by a single MHC type, because its avidity is higher than a threshold T_1, or negatively selected because its avidity exceeds a higher threshold T_2, respectively (see Figure 9.1). By this definition, thymocytes can only be negatively selected by MHC molecules by which they are also positively selected [Meerwijk et al. 1997], i.e. $n < p$. Since T cell clones need to be positively

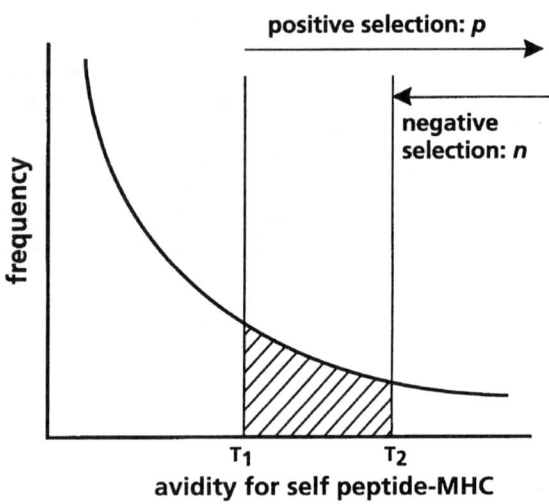

Fig. 9.1. Positive and negative selection according to the avidity model [Janeway & Katz 1984]. The curve depicts the distribution of thymocyte avidities for self peptide–MHC complexes. In our model, the chance p to be positively selected by a single MHC type is the chance that the avidity between the thymocyte T cell receptor and any of the self peptide–MHC complexes exceeds threshold T_1. Thymocytes with avidities for self peptide–MHC complexes exceeding the upper threshold T_2 are negatively selected (with chance n per MHC type).

selected by at least one of the MHC molecules, and to avoid negative selection by all of the MHC molecules, the number of clones in the functional repertoire R can be expressed as

$$R = R_0 \left((1-n)^M - (1-p)^M \right) . \tag{9.1}$$

The functional repertoire R thus contains all T cell clones that fail to be negatively selected, minus the ones that also fail to be positively selected by any of the M different MHC molecules of the host [Borghans et al. 2003].

Experimental estimates for the parameters of this model have recently become available. In mice, around 3% of the T cells produced in the thymus end up in the mature T cell repertoire [Scollay et al. 1980, Egerton et al. 1990, Shortman et al. 1991, Merkenschlager et al. 1997], and at least 50% of all *positively selected* T cells have been shown to undergo negative selection in the thymus [Meerwijk et al. 1997].

Thus, 94% of all thymic T cells fail to be positively selected by any of the MHC molecules in the host [Meerwijk et al. 1997]. We have used these estimates to calculate the chances p and n of a T cell clone to be positively or negatively selected by a single type of MHC molecule. Taking into account that inbred mice are homozygous and therefore express 3 types of class I and 3 types of class II MHC molecules, p and n follow from: $(1-p)^6 = 0.94$ and $(1-n)^6 = 0.97$. This yields $p = 0.01$ and $n = 0.005$ [Borghans et al. 2003]. Both class I and class II MHC molecules are incorporated in these calculations, because positive selection and at least part of negative selection take place at the double positive (DP) stage [Kisielow et al. 1988, Baldwin et al. 1999], when thymocytes express both CD4 and CD8 coreceptors.

Using these experimental estimates, we found that the number of clones in the functional T cell repertoire R increases with the number of different MHC molecules M in an individual until $M = 140$ (see Figure 9.2). In other words, the size of the functional T cell repertoire would increase if the MHC diversity M were to exceed its normal value of ten to twenty in heterozygous individuals. The intuitive reason is that only a very small part of the T cell repertoire has sufficient avidity for any of the self peptides presented by a single MHC type to be positively selected by that MHC. As long as additional MHC types positively select parts of the T cell repertoire that hardly overlap, negative selection only eliminates T cells that were not expected to be positively selected in the absence of those MHC molecules. A net negative effect of MHC diversity on the size of the functional T cell repertoire is only attained once the individual MHC diversity is so large that thymocytes are selected by multiple MHC types, i.e. when $M > 140$.

The finding that the expression of extra MHC molecules would increase the diversity of the T cell repertoire is a consequence of the current consensus that positive selection is a strong bottleneck [Von Boehmer 1994, Fink & Bevan 1995, Meerwijk et al. 1997, Surh & Sprent 1994]. This explains why previous studies claimed a low optimal number of MHC types due to negative selection in the thymus [Takahata 1995, Celada & Seiden 1992, De Boer & Perelson 1993, Nowak et al. 1992]. These models either did not at all account for positive selection [De Boer & Perelson 1993], or involved too stringent negative selection [Takahata 1995, Nowak et al. 1992], because T cells with avidities too low to be positively selected by a particular MHC molecule could nevertheless be negatively selected by that same MHC molecule. The latter is in disagreement with the avidity model depicted in Figure 9.1, and would lead to a net decrease of the functional T cell repertoire at MHC diversities as low as one to two MHC types per individual [Borghans et al. 2003].

Summarizing, if current estimates of positive and negative selection are correct, the consensus explanation that the MHC diversity per individual is limited to avoid repertoire depletion [Matzinger et al. 1984, Vidović & Matzinger 1988, Parham et al. 1989a, Nowak et al. 1992, De Boer & Perelson 1993, Takahata 1995, Janeway & Travers 1997, Cohn 1985, Celada & Seiden 1992, Lawlor et al. 1990] is untenable. What remains is the question what can explain the limited number of different MHC molecules per individual. First of all, it may simply be sufficient to have a few MHC types per individual. Since many peptides can be generated from a single pathogen, even a low individual MHC diversity gives a good chance to present and respond to

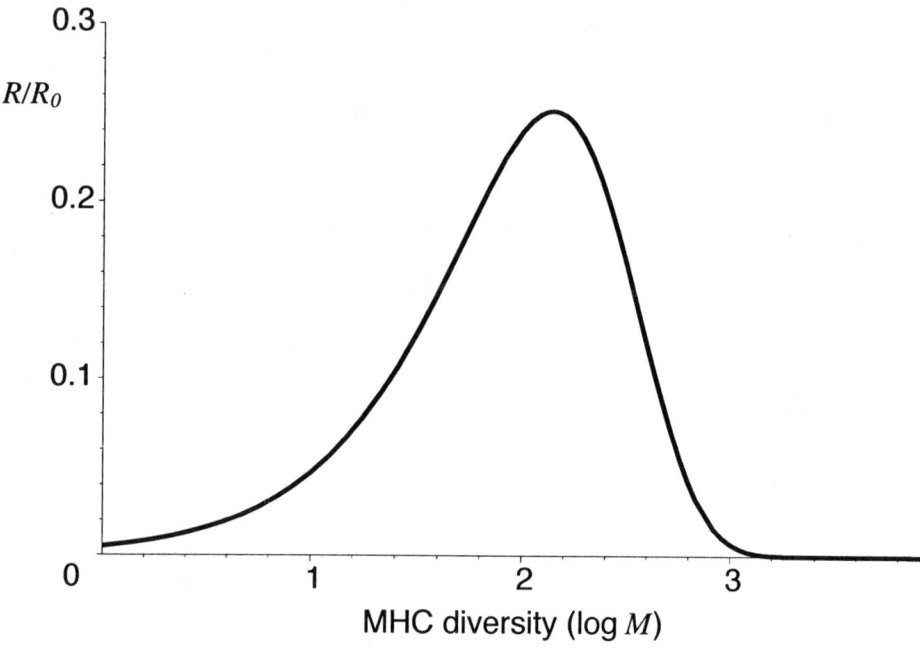

Fig. 9.2. The size of the T cell repertoire as a function of MHC diversity. The number of clones in the functional repertoire R is plotted as a fraction of the total initial lymphocyte repertoire R_0. Parameters are: $p = 0.01$, and $n = 0.005$. Note that at $M = 6$ indeed 3% of the initial T cell repertoire ends up in the functional lymphocyte repertoire [Egerton *et al.* 1990]. From: [Borghans *et al.* 2003].

pathogens [Borghans *et al.* 2003]. There could thus be no selection for more MHC diversity per individual. Another possibility is that the number of different MHC molecules per individual is limited to avoid the induction of inappropriate, cross-reactive immune responses (see also [Borghans *et al.* 1999, Borghans & De Boer 2001, Borghans & De Boer 2002]), such as anti-viral responses that cause autoimmunity as a side effect [Zhao *et al.* 1998, Bachmaier *et al.* 1999]. More MHC diversity than strictly required to ensure the presentation and recognition of pathogens, would increase the risk to induce such undesirable responses (see also [Apanius *et al.* 1997, Borghans & De Boer 2001]). Alternatively, it has been proposed that the limited MHC diversity per individual helps to induce effective immune responses by focusing the T cell repertoire at a few epitopes only [van den Berg & Rand 2003]. Indeed, if all cells were to express a great variety of MHC molecules at the cell surface, the concentration of any MHC–peptide ligand might be too low to induce an effective immune response. Finally, we emphasize again that the high optimum that we find (Figure 9.2) is due to fact that positive selection is the major bottleneck for lymphocyte selection according to the parameters that we have taken from the literature. [Huseby *et al.* 2005] recently suggested that positive selection in the thymus is much less restrictive, and that negative selection is the major bottle-neck.

If this turns out to be true the optimum number of MHC molecules could be much lower [De Boer & Perelson 1993].

9.3 Heterozygote advantage

Even though the MHC diversity within individuals is quite limited, the chance that two individuals are MHC identical is extremely small. The mechanisms underlying the high population diversity of MHC molecules have been debated for decades. The discussion basically centres around two opposing views. According to one, the large degree of MHC polymorphism is due to selection favoring MHC heterozygous hosts [Doherty & Zinkernagel 1975, Hughes & Nei 1988, Hughes & Nei 1989, Takahata & Nei 1990, Hughes & Nei 1992, Jeffery & Bangham 2000, Maruyama & Nei 1981, Hughes & Yeager 1998]. Although there is general agreement upon the significance of heterozygote advantage, others have argued that the impact of heterozygote advantage is insufficient to explain the large MHC diversity observed in nature [Lewontin et al. 1978, Aoki 1980, Parham et al. 1989b, Lawlor et al. 1990, Wills 1991]. We have developed a population genetics model to study the degree of polymorphism heterozygote advantage can lead to [De Boer et al. 2004]. In contrast to previous theoretical models for heterozygote advantage, the model is allele-based in the sense that the fitness of a host is determined by the fitness contributions of individual MHC alleles. This change of approach appeared to be an important improvement over previous models, leading to a complete revision of previous conclusions.

Consider a population of n different MHC alleles at a single locus, each characterized by a parameter f_i, representing the fitness contribution of the allele to its host. One can think of this fitness parameter as the fraction of pathogens the MHC allele can provide protection to. The frequency of the MHC alleles in the population is denoted by p_i. For simplicity, we define the fitness of a homozygote as $f_{ii} = \beta + f_i$, and the fitness of a heterozygote as $f_{ij} = \beta + f_i + f_j$, where β denotes a basis fitness parameter independent of the MHC. In fact, this additive fitness definition only holds if there is no overlap between the pathogens alleles i and j provide protection to. Refinement of this fitness definition accounting for such overlaps complicates the mathematical analysis but does not lead to qualitatively different results [De Boer et al. 2004].

In a population with n different MHC alleles present at random frequencies, the allele frequencies change according to the fitness contributions f_i of the different alleles. Alleles contributing little to the hosts' fitness attain low frequencies, while useful MHC alleles increase in frequency. At steady state, the marginal fitnesses $w_i = \sum_{j=1}^{n} p_j f_{ij}$ should be identical for all alleles i. MHC alleles with high fitness contributions, present at high frequencies, attain the same marginal fitness as rare MHC alleles, because they occur more often in MHC homozygous hosts [Apanius et al. 1997]. Thanks to the heterozygote advantage, MHC alleles with relatively low fitness contributions f_i can remain in the population by "hitch-hiking" with better alleles in heterozygous hosts.

A certain minimal fitness contribution is required, however, for an MHC allele to be maintained in the population. Let all n MHC alleles be ranked such that $f_1 > ... > f_n$. In the Appendix we show that a novel allele with fitness value $f_{n+1} < f_n$ can only successfully invade into a settled steady state population of n alleles if [van Boven & Weissing 2001, Weissing & van Boven 2001]:

$$f_{n+1} > \frac{n-1}{n} \widehat{f}, \tag{9.2}$$

where $\widehat{f} \equiv n / \sum_{j=1}^{n} f_j^{-1}$ is the harmonic mean of the fitness contributions of the n established alleles. Thus, novel alleles will only be able to invade if their fitness contribution is sufficiently close to the harmonic mean fitness contribution $\widehat{f}$ of the n established alleles. For example, in a population of 19 alleles, a 20th allele can invade only if its fitness contribution exceeds 19/20 of the harmonic mean of the fitness contributions of the 19 resident alleles. According to Eq. (9.2), the critical fitness contribution of a novel allele increases with the degree of polymorphism n. Thus, the larger the MHC polymorphism is, the harder it becomes to attain an even higher degree of polymorphism.

The above findings are illustrated in Figure 9.3, where an example is shown in which the fitness contributions of the different alleles are given by $f_i = (1 - s)^{i-1}$. The parameter s defines the steepness with which the fitness contribution decreases with the allele number. Solving s from Eq. (9.2) gives the maximum s value compatible with a polymorphism of $n + 1$ alleles (see Appendix). Figure 9.3 confirms that large degrees of polymorphism require a high fitness contribution of invading alleles, and a small variation in allele fitness contributions s. Heterozygote advantage can thus only account for a very high degree of MHC polymorphism if the variation in fitness contributions amongst the MHC alleles is vanishingly small [De Boer et al. 2004].

One could argue that evolution leads to a slow accumulation of novel MHC alleles with very similar fitness contributions [Lewontin et al. 1978]. We think this is not the case, however, because MHC alleles differ greatly in their binding motifs [Barouch et al. 1995, Rammensee et al. 1999], and even small binding motif differences can lead to large differences in protection. For instance, one amino acid difference in the peptide-binding region of the DRB1*1302 allele abrogates its protection to malaria [Davenport et al. 1995]. The fact that the HLA alleles expressed in the South Amerindian population are different from those in the founder population, while North Amerindians still express the founder alleles [Parham & Ohta 1996], also suggests that different MHC alleles provide different degrees of protection against different pathogens. Similarly, [Gao et al. 2001] found that despite the small differences in binding motifs, the B*3503 allele is associated with fast progression to AIDS, while B*3501 is not. To get an idea on the extent to which MHC allele fitness contributions differ, we predicted the best binding peptide of 17 different viral proteomes for three human class I MHC alleles (Figure 9.4), using an established MHC–peptide binding prediction method [Parker et al. 1994]. The amino acid weight matrices of HLA-A*0201, HLA-A*0205, and HLA-A*3101 were downloaded from bimas.dcrt.nih.gov/molbio/hla_bind/ (April, 2002). These three alleles were chosen because their weight matrices were sufficiently detailed to allow for an almost continuous ranking of the peptides. In the weight matrices, dominant anchor residues have the highest weights, followed by auxiliary anchor residues, and

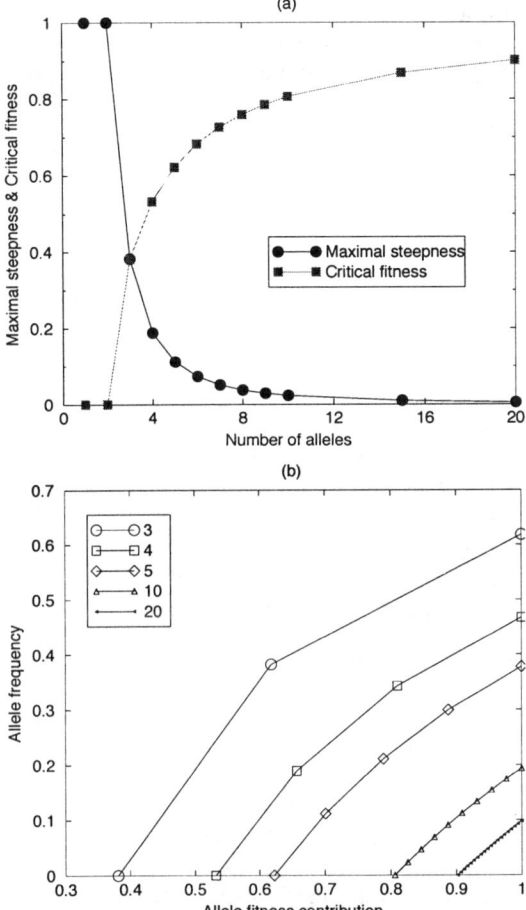

Fig. 9.3. The MHC polymorphism obtained with allele fitness contributions $f_i = (1 - s)^{i-1}$. Panel (a) gives the invasion of the $n + 1^{\text{th}}$ allele into an established polymorphism of n alleles (see Appendix). The line marked by circles depicts the maximal value of the steepness parameter s for a novel allele to invade and establish the polymorphism plotted on the horizontal axis. The line marked by squares is the critical fitness contribution of the invading allele, which approaches the harmonic mean fitness $\widehat{f}$ when $n \to \infty$. Panel (b) depicts the corresponding distributions of allele frequencies for $n+1 = 3, 4, 5, 10$ and 20 alleles, *i.e.* for $s = 0.38, 0.19, 0.11, 0.02$ and 0.005, respectively. From: [De Boer *et al.* 2004].

favorable amino acids. Unfavorable amino acids have a weight less than one. The binding score of each MHC–peptide combination is defined as $\prod_{i=1}^{9} w_i$, where w_i is the weight index of the amino acid at position i in the peptide. This definition is based on the assumption that each amino acid in the peptide contributes independently to the binding score. Since this method yields a binding score rather than an affinity, we performed a non-parametric normalization by ranking the binding score of the best binding foreign peptide among the binding scores of all 8.5×10^6 unique 9-mer peptides from the human proteome. This ranking provides an indication for the fitness contribution of an MHC allele. If many self-peptides bind better than the best-binding foreign peptide, its binding affinity to the MHC is relatively low, and the foreign peptide is likely to be out competed by self-peptides on the surface of antigen-presenting cells. The human proteome and 17 virus proteomes were downloaded from the http://www.ebi.ac.uk/genomes data base (March 2, 2002). All 9-mer peptides containing the non-standard amino acids B, X, and Z were ignored.

Our bioinformatic analysis suggests that viruses are presented very differently by different MHC alleles (Figure 9.4). Since immunity against dangerous viruses can be a matter of life and death, the differences in antigen presentation between MHC alleles can lead to large host fitness differences. Remember that even a low polymorphism of 20 MHC alleles requires that the fitness contribution of the worst allele is at least 95% of the harmonic mean of the fitness contributions of the other 19 alleles (see above). The fact that we found such large presentation differences between the three human MHC alleles in our analysis therefore makes it extremely unlikely that the fitness contributions of all MHC alleles in the human population are sufficiently similar to explain the large degree of MHC polymorphism by heterozygote selection only.

Remarkably, previous claims that heterozygote advantage suffices to explain high degrees of MHC polymorphism where also based on theoretical models [Takahata & Nei 1990, Hughes & Yeager 1998, Hughes & Nei 1992, Maruyama & Nei 1981]. These models where, however, based on random genotype fitness matrices [Lewontin et al. 1978, Maruyama & Nei 1981, Takahata & Nei 1990]. The main problem when using random fitness matrices is that the fitness of genotypes in the model becomes unrelated to the fitness contributions of the individual MHC alleles. In the model of Takahata & Nei [Takahata & Nei 1990], for example, all heterozygote fitnesses were set to one, which allows any novel allele to invade into any established polymorphism [De Boer et al. 2004]. Instead, in our model host genotype fitnesses were determined by the fitness contributions of individual MHC alleles. Since heterozygous hosts with poor alleles thus had a smaller fitness than heterozygotes with useful alleles, this gave rise to much lower degrees of polymorphism. Unless all MHC alleles in a population confer almost identical levels of protection to their hosts (which we claim not to be the case, see Figure 9.4), heterozygote advantage thus fails to explain the large degree of polymorphism of the MHC [De Boer et al. 2004].

9.4 MHC polymorphism by host–pathogen coevolution

Since heterozygote advantage is insufficient to explain the large degree of polymorphism of the MHC, additional mechanisms should be involved. Alternative mechanisms that have been proposed vary from MHC-dependent mate selection, and preferential abortion, to various pathogen-driven selection pressures (see [Apanius

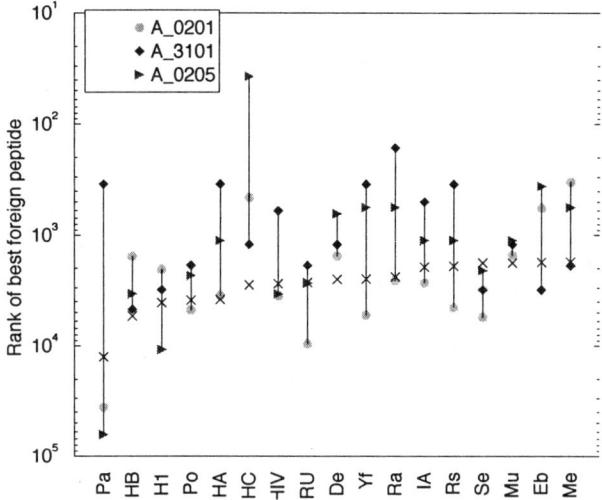

Fig. 9.4. Different HLA-A alleles provide a different level of protection to various viruses. Weight matrices of the binding motifs of the HLA-A*0201, HLA-A*0205, and HLA-A*3101 alleles were used to score the ranking of all unique 9-mers from the proteomes of various common viruses. The scores were normalized by comparing them to the predictions of all unique 9-mers from the human proteome. For each virus, the ranking of the binding score of the best-binding viral peptide in the sorted list of binding scores of all self peptides is given. For instance, the best binding peptide from the Parvovirus proteome for the HLA-A*3101 allele binds worse than 340 self peptides from the human proteome, while the best binding peptide for the HLA-A*0205 allele binds worse than more than 60,000 self peptides. This ranking provides an indication of the levels of protection provided by the MHC alleles. Since the expected ranking (denoted by stars) increases with the number of unique 9-mers in the virus proteomes, the viruses on the horizontal axis were ordered by their size as measured by the total number of unique 9-mers: Pa: *Parvovirus H1* (X01457), HB: *Hepatitis B virus* (X51970), H1: *Human T-cell leukaemia virus type I* (D13784), Po: *Human poliovirus 1* (AJ132961), HA: *Hepatitis A virus* (M14707), HC: *Hepatitis C virus* (AJ132997), HIV: *Human immunodeficiency virus 1* (AJ006287), Ru: *Rubella virus* (AF188704), De: *Dengue virus type 1* (U88536), Yf: *Yellow fever virus* (X03700), Ra: *Rabies virus* (M31046), IA: *Influenza A virus segments 1–8* (V00603, J02151, V01106, V01088, J02147, J02146, V01099, J02150), Rs: *Human respiratory syncytial virus* (AF013254), Se: *Sendai virus* (M69046), Mu: *Mumps virus* (AB04087), Eb: *Ebola virus* (AF086833), and Me: *Measles virus* (K01711), where GenBank accession numbers are given in brackets.

et al. 1997] and [Penn 2002] for extensive reviews). We have developed a computer simulation model to analyse the impact of frequency-dependent selection by host–pathogen coevolution [Borghans & De Boer 2001, Beltman *et al.* 2002, Borghans *et al.* 2004]. Since evolution favors pathogens that avoid presentation by the most common MHC molecules in the host population, hosts with rare MHC alleles have a higher fitness than hosts with common MHC alleles. The frequency of rare MHC alleles will therefore increase, and common MHC alleles will become less frequent, resulting in a dynamic polymorphism [Snell 1968, Bodmer 1972, Slade & McCallum 1992, Beck 1984]. Recent studies of a snail infected by a trematode parasite provide support for host–pathogen coevolution, by demonstrating that the parasite adapts to be most virulent in the dominant host genotype [Dybdahl & Lively 1998, Lively & Dybdahl 2000]. Another example is HIV-1, which has evolved protein regions that are devoid of epitopes because they lack immuno-proteasome cleavage sites or generate peptides that are poorly presented by MHC molecules [Korber *et al.* 2001a, Moore *et al.* 2002, Leslie *et al.* 2005].

Consider a population of N_{host} diploid hosts, each represented by two bit strings coding for the MHC alleles at a single locus. Pathogens are haploid and occur in 50 independent species of maximally 10 different genotypes. Each pathogen is modelled by 20 bit strings representing its dominant peptides. Both peptides and MHC molecules are 16 bits long. Peptide presentation by an MHC molecule is modelled by complementary matching. If the longest stretch of adjacent complementary bits is at least 7 bits long, the peptide is considered to be presented by the MHC molecule. With these parameters the chance that a random MHC molecule presents a randomly chosen peptide is about 5%, which is close to the experimental estimate [Kast *et al.* 1994]. Thus, hosts carrying different MHC molecules typically present different peptides of the same pathogen. Since pathogens typically have shorter generation times than their hosts, 10 pathogen generations occur per host generation. At each pathogen generation, every host interacts with one randomly chosen member of each pathogen species. The fitness of a host is defined as the fraction of pathogens it has presented during one host generation. The fitness of a pathogen is the fraction of hosts that the pathogen can infect during one pathogen generation without being presented by the host's MHC molecules. At the end of each generation, all individuals are replaced by fitness-proportional reproduction. The chance that an individual reproduces is proportional to its squared fitness divided by the sum of the squared fitnesses in the pathogen species or host population. Pathogen genotypes reproduce asexually; newborn pathogens come from parents of the same pathogen species. Newborn hosts have two parents, each of which donates a randomly selected MHC allele. The size of the host population and the number of pathogens remain constant. Mutations are modelled by generating new random bit strings. MHC molecules mutate at a frequency $\mu_{host} = 10^{-5}$ per allele per host generation. The mutation rates of the pathogen species vary between $\mu_{path} = 10^{-1}$ per peptide per pathogen generation and $\mu_{path} = \mu_{host}$. Thus, most pathogen species mutate much faster than the hosts.

To study the evolution of MHC polymorphism, all hosts were initialized with the same randomly chosen MHC allele. Pathogens were initialized fully randomly, *i.e.* each pathogen species started with 10 genotypes each consisting of 20 randomly generated peptides. A population of 1000 diploid hosts coevolving with 50 pathogen

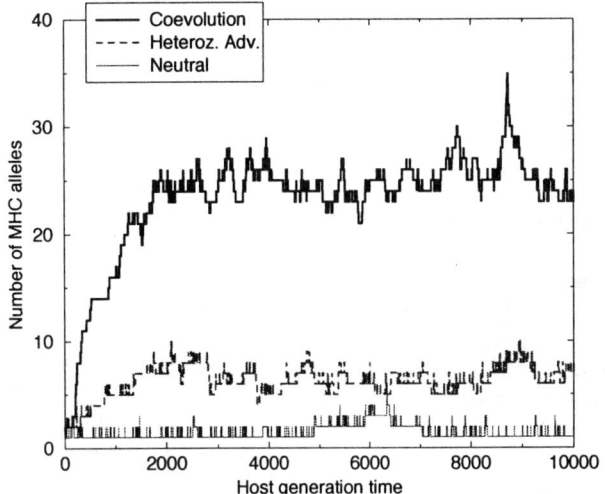

Fig. 9.5. The evolution of MHC polymorphism in host populations starting with a single MHC allele at one locus. The solid curve depicts a simulation with coevolving hosts and pathogens, the dashed curve represents a simulation of heterozygote advantage (in which the pathogens did not evolve), and the dotted curve represents purely neutral selection. From: [Borghans *et al.* 2004].

species developed a stable MHC polymorphism of approximately 27 alleles (see the solid curve in Figure 9.5). At the end of the simulation, the pathogens escaped from presentation by the MHC molecules of 45% of the heterozygous hosts, which is much more than the 16% expected for a random pathogen [Borghans *et al.* 2004]. Apparently, most of the pathogens were well adapted by being poorly presented by the 27 different MHC molecules present in the host population.

The parameter regime strongly favored MHC heterozygous hosts: a pathogen with 20 peptides had 84% chance of being presented by a host with two different MHC molecules, and only 59% in a host with a single MHC molecule [Borghans *et al.* 2004]. Nevertheless, the impact of heterozygote advantage on the MHC polymorphism was small. When pathogen selection was prevented by imposing one and the same fitness value on all pathogens, a polymorphism of only about 7 MHC alleles was attained (see the dashed curve in Figure 9.5). As the pathogens no longer adapted to the host population, this polymorphism must have been purely due to heterozygote advantage. In the absence of host and pathogen selection an even lower degree of polymorphism was attained (see the bottom line in Figure 9.5). Summarizing, host–pathogen coevolution leads to a significantly higher degree of MHC polymorphism than heterozygote advantage.

The degree of MHC polymorphism developing under host–pathogen coevolution was highly dependent on two parameters of the model. The first one is the host population size, affecting the chance that rare alleles are driven to extinction. Increasing the host population size from 1000 to 5000, almost linearly increased the degree of MHC polymorphism (see Figure 9.6a). Since a host population of 5000 individuals coevolving with pathogens developed a polymorphism of more than 50 alleles, we expect coevolution to be able to account for the naturally observed polymorphism of more than 100 alleles per locus in large natural populations [Parham & Ohta 1996]. Conversely, when there was only heterozygote advantage, the degree of polymorphism remained small when the population size was increased (see Figure 9.6a), and was mainly determined by the fitness distribution of the alleles (see Section 9.3).

Surprisingly, the second parameter affecting the degree of MHC polymorphism is the mutation rate of the pathogens. When each pathogen species had the same low mutation rate, e.g. $\mu_{path} = 10^{-6}$ for all pathogen species, the pathogens changed so slowly that the difference between coevolution and heterozygote advantage vanished (see Figure 9.6b). The hosts evolved a few MHC molecules specialized at presenting the peptides present in the pathogen population. At a larger pathogen mutation rate, e.g. $\mu_{path} = 10^{-1}$ for all pathogen species, a much higher degree of MHC polymorphism was attained. In a population of 5000 hosts nearly 100 different MHC alleles could then be found at a single locus (see Figure 9.6b). In this parameter region, the difference between the polymorphism arising under host–pathogen coevolution and under heterozygote advantage reached its maximum (see Figure 9.6b), and became similar to the simulations in which pathogen species had different mutation rates (see Figure 9.6a).

The polymorphism under heterozygote advantage increased with the pathogen mutation rate. When all pathogens had a mutation rate of 100%, coevolution and heterozygote advantage again became identical, because the pathogens were no longer selected. Under heterozygote advantage, this led to the highest possible polymorphism, because all MHC alleles have very similar fitness contributions when pathogens are random. At lower pathogen mutation rates, successful adaptation of MHC molecules to the pathogens caused differences between the MHC alleles, which reduced the degree of MHC polymorphism that was obtained under heterozygote advantage (see Section 3). Thus, these simulations confirm the conclusion that heterozygote advantage is insufficient to explain the large population diversity of MHC molecules, and show that host–pathogen coevolution gives rise to natural levels of MHC polymorphism in large host populations.

9.5 Discussion

By simulating the evolution of hosts and pathogens, we have demonstrated that a large degree of polymorphism of MHC molecules naturally arises in host populations infected by many different pathogens. The simulations confirm that there is selection favoring MHC heterozygosity [Doherty & Zinkernagel 1975, Hughes & Nei 1988, Hughes & Nei 1989, Takahata & Nei 1990, Hughes & Nei 1992]. Heterozygote advantage by itself, however, is insufficient to explain the large population diversity of MHC molecules, as it would require an unrealistic degree of similarity

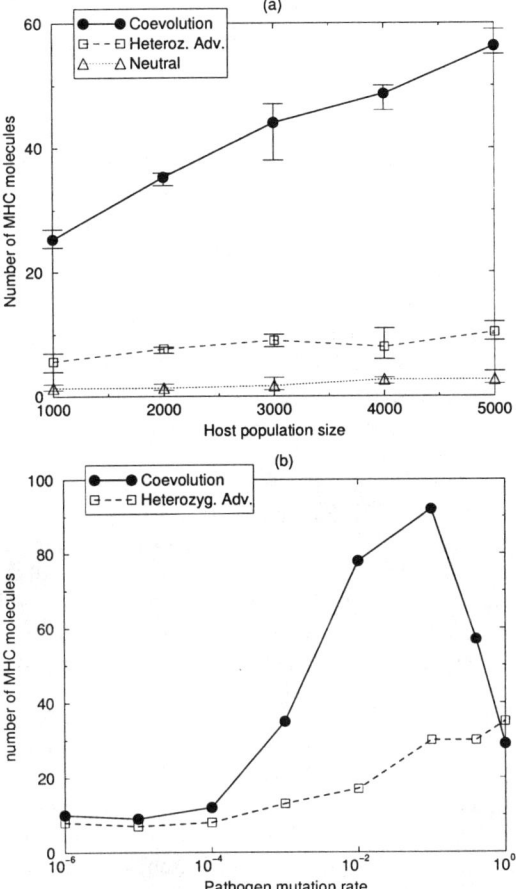

Fig. 9.6. The impact of the host population size (a) and the pathogen mutation rate (b) on the degree of MHC polymorphism at generation 10,000. Data points in panel (a) are the average of three runs, differing in random seed; error bars denote the range of the three runs. Panel (b) gives simulations with 5000 hosts, and with equal mutation rates for all pathogen species. From: [Borghans *et al.* 2004]

between MHC molecules. When hosts and pathogens coevolve, there is frequency-dependent selection in addition, favoring the expression of rare MHC molecules [Snell 1968, Bodmer 1972, Slade & McCallum 1992, Beck 1984]. Rare MHC molecules tend to provide protection against pathogens that avoid presentation by the most common MHC molecules in the population. We have shown that the MHC polymorphism arising under host–pathogen coevolution is significantly larger than the polymorphism arising under selection for heterozygosity only. In addition, in our simulations there was no explicit disadvantage of expression of many different MHC molecules per individual. If a disadvantage of a high individual MHC diversity (at multiple loci) were to be taken into account, the effect of heterozygote selection on the degree of MHC polymorphism would diminish at a sufficiently high individual diversity. Selection for expression of rare MHC molecules would, however, remain (results not shown).

Although hundreds of different MHC molecules have been observed at the population level [Parham & Ohta 1996, Vogel *et al.* 1999], individuals express only a small fraction of this MHC diversity [Paul 1999]. We have disputed the widely held view that the individual diversity of MHC molecules is limited to avoid T cell repertoire depletion during self tolerance induction [Vidović & Matzinger 1988, Parham *et al.* 1989a, Nowak *et al.* 1992, De Boer & Perelson 1993, Takahata 1995, Janeway & Travers 1997, Cohn 1985]. Using a mathematical model, we found that expression of extra MHC molecules would increase the number of T cell clones in the functional T cell repertoire, and that repertoire depletion would only occur at an unrealistically high individual MHC diversity.

The MHC is not the only gene complex that is very polymorphic. Another well-known example that is associated with immune defense to pathogens is the killer cell immunoglobulin-like receptor (KIR) gene cluster of NK cells [Carrington & Martin 2006]. The KIR gene locus has a high diversity of haplotypes expressing different and variable numbers of genes. The inhibitory KIR bind self MHC molecules to prevent the NK response to cells with normal expression levels of MHC molecules. To recognize cells that have down-regulated MHC expression, i.e., the "missing self" hypothesis [Karre 1995], an individual's diversity of KIR should be sufficient to recognize all, or most, MHC alleles that are expressed within the individual. Due to the MHC polymorphism these vary, and the polymorphism of the KIR gene tends to double the repertoire of KIRs in heterozygous individuals, which increases the chance that all MHC molecules expressed with the individual are recognized. Finally, gene polymorphisms are not restricted to the immune system. For example, G-protein coupled receptors (GPCRs) form a large protein family with numerous single nucleotide polymorphisms (SNPs) in coding regions, which are associated with disease and drug efficacy [Balasubramanian *et al.* 2005].

The MHC is polymorphic in almost all vertebrate species for which this has been studied. The "Immuno Polymorphism Database" (IPD-MHC) collects information of the MHC diversity of a large number of species [Robinson *et al.* 2005] (see [www.ebi.ac.uk/ipd/mhc/index.html]). Mate choice experiments in several species have suggested that females attempt to increase the degree of MHC heterozygosity of their offspring by choosing mates with different MHC alleles [Potts *et al.* 1991, Wedekind *et al.* 1995, Ober *et al.* 1997, Reusch *et al.* 2001], while another study suggests that MHC alone is not enough for individual recognition Hurst.pbs05. We have suggested that the major selection pressure driving MHC polymorphism of the population, and as a consequence MHC heterozygosity of the host, is the co-evolution between pathogens and their hosts. Generally, genetic variability of the host population decreases the average fitness of pathogens, i.e., this is not restricted to the MHC, and this is one of the major driving forces for the evolution of sexual reproduction [Hamilton *et al.* 1990]. It is not clear why a limited species has a much more restricted MHC diversity, e.g., cheetah [O'Brien & Yuhki 1999, Yeager & Hughes 1999], Eurasian Beaver [Ellegren *et al.* 1993], Moose [Mikko & Andersson 1995], and African mole-rates [Kundu & Faulkes 2004], although recent bottle-necking of the population may evidently play a role [Ellegren *et al.* 1993, Aguilar *et al.* 2004]. The MHC class I repertoire of chimpanzees was shown to be far less diverse than that of humans. This was proposed to be due to an "ancient selective sweep" by a wide-spread viral infection, possibly SIV [de Groot *et al.* 2002]. Bottle-necks may also play a role in explaining small ethnic human populations that sometimes express particular HLA haplotypes at unexpectedly high frequencies (see [www.ncbi.nlm.nih.gov/IEB/Research/GVWG/IHWG/ihwg.cgi] for a number of examples).

For other defence systems, *e.g.* the restriction–modification (RM) system which protects bacteria against invading genetic material, two modes of diversity have been described: an individual-based mode in which every bacterium expresses all possible RM specificities, and a population-based mode in which each bacterium expresses maximally one RM system, with the total set of RM systems being expressed at the population level [Pagie & Hogeweg 2000]. [Pagie & Hogeweg 2000] demonstrated that such a population-based mode even exists in the absence of any costs for expression of RM systems. Expressing a limited number of defence systems per individual allows individuals to be different from each other. Analogously, the population diversity of MHC molecules allows different individuals to respond differently to identical pathogens. Each host "samples" a small fraction of a pathogen's proteome for presentation that differs from host to host. This unpredictability hampers the evolution of pathogen proteomes that can successfully evade antigen presentation.

Evidence for host–pathogen coevolution has recently been found in the field of HIV. CTL escape mutants of HIV were shown to be associated with particular MHC class I alleles, suggesting viral evasion of presentation by those MHC molecules [Moore *et al.* 2002, Leslie *et al.* 2005]. On the other hand, common MHC alleles appeared to correlate with absence of viral mutants, suggesting successful previous adaptation of HIV to common MHC molecules [Moore *et al.* 2002, Leslie *et al.* 2005]. Another evidence for frequency-dependent selection on MHC alleles in man was discovered when MHC alleles were grouped together in "supertypes" that are based on similarities in their binding motifs [Sette & Sidney 1999, Lund *et al.* 2004], see Chapter 10. The frequency of different MHC supertypes in a large group of HIV-infected men was found to be inversely correlated with viral RNA loads [Trachtenberg *et al.* 2003], and common HLA supertypes were shown to be associated with a lack of CTL responses to known HIV-1 epitopes [Scherer *et al.* 2004]. One has to be cautious while interpreting these results, however, because we have recently showed that "rare and protective" HLA supertypes B58s and B27s that were found in these studies have a stronger preference for epitopes from "constrained" p24 [Von Schwedler *et al.* 2003, Leslie *et al.* 2004, Peyerl *et al.* 2004] than the non-protective supertypes B7s and A1s, which have a stronger preference for "polymorphic" Nef [Keşmir, Borghans & De Boer, in prep.]. For the most common HLA-supertype A2s the percentage of known epitopes per protein closely resembled the expected frequency distribution. Because HLA supertypes associated with slow disease progression preferentially present peptides from the most constrained parts of the HIV-1 genome, we predict that these protective alleles remain protective even in populations where they are more common, as is the case in Botswana [Novitsky *et al.* 2003]. Finally, the classification of MHC alleles into supertypes fails to fully capture their functional relationships, because different MHC alleles within one supertype may have quite different associations with disease. For example, the HLA-B*3503 allele is associated with fast progression to AIDS, while HLA-B*3501 from the same B35s superfamily is not [Gao et al. 2001].

Another example that was interpreted as viral adaptation to host antigen presentation pathways was the identification of HIV proteome regions with few immunogenic class I epitopes for a large variety of HLA alleles [Yusim et al. 2002]. As these regions contained very few human proteasome cleavage sites, (Chapter 8), it seemed that HIV was exploiting the predictability of the, hardly polymorphic, proteasome and TAP molecules [Kesmir et al. 2002]. Recent work however casts doubt on this because we fail to find evidence for an increase of HIV escape from proteasomal cleavage and TAP binding, and we think that the low density of class I epitopes is due to the hydrophilic nature of these regions [Lucchiari-Hartz et al. 2003] [Schmidt, Keşmir & De Boer, in prep.]. Due to the MHC polymorphism the particular proteasomal cleavage sites and TAP ligands that actually give rise to immunogenic peptides remain variable among different individual hosts. For a co-evolving virus like HIV the important cleavage and binding sites therefore remain unpredictable.

Acknowledgments

We thank Joost Beltman, Michiel van Boven, Franjo Weissing, Nigel Burroughs and André Noest for their excellent contributions to the papers we have reviewed here. JB acknowledges financial support by the EC (Marie Curie Fellowship, Quality of Life, contract 1999-01548). This work was financially supported by the Netherlands Organization for Scientific Research (NWO, grants 050.50.202, 916.36.003, 16.048.603).

A Appendix

Requiring that the marginal fitness $w_i = \sum_{j=1}^{n} p_j f_{ij}$ is the same for all alleles i, one obtains $f_i(1 - p_i) = f_j(1 - p_j)$ for all alleles i and j. By expressing all p_j by the same p_i and summing the n equations one obtains

$$p_i = 1 - \frac{n-1}{n} \frac{\widehat{f}}{f_i} , \qquad (9.3)$$

where $\widehat{f} \equiv n / \sum_{j=1}^{n} f_j^{-1}$ is the harmonic mean of the n allele fitness contributions. Thus, MHC alleles with a too low fitness contribution f_i will not be present at a positive frequency p_i, i.e. cannot co-exist, because $\widehat{f}/f_i$ becomes too large. Note that the basis fitness value β has canceled [van Boven & Weissing 2001].

For a novel allele with fitness contribution f_{n+1} to invade into an established polymorphism of n alleles, its marginal fitness has to exceed the marginal fitness of the other alleles [van Boven & Weissing 2001, Weissing & van Boven 2001]. Writing $w_{n+1} > w_i$ this yields

$$\sum_{j=1}^{n} p_j(f_{n+1} + f_j) > \sum_{j=1}^{n} p_j(f_i + f_j) - p_i f_i \ , \tag{9.4}$$

which can be simplified into

$$f_{n+1} > f_i(1 - p_i) \ . \tag{9.5}$$

Substituting p_i from Eq. (9.3) gives Eq. (9.2) in the text.

To obtain Figure 9.3 we substitute $f_i = \phi^{i-1}$ into Eq. (9.2) in the text, where $\phi \equiv (1 - s)$, one can test the invasion of the $n + 1^{\text{th}}$ allele, and simplify to obtain

$$\sum_{i=0}^{n} \phi^i = \frac{1 - \phi^{n+1}}{1 - \phi} > n \ , \tag{9.6}$$

which can be used to solve the critical s for invasion into any polymorphism of n alleles.

Identifying Major Histocompatibility Complex Supertypes

Pingping Guan[1], Irini A. Doytchinova[2], and Darren R. Flower[2]

[1] John Innes Centre, Norwich, NR4 7UH, UK
[2] The Jenner Institute, University of Oxford, Compton, Berkshire, RG20 7NN,
UKDarren.flower@jenner.ac.uk

Summary. Human leukocyte antigen (HLA) recognizes antigenic fragments and presents them to T cells. HLA is polymorphic. There are over 2000 different HLA alleles at present and the number is constantly increasing. However, antigen binding studies are limited to a small proportion of these alleles; the binding specificities of most alleles are unknown. Several research groups have attempted to partition different HLA alleles into groups. In this chapter previous classifications are reviewed and we present two chemometric approaches to classifying class I HLA alleles. The program GRID is used to calculate interaction energy between protein molecules and defined chemical probes. These interaction energy values are imported into another program GOLPE and used for principal component analysis (PCA) calculation, which groups HLA alleles into supertypes. Amino acids that are involved in the classification are displayed in the loading plots of the PCA model. Another method, hierarchical clustering based on comparative molecular similarity indices (CoMSIA) is also applied to classify HLA alleles and the results are compared with those of the PCA models.

10.1 Introduction

Major histocompatibility complex (MHC) molecules are polymorphic membrane glycoproteins [Zinkernagel 1986]. Human MHCs are also called human leukocyte antigen, often abbreviated to HLA [Clark & Forman 1984]. There are two classes of HLA, class I and class II. Class I HLA is present on most nucleated cells, including the surfaces of lymphocytes, which have 1000 to 10000 HLA molecules per cell [Goust 1993]. Class II HLA is mostly expressed on antigen presenting cells (APC) such as macrophages, B cells and dendritic cells. Partly as a result of their importance in mediating tissue rejection, sequencing has identified MHC proteins as amongst the most polymorphic of all human gene products. According to the international ImMunoGeneTics information system (IMGT), there are over 2000 different HLA class I and II alleles and a significant number of new alleles are discovered every year [Robinson *et al.* 2003]. In Chapter 9, Borghans *et al.* explore the nature and origin of MHC diversity in more detail.

MHCs exhibit much polymorphic amino acid variation, and seemingly trivial alterations in the identity of binding site amino acid residues give rise to differences in peptide selectivity exhibited during peptide binding. Peptide binding assays are the most widely-used way of identifying T cell epitopes and measuring the affinities of peptides binding to MHC. Such assays include direct binding and the quantitative measurement of radio- or fluorescence- labeled peptides bound to the MHC molecules [Chen & Parham 1989, Schumacher & Heemels 1990, Cerottini & Luescher 1991, Christinck & Luscher 1991, Kast & Melief 1991, Mendez-Samperio & Jimenez-Zamudio 1991, Stuber & Dillner 1995, Wauben & van der Kraan 1997, Levitsky & Liu 2000]. Several databases have been set up to store peptide binding affinity data, such as MHCPEP [Brusic *et al.* 1998], MHCBN [Bhasin & Singh 2003], and AntiJen [Blythe *et al.* 2002, McSparron *et al.* 2003, Toseland *et al.* 2005].

Many HLA alleles have been demonstrated to bind peptides with similar anchor residues [Southwood *et al.* 1998]. This has led to the concept of MHC supertypes: the idea that MHCs with distinct sequences can be classified into separate groups, each of which displays equivalent, if not necessarily identical, specificities when binding peptides. The celerity of experimental research will be greatly accelerated if one could identify a procedure able to cluster HLA alleles with similar specificities. Several research groups have sought to classify HLA alleles in this way, using a wide variety of different methods. Examples of such disparate methodologies include sequence analysis [Lawlor & Warren 1991], structural analysis [Chelvanayagam 1997], use of geometrical similarity matrix methods [Cano & Fan 1998], and motif search [Sette & Sidney 1998, Lund *et al.* 2004].

We have recently developed and applied chemometric GRID/CPCA and hierarchical clustering methods to the identification of MHC supertypes [Doytchinova *et al.* 2004b]. Within vaccinology, HLA classification, using bioinformatics methods, can potentially reduce the overall experimental burden by rending unnecessary the individual study of every allele. It can thus accelerate the discovery of both epitope-based vaccines, and other immunotherapies, that are targeted at multiple alleles. In the remainder of this chapter, we will explore attempts, both ours and those of others, to address the problem of finding and populating MHC supertypes.

10.1.1 Evolutionary Analysis

An early attempt to classify MHC molecules is from protein sequence studies [Lawlor & Warren 1991]. Lawlor compared the sequences of 14 gorilla class I MHC alleles with HLA-A, B and C alleles in human and MHC in chimpanzees. Sequences of human, gorilla and chimpanzee MHC alleles are similar but not identical, as most of the polymorphic residues appear in the same region. Also genes at A, B and C locus of gorilla and chimpanzee MHCs are similar to HLA-A, HLA-B and HLA-C, respectively. Phylogenetic trees are generated for A, B and C genes and it is found that HLA-A alleles are divided into five families: A2, A3, A9, A10 and A19. Two divergent groups of HLA-C alleles are found, one containing Cw*0701 and Cw*0702, the other with Cw*0101-Cw*0601 and Cw*1201. HLA-B is the most polymorphic locus in the human HLA genes and no consensus group is found in the study. Based on Lawlor's research, Jakobsen et al. aligned DNA and protein sequences of the HLA-A alleles. The DNA alignment showed that family signatures are not focused on one region but are distributed throughout the sequence. The protein sequence alignment revealed that position 62, 97 and 114 in the binding site are important in the classification [Jakobsen & Gao 1998].

Another HLA grouping based on evolutionary analysis was undertaken by McKenzie et al. in 1999 [McKenzie & Pecon-Slattery 1992]. In their study, phylogenetic trees were built using three methods: maximum parsimony, distance-based minimum evolution and maximum likelihood. Different classifications were carried out, based on either whole protein/nucleotide sequence, sequence of the binding site, or sequence excluding the binding site. Two clusters were found for HLA-A class: one with A1, A3, A9, A11, A36, A*8001 and some of the A19 and the other with A2, A10, A28, A*4301 and the other A19 members. HLA-B and HLA-C did not form any consistent clusters.

10.1.2 Structural analysis

The binding of peptides to MHC molecules is influenced by the interactions between the side chains of bound peptides and the binding pockets within the peptide binding site. In contrast to data driven models, which rely on the accumulation of significant quantities of binding data, an important approach seeks a structural understanding of peptide binding by analysing the structure of MHC receptor binding sites. These allow connections to be identified between different MHC alleles at the functional level. Any significant similiarity apparent between binding sites should also be mirrored in the overall peptide selectivities exhibited by different MHCs. Comparative investigation of such relationships should allow the prediction of similarities in peptide selectivity and the effective grouping of different alleles.

Kurata and Berzofsky studied the interaction of peptide analogs with the MHC binding site and their comcomitant interactions with the T cell receptor (TCR). It was identified that the same peptide can bind to class II allele I-Ed in more than one conformation. Moreover, the change in peptide conformation did not affect the recognition by T cells, indicating that the TCR may interact with different regions of the peptide in different conformations [Kurata & Berzofsky 1990]. Similarly, Gopalakrishnan and Roques simulated the interactions between a peptide and the H-2Kd binding site using the molecular dynamics program AMBER. They found that the binding orientation of the peptide may be dependent on the sequence and structure of the peptide and may be allele specific [Gopalakrishnan & Roques 1992].

In 1996, Chelvanayagam studied binding pockets and grouped HLAs according to the amino acid composition in each pocket [Chelvanayagam 1997]. HLA molecules within one group have the same or similar amino acids in a particular binding pocket are expected to bind to the same peptide. The analysis was used to classify HLA molecules that have not been studied experimentally and thus to predict their binding motif. Although classified separately, groups of HLA-B and C molecules share the same binding specificity with HLA-A if they have the same amino acids in the binding site. The drawback to this form of classification is that since the classification is done according to the residues surrounding one position of the peptide, for a nonamer peptide, the HLA alleles are classified nine times and the same allele is often found in different groups in different classifications. A similar study has been carried out by Zhang et al., in which the binding pockets of class I MHC are classified into families by modelling the structures of MHC-peptide complexes using crystal structures as templates. Five families were defined according to specificities in the pocket B, and three families were defined based on specificities inside pocket D. Three more families were also defined for alleles with a joint specificity of pocket C and D [Zhang et al. 1998].

10.1.3 Geometrical similarity matrix

Cano et al. clustered the HLA-A and HLA-B alleles by constructing similarity matrices [Cano & Fan 1998]. MHC molecules were compared in a geometric space, where each amino acid occupied one dimension. The similarities among chemical properties of the twenty amino acids such as polarity and charges were compared and the results were stored in an amino acid similarity matrix. Another reference matrix, the binding affinity matrix was generated by calculating the flexibility of each amino acid side-chain at each position of the peptide. The similarity among MHC alleles was measured using both experimental peptide elution data and by comparing the alleles using the similarity matrix. The method identified three clusters as listed in table 10.4.

10.1.4 Sequence and binding motif approach

Another way of classifying HLA molecules is to group alleles with similar binding motifs together. Class I HLA molecules have been classified into superfamilies b Sette and Sidney using this approach [Sidney et al. 1996a]. Sidney et al. defined fou supertypes by examining reported cross-reactive epitopes [Sidney et al. 1996a]. They then compared the sequences corresponding to the MHC binding pockets B and F. Experimentally confirmed binding motifs of the alleles were also examined, and those with similar motifs are grouped into one supertype [Southwood et al. 1998]. The supertypes identified in the paper are listed in table 10.4. The same group later published review papers in which the four supertypes were revised. A*0207 was added to the A2 supertype and B*1508 and B*5602 were added to the B7 supertype[Sette & Sidney 1998, Sette et al. 1989]. Recently A*2902 and A*3002 are added to the A1 supertype [Sidney *et al.* 2005]. Sette and Sidney carried out further analysis in 1999 and defined a total of nine supertypes including the previously defined supertypes [Sette *et al.* 1989]. The nine supertypes were estimated to cover 99% of the world population (table 10.4) [Sette & Livingston 2001]. Table 10.4 lists the supertypes and alleles within each supertype.

Based on Sette's study, [Lund *et al.* 2004] classified HLA-A and B molecules using specificity matrices. The nonamer ligands of all HLA-A and B molecules were collected from SYFPEITHI and MHCPEP and were aligned. The frequencies of each amino acid at each position were summarised as a matrix, and this was used for a cluster analysis. The resulting HLA superfamilies were organised into a consensus tree. In their results, the A26 alleles were separated from the A1 cluster described by Sette, and a new B8 superfamily was defined. The other superfamilies were the same.

It should be noted that class II HLA molecules have also been classified using a sequence approach. Chelvanayagam defined the HLA-DR roadmap by allele binding specificities and the polymorphic residues inside the binding site. The important residues were identified by studying the crystal structures of known HLA-DR-peptide complexes [Chelvanayagam 1997]. HLA-DP [Castelli & Buhot 202] and DQ [Baas & Gao 1999] supertypes have also been defined based on a combination of binding studies to define motifs together with structural modelling of the peptide-MHC complexes. Reche and Reinherz used multiple sequence alignment to find important residues in 774 class I and 485 II HLA molecules. Consensus sequence patterns were obtained for the binding sites of HLA-A, B, C, DP, DR and DQ groups [Reche & Reinherz 2003].

10.2 GRID/CPCA AND Hierarchical Clustering

Class I HLA supertypes have been defined by Doytchinova et al. using GRID/CPCA combined with hierarchical clustering based on comparative molecular similarity indices (CoMSIA) fields. The GRID program identifies the energetically favoured or disfavoured regions on molecules with known three-dimensional structures. Many molecules can be included in one calculation [Cruciani & Watson 1994]. A selection of chemical probes is included in the program; each probe represents atoms or functional groups with different properties. GRID calculates the interaction energy between selected chemical probes and each of the molecules. Molecular interaction fields (MIFs) between different chemical probes and a set of different HLA proteins were calculated in GRID, and these were used to build PCA and CPCA models in GOLPE.

The program Generating Optimal Linear Partial least square Estimations (GOLPE) [Cruciani & Watson 1994] has one module for PCA calculation. PCA decomposes a matrix X into two smaller matrices: the scores matrix T and the loading matrix P', which explain the overall variance of the X matrix. The scores matrix contains a few variables M, that is, the principal components (PC), which can be used to describe the observations. The loading matrix reveals the relationship between the variables in the original matrix and the principal components. Plots of the observations in the multidimensional space are called the scores plot, which identifies similarities and differences within the observations and groups them accordingly, while the loading plot relates the original variables to the PCs and identifies variables that are important in distinguishing groups of observations.

When more than one probe is used in the GRID calculation, the data generated by different probes are grouped into blocks, and they are often analysed by hierarchical PCA methods such as consensus PCA (CPCA). The advantage of CPCA compared to PCA is that it compares the relative importance of each block in the calculation and makes a 'consensus' clustering of the objects. CPCA uses the same underlying principle as PCA: a CPCA model tries to explain the overall variance of the original data matrix. The algorithm used in CPCA is an adaptation of the NIPALS algorithm used in PCA [Wold & Hellberg 1987]. Like PCA, CPCA calculates the principal components and gives the scores and loading matrix. In addition, CPCA also calculates the importance of each data block. It calculates the scores and the loading matrix for each probe used, and also returns the weight matrix that can illustrate the contribution of each probe to the overall scores.

Cluster analysis is a process of grouping of observations into subsets or clusters, the grouping is dependent on the similarities between each observation. Commonly used clustering methods are hierarchical clustering and k-means clustering, etc. Hierarchical clustering based on the agglomerative method is used in HLA classification, in which observations are separated into n clusters at the beginning of the clustering, each cluster contains one observation. The distance between two clusters is proportional to the similarities of the observations. Clusters with the shortest distance are merged and the distance between the new cluster and others is computed. These steps are repeated until there is only one cluster left. The cluster analysis used is as implemented in Sybyl6.9, complete linkage clustering is used in distance computation, in which the maximum distance between data points in two clusters is used. The clustering process makes use of the five molecular interaction fields calculated by CoMSIA.

A total of 783 class I HLA sequences were found in the IMGT/HLA database and were included in the classification. The sequences were selected on the basis of the differences at protein sequence level. The classification is defined according to the scores plots of the CPCA model and the dedrograms obtained from hierarchical clustering. The scores plot showed the clustering of the HLA alleles, whereas the loading maps highlighted regions in the peptide binding site that contributed significantly in clustering different superfamilies. Amino acid fingerprints are identified from the loading plots of the CPCA models, the fingerprints are the basis of the classification and can be used for future classification.

Three HLA-A clusters are defined from the scores plot of the CPCA model and hierarchical clustering, A2, A3 and A24 (Fig 1, Fig 2). In the scores plot, the first component of the CPCA model separated A23 and most of the A24 molecules on the left, with negative PC1 scores, from the rest of the HLA-A molecules. The second principal component separated the HLA-A*1, A*11, A*25, A*26, A*29, A*03, A*31, A*32, A*33, A*34, A*36, A*66, A*68 and A*74 families with positive PC2 scores from the others. Therefore, the CPCA analysis revealed three clusters as demonstrated in the 3D scores plot: the A3 cluster on the top right of figure 10.4, including the alleles A*01, A*03, A*11, A*25, A*26, A*29, A*30, A*31, A*32, A*33, A*34, A*36, A*4301, A*66, A*74 and A*8001. Most of the A*68 alleles (except A*6802 and A*6815, which were in the A2 cluster) were also included in the A3 family. The A24 cluster is on the top left of the figure including the A*23 and A*24 alleles. The A2 cluster is at the bottom of the figure, with most of the A*02 alleles. Other alleles in the A2 cluster were A*57, A*6802, A*6815, A*6823 and A*6901 (Table 10.4).

Hierarchical clustering analysis using CoMSIA fields also defined three clusters (Fig. 2). The cluster on the left includes HLA alleles A*02, A*25, A*26, A*3401, A*3405, A*4301, A*66, A*6802, A*6815, A*6823 and A*6901. This cluster was the A2 cluster. The A24 cluster was well distinguished and included A*23 and A*24 alleles. Finally, the A3 cluster included A*01, A*03, A*11, A*29, A*30, A*31, A*32, A*33, A*36. Some A*34 and A*68 alleles, A*74 and A*8001 were also in this cluster.

The loading plot of the HLA-A model highlighted position 9, 97, 114 and position 116 (Fig 10.3). Sequence alignment of HLA-A molecules showed that most of the A24 alleles had dominant polar amino acid Ser at position 9, while the A3 molecules had aromatic amino acids Tyr or Phe at position 9.

The scores plot of the HLA-B CPCA model reveals that the HLA-B molecules are divided into three clusters (Fig 10.4, Table 10.4): B7 (B*07, B*08, B*14, some B*15, B*18, B*35, B*3705, B*3904, B*41, B*42, B*45, B*48, B*50, B*55, B*56, B*6701, B*6702, B*7301, B*78, B*81, B*82 and B*83), which is on the left of the Y axis, B27 (B*27, B*37, B*38, B*4013, B*4019 and B*4028) in the top right corner of the plot, and B44 (B*13, B*44, B*47, B*49, B*51, B*52, B*53, B*5607, B*57, B*58 and B*5901). Similar clusters are found using hierarchical clustering method, in which three clusters (B7, B27 and B44) are identified (Fig 5).

The PC1 loading plot showed that two areas were important in the classification (Fig 10.6). Position 63 and 66 were inside pocket A and B. Position 66 was conserved while position 63 was polymorphic with two amino acid variations Glu and Asn. The other important area in the loading plot was around position 77 and 81 in the pocket F. Asn, Ser and Asp were found at position 77, and Leu and Ala at position 81.

Results the HLA-C model is in figures 10.7 and 10.8, in which HLA-C molecules were divided into two clusters. Cw*01, Cw*03, Cw*07, Cw*08, Cw*12 and Cw*16 are grouped into one cluster, and Cw*02, Cw*03, Cw*04, Cw*05, Cw*06, Cw*15, Cw*17 and Cw*18 are in the second cluster. Some of the Cw*03, Cw*07 and Cw*12 are also grouped into the second cluster. The first cluster is named C1 and the second cluster is named C4. The result from hierarchical clustering gave nearly identical groups, with only eight amino acids mis-placed Cw*0308, Cw*0310, Cw*0701, Cw*0706, Cw*0716, Cw*0718, Cw*1208 and Cw*1404 (Table 10.4).

The PC2 loading plots showed that positions 70, 74, 77 and 81 of the HLA-C molecules are involved in the classification (Fig 9). Among the HLA-C molecules, only position 77 was polymorphic. The amino acids presented at this position were Ser and Asn. The molecules in the C4 class all have Asn at position 77. The ones in the C1 cluster, on the other hand, all have serine at this position. As Asn is more polar than serine, they are more favoured for interaction with polar probes and hydrogen-bond formation.

Class I HLA classification using GRID/CPCA and hierarchical clustering based on CoMSIA fields exhibit, on average, a 77% consensus. HLA-A classification by both methods was 88% identical. HLA-B classification by the two methods gave a slightly lower consensus (68%), which may be because the group had the largest number of molecules among the three (447 HLA-B alleles) and the binding site consisted of more amino acids. The classification of the cluster B27 was debatable, as most of the molecules in the B27 cluster, as defined by hierarchical clustering, were in the B7 cluster in the CPCA model. The HLA-C classification gave the best agreement using the two methods (93% consensus). Only 8 molecules were classified into different subtypes by the two methods. Molecules that have been classified into different clusters by the two methods were considered as outliers as it was not possible to classify them properly into clusters. They require future re-classification using other, more sensitive techniques. A closer look at the protein sequence level showed that these outliers do not significantly resemble the classified alleles. For example, A*2501 - A*2503 alleles had Tyr at 9 and Asp at 116, which were identical as A*11 alleles, but they also had Glu at position 114 like the A*31 and A*32 alleles.

The GRID/CPCA procedure grouped all class I HLA-A, B and C alleles into several supertypes. Of these alleles, A*0201, A*0202, A*0204, A*0206 and A*0207 had been grouped into the A2 supertype by binding studies [del Guercio & Sidney 1995, Southwood et al. 1998, Sudo & Kamikawaji 1995, Sidney et al. 1996a, Sidney et al. 1996b] and motif studies [Rammensee et al. 1999]. All these alleles were grouped into the A2 supertype in the GRID/CPCA study with the exception of A*0204, which, like the A3 alleles, possessed Met at position 97 and was classified as belonging to the A3 family. A*0204 differed from A*0201 by having one amino acid mutation Arg -> Met at position 97. Met97 is inside pocket F. The side chain of Met97 is smaller compared with Arg, therefore increasing the volume of pocket F. However, the A*0204 binding motif (L_2L_9) was closer to A*0201, therefore it was possible that A*0204 is an outlier from the A3 superfamily. The previously classified A2 supertype also included A*6801 and A*6901, which were in the A2 superfamily in the present study.

Apart from the A2 supertype, other HLA-A supertypes are less well studied. There were three more HLA-A families in Sette's classification, the A1 superfamily (A*0101, A*2501, A*2601, A*2602 and A*3201), the A3 superfamily (A*0301, A*1101, A*3301, A*3101 and A*6801) and A24 superfamily (A*2301, A*2402, A*2403, A*2404, A*3001, A*3002, A*3003). The A1 and A3 families were grouped into the A3 superfamily in the GRID/CPCA analysis. The A*23 and A*24 alleles were in the A24 superfamily, but A*3001, A*3002 and A*3003 were placed in the A3 superfamily. Our work was also compared with the classification by Lund et al., which produced a set of five distinct HLA-A clusters (A1, A2, A3, A24, A26) using both motif information and binding site structure analysis [Lund *et al.* 2004]. The A1, A3 and A26 cluster in Lund's classification were grouped into the A3 superfamily in the present classification, although the A2 and A24 families in the two analyses were in good agreement.

HLA-B7 (B*07, B*35, B*51, B*53,B*54, B*55, B*56, B*67, B*78), B27(B*1401-02, B*1503, B*1508, B*1509, B*1510, B*1518, B*2701-08, B*3801, B*3802, B*3901-04, B*4801-02, B*7301) and B44 (B*37, B*4001-2, B*4006, B*41, B*44, B*47, B*49, B*50) families have been previously classified and tested in many binding experiments [Southwood *et al.* 1998, Sidney *et al.* 1996a, Doolan & Hoffman 1997, Lamas *et al.* 1998, Sidney *et al.* 2003]. Most of the B7 alleles in Sette's classification were in the B7 cluster defined by GRID/CPCA, apart from B*51 and B*53, which were in the B44 cluster. Alleles in the B7 and B44 family of Sette's classification were found scattered within the B7, B27 and B44 superfamilies in the present analysis. In Sette's classification two more clusters B58 (B*1516-17, B*5701-02, B*5702, B*5708) and B62 (B*1301, B*1302, B*1501, B*1502, B*1506, B*1512-14, B*1519, B*1521, B*4601, B*4652) were defined. Molecules in the B62 cluster of Sette's classification were located in either the B7 or the B44 superfamilies in the GRID/CPCA analysis. The B58 cluster in Sette's classification can be found in the B44 cluster in the present study. Compared with Lund's classification (B7, B8, B27, B44, B58, B62), the B8 cluster was included in the B7 supertype and alleles in the B58 and B62 cluster were in the B7 or B27 cluster in the current analysis.

Although there is no previous HLA-C classification available for comparison, we can nonetheless make the interesting observation that the NK cell inhibitor receptor KIR2DL can be divided into two groups based on their HLA-C specificity. KIR2DL1 recognised HLA-Cw*2, Cw*4, Cw*5 and Cw*6, all of which possessed Asn77, whereas KIR2DL2 recognised HLA-Cw*1, Cw*3, Cw*7 and Cw*8, which had Ser at position 77 [Fan & Long 2001]. The specificity of KIR2DL was in agreement with our HLA-C classification, which suggested that position 77 was important in substrate binding: HLA-C molecules with the same residue at position 77 tend to share the same specificity.

A hierarchical clustering study based on HLA binding pockets has also been carried out, in which HLA-A molecules are classified according to molecular specificities of each of the six binding pockets. Three clusters have been defined according specificities of pocket A (table 10.4. The first cluster is consisted of A1 (A*01 and A*11), A*0208, A*16, A*20, A*29 and A*56, most of the A3 (A*03, A*30 A*31 A*32 and A*36), A*6810, A*6813, A*6814 alleles, A*7401-09 and A*8001. The second cluster includes A*25, A*26, A*33, A*34 and most A*68 alleles. Most of the A*02 alleles are present in the third cluster together with A*23, A*24, A*29, A*4301, A*6601, A*6604, A*6801-09, A*6815-23, A*6901. Two residues lining pocket A are identified to be important in the classification: position 63 and 66. Alleles in the first and second clusters all have polar amino acid Asn at position 66, while alleles in the third cluster have basic amino acid Lys at this position. Alleles in the first cluster have acidic Glu63 but those in cluster 2 have Asn63.

Two clusters are identified from the hierarchical clustering based on CoMSIA fields of pocket B (table 10.4). Cluster B1 has A*01, A*0201-33, A*0236-60, A*23, A*24, A*25, A*26, A*30, A*31, A*32, A*33, A*4301, A*6803-05, A*74 and A*8001. The second cluster includes A*0234-35, A*2424, A*29, A*34, A*0301-10, A*1101-04, A*6601-01, A*6601-02, A*6604, A*68 and A*6901. The classification is based on a single amino acid at position 70. Alleles with basic amino acid His70 are in cluster 1, while those with Gln70 are in cluster 2.

The HLA-A alleles are separated into two clusters according to specificities in pocket C (table 10.10). The first cluster include mainly A*01, A*0211, A*0235, A*0248, A*03, A*11, A*23, A*24, A*25, A*26, A*29, A*30, A*31, A*32, A*34, A*36, A*66, A*6801-04, A*6806-14, A*6816-19, A*6821-23, A*6901 and A*7401-05, A*7408-09, while most of A*02, A*1106, A*2428, A*2430, A*2603, A*2606, A*3009, A*6805, A*6815, A*6820, A*7406 and A*8001 are in the second cluster. One main feature of pocket C is position 74, alleles within cluster C1 all possess acidic amino acid Asp74, whereas alleles within cluster C2 have basic His74.

There are four clusters defined for pocket D (table 10.11). Cluster D1 includes all A*01 except A*0106, A*0249, A*1108 and A*36. The second cluster also has a small group of alleles including A*0310, A*1101-07, A*2417, A*2905, A*3402 and A*6801, A*6803-05, A*6807-23. The third cluster is consisted of A*0106, A*0201-40, A*0242-48, A*0250-51, A*0253-60, A*0301, A*0304-09, A*2301-09, A*2402-16, A*2418-38, A*2901-07, A*3103-06, A*3204, A*3402-04, A*6802, A*6808, A*6815, A*6901 and A*8001. The fourth cluster includes A*25, A*26, A*3004, A*3006, A*3202, A*3401 and A*3405,A*4301 and A*66. A*0241, A*0252, A*3001-03, A*3007-12, A*3101-02, A*3105, A*3107-09, A*32, A*33 and A*74 are grouped in the fifth cluster. Amino acids fingerprint for this classification is consisted of position 114 and 156. Alleles with basic amino acids such as Arg or His at position 114 are grouped into the first three cluster and alleles with Phe or Gln are in the last cluster. The first three clusters are further separated by polymorphism at position 156. Alleles with Arg, Gln/Trp and Leu are grouped into cluster 1, 2 and 3, respectively.

Four clusters are identified for pocket E (table 10.12). The first cluster is composed of serotype A*01, A*03 (A*0302, A*0307, A*0310), A*1101-02, A*1104-07, A*1109-14, A*2612, A*2618, A*29, A*31, A*32, A*33, A*36, A*6801, A*6803-05, A*6808-11, A*6813, A*6814, A*6816, A*6818-23, A*74 and A*8001. The second cluster is consisted of A*0301, A*0304-06, A*0308-09, together with A*1103, A*1108, A*2504, A*2608, A*2905, A*3204 and A*34. The third cluster includes most of A*02, A*23, A*24, A*30, A*6802, A*6806, A*6807, A*6815, A*6817 and A*6901. The last cluster has A*0203, A*0213, A*0226, A*0238, A*2418, A*25, A*26, A*3401, A*3405, A*4301 and A*66. The classification can be explained by two amino acids at position 116 and 152. Alleles with acid Asp at position 116 are grouped in cluster 1 and 2, while alleles in cluster 3 and 4 have bulky amino acid His or Tyr.

Only two clusters are found for pocket F (table 10.4). The first one consists of A*01, A*0301-10, A*11, A*2417, A*2501-04, A*26, A*29, A*31, A*32, A*33, A*34, A*36, A*4301, A*6801, A*6803, A*6808-14, A*6816, A*6818-23, A*74 and A*8001, and the second cluster has A*02, A*23, A*24, A*2602, A*30, A*6802, A*6806-

07, A*6815, A*6817 and A*6901. One position is identified to be important in the clustering process, position 116. Alleles in the first cluster possess negatively charged amino acid Asp at position 116, while those in the second cluster have aromatic Tyr at this position.

Compared with classifications on the whole binding site, the pocket classification considers one pocket at a time, therefore one allele may be classified into different groups in different classifications. For example, pocket B of some A*02 and A*03 alleles favour aliphatic amino acids therefore they are in the same group. However, pocket F of the A*03 alleles favours charged amino acids while A*02 alleles accept small aliphatic amino acids and they are in different clusters. In contrast to the three amino acids fingerprint from whole binding site classification, eight amino acids (position 63, 66, 70, 74, 114, 116, 152, 156) are identified in the pocket classification, indicating that more amino acids of the binding site are important in peptide specificity. However, as peptide binding motifs are only available for a small group of alleles, therefore the current classifications can not be validated.

10.3 Class II HLA Classification

Class II HLA alleles have also been classified by clustering. Doytchinova et al. applied hierarchical clustering using CoMSIA fields and non-hierarchical clustering based on z-scores. The hierarchical clustering follows the same procedure as class I classification described above. Nonhierarchical clustering uses five z descriptors to describe hydrophobicity, steric bulk, polarity and electronic effects of the HLA molecules. K-means clustering is applied to the set and the initial number of k seeds is equal to the number of clusters obtained from the hierarchical clustering. The known crystal structures of class II HLA are used as templates in homology modelling. Like the class I HLA classification, the class II alleles are grouped into twelve families. HLA-DR alleles are classified into DR1, DR3, DR4, DR5 and DR9 supertypes (Table 10.4). Hierarchical clustering groups DRB1*01-11, DRB1*1501-11 and DRB1*1601-08 in DR1, DRB1*0701-07, DRB1*0301-25, DRB3*0101-10, DRB3*0301-03, DRB1*0422 and DRB1*1107 in DR3, DRB1*0401, 03-48, DRB1*1113, 17, 26, 34, 42, DRB1*1309, DRB1*1401-48, DRB1*1001, DRB4*0101-06 in DR4, DRB1*0402, 12, 15, 25, 36, 37, 47, DRB1*1101-47, DRB1*1201-09, DRB1*1301-62, DRB1*1403, 16, 22, 25, 27, 40, DRB1*0801-25 in DR5, DRB1*0901-02, DRB5*0101-12, DRB5*0202-05 in DR9. Nonhierarchical clustering classifies DRB1*01-11, DRB1*1501-11 and DRB1*1601-08 in DR1, DRB3*0101-10, DRB3*0201-18, DRB3*0301-03, DRB1*1333, DRB1*1447 in DR3, DRB1*0401, 03-48, DRB1*1107, 13, 17, DRB1*1401-48, DRB3*0215, DRB1*1001, DRB4*0101-06, DRB1*0301-25 in DR4, DRB*0402, 15, 25, 36, 47, DRB1*1101-47, DRB1*1201-09DRB1*1301-62, DRB1*1403, 16, 17, 21, 22, 24, 25, 27, 29, 30, 37, 40, 41, 48, DRB1*0801-25, DRB5*0101-12, DRB5*0202-05 in DR5, DRB1*0901-02, DRB1*0701-07 in DR9. The DQ alleles are divided into DQ1 (DQB1*0501-03, DQB1*0601-21), DQ2 (DQB10201-03) and DQ3 (DQB1*0301-13, DQB1*0401, DQB1*0402 and DQA1*0301-03) (table 10.4). Four families are discovered for HLA-DP alleles, they are DPw1, DPw2 (DPB1*0201, DPB1*0202, DPB1*32, 33, 41, 46,

47, 48, 71, 81, 86, 95), DPw4 (DPB1*0401 and 0402, DPB1*15, 18, 23, 24, 28, 34, 39, 40, 49, 51, 53, 59, 60, 62, 66, 72, 73, 74, 75, 77, 80, 83, 94, 96, 99) and DPw6 (Table 10.4). In hierarchical clustering, DPw1 cluster includes DPB1*0101, 0301 and 0501, DPB1*14, 20, 25, 26, 27, 31, 35, 36, 38, 45, 50, 52, 56, 57, 63, 65, 67, 68, 70, 76, 78, 79, 84, 85, 87, 89, 90, 91, 92, 97 and 98. DPw6 cluster has DPB1*0601, and DPB1*08, 09, 10, 11, 13, 16, 17, 19, 21, 22, 29, 30, 37, 44, 54, 55, 58, 69, 88, 93. Nonhierarchical clustering classifies DPB1*0101, DPB1*0301, DPB1*11, 14, 25, 26, 31, 35, 45, 50, 52, 56, 57, 65, 67, 68, 69, 70, 76, 78, 79, 84, 90 and 92 into DPw1 and DPB1*0501, 0601, DPB1*08, 09, 10, 13, 16, 17, 19, 20, 21, 22, 27, 29, 30, 36, 37, 38, 44, 54, 55, 58, 63, 85, 87, 88, 89, 91, 93, 97 and 98.

HLA-DR classification is due to polymorphism at position 9, 70 and 74 of the beta chain. Alleles with Trp9 are found in DR1 and those with Lys/Gln9 are in DR9. Alleles with Glu9/Asp70 are grouped in DR5. Alleles that have the combination of Glu9/Gln70 and Gln/Arg74 are in DR3 and those with Glu9, Gln/Arg70 and Glu/Ala74 are in DR4. Only two positions are responsible for HLA-DQ classification, position 71 and 86 of the beta chain. All alleles with Val86 are grouped in DQ1 cluster, while those with Glu86/Lys71 are in DQ2 and those with Glu86 and Thr/Asp71 are in DQ3. The DP classification is mainly based on polymorphism at position 69 and 84 of the beta chain. Alleles with Asp84 are grouped into DPw1/2 and those with Gly/Val84 are in DPw4/6. Alleles in DPw1/2 are separated by amino acid differences at position 69, those with Lys69 are grouped into DPw1 and those with Glu69 are in DPw2. DPw4 and DPw6 are separated by Lys69 and Glu69, respectively. The classification by hierarchical and non hierarchical clustering have a consensus of more than 85%.

A possible limitation of the GRID/CPCA technique is that it relies on accurate molecular structures. As the number of unique HLA sequences greatly exceeds the number of unique solved MHC crystal structures, protein structures used in these studies have been derived by homology modelling. Although HLA molecules are structurally similar, there may be some differences in the binding site conformation, and potentially this limitation is confounding. However, compared with HLA classifications based on peptide binding motifs, chemometric methods have some advantages (table 10.4). GRID/CPCA and hierarchical clustering are more flexible as they only require the sequence information of molecules, therefore all the HLA molecules available, whether or not they have been studied experimentally, can be classified. In contrast, the motif-based method can only classify that small number of HLA molecules with sufficient binding data. Most of the motifs include only anchor residues of the peptide, therefore only part of the peptide binding site interaction is studied. GRID/CPCA method takes the whole binding site into consideration and identifies important positions involved in the classification. Also, motif based classifications use a haphazard mismash of differently derived experimental binding data, which may be biased and inconsistent. GRID/CPCA classification only uses sequence information, albeit manifest as homology modelled 3D structures, and thus minimises data inconsistency.

10.4 Discussion

As we have seen, HLA alleles can be classified into supertypes using only their sequence information. Some have sought insights into MHC supertypes from a sequence perspective, others from the perspective of structural data. The classification we outlined here identifies crucial, cluster determining differences at several important positions in the binding site. These positions are the HLA 'fingerprints'. The HLA-A fingerprint includes position Phe/Tyr9, Arg97, His114 for A2 supertype, Ser9 and Arg97 for A24 and Ser/Thr9, Ile/Arg97, Glu114 and Asp116 for A3 supertype. The HLA-B fingerprint is Asn63 and Leu81 for B7, Glu63 and Leu81 for B27 and Ala81 for B44. The HLA-C fingerprint is Ser/Gly77 for C1 and Asn77 for C4 supertype. The important positions for Class II DR supertype classification are position 9, 70 and 74, and position 71 and 86 are identified as the fingerprint for DQ clusters. These HLA fingerprints enable us to group any new HLA molecules into supertypes, accelerating HLA function characterisation and help to define the peptide binding motif for the molecule. Also, the HLA supertype classification allows immunologists to use similarities in sequence and structure to make educated guesses about peptide binding specificity that will help in identifying good MHC binders and testable potential epitopes.

The veracity and pace of vaccine identification would be enhanced greatly were one able to group HLA alleles into effective supertypes. An accurate and sufficiently extensive classification would render experimental work much more efficient, since one could look at a few supertypes rather than at thousands of separate alleles. This would thus greatly expedite the discovery of epitope based vaccines targeted at multiple alleles. Supertype definitions have already shown utility in epitope based vaccine discovery. Epitopes taken from hepatitis B virus infected patients have been shown to cross react with alleles in the A2, A3 and B7 superfamilies [Bertoni & Sidney 1997]. Epitopes isolated from Epstein-Barr virus reacted with several alleles of the B*44 family [Khanna & Burrows 1997]. Epitopes have been shown to cross-react with the A24 family [Burrows & Elkington 2003]. Many viral and tumour antigen derived vaccine candidates have also been shown to be able to bind multiple alleles. Sette et al. predicted 223 potential cancer peptides of CEA, Her-2/neu, P53 and MAGE antigens using a T cell epitope prediction algorithm, among which 115 were cross-reactive with peptides of the A2 supertype. 43 peptides were tested for immunogenecity and 73% were positive [Sette & Livingston 2001]. Recently a protein sequence scan has been carried out to search for T cell epitopes within the sequence of the SARS virus, based on the nine HLA supertypes in Sette's analysis [Sylvester-Hvid et al. 2004]. Fifteen predicted epitopes for each supertype were identified and tested experimentally: 75% of the predicted epitopes were found to be high affinity peptides ($IC_{50} < 500nM$) and about 112 candidate epitopes were obtained.

All supertypes are theoretically derived, even the *experimental* supertypes promulgated by Sette. His supertypes were derived from the comparison of *binding motifs*. Motifs are, at best, an inadequate description of peptide specificity. Possessing a certain verisimilitude, they can only give rise to a partial and largely incomplete definition of supertypes, limited by the lack of data for most MHC molecules. Structural supertypes represent an encouraging solution to this problem, unencumbered by limitations imposed by the availability of binding data. Modern methods in particular, such as the GRID/CPCA method we outline here, allow us to propose supertype definitions solely based on sequence and structural data. In one seamless movement we can preogress from HLA sequencing to structure to supertype classification to binding specificity to epitope prediction. The clinical potential of such a process are tantalizing. Moreover, the same fundamental methodology can be used to address the issue of identifying benign HLA mismatches in tissue rejection, such as kidney transplants, bone marrow donation, and the like. Such problems require a robust, reliable and, preferably, transparent measure of structural similarity between HLA molecules in order to suggest which pairs of alleles will present the same peptides or be equally invisible to antibody surveillance. The GRID/CPCA method offers the possibility of effectively addressing all these problems and many more. All that is required is the requisite investment of time and resource in order to realize this potential coupled, of course, to the willingness of experimentalists to exploit these techniques.

Clusters	Alleles
1	HLA-A3, HLA-A11, HLA-31, and HLA-33
2	HLA-B7, HLA-B35, HLA-B51, HLA-B53 and HLA-B54
3	HLA-A29, HLA-B44 and HLA-B61

Table 10.1. Three clusters are identified by Cano's geometrical similarity matrix analysis

Supertype	Alleles
A2	A*0201-06, A*6802, A*6901
A3	A*0301, A*1101, A*3101, A*3301, A*6801
B7	B*0702-5, B*3501-3, B*5101-5, B*5301 B*5401, B*5501-2, B*5601, B*6701and B*7801
B44	B37, B41, B44, B45, B47, B49, B50, B60, B61

Table 10.2. The four supertypes defined by Sette's group.

Supertype	MHC alleles
A1	0101, 2501, 2601, 02, 2902, 3001, 3201
A2	0201-07, 6802, 6901
A24	2301, 2402-04, 3001-03
A3	0301, 1101, 3101, 3301, 6801
B44	37, 4001,4002 4006, 41, 44, 45, 47, 49, 50
B27	1401 − 02, 1503, 09, 10, 18, 2701 − 08, 3801, 02, 3901 − 04, 4801, 02, 7301
B7	07, 35, 51, 53, 54, 55, 56, 67, 78
B58	1516, 17, 5701, 02, 58
B62	1301 − 02, 1501, 02, 06, 12, 13, 14, 19, 21, 4601, 52

Table 10.3. Nine supertypes defined by Sette and Sidney

Supertype	Motif-based	Hierarchical clustering	Consensus PCA	Fingerprint
A1	0101 2501 2601,02 3201			
A2	0201-07 6802 6901	0201 – 60 2501 – 04 2601 – 18 3401, 05 4301 6601 – 04 6802, 15, 23 6901	0201 – 60 without 04, 17, 57 6802, 15, 23 6901	Tyr9/Phe9 Arg97 His114 and Tyr116
A24	2310 2402-04 3001-03	2301-09 2402 – 38	231-09 2402 – 38	Ser9 Met97
A3	0301 1101 3101 3301 6801	0101 – 09 0301 – 10 1101 – 14 2901 – 07 3001 – 12 3101 – 09 3201 – 07 3402 – 04 3601 – 04 6801 – 22 without 02, 15 15 7401 - 09 8001	0101 - 09 0301 – 10 1101 – 14 2501 – 04 2601 – 18 2901 – 07 3101 – 09 3201 - 07 3301 – 06 3401 – 05 3601 – 04 4301 6601 - 04 6801 – 23 without 02, 15 7401 – 09 8001	Tyr9/Phe9/Ser9 Ile97/Met97 Glu114 and Asp116

Table 10.4. A list of HLA alleles included in each cluster in the scores plot. For simplicity only the beginning and the end of the alleles were listed. For example, A*0201 – 60 meant that all sixty alleles from A*0201, A*0202, A*0203 ... to A*0260 were included in the cluster, etc. The amino acids used to define each cluster are shown in the last column.

Supertype	Motif-based	Hierarchical clustering	Consensus PCA	Fingerprint
B44	37	0802	0802	Ala81
	40012	1301 – 1311 without 09	1301 – 1311 without 09	
	4006	1513, 16, 17, 23, 24, 36, 67	1513, 16, 17, 23, 24, 36, 67	
	41	1809	1809	
	44	2701, 02	3805	
	45	3801 – 3809 without 03		
	47	4013, 4019		
	49	4402 – 4433 without 09, 31		
	50	4704	5101 – 34	
		4901 – 03	5201 – 05	
		5101 – 34	5301 – 09	
		5201 – 05	5607	
		5301 – 09 without 03, 05		
		5701 – 09	5701 – 09	
		5801 – 07	58 – 07	
		5901		
B27	1401 – 02	0713	0727	Glu63
	1503, 09, 10, 18	1309	2701 – 25 without 08, 12, 18	Leu81
	2701 – 08	1501 – 1575 without these in B7	3701 – 04	
	3801. 02	and B44	3801 – 09	
	3901 – 04	1812	4013, 19, 28	
	4801, 02	2703 – 2725		
	7301	3513, 16, 28		
		3701 – 05		
		3803		
		3902, 08, 13, 22, 23		
		4001 – 44 without 08, 13, 19, 25		
		4101 – 06		
		4409, 31		
		4501 – 06		
		4601, 02		
		4701 – 03		
		4801 – 07		
		5001 – 04		
		5608		
		6702		
		7805		

Table 10.5. A list of the HLA-B molecules in the scores plot (Part 1).

Supertype	Motif-based	Hierarchical clustering	Consensus PCA	Fingerprint
B7	07	0702 – 31 without 13	0702 – 31 without 0727	Asn63
	35	0801 – 17 without 02	0801 – 17 without 02	Leu81
	51	1401 – 06	1309	
	53	1502, 08, 09, 10, 11, 15, 18, 21, 29, 37,	1401 – 06	
	54	44, 51, 52, 55, 64, 72	1501 – 75 without 13, 16, 17	
	55	1801 – 18 without 09, 12	23, 24, 36, 67	
	56	2723	1801 – 18 without 09	
	67	3501 – 45 without 13, 16, 28	2708, 12, 18	
	78	3901 – 27 without 02, 08, 13, 22,	3501 – 45	
		23	3705	
		4008, 25	3904	
		4201 – 04	4101 – 06	
		4806	4201 – 04	
		5303, 05	4409	
		5401, 02	4501 – 06	
		5501 – 12	4601, 02	
		5601 – 11 without 08	4702	
		6701	4801 – 07	
		7301	5001 – 04	
		7801- 04	5401, 02	
		8101	5501 – 10	
		8201, 02	5601 – 11 without 5607	
		8101	6701, 02	
			7301	
			7801 – 05	
			8101	
			8201, 02	
			8301	
B58	1516, 17			
	5701, 02			
	58			
B62	1301 – 02			
	1501, 02, 06, 12,			
	13, 14, 19, 21			
	4601			
	52			

Table 10.6. A list of the HLA-B molecules in the scores plot (Part 2).

Supertype	Motif-based	Hierarchical clustering	Consensus PCA	Fingerprint
C1	No data	0102 – 09 0302 – 16 without 07, 08, 10, 15 0702 – 18 without 01, 06, 07, 09, 16, 18 0801 – 09 1202 – 07 without 04, 05, 08 1402 – 05 1601, 04	0102 – 09 0302 – 16 without 7, 15 0701 – 18 without 07, 09 0801 – 09 1202 – 08 without 04, 05 1402 – 05 1601, 04	Ser77/Gly77
C2	No data	0202 – 06 0307, 08, 10, 15 0401 – 10 0501 – 06 0602 – 09 0701, 06, 07, 09, 16, 18 1204, 05, 08 1502 – 11 1602 1701 – 03 1801, 02	0202 – 06 0307, 15 0401 – 10 0501 – 06 0602 – 09 0707, 09 1204, 05 1404 1502 – 11 1602 1701 – 03 1801, 02	Asn77

Table 10.7. A list of the HLA-C molecules in each cluster. The important residues in defining the clusters were listed in the last column.

Cluster	Alleles	Fingerprint
A1	0101-09 0208 0216 0220 0229 0256 0301-10 1101-09 1112-14 2309 3001-12 3101 3103-06 3019 3201-04 3206-07 3601-04 6002-03 6810 6813-14 7401-09 8001	Asn66 Glu63
A2	0255 1110-11 2424 2501-04 2601-08 3301 3303-06 3402-04 4301 6601 6604 6801-09 6815-23 6901	Asn66 Asn63
A3	0201-07 0209-19 0221-28 0230-54 0257-60 2301-06 2402-38 2607 2901-07 3007 3102 3107-08 3205 3401 3405 7409	Lys66

Table 10.8. Supertypes defined by pocket analysis: pocket a

Supertype	alleles	Fingerprint
B1	0101-08 0201-33 0236-60 2301-06 2401-23 2425-38 2501-04 2601-08 3002-10 3012 3101-09 3201-07 3301-06 4301 6803-05 7401-09 8001	His70
B2	0234-35 2424 2901-07 3001 3011 3401-05 0301-10 1101-14 6601-02 6604 6801-23 6901	Gln70

Table 10.9. Supertypes defined by pocket analysis: pocket b

Supertype	Alleles	Fingerprint
C1	0101-09 0211 0235 0248 0301-10 1101-05 1107-14 2301-09 2402-27 2429 2431-38 2501-04 2601-18 2901-07 3001-12 3101-09 3201-07 3401-05 3601-04 4301 6601-04 6801-04 6806-14 6816-19 6821-23 6901 7401-05 7408-09	Asp74
C2	0201-10 0212-34 0236-47 0249-60 1106 2428 2430 2603 2606 3009 6805 6815 6820 7406 8001	His74

Table 10.10. Supertypes defined by pocket analysis: pocket c

Supertype	Alleles	Fingerprint
D1	0101-03 0107-08 0249 1108 3601-04	Arg/His114 Arg156
D2	0310 1101-07 1109-14 2417 2905 3402 6801 6803-05 6807-23	Arg114 Gln/Trp156
D3	0106 0201-40 0242-48 0250-51 0253-60 0301 0304-09 2301-09 2402-16 2418-38 2901-07 3103-06 3204 3402-04 6802 6808 6815 6901 8001	Arg114 Leu156
D4	2501-04 2601-18 3004 3006 3202 3401 3405 4301 6601-04	Phe/Gln114 Gln/Trp156
D5	0241 0252 3001-03 3007-12 3101-02 3105 3107-09 3201-03 3205 3207 3301-06 7401-09	Phe/Gln114 Leu156

Table 10.11. Supertypes defined by pocket analysis: pocket d

Supertype	Alleles	Fingerprint
E1	0101-09 0302 0307 0310 1101-02 1104-07 1109-14 2612 2618 2901-04 2906-07 3101-09 3201-03 3205-07 3301-06 3601-04 6801 6803-05 6808-11 6813-14 6816 6818-23 7401-09 8001	Asp116 Val/Ala152
E2	0301 0304-06 0308-09 1103 1108 2504 2608 2905 3204 3402-04	Asp116 Glu152
E3	0201-02 0204-12 0214-25 0227-37 0239-60 2301-09 2402-17 2419-38 3001-12 6802 6806 6807 6815 6817 6901	Tyr/His116 Val152
E4	0203 0213 0226 0238 2418 2501-03 2601-11 2613-17 3401 3405 4301 6601-04	Tyr/His116 Glu152

Table 10.12. Supertypes defined by pocket analysis: pocket e

Supertype	Alleles	Fingerprint
F1	0101-09 0301-10 1101-14 2417 2501-04 2601-09 2901-07 3101-09 3202-07 3301-06 3401-05 3601-04 4301 6601-04 6801 6803 6808-14 6816 6818-23 7401-09 8001	Asp116
F2	0201-60 2301-06 2309 2402-38 2602 3001-12 6802 6806-07 6815 6817 6901	Tyr116

Table 10.13. Supertypes defined by pocket analysis: pocket f

Supertype	Hierarchical clustering	Non-hierarchical clustering	Common alleles	Fingerprint	Known supertypes[a]	Known motifs p1 p 4 p6 p7 p9
DR1	DR1 (DRB1*0101–11) DR2 (DRB1*1501–11, DRB1*1601–08) DR7 (DRB1*0701–07)	DR1 (DRB1*0101–11) DR2 (DRB1*1501–11, DRB1*1601–08) –	11 13 7 –	Trp$^{9\beta}$		DRB1*0101 (9, 50-52) YFW LA AG – LA DRB1*1501 (53, 54) LVI FYI – IL GSP
DR3	DR3 (DRB1*0301–25) DR52 (DRB3*0101–10, DRB3*0201–18, DRB3*0301–03) DRB1*0422 DRB1*1107	– DR52 (DRB3*0101–10, DRB3*0201–18, DRB3*0301–03) DRB1*1333 DRB1*1447	– 10 18 3 – –	Glu$^{9\beta}$ Gln$^{70\beta}$ Gln/Arg$^{74\beta}$	DR RSP "R"	DRB1*0301 (55-57) LIF D KR – YLF MV
DR4	DR4 (DRB1*0401, 03–48 without the alleles from DR5 supertype) DR5 (DRB1*1113, 17, 26, 34, 42) DR6 (DRB1*1309, DRB1*1401–48 without the alleles from DR5 supertype) – DR10 (DRB1*1001) DR53 (DRB4*0101–06) –	DR4 (DRB1*0401, 03–48 without the alleles from DR5 supertype) DR5 (DRB1*1107, 13, 17) DR6 (DRB1*1401–48 without the alleles from DR5 supertype) DRB3*0215 DR10 (DRB1*1001) DR53 (DRB4*0101–06) DR3 (DRB1*0301–25)	38 2 31 – 1 5 –	Glu$^{9\beta}$ Gln/Arg$^{70\beta}$ Glu/Ala^{74}	DR RSP "A"	DRB1*0401 (58-60) FY no RK NS pol[b] pol W chg[c] ali[d] ali K DRB1*0404 (61) VIL no RK NT pol pol chg ali ali K DRB1*0405 (61-63) FY VIL NS pol DE chg ali

Table 10.14. DR supertypes and fingerprints. Results of The content of hierarchical and non-hierarchical clustering are compared and fingerprints are defined (Part 1).

Supertype	Hierarchical clustering	Non-hierarchical clustering	Common alleles	Fingerprint	Known supertypes[a]	Known motifs p1 p 4 p6 p' p9
DR5	DR4 (DRB1*0402, 12, 15, 25, 36, 37, 47) DR5 (DRB1*1101– 47, DRB1*1201– 09) DR6 (DRB1*1301– 62, DRB1*1403, 16, 22, 25, 27, 40) DR8 (DRB1*0801– 25) – –	DR4 (DRB1*0402, 15, 25, 36, 47) DR5 (DRB1*1101– 47, DRB1*1201– 09) DR6 (DRB1*1301– 62, DRB1*1403, 16, 17, 21, 22, 24, 25, 27, 29, 30, 37, 40, 41, 48) DR8 (DRB1*0801– 25) DR51 (DRB5*0101 – 12, DRB5*0202 – 05)	5 42 9 61 6 25 – –	$Glu^{9\beta}$ $Asp^{70\beta}$	DR RSP "D"	DRB1*0402 (61) VIL no DE NQ RK pol ali H DRB1*1101 (59, 64) WYF LVI RK – – DRB1*1201 (51) IL LMN VYF – YFM
DR9	DR9 (DRB1*0901, 02) DR51 (DRB5*0101 – 12, DRB5*0202 – 05) –	DR9 (DRB1*0901, 02) – – DR7 (DRB1*0701– 07)	2 – –	$Lys/Gln^{9\beta}$		DRB5*0101 (53, 54) FY QV – – RK
Sum	347	347	285 (82%)			

Table 10.15. DR supertypes and fingerprints. Results of The content of hierarchical and non-hierarchical clustering are compared and fingerprints are defined (Part 2).

Supertype	Hierarchical clustering	Non-hierarchical clustering	Common alleles	Fingerprint
DQ1	DQB1*0501–03 DQB1*0601–21	DQB1*0501–03 DQB1*0601–21	45 300	$Val^{86\beta}$
DQ2	DQB1*0201–03	DQB1*0201–03	45	$Glu^{86\beta}$ $Lys^{71\beta}$
DQ3	DQB1*0301–13 DQB1*0401, 02	DQB1*0301–13 DQB1*0401, 02	195 30	$Glu^{86\beta}$ $Thr/Asp^{71\beta}$
$DQA1*03^c$	DQA1*0301-03	–	–	$Arg^{53\alpha}$
Sum	738	738	615 (83%)	

Table 10.16. HLA-DQ supertypes and fingerprints. Results of The content of hierarchical and non-hierarchical clustering are compared and fingerprints are defined.

Supertype	Hierarchical clustering	Non-hierarchical clustering	Common alleles	Fingerprint
DPw1	DPw1 (DPB1*0101) DPw3 (DPB1*0301) DPw5 (DPB1*0501) DPB1*14, 20, 25, 26, 27, 31, 35, 36, 38, 45, 50, 52, 56, 57, 63, 65, 67, 68, 70, 76, 78, 79, 84, 85, 87, 89, 90, 91, 92, 97, 98	DPw1 (DPB1*0101) DPw3 (DPB1*0301) – DPB1*11, 14, 25, 26, 31, 35, 45, 50, 52, 56, 57, 65, 67, 68, 69, 70, 76, 78, 79, 84, 90, 92	12 12 240	$Asp^{84\beta}$ $Lys^{69\beta}$
DPw2	DPw2 (DPB1*0201 and 0202) DBP1*32, 33, 41, 46, 47, 48, 71, 81, 86, 95	DPw2 (DPB1*0201 and 0202) DBP1*32, 33, 41, 46, 47, 48, 71, 81, 86, 95	24 120	$Gly/Val^{84\beta}$ $Glu^{69\beta}$
DPw4	DPw4 (DPB1*0401 and 0402) DPB1*15, 18, 23, 24, 28, 34, 39, 40, 49, 51, 53, 59, 60, 62, 66, 72, 73, 74, 75, 77, 80, 83, 94, 96, 99	DPw4 (DPB1*0401 and 0402) DPB1*15, 18, 23, 24, 28, 34, 39, 40, 49, 51, 53, 59, 60, 62, 66, 72, 73, 74, 75, 77, 80, 82, 83, 94, 96, 99	24 312	$Gly/Val^{84\beta}$ $Lys^{69\beta}$
DPw6	– DPw6 (DPB1*0601) DPB1*08, 09, 10, 11, 13, 16, 17, 19, 21, 22, 29, 30, 37, 44, 54, 55, 58, 69, 88, 93	DPw5 (DPB1*0501) DPw6 (DPB1*0601) DPB1*08, 09, 10, 13, 16, 17, 19, 20, 21, 22, 27, 29, 30, 36, 37, 38, 44, 54, 55, 58, 63, 85, 87, 88, 89, 91, 93, 97, 98	- 12 216	$Asp^{84\beta}$ $Glu^{69\beta}$
Sum	1140	1140	972 (85%)	

Table 10.17. HLA-DP supertypes and fingerprints. Results of The content of hierarchical and non-hierarchical clustering are compared and fingerprints are defined.

Motif based	GRID/CPCA approach
Considers part of the binding site	Considers the whole binding site
Requires the binding motif of the alleles	Sequence information only
Can only classify alleles with known motifs	Able to classify all HLA alleles

Table 10.18. The advantages of chemometric methods over motif based classification.

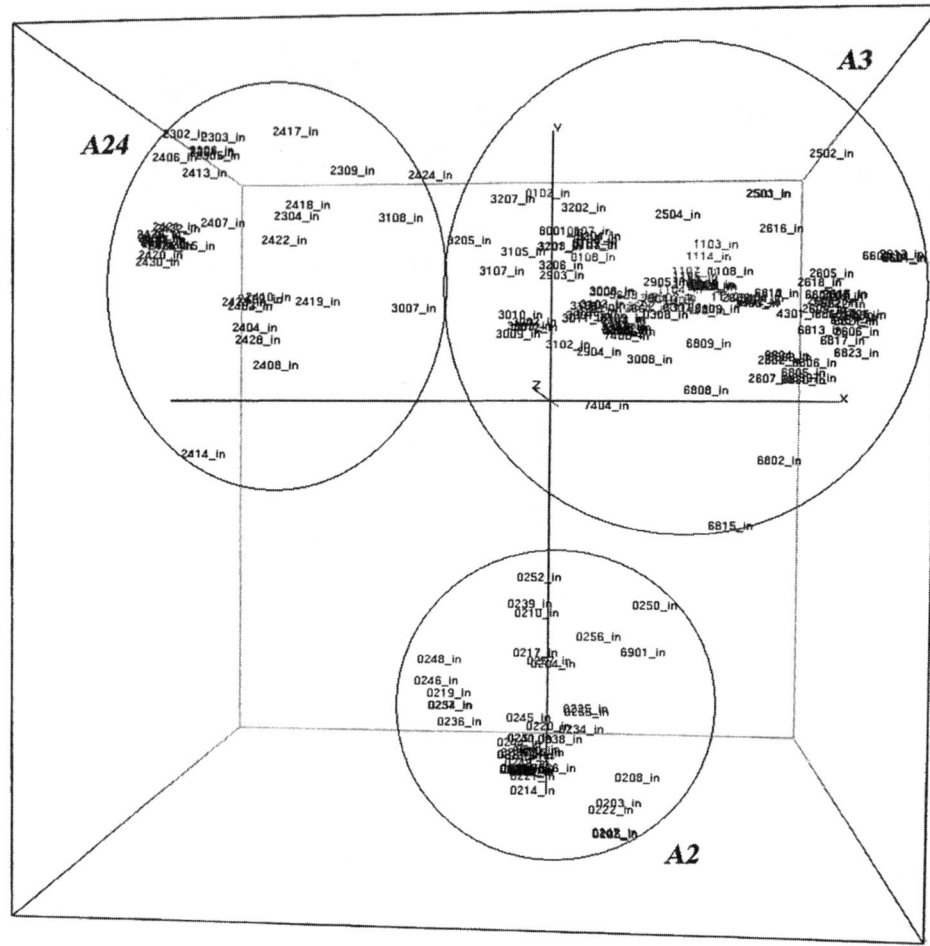

Fig. 10.1. The 3D scores plot of the CPCA analysis for HLA-A molecules. The A24 cluster is on the top left of the plot, the A3 cluster is on the top right of the plot and the A2 cluster is below the X axis.

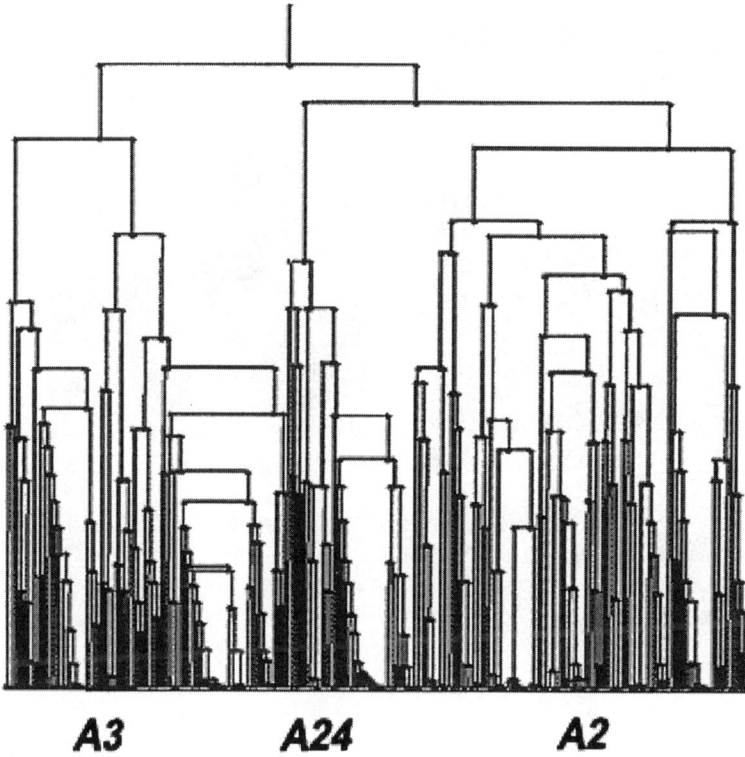

A3   **A24**   **A2**

Fig. 10.2. The HLA-A classification defined by hierarchical clustering. A hierarchical tree was built for the 229 HLA-A alleles. Each leaf represented one allele. The results of the clustering were similar to that of the GRID/CPCA analysis, the three clusters were defined in both experiments: A2, A3 and A24.

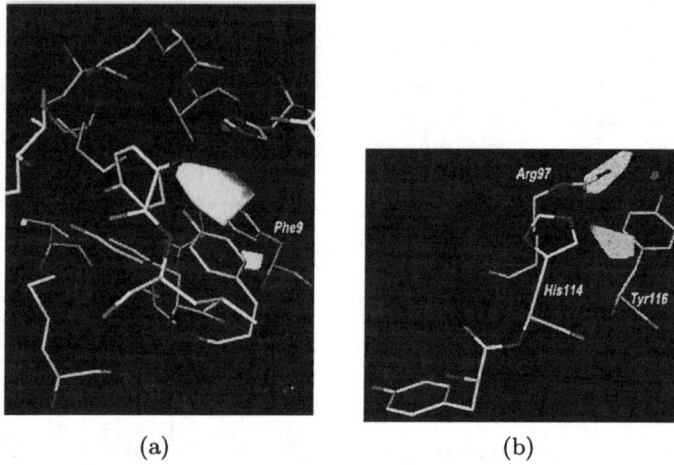

(a)   (b)

Fig. 10.3. The loading plot of the HLA-A CPCA model. The binding site of A*0201 is used in the plot to display the positions of the amino acids. There were two important interactions in the plot. The hydrophobic interaction is favoured at position 9 (a), and disfavoured around position 97, 114 and 116 (b).

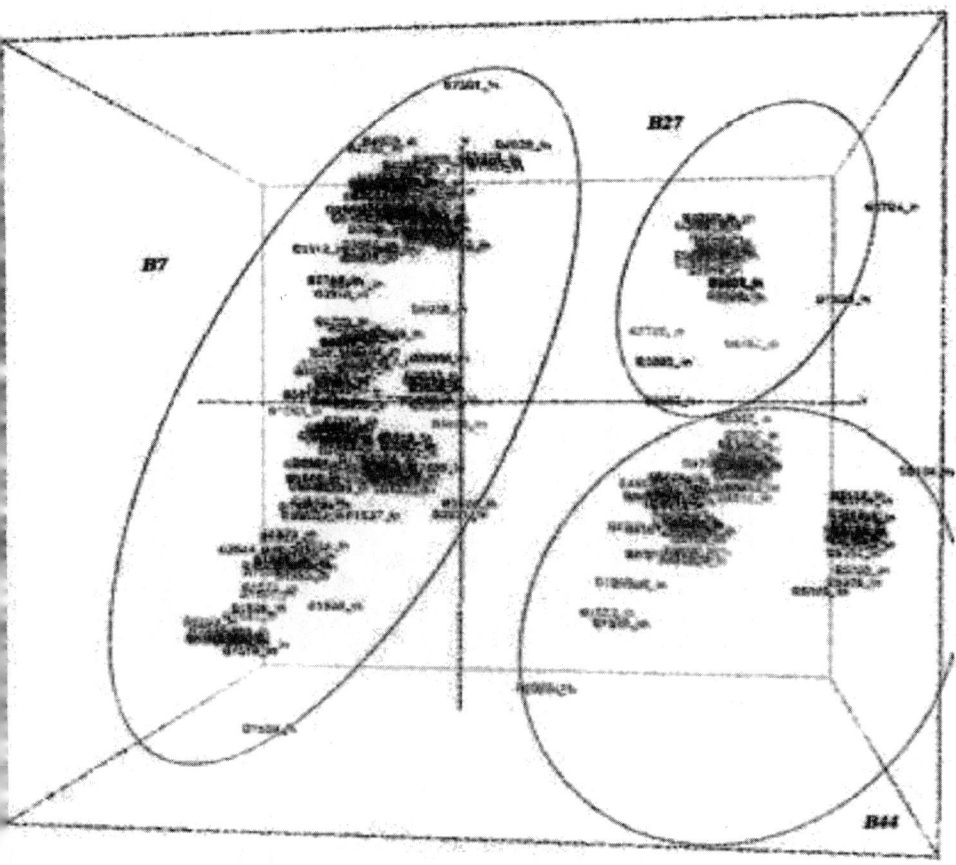

Fig. 10.4. The 3D scores plot of the CPCA analysis for HLA-B molecules. Three clusters were identified in the plot: B7, B27 and B44.

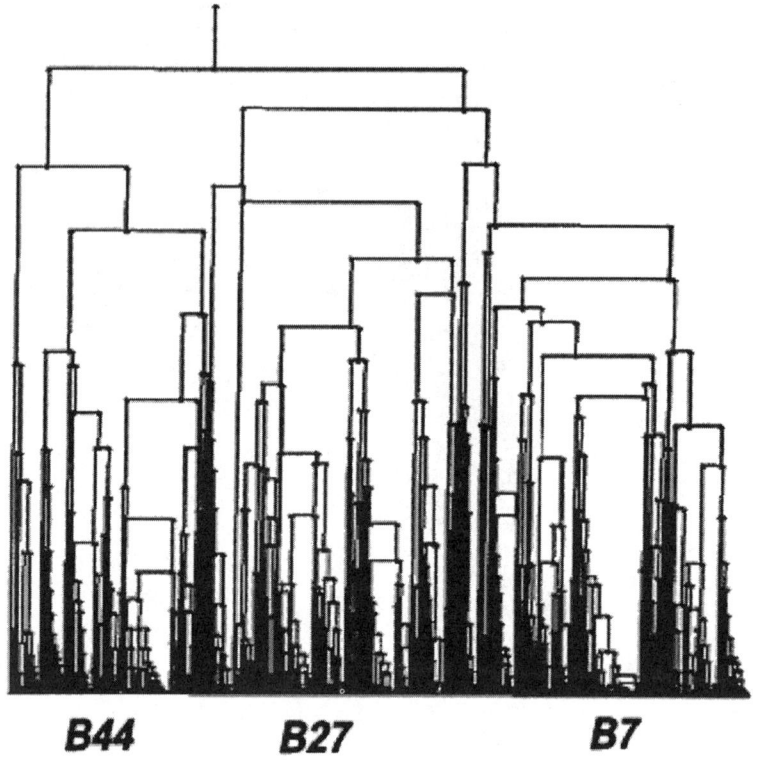

Fig. 10.5. HLA clusters produced using hierarchical clustering. A hierarchical tree was produced for the 447 HLA-B alleles. Each leaf represents one allele.

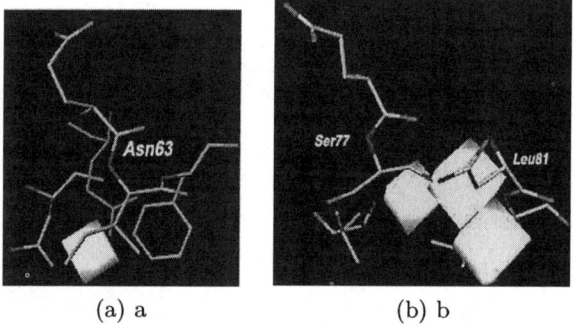

(a) a   (b) b

Fig. 10.6. Loading plot of the CPCA model for the HLA-B superfamilies classification. Part of the B*0801 binding site is shown in the plot. The hydrophobic interaction is found around position 63 and 66 (a), 77 and 81 (b)

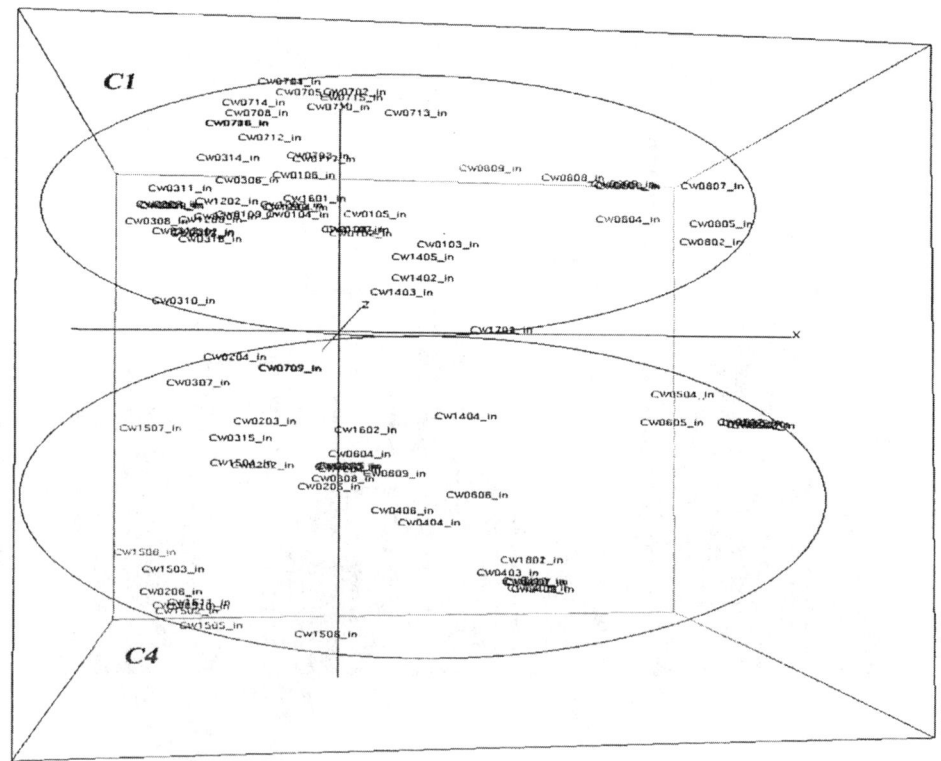

Fig. 10.7. The 3D scores plot of the HLA-C CPCA analysis. Two clusters were displayed in the plot. The main cluster above the X axis had many C1 molecules and was named the C1 cluster. The cluster below the X axis had lots of C4 molecules and was named the C4 cluster.

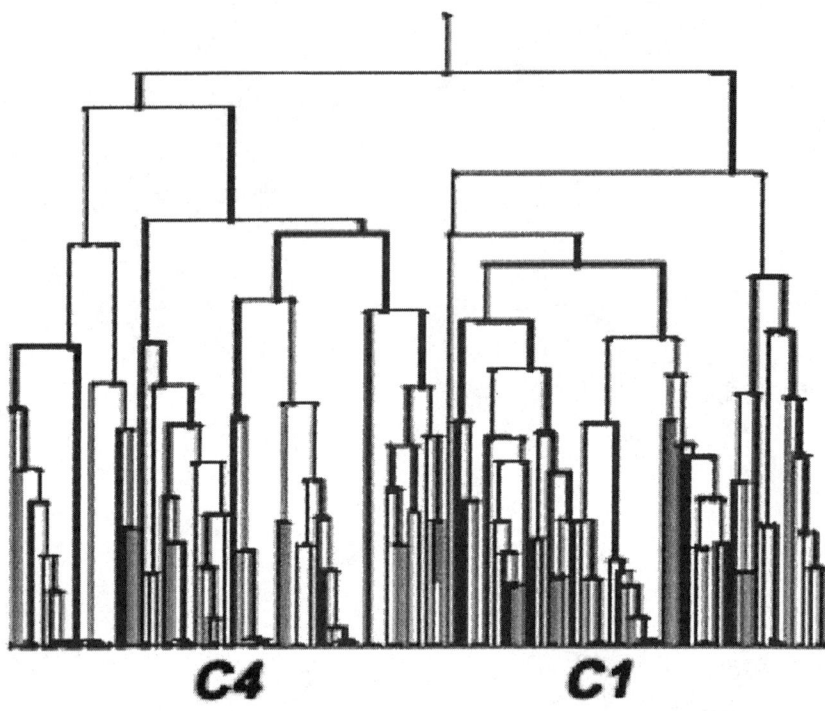

Fig. 10.8. The hierarchical tree obtained from hierarchical clustering, in which the HLA-C molecules were classified into C1 and C4 clusters. Each leaf represented one HLA-C allele. Results of the analysis were in accordance with the GRID/CPCA classification.

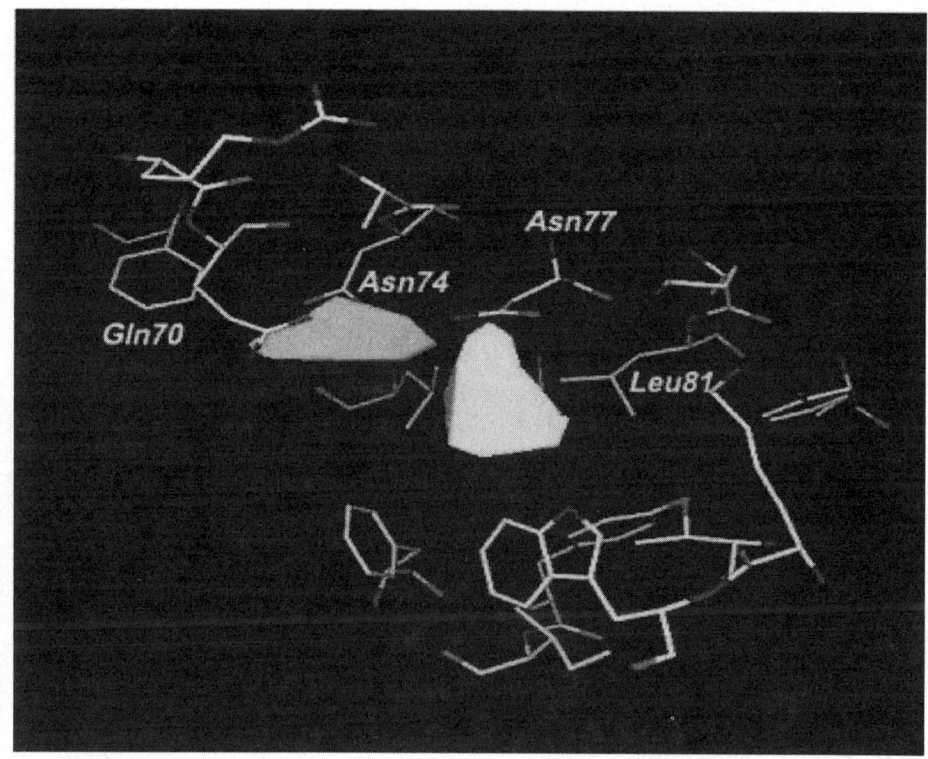

Fig. 10.9. The loading plot of the HLA-C CPCA model for the water probe. The binding site of Cw*0401 is shown in the plot. The highlighted area is around position 70, 74 and 81.

Biomolecular Structure Prediction Using Immune Inspired Algorithms

Vincenzo Cutello and Giuseppe Nicosia

Department of Mathematics and Computer Science
University of Catania
V.le A. Doria, 6 - 95125 Catania Italy
{cutello, nicosia}@dmi.unict.it
www.dmi.unict.it/~ {cutello,nicosia}

Summary. Proteins are responsible for almost all functions in a living organism: such as enzymatic catalysis, storage and transport of material, RNA editing, organ development, antibodies and more. The Protein Structure Prediction problem (PSP) is concerned with the prediction of the 3D structure of a protein given its sequence of amino acids. If we know the protein structure we can infer the protein function. One computational approach for predicting the 3D structure of a protein is related to the minimisation of the energy function derived from physico-chemical and statistical considerations. The Chapter proposes the use of hybrid Immune Algorithms to face the PSP as a single objective optimisation problem, the classical approach, and as a multi-objective optimisation problem, the innovative approach. This Chapter presents a link between the world of Artificial Immune Systems and the application to real biological problems.

11.1 Introduction

Understanding the action of proteins is the key to understanding the spark of life itself. The information contained in the primary sequence is known to be sufficient to completely determine the geometrical three-dimensional structure of the protein, at least for simpler proteins that are observed to reliably refold when denatured *in vitro*.

This chapter proposes the use of immunological inspired algorithms (as discussed in Chapter 3 in this book, based on the clonal selection principle, to the application to biological problems, a sort of biological recursive process, if you will. In particular, the biomolecules structure prediction problem is faced both as a single objective optimisation problem, the classical approach, and as a multi-objective optimisation problem, the innovative approach. These two distinct optimisation problems have been tackled by designing algorithms inspired by the inner workings of the biological immune system. Both approaches are very effective and efficient both in terms of quality solution and computational cost.

This Chapter presents a link between the world of Artificial Immune Systems and the application to real biological problems.

11.2 Biomolecular Structure Prediction

The explosion of research in molecular biology has been made possible by the fundamental discovery that hereditary information is stored and passed on in the simple, one-dimensional (1D) sequence of DNA base pairs [Watson & Crick 1953]. The connection between heredity and biological function is made through the transmission of this 1D information, through RNA, to the protein sequence of amino acid. The information contained in this sequence is known to be sufficient to completely determine the geometrical three-dimensional (3D) structure of the protein, at least for simpler proteins which are observed to reliably refold when denatured *in vitro*, i.e., without the aid of any cellular machinery such as chaperones or steric constrains due to the presence of a ribosomal surface [Anfinsen 1973].

Folding to a specific structure is typically a prerequisite for a protein to function. Further understanding of the molecular description of life requires answering the deceptively simple question of how the 1D sequence of amino acids in a protein chain determines its 3D folded conformation in space, or more precisely, the set of *near native conformations*.

There are many large biological molecules, including: nucleic acids, carbohydrates, lipids and proteins. While each play a vital and interesting part in life, there is something special about proteins. Indeed, of all the components that make up life, almost all but proteins are relatively inert and are generally, the substrates that are chopped changed by the action of proteins. In doing this, proteins do not act using some abstract *bulk property*, unlike lipids and carbohydrates, proteins act like *individual agents* that latch onto their 'objectives', the substrates, and cut and change them. Indeed, when located across a lipid membrane, they are also quite good at opening and shutting 'trapdoors'.

Chemical structure of protein

Protein is a polymer consisting of a long chain of amino acids. There are 20 different amino acids, the simple building blocks of proteins (see table 11.2; the amino acid residues are usually abbreviated with three identifying letters, or one-letter code, of the corresponding amino acid).

Amino acid	3-letter	1-letter	relevant property [Fauchere & Pliska 1983]	hydrophobicity
Alanine	Ala	A	small	0.31
Serine	Ser	S	small	−0.04
Threonine	Thr	T	small	0.26
Cysteine	Cys	C	hydrophobic	1.54
Valine	Val	V	hydrophobic	1.22
Isoleucine	Ile	I	hydrophobic	1.80
Leucine	Leu	L	hydrophobic	1.70
Proline	Pro	P	hydrophobic	0.72
Phenylalanine	Phe	F	hydrophobic	1.79
Tyrosine	Tyr	Y	hydrophobic	0.26
Methionine	Met	M	hydrophobic	1.23
Tryptophan	Trp	W	hydrophobic	2.25
Asparagine	Asn	N	polar	−0.60
Glutamine	Gln	Q	polar	−0.64
Histidine	His	H	polar	0.13
Aspartic acid	Asp	D	acidic, negatively charged	−0.77
Glutamic acid	Glu	E	acidic, negatively charged	−0.64
Lysine	Lys	K	basic, positively charged	−0.99
Arginine	Arg	R	basic, positively charged	−1.01
Glycine	Gly	G		0.0

Table 11.1. The amino acids

Under the influence of RNA containing the genetic information coding for the amino acid sequence, amino acids polymerise in a specific sequence to a chain forming the *primary structure* of the protein.

Each amino acid consists of a central carbon atom, the alpha carbon (C_α), bounded to an amino group (NH_2), a carboxyl group $(COOH)$ and a side chain. Each amino acid has a different side chain which is uniquely responsible for the characteristics of the amino acids like shape, size, and polarity. The side chains have $1 - 18$ atoms.

Amino acids are linked to each other by means of peptide bonds. They result in a constrained backbone of C_α, the repeating $-NC_\alpha C\prime-$ is called the *backbone* of a protein.

The interatomic forces bend and twist the chain in a way characteristic for each protein. These interatomic forces cause the protein to curl up into a specific 3D geometric configuration, the *folded state* of the protein. This configuration and the chemically active groups on the surface of the folded protein determine its biological function [Whisstock & Lesk 2003].

From 1D to 3D information

Although this problem is now considered central to molecular biology, only after about 45 years of study is it beginning to yield to the combined efforts of molecular biologists, chemists, physicists, computer scientist and mathematicians [Tramontano 2006].

The difficulty of the problem lies in the unfamiliar nature of *1D to 3D information transcription in going from protein sequence to protein structure*: the information processing cannot go in a sequential symbol by symbol fashion, but must operate simultaneously using remote parts of the sequence, and hence is essentially a *non-local, collective process* rather than the trivial translation of a message.

Initial progress was primarily descriptive. Proteins were observed to be covalently bonded, linear polymers with a specific primary sequence of side-chains or amino acids attached at regular intervals, constituting the 1D information pattern. While a given sequence uniquely determines a folded structure, a given structure may be highly degenerate in the sequences that fold to it, or can be 'designed' for it. Finding these sequences is known as the *inverse protein folding* or design problem [Bowie *et al.* 1991]. It is not uncommon for sequences with less than 50% identity of amino acids to fold to the same or similar structure, with nearly the same folding rate at a given stability.

Allowed regions

A pioneer analysis showed that the space of sterically allowed, local rotational angles of the backbone chain was quite restricted [Ramachandran & Sassiekharan 1968]. One of the allowed regions corresponded to the α-helix, another to a pleated 2D structure of parallel or anti-parallel strands (β−sheets). These structures can persist indefinitely in effectively infinite protein structures such as wool (the α−helix) or silk (the β−sheet), but in globular proteins they are broken up by turns of dense, semi-rigid random coil. The secondary structural elements of a protein are determined through the *collective interactions* of the elements with the rest of the molecule: the identity of an amino acid does not, by itself exclusively determine what secondary structure it will be found in.

Collective versus Local interactions

The formation of the folded structure is governed by the *collective effects of non-covalent interactions* in essentially the whole molecule: a theory that is local cannot solve the protein folding problem [Ngo *et al.* 1994]. This is fundamentally different from the *self-organization* occurring in biological systems without fixed disorder. For example, the hierarchical helix-formation involved in DNA folding, where the disorder in the sequence is suppressed by complementary base pairs, and the *folding (supercoiling) mechanism is local*. Collective interactions slow folding, whilst at the same time enhancing stability by involving non-local parts of the chain in the folding nucleus.

Cooperative interactions

The interactions stabilizing the native structure tend to also be *cooperative*: the energetic gains in forming native structure are achieved only when several parts of the protein are in spatial proximity and interacting [Perutz 1970]. There is a larger entropic barrier for this kind of process than for one involving simple pair interactions. In forming a protein which is compact overall, the elements of secondary structure must themselves be packed together into what is called tertiary structure. β−sheets can only be oriented in certain ways when stacked on top of each other.

Levinthal's paradox

The observation mentioned earlier, that proteins can fold reversibly in vitro without any external cellular machinery, means that the folding mechanism can be theoretically and experimentally studied for a single isolated protein molecule *interacting with solvent*. The reversible in vitro folding of a single protein means that the protein in the native state is *thermodynamically stable*, and therefore that the native state has the global minimum free energy of all kinetically accessible structures [Levinthal 1969].

11.2.1 Intractability of Protein Structure Prediction

Energy Landscapes

The folded state is a *small ensemble of conformational structures* compared to the conformational entropy present in the unfolded ensemble. Considering a coarse-grained level of description, the *folded structure* must then have the lowest internal energy of all kinetically accessible conformational structures. Internal energy is defined here as the *free energy* of a single backbone conformation. We can define the energy landscape for this system as a mapping of the chain conformation to its internal energy, along with rules defining what configurations are accessible from a given configuration and how the protein can move between them.

Following to the principles of thermodynamics, if a system has n degrees of freedom

$$\phi = [\phi_1, \phi_2, \ldots, \phi_n]$$

the *stable state of the system* can be found by determining the set of values

$$\phi^* = [\phi_1^*, \phi_2^*, \ldots, \phi_n^*]$$

that gives the minimum value of the free energy function

$$F(\phi) = F(\phi_1, \phi_2, \ldots, \phi_n),$$

when explored over all possible values of ϕ. Such functions $F(\phi)$ are called *energy landscapes*. For example, for the protein folding problem, ϕ could be the *backbone and side-chain torsion angles*.

Hence, a energy landscape is a network of all conformational states of a protein, with an internal free energy associated with each conformation, and with the connectivity of the network specified or assumed implicitly.

Impossibility and intractability

Finding the native structure is analogous to a "drunken golfer finding the hole on a vast green", and it is useful to think of folding in this scenario as occurring on an energy landscape with a golf-green topography: the system obtains no energetic gain from ordering any residues until it stumbles upon the complete native structure, thus sampling is unbiased. Problems with energy landscape of this nature (the golf course), when viewed as optimisation problems to find the ground state, have been shown to be NP complete [Baum 1986], a computational problem which is both NP (verifiable in non-deterministic polynomial time) and NP-hard (any NP-problem can be translated into this problem).

Hence, as one would suspect, finding a conformation that minimizes energy landscapes is very hard. There are Protein Structure Prediction (PSP) *non-discrete models* formally defined with NP hardness results, for example, the Ngo and Marks' PSP model [Ngo & Marks 1992]. In this non-discrete model, the protein is described by the complete list of the atoms in the molecule, their connectivity, bond lengths and angles and force constants between all pairs of atoms. The energy function is a *non-convex function* (obtained by summing the contributions of local and non-local interactions); to find a conformation that minimizes this kind of function is NP hard [Ngo & Marks 1992]. To show the hardness the authors devised a reduction from the *Partition problem*. No approximation algorithms are known for this PSP non-discrete model.

11.3 PSP as a Single-Objective Optimisation Problem

The most difficult task when using evolutionary algorithms, or any other type of stochastic search, for the PSP problem, is to come up with "good" *representation* of the conformations, and a *objective function* for evaluating conformations.

Few conformation-representation are commonly used: all-atom 3D coordinates, all-heavy-atom coordinates, backbone atom coordinates + side chain centroids, C_α coordinates, backbone and side chain torsion angles. Some algorithms use multiple representations and move between them for different purposes.

In this Chapter, we use an internal coordinate representation, torsion angles, each residue type requires a fixed number of torsion angles to fix the 3D coordinates of all atoms. Bond lengths and angles are fixed at their ideal values. The degrees of freedom in this representation are the backbone and side chain torsion angles (ϕ, ψ, ω, and χ_h). The number of χ angles depends on the residue type.

CHARMM potential energy function

In order to evaluate the structure of a molecule we use energy functions. In particular, we use classical physics to come up with energy functions. Sometimes called potential energy functions or force fields, these functions return a value for the energy based on the conformation of the molecule. They provide information on what conformations of the molecule are better or worse. The lower the energy value, the better the conformation should be.

In this Chapter, in order to evaluate the conformation of a protein, we use the CHARMM (version 27) energy function. CHARMM (Chemistry at HARvard Macromolecular Mechanics) [Brooks *et al.* 1983, MacKerell Jr. *et al.* 1998] is a popular all-atom force field used mainly for the study of macromolecules. It is a composite sum of several molecular mechanics equations: stretching, bending, torsion, Urey-Bradley, impropers, van-der-Walls, electrostatics. The CHARMM energy function has the form:

$$E_{charmm} = \underbrace{\sum_{bonds} k_b(b - b_0)^2}_{E_1} + \underbrace{\sum_{UB} k_{UB}(S - S_0)^2}_{E_2}$$

$$+ \underbrace{\sum_{angles} k_\theta(\theta - \theta_0)^2}_{E_3} + \underbrace{\sum_{torsionss} k_\chi[1 + \cos(n\chi - \delta)]}_{E_4}$$

$$+ \underbrace{\sum_{impropers} k_{imp}(\phi - \phi_0)^2}_{E_5}$$

$$+ \underbrace{\sum_{nonbond} \varepsilon_{ij}\left[\left(\frac{Rmin_{ij}}{r_{ij}}\right)^{12} - \left(\frac{Rmin_{ij}}{r_{ij}}\right)^6\right]}_{E_6}$$

$$+ \underbrace{\frac{q_i q_j}{er_{ij}}}_{E_7} \tag{11.1}$$

where, in order, b is the bond length, b_0 is the bond equilibrium distance and k_b is the bond force constant; S is the distance between two atoms separated by two covalent bonds (1,3 distance), S_0 is the equilibrium distance and k_{UB} is the Urey Bradley force constant; θ is the valence angle, θ_0 is the equilibrium angle and K_θ is the valence angle force constant; χ is the torsion angle, k_χ is the dihedral force constant, n is the multiplicity and δ is the phase angle; ϕ is the improper angle, ϕ_0 is the equilibrium improper angle and k_{imp} is the improper force constant; ε_{ij} is the Lennard Jones (LJ) well depth, r_{ij} is the distance between atoms i and j, $Rmin_{ij}$ is the minimum interaction radius, q_i is the partial atomic charges and e is the dielectric constant. Typically, ε_i and $Rmin_i$ are obtained for individual atom types and then combined to yield ε_{ij} and $Rmin_{ij}$ for the interacting atoms via combining rules. In CHARMM, ε_{ij} values are obtained via the geometric mean $\varepsilon_{ij} = sqrt(\varepsilon_i * \varepsilon_j)$, and $Rmin_{ij}$ via the arithmetic mean, $Rmin_{ij} = (Rmin_i + Rmin_j)/2$.

Finally the energy function CHARMM (equation 11.1) represents our minimization *objective*, the torsion angles of the protein are the *decision variables* of the optimisation problem, and the constraint regions are the *variable bounds*.

To evaluate the CHARMM energy function we use routines from TINKER Molecular Modeling Package [Huang et al. 1999]. First, the protein structure in internal coordinates (torsion angles) is transformed in cartesian coordinates using the PROTEIN routine. The conformation is then evaluated using the ANALYZE routine, that gives back the CHARMM energy potential of a given protein structure.

The metrics: DME and RMSD

To evaluate how similar the predicted conformation is to the native one, we employ two frequently used metrics: Root Mean Square Deviation (RMSD) and Distance Matrix Error (DME). RMSD is calculated by the formula:

$$RMSD = \sqrt{\frac{\sum_{i=1}^{n} |r_{ai} - r_{bi}|^2}{n}} \qquad (11.2)$$

where r_{ai} and r_{bi} are the positions of atom i of structure a and structure b, respectively, and where structures a and b have been optimally superimposed. Fitting was performed using the McLachlan algorithm [McLachlan 1982] as implemented in the program ProFit [Profit Program].

DME is calculated by the formula:

$$DME = \frac{\sqrt{\sum_{i=1}^{n} \sum_{j=1}^{n} (|r_{ai} - r_{aj}| - |r_{bi} - r_{bj}|)^2}}{n} \qquad (11.3)$$

which does not require the *superposition of coordinates*. For a particular pair of structures, the RMSD, which measures the similarity of atomic positions, is usually larger than DME, which measures the similarity of *inter-atomic distances*.

11.3.1 The Immune Algorithm for the PSP

The scientific discipline of the Artificial Immune Systems (see Chapter 3 of this book) appears to be a powerful computing paradigm as well as a prominent apparatus for improving understanding of biological data and systems. In this Chapter we describe a class of Immune Algorithms based on *clonal selection principle* (see Chapter 3) using new immunological operators, *aging operators*, and particular mutation operators (the *hypermutation and hypermacromutation operators*) to face the Protein Structure Prediction problem for real proteins [Nicosia 2004].

The IA encodes each structure of a given protein as a set of torsion angles. In the real code representation, each B cell at time step t is a vector of real variables: $x^t = (x_1, x_2, \ldots, x_n) \in \Re^n$ where $-\pi \leq x_i \leq \pi \quad \forall i = 1, \ldots, n$.

Two new hypermutation and hypermacromutation operators were used together in the immune algorithm for the PSP. Usually, variation operators mutate a individual by adding a Gaussian distributed random vector of mean zero and predefined deviation [Fogel 2000] to it as follows: $x^{(t,\prime)} = x^t + u$ where the mutation vector u is computed from $u = (u_1, u_2, \ldots, u_n)$, $u_j = N_j(0, \sigma^t)$ where σ^t is the standard deviation over the entire population at generation t.

The new hypermutation operator performs a local search of the conformational space. It will perturb all torsion angles $(\phi, \psi, \omega, \chi_h)$ of a randomly chosen residue with the law: $x_i^{(t,\prime)} = x_i^t + N(0, \sigma^t)$, where x_i^t is the i-th torsion angle of a residue of B cell receptor x at time step t, and $N(0, \sigma^t)$ is a real number generated by a Gaussian distribution of mean $\mu = 0$ and standard deviation σ^t. Starting at initial generation with $\sigma^{t=0} = initial\_sigma$, a user defined parameter (for all the simulations performed in this Section we set $initial\_sigma = 5$), the standard deviation at each generation t varies with the following law $\sigma^{t+1} = \sigma^t - step$ (typically $step = 0.5$)

if the average fitness of the population at time step t decreases of one-order magnitude. This is a simple strategy to allow for quick convergence and to explore more accurately the funnel landscape. In fact using a fixed σ value requires more generations, and does not allow the algorithm to reach near native conformations.

The new hypermacromutation operator may change the conformation dramatically. When this operator acts on a peptide chain, all the values of the backbone and side chain torsion angles of a randomly chosen residue are reselected from their corresponding constrained regions.

Immune Algorithm($ProteinSequence, d, dup, \tau_B, initial\_sigma, step, I$)
1. $Nc := d * dup;$
2. $t := 0;$
3. **If** I $P^{(t)} :=$ Random_Initial_Pop($ProteinSequence$);
4.   **else** $P^{(t)} :=$ DIRECT($ProteinSequence$);
4. Evaluate($P^{(0)}$);
5. **while** ($\neg$ Termination_Condition()) **do**
6. $P^{(clo)} :=$ Cloning ($P^{(t)}, Nc$);
7. $P^{(hyp)} :=$ Hypermutation ($P^{(clo)}$);
8.   Evaluate ($P^{(hyp)}$);
9. $P^{(macro)} :=$ Hypermacromutation (P^{clo});
10.   Evaluate ($P^{(macro)}$);
11. $(P_a^{(t)}, P_a^{(hyp)}, P_a^{(macro)}) :=$ Aging($P^{(t)}, P^{(hyp)}, P^{(macro)}, \tau_B$);
12. $P^{(t+1)} := (\mu + \lambda)$-Selection ($P_a^{(t)}, P_a^{(hyp)}, P_a^{(macro)}$);
13.   **if** (averageFitness($P^{(t+1)}$) decreases $1-$order magnitude)
14.   **then** $\sigma^{t+1} = \sigma^t - step$
15. $t := t + 1;$
16. **end_while**

Table 11.2. Immune Algorithm for real PSP problem [Nicosia 2004]

The immune algorithm designed in this Section can use two different initialisation procedures for the first population of conformations; random initialisation of the population, the standard approach, and an initial population of conformations given by a global optimisation procedure. In the first method, the protein sequences are conformations of torsion angles randomly selected from their corresponding constrained regions. In the latter, the initial population is constituted by a set of conformations created by the DIviding RECTangles (DIRECT) global optimisation algorithm [Jones et al. 1993]. The DIRECT algorithm does not use gradient information to avoid the local minima inherent in the search space. It can be defined as a pattern search method that works by sampling the search space by dividing rectangles. It uses a set of exploratory moves to consider the behaviour of the objective function at the centres of rectangles. DIRECT balances local and global search by selecting *potentially optimal rectangles* to be further explored. We are interested in using such an interesting feature of the DIRECT algorithm. DIRECT procedure produces an initial population of promising candidate solutions inside a potentially optimal rectangles of the funnel landscape of the PSP. This scheme reduces the number of fitness function evaluations of the overall search process to the lowest energy value. This is the first known application of DIRECT algorithm to the protein structure prediction problem.

Considering the two different initialisation procedures (devised by boolean variable I) for the population at initial generation ($t = 0$), we have the IA starting from a random initial population, and the IA-DIRECT approach for the IA starting from a population of promising protein conformations created by the global optimiser DIRECT.

For both versions, the main loop of the algorithms is the same. From the current population, a number dup of clones will be generated, producing the population $Pop^{(clo)}$, which will be mutated into $Pop^{(hyp)}$ by the new hypermutation operator, and into $Pop^{(macro)}$ by the new hypermacromutation operator. The other steps of the algorithm remaining unchanged. Table 11.3.1 shows the pseudo-code of the Immune Algorithm for the structure prediction of real proteins.

Residue	ϕ	ψ	ω	χ_1	χ_2	χ_3	χ_4
Tyr	-86	156	-177	-173	79	-166	
Gly	-154	83	169				
Gly	84	-74	-170				
Phe	-137	19	-174	59	95		
Met	-164	160	-180	53	175	-180	-59

Table 11.3. Global minimum energy structure of Met-enkephalin obtained by Scheraga and Z. Li [Li & Scheraga 1988]

11.3.2 Results on Met-enkephalin peptide and 1ZDD protein

Met-enkephalin peptide

The Immune Algorithm has been applied to determine the three dimensional structure of the pentapeptide Met-enkephalin. It is a very short polypeptide, only 5 amino acids, and twenty-four variable backbone and side-chain torsion angles ($n = 24$). From a optimisation point of view the Met-enkephalin polypeptide is a paradigmatic example of *multiple-minima problem*. It is estimated to have more than 10^{11} locally optimal conformations. Nevertheless, the Met-enkephalin (1MET) has defined structure (see table 11.3.1), and an apparent global minimum (with conformational energy of $-12.90 kcal/mol$ based on ECEPP/2 routine) first located by a Monte Carlo-minimization method in 4 hours only, in 1987 by Scheraga and Purisima [Purisima

& Scheraga 1987]. For all these reasons, this peptide is an obvious "test bed", for which a substantial amount of in silico experiments has been done [Purisima & Scheraga 1987, Li & Scheraga 1988, Kaiser *et al.* 1997, Bindewald *et al.* 1998]. For frequent convergence to the global minimum (table 11.3.1), in [Li & Scheraga 1988] the authors set a maximum number of energy function evaluations $T_{max} = 10^6$.

Using the IA-DIRECT approach, the initial population of 10 individuals was generated by global optimisation procedure, and run for $T^1_{max} = 30275$ energy function evaluations. After T^1_{max} energy function evaluations, the DIRECT procedure does not improve the detection of good conformations. The immune algorithm starting from this *ad hoc* initial conformation was executed for 1500 generations with a population size of $d = 10$ individuals, duplication parameter $dup = 2$, expected life time parameter $\tau_B = 5$, for a maximum number of energy function evaluation equal to $T^2_{max} = 30000$. Hence, considering the computational costs of the DIRECT procedure and the IA we have a overall $T_{max} = T^1_{max} + T^2_{max} = 60275$ maximum number of energy function evaluations for frequent convergence to the best energy value close to the global minimum. The best conformation with the lowest energy value of $-20.56 kcal/mol$ obtained by IA-DIRECT is reported in table 11.3.2.

Residue	ϕ	ψ	ω	χ_1	χ_2	χ_3	χ_4
Tyr	0.95	136.67	-169.44	-166.93	-108.85	12.00	
Gly	-181.69	-36.08	-188.63				
Gly	-180.00	-42.48	-175.28				
Phe	-120.00	-47.03	-187.71	-59.49	-65.78		
Met	-120.00	-16.73	-174.28	-65.90	-177.93	179.07	-83.74

Table 11.4. Global minimum energy structure of *1MET* obtained by the IA

Figure 11.1 shows the RMSD and energy values of the protein conformations obtained by the Immune Algorithm using the procedure DIRECT to generate the initial B cell population. The figure shows a large cluster of conformations around energy value $2.885 kcal/mol$ with an individual conformation with $RMSD = 2.835$, and energy equal to $-20.47 kcal/mol$, the best conformation obtained by designed IA. Figure 11.1 displays a good correlation between RMSD and energy value E, suggesting that minimizing E of a small protein using Immune Algorithm will tend to drive the conformation toward the true structure.

Figure 11.2 shows the superimposition of the conformation obtained by IA and the Scheraga's conformation using a full atom resolution. Computing the RMSD and DME metrics with respect the C_α atoms, and all the atoms of the two structures, we have the following results: $RMSD = 2.835$Å, and $RMSD_{c_\alpha} = 0.501$Å, while in terms of DME measure $DME = 2.142$Å and $DME_{c_\alpha} = 0.468$Å.

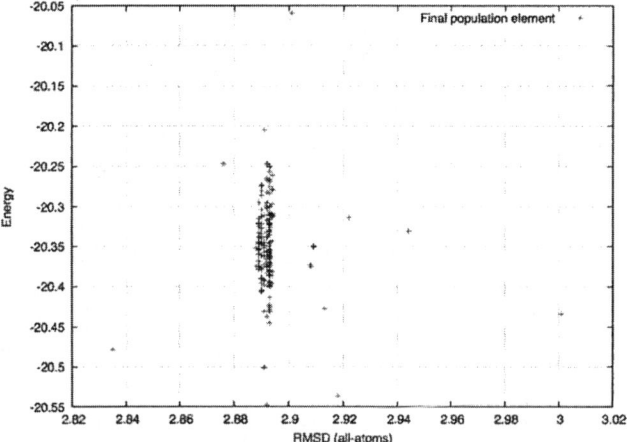

Fig. 11.1. RMSD values versus Energy values of protein conformations obtained by the Immune Algorithm

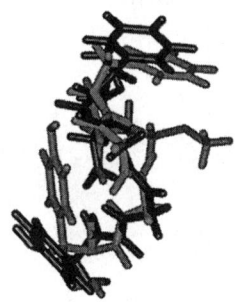

Fig. 11.2. In the plot the superimposition of the 1MET protein conformation obtained by IA and the optimal conformation obtained by Scheraga [Rabow & Scheraga 1996]; the RMSD of the two structure is 2.835Å, the $RMSD_{c_\alpha}$ is 0.490Å. In terms of DME measure for the two structures we have $DME = 2.211$Å and $DME_{c_\alpha} = 0.454$Å

Table 11.3.2 shows the comparisons of the two designed immune algorithm versions with other folding algorithms. For each algorithm, the table reports the energy function used, the mean and standard deviation energy values, and the best RMSD measure with respect the Scheraga's conformation, the accepted optimal conformation. It is important to note that the optimal conformation for ECEPP/2 and CHARMM energy functions are different. From a structural similarity point of view, it is more significant to consider the RMSD. In terms of RMSD value, the IA-DIRECT obtains

Algorithm	Energy funct.	Energy ($kcal/mol$)	RMSD
Scheraga's MC [Li & Scheraga 1988]	ECEPP/2	−12.90	n.a.
IA-DIRECT	CHARMM27	−20.47 ± 1.54	**2.835Å**
REGAL Tight constr. [Kaiser et al. 1997]	CHARMM	−23.55 ± 1.69	3.23Å
IA	CHARMM27	−19.92 ± 2.87	**3.30Å**
Lamarkian [Kaiser et al. 1997]	CHARMM	−28.35 ± 1.29	3.33Å
Baldwinian [Kaiser et al. 1997]	CHARMM	−22.57 ± 1.62	3.96Å
REGAL Loose constr. [Kaiser et al. 1997]	CHARMM	−22.01 ± 2.69	4.25Å
SGA [Kaiser et al. 1997]	CHARMM	−22.58 ± 1.57	4.51Å
REGAL [Kaiser et al. 1997]	CHARMM	−24.92 ± 2.99	4.57Å
standard GA [Bindewald et al. 1998]	ECEPP/2	−3.17 ± 0.37	
GA with sterical constraint [Bindewald et al. 1998]	ECEPP/2	−2.35 ± 0.33	

Table 11.5. Immune Algorithms versus Folding algorithms for *1MET*

the structure more similar to the the accepted optimal conformation of Scheraga with a computational cost of $T_{max} = 60275$ energy function evaluations. The IA with a random initial population reaches a conformation with the best RMSD of 3.30 with 150000 energy function evaluations. The algorithms designed in [Kaiser et al. 1997] were allowed to reach 150000 evaluations, while in [Bindewald et al. 1998] this information is not available.

Considering both versions of the IA algorithm, we can claim that they are competitive and effective search methods in the conformational search space of real proteins. In particular, the IA-DIRECT approach is very effective and efficient in terms of quality solution and computational effort.

 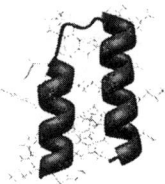

Fig. 11.3. Predicted (left plot) and native (right plot) conformations for 1ZDD protein ($DME_{C_\alpha} = 1.499$Å, $RMSD_{C_\alpha} = 2.220$Å)

Disulfide-Stabilized Mini Protein A Domain (1ZDD)

1ZDD is a two-helix peptide of 34 residues [Starovasnik *et al.* 1997]. For this protein the secondary structure constraints was predicted by the SCRATCH prediction server [Pollastri *et al.* 2002]. The best computed structure matches the crystal structure with $DME_{C_\alpha} = 1.499$Å and $RMSD_{C_\alpha} = 2.220$Å(see figure 11.3) and the energy of the structure is -1037.831 $kcal/mol$.

11.4 PSP as a Multi-Objective Optimisation Problem

In this second part of Chapter, we investigate the applicability of a multi-objective formulation of the Protein Structure Prediction to medium size protein sequences ($46 - 70$ residues). In particular, we introduce a modified version of Pareto Archived Evolution Strategy (PAES) [Knowles & Corne 1999] which makes use of immune inspired computing principles and which we will denote by "I-PAES". Experimental results on the test bed of five proteins from PDB show that I-PAES is better than (1+1)-PAES both in terms of best solution found and convergence. Moreover, I-PAES is very competitive with other single-objective evolutionary algorithms and multi-objective evolutionary algorithms proposed in literature, both in terms of minimal energy value found, and RMSD and DME metrics.

When an optimisation problem involves more than one objective function, the task of finding one (or more) optimum solution, is known as multi-objective optimisation. In general, most bioinformatics problems involve multiple, conflicting, and sometime noncommensurate objectives. The PSP naturally involves multiple and conflicting objectives [Cutello *et al.* 2006]. Different solutions (the 3D conformations) may involve a *trade-off* (the conflicting scenario in the funnel landscape) among different objectives. An optimum solution with respect to one objective may not be optimum with respect to another objective. As a consequence, one cannot choose a solution which is optimal with respect to only one objective. In general, in problems with more than one conflicting objective, there is no single optimum solution. Instead, there exist a number of solutions which are all optimal: the *Optimal Pareto front* and the *Observed Pareto front*. This is the fundamental difference between a single-objective and multi-objective optimisation task. Hence, for a multi-objective optimisation problem we can define the following procedure:

1. find the Observed Pareto front with a wide range of values for objectives;
2. choose one of the obtained solutions using some "higher-level information".

Lamont *et al.* [Day *et al.* 2001] reformulated PSP as a multi-objective optimisation problem and used a multi-objective evolutionary algorithms (MOFMGA) for the structure prediction of two small peptide sequences: [Met]-Enkephalin (5 residues), Polyalanine (14 residues). In this Chapter we investigate for the first time the applicability of such a multi-objective approach to medium size protein sequences.

Constraints

The size of the conformational space, backbone torsion angles are bounded in regions derived from secondary and super-secondary structure prediction (table 11.4). These

Secondary and Super-secondary Structures ϕ		ψ
H (α-helix)	$[-75°, -55°]$	$[-50°, -30°]$
E (β-strand)	$[-130°, -110°]$	$[110°, 130°]$
a	$[-150°, -30°]$	$[-100°, 50°]$
b	$[-230°, -30°]$	$[100°, 200°]$
e	$[30°, 130°]$	$[130°, 260°]$
l	$[30°, 150°]$	$[-60°, 90°]$
t	$[-160°, -50°]$	$[50°, 100°]$
undefined	$[-180°, 0°]$	$[-180°, 180°]$

Table 11.6. Corresponding regions of the Secondary and Super-secondary Structure Constraints

regions are the constraints of the protein structure prediction problem, they define the *feasible region* of the multi-objective optimisation. Side chain torsion angles are constrained in regions derived from the backbone-independent rotamer library of Roland L. Dunbrack [Dunbrack & Cohen. 1997]. *Super-secondary structure* is defined as the combination of two secondary structural elements with a short connecting peptide between one to five residues in length. A short connecting peptide can have a large number of conformations. They play an important role in defining protein structures. The conformations of the residues in the short connecting peptides are classified into five major types, namely, a, b, e, l, or t [Sun & Jang 1996] each represented by a region on the ϕ-ψ map. Sun et al. [Sun et al. 1997] developed an artificial neural network method to predict the 11 frequently occurring supersecondary structure:

$$\text{H-}b\text{-H, H-}t\text{-H, H-}bb\text{-H, H-}ll\text{-E, E-}aa\text{-E, E-}ea\text{-E,}$$

$$\text{H-}lbb\text{-H, H-}lba\text{-E, E-}aal\text{-E,E-}aaal\text{-E, and H-}l\text{-E,}$$

where H and E represent α-helix and β-strand, respectively. Side chain constraint regions are of the form: $[m-\sigma, m+\sigma]$; where m and σ are the mean and the standard deviation for each side chain torsion angle computed from the rotamer library. Under these constraints the conformation is still highly flexible and the structure can take on various shapes that are vastly different from the native shape.

Figure 11.4 shows predicted and native secondary and super-secondary structure of the five proteins used in the present Chapter as test bed.

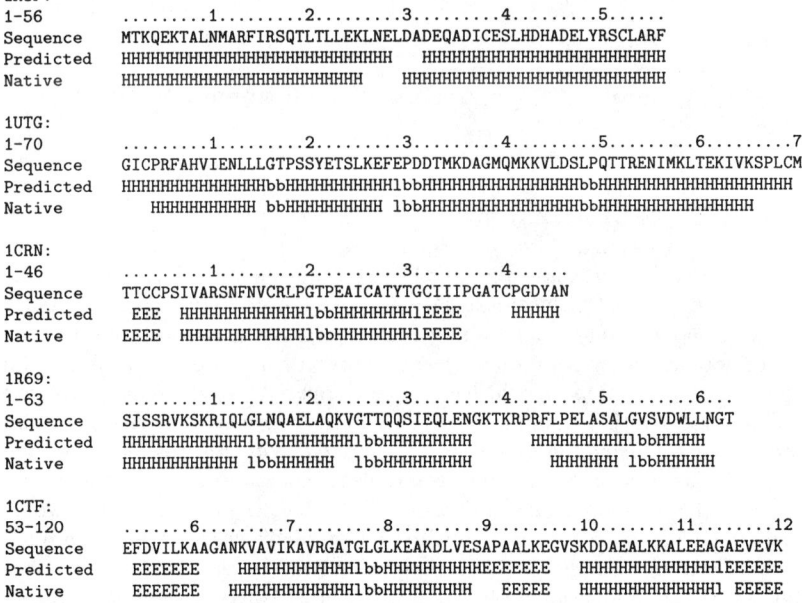

Fig. 11.4. Comparison of the predicted secondary and super-secondary structures (predicted) and the X-ray-elucidated secondary and super-secondary structure (native) of $1ROP$, $1UTG$, $1CRN$, $1R69$, and $1CTF$

Multi-objective formulation

The energy function CHARMM (11.1) is decomposed in two partial sums: *bonded* and *non-bonded* atom energies, following the definition (11.1):

$$f_1 = E_{bond} = \sum_{k=1}^{5} E_k \tag{11.4}$$

$$f_2 = E_{non-bond} = \sum_{k=6}^{7} E_k \tag{11.5}$$

These two functions represent our minimization *objectives*, the torsion angles of the protein are the *decision variables* of the multi-objective problem, and the constraint regions are the *variable bounds*. To evaluate how similar the predicted conformation is to the native one, we employ two frequently used metrics: Root Mean Square Deviation (RMSD) and Distance Matrix Error (DME) on C_α atoms (see Section 11.3 for the definition of these metrics).

11.4.1 The Immune Pareto Archived Evolutionary Strategy Algorithm

The algorithm PAES (Pareto Archived Evolutionary Strategy) was proposed for the first time by Knowles and Corne in 1999 [Knowles & Corne 1999]. PAES is a multi-objective optimiser which uses a simple (1+1) evolution strategy. Nonetheless, it is capable of finding diverse solutions in the Observed Pareto front because it maintains an archive of nondominated solutions which it exploits to accurately estimate the quality of new candidate solutions. At any iteration t, there is a candidate solution c_t and a mutated solution m_t which must be compared for dominance. Acceptance is simple if one solution dominates the other. In case neither solution dominates the other, the new candidate solution is compared with the reference population of previously archived nondominated solutions. If comparison to the population in the archive fails to favour one solution over the other, the tie is split to favour the solution which resides in the least crowded region of the space. A maximum size of the archive is always maintained. The crowding procedure is based on recursively dividing up the M-dimensional objective space in 2^d equal-sized hypercube, where d is a user defined depth parameter. The algorithm proceeds until a fixed number of *iterations* is reached.

I-PAES is a modified version of (1+1)-PAES[Knowles & Corne 1999, Knowles & Corne 2000] with a different solution representation (polypeptide chain) and immune inspired operators: *cloning* and *hypermutation* [Cutello & Nicosia 2004]. Hypermutation can be seen as a search procedure that leads to a fast maturation during the affinity maturation phase. The clonal expansion phase triggers the growth of a new population of high-value solutions centered on a higher affinity value. The algorithm starts by initialising a random conformation. From the protein sequence we generate

```
I-PAES(dup, depth, archive_size, objectives)
1. t := 0;
2. Initialize(c);                            /*Generate initial random solution*/
3. Evaluate(c);                              /*Evaluation of initial solution*/
4. AddToArchive(c);                          /*Add c to archive*/
5. while(not(Termination()))

            /* START IMMUNE PHASE*/
6.      Pop^clo := Cloning(c, dup);          /*Clonal expansion phase*/
7.      Pop^hyp := Hypermutation(Pop^clo);   /*Affinity maturation phase*/
8.      Evaluate(Pop^hyp);                   /*Evaluation phase*/
9.      m := SelectBest(Pop^hyp);            /*Selection phase*/
            /* END IMMUNE PHASE*/

            /* START (1+1)-PAES */
10.     if(c dominates m) discard m;
11.     else if(m dominates c)
12.             AddToArchive(m);
13.             c := m;
14.     else if(m is dominated by any member of the archive) discard m;
15.     else test(c, m, archive_size, depth) to determine which becomes
16.             the new current solution and whether to add m to the Archive;
17.     t := t + 1;
18.     endwhile
```

Table 11.7. I-PAES for PSP as Multi-objective Optimisation [Nicosia 2004]

a random conformation in torsion angles. The torsion angles (ϕ, ψ, χ_i) are selected randomly from the constraint regions derived from the super-secondary structure predicted with the Artificial Neural Network method developed by Sun *et al.* [Sun *et al.* 1997].

After that, the energy of the conformation (a point in the landscape) is evaluated using routines from TINKER Molecular Modeling Package [Huang *et al.* 1999]. First, the protein structure in torsion angles is transformed in cartesian coordinates using the PROTEIN routine. Then the conformation is evaluated using the ANALYZE routine, that gives back the CHARMM energy potential of the structure.

At this point, we have the main loop of the algorithm. From the current solution, a number δ of clones will be generated, producing the population (Pop^{clo}) which will be mutated into (Pop^{hyp}) and then evaluated reselecting the best one. From this moment on, the algorithm proceeds following the standard structure of (1+1)-PAES. Table 11.4.1 shows the pseudo-code of *I-PAES* for the protein structure prediction as a Multi-Objective Optimisation problem.

Hypermutation operators

Two kinds of mutation operators were used together in the affinity maturation phase (line 7 of I-PAES). The first mutation operator, mut_1, may change the conformation dramatically. When this operator acts on a peptide chain, all the values of the backbone and side chain torsion angles of a randomly chosen residue are reselected from their corresponding constrained regions (see table 11.4). The second mutation operator, mut_2, performs a local search of the conformational space. It will perturb all torsion angles (ϕ, ψ, χ_h) of a randomly chosen residue with the law: $\mathbf{x}^{(t,\prime)} = \mathbf{x}^t + \mathbf{u}$ where the mutation vector $\mathbf{u}$ is computed from $\mathbf{u} = (u_1, u_2, \ldots, u_n)$, $u_j = N(0, \sigma)$ where $\mathbf{x}^{(t)}$ is a B cell receptor (a conformation) at time step t, and $N(0, \sigma)$ is a real number generated by a Gaussian distribution of mean $\mu = 0$ and standard deviation $\sigma = 3$. In contrast with the hypermutation operator used in the previous Section (Sect. 11.3.1), in this new hypermutation operator σ is fixed for all the searching process. Preliminary experimental results give good convergence performance using a predefined σ value. The first half of Pop^{clo} is mutated using mut_1 and the second half using mut_2.

Two mutation rates are studied. The first one is a static scheme where each clone is mutated only once using one of the two possible mutation operators. We call I-PAES(1-mut) the algorithm version that uses this mutation rate.

The second mutation rate instead is similar to the scheme presented in [Cui et al. 1998]. The number of mutations decreases as the search method proceeds following the law: $M = 1 + \frac{L}{k} \times e^{\left(\frac{-t}{\gamma}\right)}$ where L is the number of residues, k is set to 6 for mut_1 and 4 for mut_2 and γ controls the shape of the mutation rate. We call I-PAES(M) the algorithm version that uses this second mutation rate.

11.4.2 Immune PAES dynamics on 1CRN protein

To show the dynamic behaviour of the algorithm, we present in figure 11.5 two typical plots produced by a generic run of I-PAES on *Crambin* (*1CRN*). Crambin is a protein of 46 amino acids with two α-helix and a pair of β-strands. The predicted supersecondary structures (see figure 11.4) are:

$$\beta - loop - \alpha - lbb - \alpha - l - \beta - loop - \alpha.$$

The top plot of figure 11.5 shows best solution energy versus iterations. The non-bound term endures strong fluctuations, and this happens because it includes van-der-Walls interaction energy. During the folding process, it is highly probable that conformations with atom clashes can be created which will produce high van-der-Walls energy levels.

The bottom plot of figure 11.5 shows variations of the archive size at each iteration. Archive size falls back when the algorithm finds a solution which is better, in terms of dominance, than a number of solutions in the archive, which are all eliminated in favour of the new solution.

The figure 11.6 shows the Pareto front of the last archive for 1CRN best run. In figure 11.6, the nondominated solutions are subdivided into three clusters which correspond to the three regions of the Pareto front.

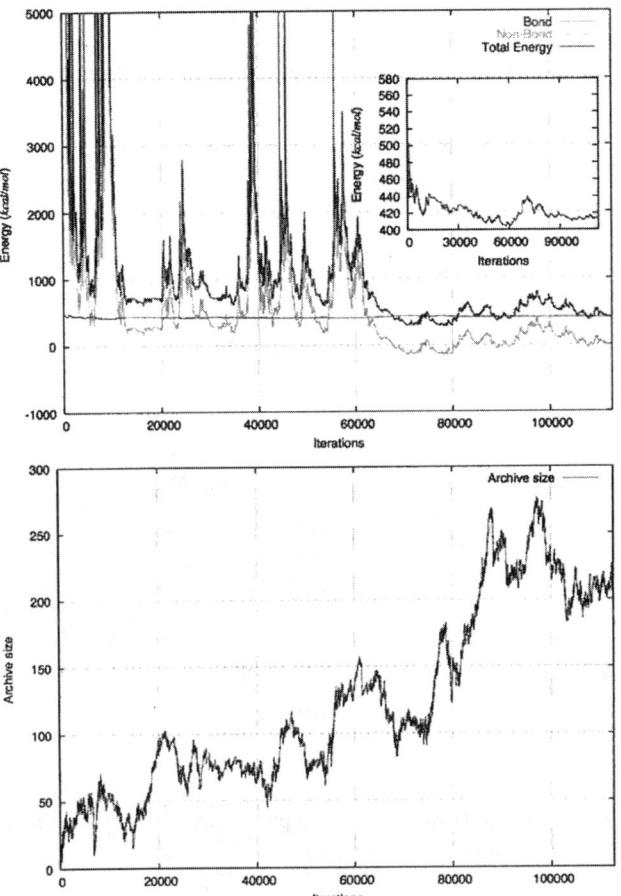

Fig. 11.5. Best energy value (top plot), and Archive size (bottom plot) versus iterations for *1CRN*

In table 11.4.2 we compared I-PAES and its results to other works in literature and others MOEA's, in particular NSGA2 [Deb *et al.* 2002], that we implemented and tested on protein structure prediction of real proteins.

In [Cooper *et al.* 2003] the best RMSD found for 1CRN, using a Hill-climbing genetic algorithm, is 5.6Å and with average number of evaluation to solution (*AES*) equal to 5×10^6 fitness function evaluations. In this case, our method performed better both in terms of best solution found and time efficiency. Inspecting the results reported in table 11.4.2, both versions of I-PAES outperform the good RMSD value obtained by GA designed by Dandekar and Argos [Dandekar & Argos 1996] which uses 4×10^5 fitness function evaluations. Moreover, both Dandekar *et al.* [Dandekar & Argos 1996] and Corne *et al.* [Cooper *et al.* 2003] used a reduced representation for the protein where side chain atoms are not included and are represented only by the

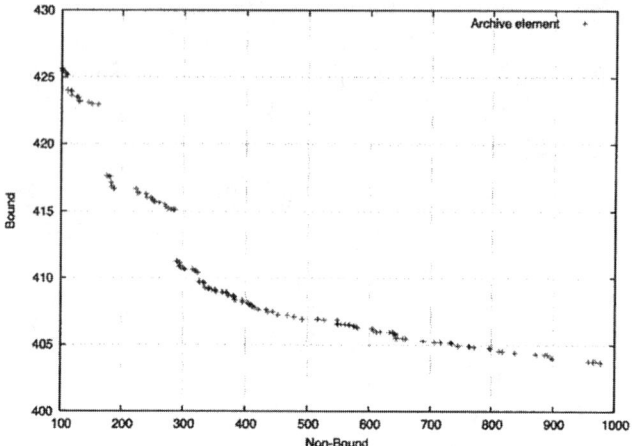

Fig. 11.6. Pareto front for *1CRN*

Algorithm	RMSD	AES
I-PAES(*M*)	**4.316Å**	**2.255 × 10⁵**
I-PAES(1-mut)	4.731Å	2.255 × 10⁵
Dandekar and Argos' GA[Dandekar & Argos 1996]	5.4Å	4 × 10⁵
Hill-climbing GA [Cooper *et al.* 2003] (with hydrophobic term)	5.6Å	5 × 10⁶
(1+1)-PAES	6.181Å	2.255 × 10⁵
NSGA2 (with high-level operators)	6.447Å	2.5 × 10⁵
Hill-climbing GA [Cooper *et al.* 2003] (without hydrophobic term)	6.8Å	10⁶
NSGA2 (with low-level operators)	10.34Å	2.5 × 10⁵

Table 11.8. Best results between (1+1)-PAES, I-PAES, NSGA2, Hill-climbing GA [Cooper *et al.* 2003] and Dandekar and Argos' GA[Dandekar & Argos 1996] on *1CRN*

protein mainchain. The ϕ and ψ torsion angle values are taken from a set of seven and four possible standard conformations in known tertiary structures respectively. In this way, the space of conformations is reduced but the model is really rigid.

Two possible versions of NSGA2 were implemented. The first one uses standard low-level operators (SBX crossover and polynomial mutation), and the protein is considered as a long sequence of torsion angles. The second one uses high-level operators (naive crossover and the scheme of mutation used by I-PAES). In this case, the protein is manipulated at the amino acid level. The dimension of the population is 300. The better performance of the high-level version is very clear. Table 11.4.2 shows the comparison between I-PAES, (1+1)-PAES, NSGA2, Hill-climbing GA [Cooper *et al.* 2003] and Dandekar *et al.*'s GA [Dandekar & Argos 1996] on 1CRN.

11.4.3 Experimental Results

The simplicity of having effectively only two secondary structures is that there are only three (pairwise) combinations of them that can be used to construct proteins; so giving the three major structural classes:

1. $\alpha-$helix with $\alpha-$helix,
2. $\alpha-$helix with $\beta-$sheet, and
3. $\beta-$sheet with $\beta-$sheet.

Incorporation of a β-sheet, however, imposes a long-range constraint across the structure. The β-sheet has free hydrogen bonds on its two edges, which consequently prevents the sheet from terminating in the hydrophobic core. This divides the core into two and, if considered more generally, imposes a layered structure onto the further arrangement of secondary structures in the protein.

All-α proteins

The all$-\alpha$ protein class is dominated by small folds, many of which form a simple bundle with helices running up then down. The interactions between helices are not discrete (in the way that hydrogen bonds in a $\beta-$sheet are either there or not) which makes their classification more difficult. Set against this, however, the size of the $\alpha-$helix (which is generally larger than a $\beta-$strand) gives more interatomic contacts with its neighbours (relative to the a $\beta-$strand) allowing interactions to be more clearly defined.

All$-\beta$ proteins

The all$-\beta$ proteins are often characterized by the number of $\beta-$sheets in the structure and the number and direction of $\beta-$strands in the sheet. This leads to a fairly rigid classification scheme which can be sensitive to the exact definition of hydrogen bonds and $\beta-$strands. Being less rigid than an $\alpha-$helix, the $\beta-$sheets can be relatively distorted often with differing degrees of twist of fragmented or extra strands on the edges of the sheet. Various patterns can be identified in the arrangement of the β-strands, often giving rise to the identification of recurring motifs.

$\alpha - \beta$ proteins

The $\alpha - \beta$ protein class can be subdivided roughly into proteins that exhibit a mainly alternating arrangement of $\alpha-$helices and $\beta-$strands along the sequence

Protein	Algorithm	min (kcal/mol)	mean (kcal/mol)	σ (kcal/mol)
1ROP(56 aa)	I-PAES(M)	-526.9542	-417.4685	98.2774
class: α	I-PAES(1-mut)	**-661.4819**	**-554.9819**	**82.9940**
energy: *-667.0515 kcal/mol*	(1+1)-PAES	2640.7719	833976.1875	1497156.7511
1UTG(70 aa)	I-PAES(M)	357.9829	619.8551	174.8500
class: α	I-PAES(1-mut)	**282.2497**	**511.4623**	**142.1591**
energy: *202.7321 kcal/mol*	(1+1)-PAES	7563.0714	53937.0271	55304.4139
1CRN(46 aa)	I-PAES(M)	410.0382	464.2972	**42.4524**
class: $\alpha + \beta$	I-PAES(1-mut)	**232.2967**	**357.2083**	75.9134
energy: *-142.4612 kcal/mol*	(1+1)-PAES	1653.9359	27995.0374	43275.1845
1R69(63 aa)	I-PAES(M)	264.5602	397.6853	74.9013
class: α	I-PAES(1-mut)	**211.2640**	**290.0966**	**46.3440**
energy: *-676.5322 kcal/mol*	(1+1)-PAES	9037.8915	2636441.5872	4462510.0991
1CTF(68 aa)	I-PAES(M)	**218.9968**	281.27994	64.3010
class: $\alpha + \beta$	I-PAES(1-mut)	71.5572	**161.4119**	**48.8140**
energy: *230.0890 kcal/mol*	(1+1)-PAES	1424.3397	52109.3556	44669.0231

Table 11.9. Results of application of I-PAES (both versions) and (1+1)-PAES to the test bed of five proteins used in [Cui *et al.* 1998]. For each algorithm, the minimum, the mean and the standard deviation on ten independent runs of best energy values are reported. For each protein were reported, the Protein Data Bank (PDB) identifier, the length and the approximate class (α-helix, β-sheet)

and those that have more segregated secondary structures. The former class includes structures in which the secondary structures are arranged in layers and those that form a circular of barrel-like arrangement. Recurring folds can also be identified in the latter type.

Comparisons

To assess the quality of the prediction ability of I-PAES, hence we need to face protein instances that belong to all the above-cited classes of proteins.

Tables 11.9 and 11.4.3 show results applying I-PAES(M), I-PAES(1-mut) and (1+1)-PAES to the five PDB proteins on ten independent runs. Inspecting the data reported in data one can seen as the five protein instances have been sorted by classes: α proteins, β proteins, and $\alpha + \beta$ proteins.

We set the maximum number of fitness function evaluations ($Tmax$) to 2.255×10^5 (so to compare it to [Cui *et al.* 1998]), a minimal duplication value ($\delta=2$), archive size of 300 and *depth* = 4. The mutation operator used for (1+1)-PAES is the local mutation mut_2, since PAES is based on a local search strategy. From table 11.9 is clear that both versions of I-PAES perform better than (1+1)-PAES. Minima energy values obtained by I-PAES are closer to those of the native structure. Moreover, the high value of the standard deviation for (1+1)-PAES shows worse convergence than I-PAES.

Protein	Algorithm	min		mean		σ	
		DME(Å)	RMSD(Å)	DME(Å)	RMSD(Å)	DME(Å)	RMSD(Å)
1ROP(56 aa)	I-PAES(*M*)	**1.684**	**3.740**	4.444	6.462	2.639	2.661
class: α	I-PAES(1-mut)	2.016	4.110	**3.405**	**5.592**	**1.036**	**1.128**
	(1+1)-PAES	4.919	6.312	9.465	10.111	3.866	3.468
	GA[Cui *et al.* 1998]	1.48	-	-	-	-	-
1UTG(70 aa)	I-PAES(*M*)	**3.474**	**4.272**	5.417	7.404	1.484	2.330
class: α	I-PAES(1-mut)	4.498	5.117	**5.221**	**6.351**	**0.817**	**1.066**
	(1+1)-PAES	4.708	6.047	6.637	8.936	1.242	1.647
	GA[Cui *et al.* 1998]	3.47	-	-	-	-	-
1CRN(46 aa)	I-PAES(*M*)	**3.436**	**4.316**	**5.057**	5.874	1.278	0.960
class: α + β	I-PAES(1-mut)	4.137	4.731	5.156	**5.817**	**0.758**	**0.726**
	(1+1)-PAES	4.676	6.181	6.700	7.778	2.164	1.404
	GA[Cui *et al.* 1998]	2.73	-	-	-	-	-
1R69(63 aa)	I-PAES(*M*)	**4.091**	**5.057**	7.867	9.630	0.815	0.911
class: α	I-PAES(1-mut)	5.932	8.425	**7.218**	**9.557**	**0.669**	**0.551**
	(1+1)-PAES	5.167	7.599	7.589	9.607	2.544	1.809
	GA[Cui *et al.* 1998]	4.48	-	-	-	-	-
1CTF(68 aa)	I-PAES(*M*)	**6.822**	**10.121**	10.773	13.559	1.351	0.727
class: α + β	I-PAES(1-mut)	8.081	10.691	**9.192**	**11.303**	**0.988**	**0.468**
	(1+1)-PAES	9.609	12.092	10.534	12.957	**0.936**	0.832
	GA[Cui *et al.* 1998]	4.00	-	-	-	-	-

Table 11.10. Results of application of I-PAES (both versions) and (1+1)-PAES to the test bed of five proteins used in [Cui *et al.* 1998]. For each algorithm, the minimum, the mean and the standard deviation on ten independent runs for DME and RMSD values are reported. For each protein we report, the PDB identifier, the length and the approximate class (α-helix, β-sheet).Also shown is the best DME value obtained for each protein respect the GA proposed in [Cui *et al.* 1998] (no RMSD values were presented by the authors)

Fig. 11.7. 1ROP: predicted structure (left plot) with DME = 1.684Å and RMSD = 3.740Å, native structure (right plot)

In table 11.4.3 the best DME and RMSD values are always obtained with I-PAES(*M*), although I-PAES(1-mut) reaches better energy values. The conformations in figures 11.7, 11.8, and 11.9 show the predicted structures [Huang *et al.* 1996] (in wire-style) calculated using I-PAES algorithms versus the native structure for the five proteins examined.

Fig. 11.8. 1UTG: predicted structure (left plot) with DME = 3.474Å and RMSD = 4.272Å, native structure (right plot)

Fig. 11.9. 1CRN: predicted structure (left plot) with DME = 3.436Å and RMSD = 4.316Å, native structure (right plot)

Table 11.4.3 shows the best DME values obtained by the genetic algorithm proposed in [Cui *et al.* 1998] (no values were presented by the authors) on the same protein test bed. Results obtained by GA [Cui *et al.* 1998] are comparable to those obtained I-PAES. I-PAES did not perform well for 1CTF where GA[Cui *et al.* 1998] reached a better DME value.

11.5 Conclusions

Experimental results on Met-Enkephalin peptide and Disulfide-Stabilized Mini Protein A Domain (1ZDD) show how the designed Immune Algorithm for single-objective optimisation is a very competitive and effective search method in the conformational search space of real proteins. In particular, the IA-DIRECT approach is very efficient in terms of quality solution and computational cost comparing the results of the current state-of-art algorithms. A short protein sequence, the Met-Enkephalin peptide (a well-known test bed), has been guided by Immune Algorithm to folds resembling its crystal structure with deviation from observed structure of $2.8 - 2.9$Å for C_α atoms. At our knowledge, the immune algorithm designed obtained the best result for the *1MET* peptide. The good correlation between RMSD and energy value E, suggesting that minimizing the energy function of a small protein using Immune Algorithm will tend to drive the conformation toward the true structure. Moreover, the PSP has been faced as a multi-objective optimisation problem using an hybrid immune algorithm.

The global fold of the test bed proteins have been computed from their sequence, with deviations from crystal structure of $1-4$ DME Å for C_α atoms. We proposed a modified version of the algorithm PAES that uses immune inspired principles (Clonal Expansion and Hypermutation operators) as a new search method for PSP. The obtained algorithm, denoted by I-PAES, has better performance than the standard (1+1)-PAES for PSP both in terms of best energy and metrics (DME, RMSD) solutions. Moreover, I-PAES has better convergence than PAES as shown by the smaller values of the standard deviation in 10 independent runs. For the first time, a multi-objective approach was used to fold medium size proteins (46-70 residues), and the results are comparable and sometimes better than other approaches in literature. Lamont *et al.* were the first to study PSP as multi-objective problem, but in their work [Day *et al.* 2001] was related only to two peptide sequences (5 and 14 residues). Experimental results on 1CRN protein show also a better performance of the PAES algorithms (I-PAES and (1+1)-PAES) with respect to genetic algorithms [Dandekar & Argos 1996], hill-climbing genetic algorithms [Cooper *et al.* 2003], and NSGA2 multi-objective optimisation evolutionary algorithm with low and high level representations. Our prediction algorithms, based on general full-atom potential energy models, are expanded to incorporate secondary and supersecondary structure information into the search process. We contrast our method with the state-of-art prediction algorithms obtaining good results in terms of energy values and RMSD and DME values.

Finally, the experimental results of the Chapter indicates that Multi-Objective Optimisation approach using real-valued evolutionary algorithms for determining the protein structure of real proteins has excellent potential to determine and investigate the Observed Pareto front in order to select a *stable fold protein* near native conformation, under biological conditions, satisfying the objectives as "best" possible.

Acknowledgements

The authors want to thanks G. Narzisi for his support providing the images reported in the manuscript.

Part III

How Natural and Artificial Immune Systems
Interact with the World

12

Embodiment

Susan Stepney

Department of Computer Science, University of York, York, YO10 5DD, UK.
susan@cs.york.ac.uk

Summary. *Embodiment* may help to reduce the computational burden on a system, by transferring some of that burden to the complex embodying environment. Embodiment can be viewed as a property not just of situated material systems, but of any suitably complex system engaged in a complex intertwined feedback relationship with its suitably complex environment. Various features and requirements of embodiment are examined in the context of natural and of artificial immune systems. This leads to a set of suggested *design principles* for engineering embodied systems and their environments.

12.1 Embodiment : what is it?

12.1.1 Embodied in an environment

Consider a system that can sense and manipulate its environment, with its internal state depending on what it senses, and its manipulations depending on its state. Those manipulations then change the environment, and hence what is subsequently sensed, and so change the system's subsequent state and its further manipulations. This produces a complex dynamical stigmergic feedback process: the system is *embodied* in its environment.

Biological immune systems, like all biological processes, are made from physical material, they are situated in and interact with a physical environment, and they are constrained by physical laws such as conservation of matter and energy. Artificial immune systems (AIS) and other software systems, on the other hand, are informational, exist in the *virtual* world of the computer, and labour under no such constraints. Does this difference matter?

A new generation of roboticists [Brooks 1991a, Brooks 1991b], AI researchers [Maturana & Varela 1980, Varela *et al.* 1991, Simon 1996], psycholinguists [Lakoff & Johnson 1980, Lakoff 1987, Lakoff & Núñez 2000], and cognitive philosophers [Clark 1997] insist that it does. They argue that the rich dynamical interaction between a system embodied in its complex physical environment crucially gives something to the system not achieved by pure virtual or symbolic or simulated inputs alone.

Much of Brooks' robotics work has focused on the use of the environment as a resource; rather than requiring the robot to abstract data from the world, to form some impoverished model, [Brooks 1991b] exhorts the robot's designer to *"use the world as its own model"*.

As an illustration of the importance of a complex environment, Herbert Simon invites us to imagine an ant walking on the beach:

> Viewed as a geometric figure, the ant's path is irregular, complex, hard to describe. But its complexity is really a complexity in the surface of the beach, not a complexity in the ant ... *The apparent complexity of its behaviour over time is largely a reflection of the complexity of the environment in which it finds itself.*   [Simon 1996]

Clark also emphasises the crucial involvement of the environment:

> the deeply misguided vision of the environment as little more than the stage that sets up a certain problem. ... the environment [is] a rich and active resource—a partner in the production of adaptive behavior.
>   [Clark 1997]

So, the environment itself is a resource, and embodiment may help to reduce the computational burden on the system itself, with some, or much, being obtained "for free" from the complex embodying environment. Of course, nothing is really for free, and the requirement for embodiment puts some interesting constraints on the design and deployment of embodied systems (later).

12.1.2 Coupled to the environment

But what precisely *is* such "embodiment"? [Kushmerick 1997, Quick & Dautenhahn 1999, Quick *et al.* 1999, Quick *et al.* 2000] note that these roboticists and AI researchers rarely bother to define the term beyond saying something like *having some (physical) body interacting with some (physical) environment*. They note that this (lack of) definition, along with the accompanying assumption of *material* situatedness, makes it particularly difficult to use the concept of embodiment when talking about Artificial Life, or virtual artefacts such as software agents.

Quick *et al.* are keen to develop a definition that makes as few assumptions as possible, in particular, one that has no requirement for the embodied system to be in some sense intelligent or cognitive or neuronal. So they turn to the concept of "*structural coupling*" [Maturana & Varela 1980], and the key idea of

> non-destructive perturbations between a system and its environment, each having an effect on the dynamical trajectory of the other, and this in turn effecting the generation of and responses to subsequent perturbations.
>
> [Quick & Dautenhahn 1999]

[Clark 1997] dubs such close structural coupling "*continuous reciprocal causation*". These ideas of dynamical trajectories, attractors, and bifurcations, which encompass both situated continuous physical processes *and* abstract discrete computations (captured in the appropriate phase or state space), are also important, implicitly or explicitly, in the writings of [Maturana & Varela 1980, Varela *et al.* 1991, Kelso 1995, Chiel & Beer 1997].

Focusing on this coupling between the system X and its environment E, [Quick & Dautenhahn 1999, Quick *et al.* 1999, Quick *et al.* 2000] offer the following definition of embodiment:

> *A system X is embodied in an environment E if perturbatory channels exist between the two. That is, X is embodied in E if for every time t at which both X and E exist, some subset of E's possible states with respect to X have the capacity to perturb X's state, and some subset of X's possible states with respect to E have the capacity to perturb E's state.*

They suggest that this definition in terms of coupling can be used as a basis to *quantify* the degree of embodiment, in terms of measures of the perturbatory bandwidths and modalities, size of affected state subspaces, size of effect on state spaces, scope for variation in behaviour, structural plasticity, computational power in the interaction, and so on. They conclude that for rich embodiment one needs *both* a large perturbatory bandwidth (rich sensors and actuators to enable complex couplings with the environment) *and* large scope for variation in behaviour (complex internal dynamics for the perturbations to work on).

And most importantly for the arguments in this chapter, they note that their definition is "ontologically neutral": it is applicable to systems embodied in software (virtual) environments as well as to those embodied in material (physical) environments.

In summary, some authors argue that embodiment is essential for certain kinds of systems, and others argue that it is a property not just of situated material systems, but of any suitably complex system engaged in a complex intertwined feedback relationship with its suitably complex environment.

One feature of embodiment is that, because of the complex feedback relationship, an embodied system cannot be fully analysed in isolation: it can be analysed only in the context of the environment in which it is embodied. The fact that this environment is usually open (unbounded) has interesting consequences for such analysis.

12.1.3 Some terminology

We follow Quick *et al.*'s definitions, and so have a computational system embodied in an environment (figure 12.1).

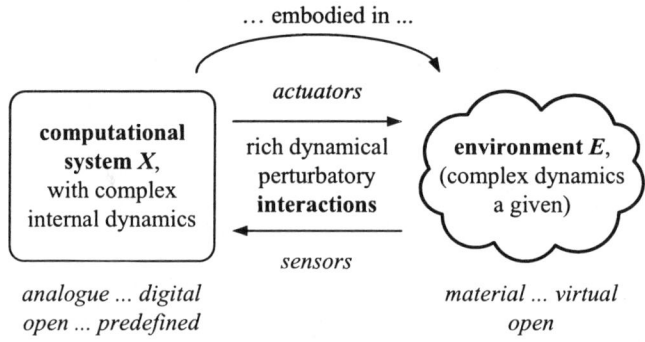

Fig. 12.1. A computational system embodied in an environment

The **computational system** itself *necessarily* runs on some form of physical hardware platform, but that hardware does not constitute the embodiment: it is the coupling with the environment that provides the embodiment. (The boundary between the computational hardware and its environment may be drawn in different places, depending on the analysis being performed. For example, in a robot, it might sometimes be useful to consider the robot body as part of the environment, and sometimes as part of the computational system.) The computational hardware platform may be analogue, digital, or a hybrid. The computational processes may be predefined (fixed code running on fixed hardware) or open (which could include dynamic binding, self-modifying code, self-reconfiguring hardware, and so on).

Following Quick *et al.*, we require that the computational system has a suitably rich complex internal dynamics. While most computational systems do indeed have a complex internal dynamics, this may not be "suitably rich", rather, it tends instead be distressingly fragile and impoverished. Engineering a *suitable* dynamics for exploiting embodiment may be non-trivial.

The **environment** in which the system is embodied may be material (for example, the physical world in which a robot is situated), or virtual (for example, the Internet). Again, a virtual environment necessarily has some underlying physical (hardware) implementation, but its behaviour may be abstractable from the details of the particular implementation in terms of certain logical properties only. The environment is required to be *open*, hence its behaviour cannot be captured in some predefined description. A physical environment is always open: the world does what the world does, and any model we have of it is just that: a model, that abstracts away from some details. A virtual environment may also be open: for example, the capabilities and topology of the Internet cannot be captured in some predefined description, but are changing and growing in unpredictable ways on a daily basis.

The environment may be relatively passive, changing only in response to the system's actuation, or it may be following its own physical laws over time, so the effect of actuations decay or grow, or it may contain other embodied computational systems (a *social* environment), each capable of being sensed, each acting on the rest of the environment, and each altering its state based on what it senses.

The **embodiment** is provided by the coupling between the computational system and its environment, and is a rich complex feedback process. Inputs to the system from the environment are through its **sensors**, and it outputs to the environment through its **actuators**.

The effect of an actuation might be minimal, merely change the system's location or orientation in the environment, thereby changing what it senses. Such actuation is perturbing the environment in a *relativistic* sense only: we can choose to say that the actuation causes the environment to move relative to the system.

A more fully embodied effect of the actuation is to alter the environment in some explicit stigmergic way, such as by making or erasing a mark, thereby transferring a memory burden to the environment, or by building or changing a structure. Subsequent sensing, by this system or by other systems in the environment, may perceive these alterations, and thereby alter behaviours.

If the environment adapts in some way to the embodied system's change, it needs to be aware of this change, unlike in the case of mere relativistic movement. For example, a virtual environment might allocate more resources to the region that the system is currently inhabiting.

A *materially embodied system* is a computational system embodied in a physical environment; a *virtually embodied system* is one embodied in a virtual environment (that is, the terms *material* and *virtual* refer to properties of the environment, not of the computational system).

12.2 Rich dynamics

12.2.1 Physical and virtual constraints

Laws. A physical structure (here, either system X or environment E) labours under the *physical constraints* of the material from which it is constituted. These constraints include such things as: basic laws of physics, from speed of light constraints to energy density constraints; physical properties of the material, its strength and resistance; natural length scales and timescales governing the dynamics and the attractor structure of the phase space.

A virtual structure similarly labours under various *computational constraints*. The most obvious of these is *computability*. If the virtual structure in question is a Turing-equivalent machine, then the Church-Turing thesis states that it is limited to performing effectively computable functions. Some argue that this computability constraint applies to *all* physical and virtual systems; others disagree. For a good discussion of this point, see [Copeland 2002].

Questions of computability notwithstanding, the issue of *computational complexity* is crucial for virtual systems. Certain computations can be performed in principle, but in practice they take too long (longer than the age of the universe, say). This is a direct analogue of the speed of light constraint in physical systems: in principle any destination is reachable if one travels for long enough, but it may take an infeasibly long time to travel there. A class of problem whose solution time scales at most polynomially with the problem size is classed as *efficient*; one that scales exponentially with problem size is infeasible. It should be noted, however, that feasibility is a technical "worst case" measure: even if a *class* of problems is infeasible, particular *instances* of that problem may be relatively simple, and even if an exact solution is infeasible to find, a good enough approximate solution may be feasible. AIS are one bio-inspired approach to providing feasible approximate solutions to a subset of infeasible exact problems.

Even if one has efficient computation, there are further constraints when considering an embodied system. The computational system is coupled with the environment, and the environment is acting and reacting on certain timescales, constrained by the relevant laws. So it behooves the computational system to act and react on appropriately similar timescales, which puts constraints on its implementation technology. Speed matters. A Turing machine built from beer cans, or John Searle locked in his Chinese Room [Searle 1980], would not be able to constitute appropriate *embodied* intelligences with respect to our everyday environment, since they would be acting on glacially slow timescales, and could not engage in rich perturbatory interactions on our timescales.

One of the tenets of embodiment, certainly from an artificial *intelligence* point of view, is that an embodied system transfers some of its computational burden to the (much larger, much richer) environment. Thus new classes of problems may become feasible for it to solve, and others may even disappear altogether [Kushmerick 1997].

Initial conditions. Physical structures are constrained by their history and dynamics: there may be certain physical states that, although potentially physically realisable, are unreachable from the system's starting state under the physical laws of the system. [Goodwin 1994] argues that such physical constraints and processes play the major role in the evolutionary and developmental dynamics of an organism, and that genetic variation provides a relatively small modulation to this.

Virtual structures are similarly constrained by their initial conditions and the dynamics of the computation. For example, it has been found, in the context of genetic programming, that there are some potential solutions that cannot be found by an evolutionary search process, given a certain class of move function [Daida *et al.* 2003a, Daida *et al.* 2003b, Daida & Hilss 2003]. Analogous constraints are doubtless features of other bio-inspired algorithms, such as AIS, that develop from some initial population using their own move functions, although the precise forms of these constraints are yet to be uncovered.

Viability. Natural biological systems are constrained by viability: individuals must be viable at all stages of their life, and members of species must be viable at all stages of their evolutionary history. Additionally, individuals have many functions that are necessary simply for maintaining their life.

Artificial systems (virtual or physical) potentially suffer fewer such constraints. They need not be viable until their construction is complete. Also, they often need have little or no functionality beyond that needed to support their primary purpose: any resources they need merely to "survive" are usually supplied by some external agency.

As these artificial systems become more complex, however, and particularly as designers look to biology for inspiration, some of the constraints of natural systems are in turn added to artificial ones: they may be artificially evolved and grown, with some requirement for intermediate viability, and they may need to compete for resources with other artefacts.

An AIS certainly has a competitive element to it, in that the individual elements within the system may compete with other such elements; the overall system, however, is usually situated in a more computationally traditional non-competitive environment.

12.2.2 Natural physical richness

As argued above, physical and virtual systems both labour under certain constraints, and it might therefore seem that (physical) embodiment adds nothing to the equation. However, physical systems are essentially rich, whereas virtual systems tend to be *impoverished* unless specially designed for richness. It is here that the effortless richness of physical embodiment can offer new opportunities, and where analogous properties may need to be designed in to virtual embodiment.

Physical systems can exploit continual novelty. It is in this sense that they are *rich*: they can, and often do, use *any* feature of the real world to perform their task, not just the ones abstracted out for analysis in the mathematical or computational model. Physical systems can move outside the model, and evolve new representations. *"Evolution tends to produce designs that take full advantage of the available freedom"* [Beer 1995]. Computational systems embodied in a physical environment can exploit this richness.

The classic work that demonstrates such physical richness is [Thompson & Layzell 1999], who use a genetic algorithm to evolve a two-frequency discrimination algorithm running on an unclocked Field Programmable Gate Array. The resulting solution circuits perform their task, but are bizarrely inexplicable in operation. In particular, some solutions have unconnected components, yet if these apparently irrelevant components are removed, the circuit fails to operate. Also the circuits are not *portable*: they cannot be moved to other places on the chip's array, or to other chips, or run at different temperatures. The circuits appear to be exploiting *extralogical* properties of the chips, such as capacitances between components (even ones not directly connected); these properties are not controlled by design or manufacture to be the same in all places or at all temperatures.

Even with a conventional digital microprocessor, such extra-logical properties can be important. For example, recent breakthroughs in cryptanalysis exploit extra-logical *side-channels*, for example using timing or power measurements in correlation with the computation being performed, in order to break the cryptographic systems. See, for example [Kocher 1996, Kocher *et al.* 1999, Clark *et al.* 2005b].

12.2.3 Achieving virtual richness

Physical systems do not need to develop any special mechanisms to obey the laws of nature: they just *naturally* follow such laws, and can evolve to exploit these laws. In contrast to physical environments with their open range of extra-logical properties occurring "for free", most virtual environments are severely constrained and impoverished. Typically they are *closed* systems with a pre-defined finite discrete logical representation, and cannot move out of this (or if they *do* move out, it constitutes an error).

The Internet/Web provides an interesting *open* virtual environment, in that it is constantly changing and growing. [Quick *et al.* 2000] describe *Phenomorph*, a system designed to be embodied in the open virtual environment of the Web. The system's *sensors* parse pages for keywords while its *actuators* select and navigate links. The sensory input alters the system's behaviour (in a way inspired by the locomotive behaviour of *E. coli*), which alters the actuator's choice of link to follow, which affects where the system is located in the environment.

The work is an initial attempt to explore the question: Can a purely virtual embodied system (that is, a system embodied in a virtual environment) experience the same degree of richness as a *physically* embodied system?

The answer appears to be a qualified 'yes', provided that the system and its virtual environment are designed to achieve a *suitably complex dynamics*, and that the environment is sufficiently open, allowing it to remain *far from equilibrium*.

12.2.4 Suitably complex dynamics

It is not sufficient for the system or the environment merely to have a *large* state space, or phase space: that space must also have a "suitably complex dynamics", that is, have a complex structure, neither too regular, not too random. The rich perturbatory interactions then make suitably complex changes to this structure.

The dynamics is described in terms of the trajectory that the system follows through its relevant state space. This trajectory is governed by the attractor structure of the state space. Inputs can change the system, possibly by altering the values of parameters describing the space, thereby altering its attractor structure, for example by moving attractors, or causing attractors to merge or bifurcate. [Beer 1995] provides a concise overview of the relevant dynamical systems theory.

A dynamics is suitably complex when it results in complex *emergent properties*, which may be identified with attractors or with other complex structures in the dynamics. These emergent properties are new higher level properties (patterns, agents) in space and time, and they have their own structure and dynamics, their own higher level state space, trajectories, and attractors. This higher level state space can then support the emergence of still higher level patterns, and so on. In an *open* system, where arbitrarily many levels of patterns can emerge, it is impossible to pre-define all the state space: the higher level state spaces emerge along with the patterns [Kauffman 2000], resulting in the potential for constant novelty.

Some authors argue that to achieve such richness, the state space must be continuous, rather than discrete: "*It is my belief that the versatility and robustness of animal behavior resides in the rich dynamical possibilities of continuous state spaces*" [Beer 1995]. However, research on Cellular Automata and other similarly "simple" systems shows that these discrete spaces can nevertheless have amazingly rich dynamical possibilities. See, for example, [Gardner 1970, Langton 1991, Wuensche & Lesser 1992, Wolfram 1994, Wuensche 2002].

It is an open question whether finite discrete systems can provide a sufficient richness of multiple levels of emergent structures, or whether continuous systems are qualitatively different. There has been much debate, and no doubt will be much more, on both sides of the argument.

Fractal proteins [Bentley 2004] provide one form of computational richness. The aim is to achieve the "complexity, redundancy and richness" of natural protein systems, without using (or simulating) the actual real-world mechanisms. Each fractal protein is a triplet of real numbers that encodes a small square patch of the Mandelbrot set centred at (x, y) and with size z. The rich and diverse shapes of natural proteins are mirrored in the complex and diverse shapes of these patches of the Mandelbrot set. Natural protein interactions are mirrored by intersecting fractal patches to measure their *affinity*, or how closely they match. An evolutionary or other search process searches for sets of triplets that exhibit appropriate dynamics under such intersection. The aim is for a sufficiently complex artificial chemistry, provided by the complexity of the highly non-linear Mandelbrot set.

Fractal proteins were originally designed to "provide a rich medium for evolutionary computation" of artificial gene regulatory networks (GRNs) [Bentley 2004]. Although originally designed for GRNs, fractal proteins, or related concepts, may also provide a rich medium for other bio-inspired processes, including immune ones. Antigens are a kind of protein, and [Bentley & Timmis 2004] use "fractal antigens" in the context of an artificial immune network, to represent the shape space and affinity functions.

The scheme works well in practice, but as used it may not be achieving its full potential. The search process as described [Bentley 2004] modifies the z parameter (patch size) by a random *additive* rather than *multiplicative* factor, making "deep zooming" (very small values of z) unlikely, and hence not exploiting the deep self-similar nature of the fractal. It also uses a relatively coarse-grained sampling grid on the patches to calculate their affinity. However, there is no reason in principle why a more fine-grained approach could not be used to produce arbitrarily complex chemistries.

It would be interesting to explore the effects on the generated dynamics of the precise choice of fractal, and even of (co-)evolving the underlying fractal itself.

12.2.5 Far-from-equilibrium openness

A closed dissipative system will reach equilibrium: it will converge on an attractor and stay there. If that attractor is a *strange attractor* [Lorenz 1963, Strogatz 1994], the system can exhibit complex-looking behaviour, but it will still be bounded, and in a form of "steady" state. This is why we desire the environment (at least) of our embodied system to be *open*. This openness (a constant flow of matter, energy, or information through the system) allows constant novelty, by allowing the combined system to be far from equilibrium.

A far-from-equilibrium system, rather than converging, may *self-organise* [Bak 1997] to the computational *edge of chaos* [Langton 1991]. There it can form stable structures, patterns, emergent properties, that persist; yet it is simultaneously "poised" [Kauffman 1995] in that it can readily change in response to inputs. This is what we want from an adaptive learning system: the stable patterns form the memory, and the poised response forms the adaptation.

A physical environment is naturally open, with its noise, unpredictability, and complexity. Some virtual environments (such as the Internet) are also open. How can we provide other virtual environments with desirable open properties?

A source of *noise*, or randomness, to simulate richness is not suitable, because it has no underlying structure for the system to exploit (or, in the case of pseudo-randomness, it may have the wrong kind of structure). To get at least the flavour of the right *kind* of complexity [Crutchfield 1994], an otherwise impoverished virtual environment could be coupled to some suitable "edge of chaos" or other complex non-linear device. Depending on the nature of this device, it might still be difficult to achieve multiple levels of emergence. But in the short term, this is a potentially valuable approach to the openness problem.

12.3 Rich coupling

12.3.1 Co-evolution of sensors/actuators and processing elements

In addition to a suitably complex dynamics of the individual components (both system and environment), embodiment requires rich perturbatory channels between the systems. Similar to the argument about the state space, this does not merely mean a high bandwidth, it means a communication flow that affects the various dynamics in a suitably complex way.

If we want to engineer such a system, how can we find suitable state spaces and communication channels in the truly vast space of possibilities?

Species are not created with a given complement of sensors and actuators, filtering their interaction with their environment in fixed ways. These filters, their modalities, number and positioning, have evolved to suit the needs of the particular organism in its particular environment. Different species have different modalities (bats and dolphins "see" with sound; ants communicate by pheromones; certain fish and birds can usefully sense magnetic fields; etc), and different manipulatory appendages.

[Chiel & Beer 1997] discuss this cooperative coevolutionary history. [Percus *et al.* 1993] note that immune receptors and the antigens they sense have competitively coevolved. Kaufmann [2000] goes further, and suggests the reason why biological evolution works so well as a search technique in organism space is that organisms and evolution have themselves coevolved. Polani and co-workers [Polani *et al.* 2001, Klyubin *et al.* 2005] use a mutual information-theoretic approach to quantify informational and bandwidth requirements, given particular tasks or goals, which might be used to help evolve suitable sensors and actuators.

For artificial systems, it is possible for the environment to provide a system with inputs (or "insults") other than through its designed sensors, such as by impact, heat, etc. A sufficiently flexible system may be able to adapt to exploit these inputs, and an adaptive environment then learn to exploit them more explicitly. Similarly, an adaptive environment might learn to interpret some non-designed outputs in a useful manner: a sufficiently flexible system may be able to learn or evolve to modulate those outputs. These evolutions will simultaneously adapt the system's state space and internal dynamics.

So, for an artificial embodied system, it is essential to coevolve (or at the very least codesign) the system's computational engine along with its sensors and actuators, including the bandwidths and formats of the input/output data, in the context of the relevant environment. And ideally, for an adaptive and learning system, its sensors and actuators should be able to adapt as well.

This is contrary to the classical software engineering view that interfaces need to be clean, well-defined, controlled, and that the low-level implementation details of data representations (once a standard format has been agreed) are unimportant.

12.3.2 Co-development of system and environment

Embodiment can affect many processes, including, for example, the rate of evolution. Johan Metz [private communication, 2005] says that embodied development is a reason why mammalian morphology evolves much faster than the morphology of, say, indirectly developing insects. Mammals interact with and use cues from their environment during their development in the womb, and so bones and muscles, for example, can develop in a coordinated manner according to their use. On the other hand indirectly developing insects develop "ballistically" in the pupa, and so have no such environmental cues to exploit.

Indeed, [Riegler 2002] argues for a stronger definition of embodiment than Quick *el al.*'s. Not only must the perturbatory interactions with the environment exist, but, he claims, additionally "embodiment of a system is synonymous with competence in its environment". He goes on to claim that this competence cannot arise by engineered design, but requires "historical development in synchronization with their environment". He allows that artefacts may be embodied in virtual environments, but requires them to have developed their own goals, their own competences: designed systems whose main goals are those of the designer are merely "embedded", not embodied.

Although we do not adhere to Riegler's strong view about the source of the system goals, it is clear that the developmental process plays a key role in embodiment. Embodied systems do not spring into existence fully formed; they grow and adapt in an environment that, due to the close coupling, shapes, and is shaped by, that growth and adaptation. The same "seed", planted in two different environments, can develop into two quite different embodied mature forms. The growth itself is part of the adaptive process of the computational system, with its complex dynamics changing as it grows.

A requirement for embodied development is rather daunting for physical systems, but may be less so for virtual ones. Recent research on virtual developmental systems might hold a key: the system is encoded as a "seed", then "grown" into its final complex adult form. See, for example, [Prusinkiewicz & Lindenmayer 1990, Kumar & Bentley 2003]. This is relevant for embodiment when the system is coupled with the environmental during its growth [Měch & Prusinkiewicz 1996].

Such an approach may help to overcome some of Riegler's arguments against designed systems. The computational system is deployed as a "seed", and grown to maturity in its specific environment, rather than designed in an explicit fully grown form; the seed itself, however, may be designed.

As a consequence, a mature embodied system cannot be simply transplanted to a different kind of environment. It must develop and learn in the relevant environment. This has been noticed in practice, for example with work on developing an AIS for fault prediction in Automatic Teller Machines (ATMs), where an AIS trained on data from an ATM in one location is ineffective when transferred to a machine in a different location [Ayara 2005]. Hence one will not be able to develop a virtual embodied system, make multiple copies of its mature form, and then deploy them in other environments, unless those environments are sufficiently similar.

12.4 Design principles for embodied systems

[Kushmerick 1997] begins a computational analysis of embodiment that could be used for the design and characterisation of virtual as well as physically embodied systems. Although his emphasis is on *intelligent* behaviour, some of the points are relevant to AIS.

One key aspect of Kushmerick's analysis is that intelligent embodied systems (animals) have a *high bandwidth* of high quality input from their environment. Vision, in particular, is a high bandwidth channel, and involves a variety of filters (such as gaze direction and attention) to control the data flow. This does not mean that all embodied systems require visual sensors: plants, for example, are embodied, but are not renowned for their sharp eyesight. What is important is the bandwidth on the relevant timescales of interaction, which is much slower for (most) plants than for (most) animals.

Another aspect is that the sensory flow is continuous: it is always present and does not have to be requested. Once an animal has focused its visual attention on a part of its environment, it does not have to "request" or "poll" for its visual input; the channel is broadcasting continuously. [Kushmerick 1997] argues that this is an important part of the interaction that allows the computational and memory burden to be shifted to the environment (cf Brooks' "the world is its own best model"). The system lives in a sea of constantly updating data, and much of the problem of perception is what to throw away, not what to request.

[Kushmerick 1997] notes that the task that an embodied system has to perform is often more highly constrained that the general class of tasks of which it is a member, because of existing environmental constraints on the solution, or because the system itself imposes extra constraints via the coupling that reduce the number of degrees of freedom. Those constraints can then be exploited to simplify the computation needed to perform the task. (Recall that computational complexity is a worst case property of a class of problems, and individual instances may have much lower complexity.)

Finally, [Kushmerick 1997] observes that physically embodied systems are usually satisfied with approximate "good enough" solutions, rather than optimal or exact solutions, particularly when the approximate solution can be achieved with significantly reduced computational burden.

These observations, along with arguments discussed earlier, suggest some design principles for embodied systems.

1. Design the system X with sufficiently complex dynamics, that can execute this dynamics on the relevant timescale(s) of the environment E.
2. Design a sufficiently high interaction bandwidth on the relevant interaction timescale(s).
3. Ensure that input from the environment is constantly available and up to date.
4. Ensure that the system perturbs the environment, rather than being merely a passive observer.
5. Ensure that the environment has sufficiently complex dynamics.
6. Allow the system to exploit structure and constraints in the environment in order to simplify its tasks.
7. Apply embodied systems only in "softer" problem domains where approximate solutions are appropriate and acceptable.
8. Co-design the system and its interface (sensor and actuator numbers, positions, data formats, etc).
9. Design the system to develop, "grow", in the relevant environment.

These principles apply both to *material* and to *virtual* embodied systems, but the emphases and difficulties are different in each case. For example, complex dynamics may be easier to achieve for a physical than a virtual environment, whereas the co-design of sensors and actuators may be easier to achieve for virtual than for physical interactions.

Of course, *how* to design and ensure many of these things are still open research problems. Certain aspects of the "design" may need to be accomplished by some evolutionary search process. However, one point to note is that some of these design principles refer to the *environment*, not just to the computational system. This is not unexpected: we noted earlier that it it impossible to analyse an embodied system in isolation, and so it is impossible to design it in isolation. (In chapter 16 of this book, Hone and Van Den Burg analyse an isolated AIS. This does not contradict the claim here, since their AIS is not embodied: it does not have rich perturbatory interactions with its environment.)

12.5 Embodiment in the natural immune system

12.5.1 Instantiating the model of embodiment

Let us now consider the natural immune system as an embodied computational system. (The following discussion is necessarily a gross simplification of the actual biological processes, omitting many constituent agents and processes, but is sufficient to illuminate the key concepts.) We instantiate the generic embodied system of figure 12.1 with immune system concepts (figure 12.2).

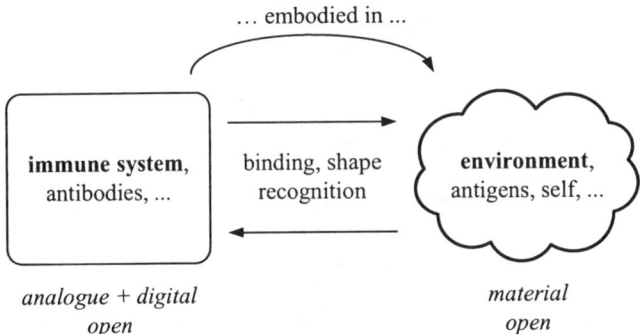

Fig. 12.2. A computational immune system embodied in an environment

The immune system contains, among other things, a population of *antibodies*, which are particular kinds of proteins. The system is hybrid: partially digital (proteins are strings of amino acids drawn from a small alphabet of possibilities, partly coded for in the DNA) and partially analogue (the proteins fold into complex three-dimensional shapes whose physical properties determine their behaviour). The system dynamics includes the modification of the population by the production, modification, and destruction of different kinds of antibodies, in response to events sensed in the environment. See, for example, [Nowak & May 2000].

The environment is the open material environment of the body, taken here as containing the antigens to be handled, and self proteins that should not be attacked. The environment is modified by the immune system, and some of its components evolve to resist this modification, some to aid it. It is open in that there is a flow of matter through it.

The interaction between the system and the environment occurs by antibodies binding to antigens. This binding depends on shape recognition (hydrogen bonds form between the molecules; if the shapes are sufficiently complementary, enough hydrogen bonds can be created to form a strong enough aggregate bond to stick the molecules together), and other environmental factors (for example, water molecules can plug gaps between the shapes and form additional hydrogen bonds). The stronger the bond, the higher the *affinity* the antibody has for the antigen.

12.5.2 Danger Theory as embodied sensing

There may be hierarchies of embodiment. Here we have described the immune system as being comprised of its cells, with the host organism and pathogens acting as its environment. The host organism itself is also a system, embodied in its own environment. We could instead consider the immune cells to be embodied in their host, with the pathogens acting as an external environment, much in the way a robot controller is considered to be embodied in the physical robot host, in the environment of the world. However, in our case, such a choice would beg the question of precisely what comprises the host organism, and what comprises the pathogenic environment.

There are several theories, all more or less contentious, of how the natural immune system determines what of its environment is host organism, and what are the pathogens. These range from externally directed self-non-self learning, to fully autopoietic ideas of self-assertion (summarised in [Bersini 2002]).

Danger Theory [Matzinger 1994a, Matzinger 2002] sits towards the middle of this spectrum, and says that the immune system reacts to danger, or rather, to damage, by reacting to certain chemicals given off by distressed cells. When it senses such a "danger signal", it reacts by associating "nearby" cells with the danger, and so they are classified as the cause of the problem, in a form of "guilt by association". Danger Theory helps to explain why an adjuvant (essentially, a poison) is a necessary part of some vaccines: it causes the required damage.

In Danger Theory, the immune system has cells with sensors that detect danger signal chemicals: specific chemicals given out by damaged or killed host organism cells. It also has sensors to detect nearby cells, and a system dynamics that correlates the inputs from these sensors. Additionally, it has actuators (further cells) that then respond to (kill) these associated cells, whether or not they subsequently occur in proximity to any danger signal.

Under this theory, the immune system has evolved its internal computational dynamics and its sensors to respond to danger signals. One can also postulate that the host organism part of the environment may have coevolved to enhance these signals, and that the pathogens have coevolved to evade the associated recognition, providing the environment with its own complex dynamics. Thus we see the evolution of a rich embodied computational system.

Ideas from Danger Theory are now being incorporated in AIS models. See, for example, [Bentley *et al.* 2005, Greensmith *et al.* 2005]

12.5.3 Shape space in the natural immune system

As noted earlier, the interaction between the immune system and its environment is in terms of shape recognition.

[Perelson & Oster 1979] introduce the concept of an abstract **shape space**, S, in which the shape of an object is represented by a single point. They take S to be an N-dimensional Euclidean vector space, $S = \Re^N$. They use the N shape parameters as a model of the relevant features of an antibody combining region, and represent these values by a point **Ab** in S. They do not identify these parameters further, beyond suggesting that they could be physical properties such as size, charge, and dipole moment. [Lapedes & Farber 2001] use multidimensional scaling techniques to derive further geometrical properties of shape space from influenza data.

[Perelson & Oster 1979] then capture how well an antibody combining region **Ab** and antigen **Ag** fit together in terms of the *distance* between the shape of the antibody and the complementary shape of the antigen, $\|\mathbf{Ab}, \overline{\mathbf{Ag}}\|$, measured using an "appropriate" metric on S. A small distance represents a good fit, and hence a high *affinity*. They explicitly decline to define this metric, its being "a complicated chemical problem", and continue their analysis in terms of the well-known Euclidean metric. (See appendix A for some other possibilities.)

What is the distribution of antibodies and antigens in this shape space? Perelson and Oster note that it is almost certainly not *random*, since the components will have been subject to negative selection (in order not to recognise *self*) and evolutionary selection respectively. It is presumably also affected by the physical constraints of embodiment: certain combinations of parameters may be physically impossible, or at least highly unlikely, to be realised.

Since nothing is known of the actual distribution, analyses tend to proceed on the minimal assumption of random distributions within some finite volume $V \subset S$ (or are parameterised by the actual, but unknown, distribution). Perelson and Oster analyse the size of the antibody repertoire needed to cover this volume, in terms of a recognition specificity $\hat{\varepsilon}$ and of the dimensionality N. Their analysis shows that higher values of N need significantly larger antibody repertoires to cover a given volume of shape space, for a given specificity. They argue that this is why recognition uses only a small portion of each antigen, to limit the complexity, and hence to limit N and the required size of the repertoire. For typical animals, they estimate that $N = 5 - 10$.

Using different techniques, [Lapedes & Farber 2001] recover $N = 5$ from various influenza data. [Smith *et al.* 1997] find that $N = 5 - 8$ is consistent with their data; using a Hamming metric (see appendix A) instead of a Euclidean metric, they find $N = 20 - 25$, for an alphabet of size $3 - 4$, is consistent with their data.

[Perelson & Oster 1979] define their recognition specificity $\hat{\varepsilon}$ as the radius of an N-sphere centred on **Ab** in S, scaled by the radius R of the full N-sphere of volume V, making it a non-dimensional quantity with $0 \leq \hat{\varepsilon} \leq 1$. The number of such spheres needed to fill V goes like $1/\hat{\varepsilon}^N$. Keeping $\hat{\varepsilon}$ fixed whilst varying N corresponds to a keeping a fixed specificity along each individual dimension whilst changing the number of dimensions. This form of scaling is (most of) the reason for the dependence of repertoire size on N: if one fixes specificity along each dimension whilst increasing N, the number of antibodies in the repertoire needed to cover the space increases.

One can look at this argument from another perspective. Consider a *fixed repertoire size* $1/\hat{\varepsilon}^N$; then as N increases, $\hat{\varepsilon}$ *also* increases. For example, for a repertoire size of 64, then $N = 1$ has $\hat{\varepsilon} = 1/64$; $N = 2$ has $\hat{\varepsilon} = 1/8$; $N = 3$ has $\hat{\varepsilon} = 1/4$; and $N = 6$ has $\hat{\varepsilon} = 1/2$. So, for a fixed repertoire size, one can trade off specificity against dimension: a larger N requires less precise discriminations along each of its individual dimensions, which might be easier to realise.

Shape space is an essentially physically embodied notion, as further demonstrated by the following quotations from [Perelson & Oster 1979] (our italics):

> multisite recognition is a more reliable method of distinguishing between *molecules* than single site recognition. This may have been an important evolutionary consideration in the selection of *weak non-covalent interactions* as the basis of *antigen–antibody bonds.*
>
> ... a large repertoire of *antibody molecules* with different *three-dimensional binding sites* specific for the *different chemical groups* ... found on *antigen molecules.*
>
> because of *physical restrictions* on the manner in which *molecules fold*, smaller regions can take on fewer shapes

Shape space as originally envisaged is a metric space, where the only property of interest is the distance between pairs of points in the space. Spaces with more structure, with a value at each point (for example a scalar value that might represent a fitness, or a vector value that might represent a force), are also of potential interest for modelling the interaction between a system and its environment. Entities can then have a dynamics, can move around in these spaces, affected by the values at various points. The values in turn can be affected by how things move, and thus provide even richer dynamics [Saunders 1993]. In Chapter 4, Lee and Perelson discuss further models of affinity, and their dynamics.

Complementarity revisited. Care needs to be taken with the various forms of complementarity used in different shape space arguments. The idea is that complementary shapes match well. However, the complementarity may be included at different points in the argument, resulting in different interpretations of the metric.

The original shape space arguments [Perelson & Oster 1979] consider a specificity "ball" around an antibody, with this ball containing the antigens recognised. The antigens are located in the shape space according to the complement of their shape. The Euclidean metric measures how well antigen and antibody match, where *small* values correspond to good matches, that is, to high affinity. So in this case, high affinity corresponds to small values of $\|\mathbf{Ab}, \overline{\mathbf{Ag}}\|_E$. [De Boer *et al.* 1992] take a slightly different interpretation: shapes have good matches in shape space when their coordinates are *"equal and opposite"* (thus putting an explicit interpretation on complementarity), so high affinity corresponds to small values of $\|\mathbf{Ab}, -\mathbf{Ag}\|_E$.

The "lock and key" metaphor [Percus *et al.* 1993] has a more direct measure of complementary shapes matching well. Here r-contiguous bits (which measures complementary contiguous subsequences) is used, and *high* values correspond to good matches. So high affinity corresponds to large values of $\|\mathbf{Ab}, \mathbf{Ag}\|_C$.

In practice, it is common for *artificial* immune system researchers to neglect complementarity when using real-valued artificial shape spaces, and instead measure direct matching, and use a large (or sometimes, small) value of $\|\mathbf{Ab}, \mathbf{Ag}\|_E$ to represent a high affinity.

Discussions of the role of complementarity and metrics in the context of artificial immune systems can be found in [Garrett 2003, Hart & Ross 2005]. It is clear that great care should be taken in choosing, and precisely documenting, what measure of matching is being used.

12.6 Embodied artificial immune systems

Engineered embodied systems must have their own suitably complex internal dynamics and complex perturbatory interactions with a suitably complex open environment. This is a potentially enormous class of systems, and so we seek inspiration from existing natural systems to help us to constrain our design space.

As discussed in Chapter 3, Artificial immune systems (AIS) are computational systems that take their inspiration from various theories of the natural immune system. However, research has tended to concentrate more on the complex internal dynamics than on the complex perturbatory interactions, and the systems are often situated in a closed and impoverished virtual environment: AIS tend not to be *embodied*. Let us look at the design principles suggested earlier, to see how they might be used to achieve (materially or virtually) embodied AIS.

Timescales. Physically embodied systems usually interact on "human" timescales (seconds, hours, days), whereas virtual systems may have to react much faster.

There may be more than one relevant timescale, for example, for immediate reaction to a novel situation, and for slower adaptation/learning of that novelty. One of the reasons that AIS have not managed to act as fully effective "immune system" defences against computer viruses may be that, although the analogy holds between the two sets of terminologies, it does not hold between timescales. A computer virus can destroy an entire network on the *same* timescale that the defences react.

Bandwidth. A physical environment naturally offers high bandwidth, so the AIS merely needs to be supplied with sensors and actuators to access the relevant portion of it. A virtual environment may need to be adapted to provide a richer source of data: not simply, say, packet headers, process ids, or return codes, but more contentful information. This may pose a problem on, say, a network where all traffic is encrypted. It is worth considering if "side channel" information (such as timing, power consumption, etc [Kocher 1996, Kocher *et al.* 1999, Clark *et al.* 2005b]) are also available, as these provide rich extra-logical resources.

Constantly available up-to-date input. The AIS should be able to directly sense the relevant data streams, rather than explicitly request data. The requirement to be up to date might imply that virtual data streams have limited temporal and spatial extent, so that they "decay" before they have become outdated.

Actuation. An embodied system perturbs its environment: it takes actions based on what it senses, and these actions in turn change the environment. The AIS must interact, and be able to change aspects of the environment (for example, by injecting packets into relevant data streams). Thus purely passive recognisers or monitors (the major application areas of AIS to date) cannot be considered to be embodied.

Sufficiently complex environmental dynamics. This should be a "given" for truly physical embodiment. However, in some cases the physical world may instead be severely impoverished, by design. For example, many robot control experiments take place in sterile featureless mazes that offer few environmental cues. These are not appropriate environments for embodiment.

A virtual environment may have to be designed explicitly to have sufficiently complex dynamics; this may be simulated by artificially hooking up the environment to an edge of chaos generator. "Sufficient" complexity allows the actuations of the system to affect the dynamics of the environment in complex ways.

Environmental structure and constraints. Exploiting these requires an understanding of the constraints (for example, restricted network topology resulting in restricted navigational opportunities) and their effects, and how they might change over time.

Approximate solutions. In most physically embodied cases, approximate solutions are acceptable, because the system is interfacing with the continuous, analogue real world, and there is always limited precision. The acceptable degree of approximation still needs to be ascertained. In the virtual world, crisp digital problems are not appropriate: this excludes many classical computer domains, from word processing to payroll systems. Even for softer domains such as pattern recognition, it requires consideration of the acceptable false positive and false negative rates.

Co-design the system with its sensors and actuators. Natural immune systems interact with the environment by shape recognition and binding, modelled by shape space. The analogue for AIS is an artificial shape space (sensing affinities), and artificial binding (actuation).

Relatively simple shape space arguments show how the choice of geometrical parameters such as representation dimensionality, alphabet size, and specificity, determine the antibody repertoire size needed to cover shape space. This is a crucial design consideration for artificial immune systems. Evolution appears to have settled on a value of $N \sim 5$ for real immune system shape space. Design choices include the actual dimensionality N, what each axis of dimension represents, the metric used to measure affinity distance, and the specificity $\hat{\epsilon}$.

The data representation also matters. For example, interpreting a b-bit string as representing a low dimensionality, large alphabet system, rather than as a b-dimensional binary hypercube, results in smoother measures [Smith *et al.* 1997], which may be beneficial to the behaviour of the system. It may be that the use of complementarity and the choice of metric (or non-metric distance measure, see appendix) have subtle effects on the underlying dynamics of the system [Hart & Ross 2005], and so should be designed into artificial systems with care.

Grow the system in the relevant environment. Essentially this affects the initial conditions of the system. A system should start as small as possible (maybe from an embryonic or an infant state), and develop in the context of its environment, learning and adapting as it goes. It should not merely accrete: the dynamics of its growth should be complex, and affected by the environmental inputs. It should be a continually developing system. This implies that the system should undergo continual online learning, and not be a system that has an initial learning phase, then a frozen deployment phase.

12.7 Conclusions

Embodiment offers substantial advantages in allowing a computational system to offload much of its computational burden to the surrounding rich environment. This advantage comes at a price, however, as the design of such a system is non-trivial.

Despite embodiment seeming to require a physical body, many of the advantages can also be achieved in systems embodied in a virtual environment, provided that the design of both system and environment supports the rich dynamics and interactions necessary.

Current artificial immune systems tend not to have the properties necessary for embodiment. Several design principles for embodied systems have been abstracted, and their application to potential embodied artificial immune systems explored. The next step is to attempt to build some fully embodied artificial immune systems according to these principles.

A Metrics

A *metric* is the mathematical abstraction of the intuitive concept of *distance*. It is a function that maps pairs of points x, y in a space S to a real number, the "distance" between them: $\|x, y\| \in \Re$. To qualify as a *metric*, the function must obey the following properties.

1. distances are not negative: $\forall x, y : S \bullet \|x, y\| \geq 0$
2. the distance from a point to itself is zero; the distance between different points is not zero: $\forall x, y : S \bullet \|x, y\| = 0 \Leftrightarrow x = y$
3. distance is *symmetric* (the distance from x to y is the same as the distance from y to x): $\forall x, y : S \bullet \|x, y\| = \|y, x\|$
4. distance obeys the *triangle inequality* (detours are further than direct routes): $\forall x, y, z : S \bullet \|x, y\| + \|y, z\| \geq \|x, z\|$

Give a prospective metric, usually the first three properties are immediate from the definition and only the triangle inequality property needs further demonstration.

There are many metrics; we consider here those most often used in the AIS literature. In what follows, we take the points in space to be vectors $\mathbf{x}$ in the N-dimensional shape space of an antibody $\mathbf{Ab}$ or antigen $\mathbf{Ag}$. A vector $\mathbf{x}$ can be written in terms of its components x_i, with $\mathbf{x} = \sum_{i=1}^{N} \hat{\mathbf{e}}_i x_i$, where the $\hat{\mathbf{e}}_i$ are a basis set of orthonormal vectors.

The **Euclidean metric**, favoured by mathematicians because of its nice analytical properties, is

$$\|\mathbf{x}, \mathbf{y}\|_E = \sqrt{\sum_{i=1}^{N} (x_i - y_i)^2}$$

For example, $\|3\hat{\mathbf{e}}_1, 4\hat{\mathbf{e}}_2\|_E = \sqrt{(3 - 0)^2 + (0 - 4)^2} = 5$

The **Manhattan metric**, favoured by computer scientists because it is simple and cheap to compute, is

$$\|\mathbf{x}, \mathbf{y}\|_M = \sum_{i=1}^{N} |x_i - y_i|$$

For example, $\|3\hat{\mathbf{e}}_1, 4\hat{\mathbf{e}}_2\|_M = |3 - 0| + |0 - 4| = 7$

The **Hamming metric** is even simpler. Let the components of $\mathbf{x}$ be drawn from some finite alphabet Σ (so $\mathbf{x} \in \Sigma^N$ rather than $\mathbf{x} \in \Re^N$). Then

$$\|\mathbf{x}, \mathbf{y}\|_H = \sum_{i=1}^{N} \delta(x_i, y_i)$$

where $\delta(x, y) = $ if $x = y$ then 0 else 1

So, for example, $\|120001, 020020\|_H = 1 + 0 + 0 + 0 + 1 + 1 = 3$. For binary alphabets (that is, Boolean vectors), the Hamming metric is the *total number of complementary bits*, and is the same value as the Manhattan metric over the N-dimensional hypercube.

The **r-contiguous bits** distance [Percus *et al.* 1993] is also applicable to Boolean vectors. It measures the *longest contiguous subsequence of complementary bits*. The definition, expressed in the Z mathematical language [Spivey 1992, Valentine *et al.* 2004], is as follows. Let s be the sequence of vector components of $\mathbf{x} - \mathbf{y}$, that is, $s = \langle|x_1 - y_1|, \ldots, |x_N - y_N|\rangle$. So, for complementary bits,

$$\|\mathbf{x}, \mathbf{y}\|_C = \max\left\{t : \mathrm{seq}\{1\} \mid t \text{ infix } s \bullet \#t\right\}$$

Consider the distance $\|110001, 010010\|_C$. The sequence $s = \langle 1, 0, 0, 0, 1, 1\rangle$. So $\|110001, 010010\|_C = \max\left\{\#\langle 1\rangle, \#\langle\rangle, \#\langle 1, 1\rangle\right\} = \max\{1, 0, 2\} = 2$.

[Percus *et al.* 1993] motivate this choice of distance in terms of the immunological *lock and key* metaphor, that the 1s and 0s of each bit string are modelling the relevant ups and downs of the key teeth and complementary lock shape, and that a contiguous run is needed to generate sufficient affinity. They also allow the shape space vectors to have non-Boolean-valued components (they find evidence for a trinary alphabet of values in natural immune systems, corresponding to positively, negatively, and neutrally charged regions), with matching being all (for complementary values) or nothing.

Formalising this, each vector component is drawn from some alphabet Σ. Each element $\sigma \in \Sigma$ has a complement $\bar{\sigma} \in \Sigma$. An element may be self complementary, $\bar{\tau} = \tau$. Complementarity is a symmetric relation: $\bar{\sigma} = \tau \Leftrightarrow \bar{\tau} = \sigma$. Now define the sequence of components as $s = \langle\delta(x_1, y_1), \ldots, \delta(x_N, y_N)\rangle$ where $\delta(x, y) = $ if $y = \bar{x}$ then 1 else 0. Then r-contiguous bits is as defined earlier, using this s. For example, consider the alphabet $\Sigma = \{+, -, 0\}$ with complementarity relation $\bar{+} = -, \bar{0} = 0$. Now consider $\|-+000+, +++-0-\|_C$, which has the sequence $s = \langle 1, 0, 0, 0, 1, 1 \rangle$, and so $\|-+000+, +++-0-\|_C = 2$.

It should be noted that r-contiguous bits does *not* form a metric (the notation we use above notwithstanding). To see this, consider the binary vectors $\mathbf{x} = 0000$; $\mathbf{y} = 1010$; $\mathbf{z} = 1111$. If we consider the usual complementarity relation, $\bar{0} = 1$, we have that $\|\mathbf{x}, \mathbf{y}\|_C + \|\mathbf{y}, \mathbf{z}\|_C = 1 + 1 < 4 = \|\mathbf{x}, \mathbf{z}\|_C$, which does not obey the triangle inequality. If we take the self-complementarity relation, $\bar{0} = 0, \bar{1} = 1$, then we have $\|\mathbf{x}, \mathbf{x}\|_C = 4 \neq 0$, which violates another of the metric conditions.

The Multi-scale Immune Response to Pathogens: *M. tuberculosis* as an Example

Denise Kirschner[1]

Department of Microbiology and Immunology, University of Michigan Medical School, Ann Arbor, MI 48109-0620 kirschne@umich.edu

Summary. The immune response occurs over multiple temporal and spatial scales. Events at the genetic level can influence events at the cellular level and finally manifest at the population scale. Through the example of the human pathogen *Mycobacterim tuberculosis* we explore immune response events over multiple scales and how bridging these scales may ultimately lead to the greatest picture of how this complex system works.

13.1 Introduction

When a pathogen invades a host, the host mounts a response that occurs at several levels of biological organization including genetic, molecular, cellular, tissue and system level. A number of host cells are called into action including antigen presenting cells (APCs) and T cells. At the body's peripheral sites, populations of resident APCs are maintained consisting primarily of macrophage and dendritic cells (DCs). These cells are among the first to encounter pathogens that breach host barriers. Foremost among their responsibilities is the presentation of peptide antigens from pathogens that are taken up at the site of infection in the form of peptide-MHC (pMHC) complexes on their cell surface. Some APCs, namely DCs, migrate to the nearest lymph node (LN) where they activate naïve T cells. Other APCs, namely macrophages, remain at the site of the infection and respond to an influx of activated CD4$^+$ T cells by increasing their presentation and microbicidal activity.

While most of these events occur at the cellular level, they are embedded in the context of multiple biological levels. The initial APC-T cell interaction occurs mainly in the specialized structured environment of the LN. The lymphatic system serves as a conduit for immune cells between tissues, LNs and organs. While the blood supplies immune cells to the LNs, the lymphatics drain the tissues, acting as the key source of antigens and DCs in most infections. Hence, both tissue- and system-level events play a role in response efficacy. At the same time, APCs may vary in their ability to perform antigen presentation due to events occurring at the molecular and genetic levels. The APC-T cell interaction depends on stable expression of pMHC complexes on the APC surface that in turn depends on pMHC binding affinity. A high degree of variability exists in the peptide-binding region of MHC throughout the human population, resulting in considerable APC heterogeneity, both within a single individual and between individuals. Antigen presentation therefore lies at the crux of the immune response, between the larger scales (tissue- and system-level) that determine its context and the small scales (genetic- and molecular-levels) that determine its constituents. In fact, susceptibility and resistance to some diseases have been linked directly to the basic genetic components underlying antigen presentation.

Certainly there has been a wealth of basic science performed at the molecular and cellular levels attempting to elucidate immunity. However, given its complexity, the multi-scale system is presently impossible to study in an experimental setting. Thus, mathematical and computational models bridging the multiple scales that encompass the immune response are necessary to help uncover mechanisms underlying the dynamics of this complex system.

Mathematical models of the host-pathogen interaction have mainly been restricted to the study of host-viral interactions. Relatively few models have explored bacterial-host interactions [Freter et al. 1983, Kirschner & Blaser 1995, Asachenkov 1994, Gordon & Riley 1992, Lipsitch & Levin 1997]. Regardless, most have focused on the single-scale of cellular-level dynamics.

We have made attempts to explore the complex system of immunity by studying the immune response to a specific pathogen. We have studied the interaction of the immune system with the intracellular pathogen *Mycobacterium tuberculosis* at a number of biological and spatial scales. Here we highlight both the biology we are addressing and the mathematical approach taken as a means for beginning to understand the integrated, multi-scale complex system know as the immune response.

13.2 *M. tuberculosis*

Tuberculosis (TB) has been a leading cause of death in the world for centuries. Today it remains the number one cause of death by infectious disease world wide - 2 million deaths per year. TB is not only one of our oldest microbial enemies, but it remains one of the most formidable: An estimated one third of the world population has latent TB—2 billion people. Thus, there is a great need to elucidate the mechanisms of TB disease progression. There are 2 major infection outcomes for TB–latency and active disease; the ability to clear TB has not been demonstrated, although only a subset (~30%) become initially infected upon exposure [Styblo et al. 1969] suggesting some (perhaps most) are able to clear upon initial infection.

Reactivation can occur in latent infection, although we do not discuss this here for brevity (see [Singer & Kirschner 2004] for more information). Key issues are to understand immune mechanisms involved in controlling infection leading to latency. To this end, elaborating the primary immune response against the causative agent, *M. tuberculosis* (Mtb), is essential to understanding the functional immune response that leads to latency.

Primary infection usually develops in the alveoli of the lung after inhaling droplets containing Mtb. The bacteria are then ingested by resident alveolar macrophages and begin to multiply [Canetti 1955]. These macrophages are poor at destroying their occupants in part because Mtb can prevent phagosome-lysosome fusion in resting macrophages [Myrvik *et al.* 1984, McDonough *et al.* 1993]. Infected macrophages may burst due to the large number of multiplying bacteria within. Infected dendritic cells or macrophages circulate out through the lymphatic ducts to the draining lymph nodes where the specific immune response is initiated. Here, CD4+ T cells are stimulated to become effector cells, most likely of the Th1 type. These and other effector cells such as CD8+ T cells and monocytes must then be recruited and migrate to the site of infection, interact with cells at the site, where they participate in the formation and function of a unique immunological structure known as a granuloma.

Granuloma formation is dependent on a number of factors, including chemokines, cytokines, cell adhesion molecules and immune effector cells. There exists a large body of literature regarding these individual elements in the immune response in TB; however, little is known about the interaction among these elements that leads to granuloma formation and function. Characterization of the immunologic factors operating during granuloma formation is likely to shed light on our understanding of host defense and pathogenetic mechanisms involved in TB. This is a daunting task as infection with Mtb triggers production of a complex set of immunologic factors, including potent pro- and anti-inflammatory cytokines and chemokines that are capable of interacting with and cross-regulating one another. These analyses are further complicated by the fact that many of the participating members of the tuberculosis immune network possess pleiotropic and often opposing functions. Mathematical models provide a framework for integration of large amounts of data into a complex system that can then be analyzed, and thus is currently the only integrative approach for studying complex biological systems.

13.2.1 Immune cells participating in the immune response to *M. tuberculosis*

Macrophages are the preferred host cell for mycobacteria. These phagocytic cells take up *M. tuberculosis* and are unable to clear it as they normal do most other bacteria. However, if the macrophage receives appropriate cellular and cytokine signals (such as IFN-γ) within an efficient amount of time, then these macrophages can become activated and clear their intracellular load [Nathan *et al.* 1983, Flesch & Kaufmann 1990]. Otherwise, macrophages become chronically infected and will not only never be able to clear their intracellular bacteria [Armstrong & Hart 1971, Sturgill-Koszycki et al. 1994], but will eventually burst due to increasing bacterial numbers or be killed by cytotoxic T cells [Lewinsohn et al. 1998, Tan et al. 1997].

It is well established that cell-mediated immunity is essential for controlling initial as well as latent Mtb infection both in humans and murine models. CD4+ and CD8+ T cells are believed to be important in this response [Chan & Kaufmann 1995]. Support for the importance of CD4+ T cells comes from the extreme susceptibility of HIV+ subjects to acute and reactive TB. Mice deficient in CD4+ T cells succumb to fatal TB [Leveton et al. 1989, Muller et al. 1987, Tascon et al. 1998, Caruso et al. 1999]. CD4+ T cells produce cytokines, such as IFN-γ, and thus activate macrophages to eliminate intracellular Mtb [Caruso et al. 1999, Silver et al. 1998]. This is partially mediated, in mice and possibly in humans, by the production of reactive nitrogen intermediates, such as nitric oxide, produced by inducible nitric oxide synthase (NOS2) within macrophages [Chan 1993]. Mice deficient in CD8+ T cells are more susceptible to Mtb than are wild type mice [Flynn et al. 1992]. CD8+ T cells in the lungs of infected mice can produce cytokines and act as cytotoxic T cells (CTL) for infected macrophages [Dolin et al. 1994, Serbina & Flynn 1999, Serbina et al. 2000]. Mtb-specific human CD8+ T cells from tuberculosis patients have recently been reported (reviewed in [Flynn & Ernst 2000].

13.2.2 Cytokines Involved in the Response to *M. tuberculosis*

An essential cytokine in control of infection is IFN-γ; mice deficient in this gene are extremely susceptible to acute TB [Flynn et al. 1993, Cooper et al. 1993]. A consequence of the absence of IFN-γ is the lack of macrophage activation, including NOS2 production [Flynn et al. 1993, Cooper et al. 1993, Dalton et al. 1993]. IL-12 is also required for control of acute TB [Cooper et al. 1997b, Cooper et al. 1997a]. Human studies have demonstrated that mutations in genes for IFN-γ and IL-12 receptors increase susceptibility to mycobacterial infections [Ottenhoff et al. 1998]. TNF is also essential to control of both acute and chronic Mtb infection [Flynn 1995, Adams et al. 1995, Mohan et al. 2001, Bean et al. 1999]. This cytokine has effects on chemokine and adhesion molecule expression and therefore is an apparent key player in granuloma formation [Flynn 1995, Bean et al. 1999, Kindler 1989, Mohan et al. 2001]. Recently, TNF has shown to be an important cytokine in human studies (with anti-TNF treatment for arthritis), which have induced reactivation of TB [Ehlers 2003, van Deventer 2001, van Deventer 2002], as well as in mouse systems were TNF knock-out mice were highly susceptible to active TB [Mohan et al. 2001, Botha & Ryffel 2003].

13.2.3 Chemokines Involved in the Response to *M. tuberculosis*

A successful host inflammatory response to invading microbes requires precise co-ordination of myriad immunologic elements. An important first step is to recruit intravascular immune cells to the proximity of the extravascular location of infection, preparing them for the process of extravasation. This is controlled by adhesion molecules and chemokines. The field of chemokine research is expanding at a rapid rate. These molecules induce migration of various cells, including monocytes/macrophages, dendritic cells, neutrophils and leukocytes [Baggiolini 1998]. The migration of cells occurs as a result of the integration of various chemokine signals and their receptors [Foxman *et al.* 1997]. There is evidence that cytokines play both direct and indirect roles in modulating this process [Lane *et al.* 1999, Czermak *et al.* 1999, Crippen *et al.* 1998, Koyama *et al.* 1999]. Chemokines in Mtb infection have been investigated to a limited extent [Orme 1999a, Orme & Cooper 1999, Orme 1999b, Orme 1999c]. We begin to elucidate the role of chemokines in our models of the immune response to Mtb.

13.3 *In Silico* Models at Different Biological Scales

Our goal is to illustrate the application of mathematical modeling at different biological scales towards better understanding the immune response to Mtb. To this end, we present 4 distinct models. First, we study the role of antigen presentation at the intracellular level exploring processing and genetic events that are interfered with by *M. tuberculosis* to its favor. Second, we bridge two distinct biological scales: genetic level, immune system events that impact the epidemiology of TB. Next, we explore the immune response to *M. tuberculosis* using a two-pronged approach. We developed a temporal model tracking a spatially homogenous population of cells and cytokines in the lungs. This model was designed with ordinary differential equations. And lastly, we then narrowed the spatial scale to a single granuloma forming and accounted for the heterogeneous spatialization and behavior of cells on an individual level using an agent-based model.

13.3.1 Antigen Presentation and its role in *M. tuberculosis* Infection

Antigen presentation is critical to triggering an appropriate immune response. It is the process whereby peptide fragments of proteins derived from pathogens are presented on an immune cell surface signaling the presence of infection. This process occurs via two pathways. All cells of the body (except red blood cells) have the ability to process and present antigens that are derived from the cytosol. This allows for cells to signal they are infected to the immune response for clearance. This process occurs via the MHC class I presentation pathway. Other cells, termed professional antigen presenting cells, or APCs, present antigen to immune cells for activation via the MHC class II pathway. It is this route of presentation that we focus on here.

Briefly, specialized APCs, dendritic cells and macrophages, take up pathogens or other factors produced by pathogens at the site of infection. Once taken up, pathogens are sequestered into vacuoles and their proteins are processed into peptides. These peptides are bound by MHC class II (MHC II) molecules, named for the region of the genome in which they are encoded, the major histocompatibility complex. Within this region lie the most polymorphic genes in the human genome, giving rise to MHC molecules with different peptide-binding specificities. Peptide-MHC complexes (pMHC) are displayed on the surface of the APC and are recognized by the T cell receptor on T helper cells that become activated and proliferate in response For a complete treatment of T cell receptors, see Chapter 4 by Lee and Perelson in this book.

13.3.2 A Model for MHC class II Antigen Presentation During Infection with *M. tuberculosis*

While MHC II polymorphism may be the strongest genetic determinant of an antigen presentation outcome due to its effect on pMHC binding, this is by no means the only regulated step. Several critical cellular processes contribute to successful antigen presentation by APCs. These processes occur in the time frame of minutes to hours and can be stated briefly as: (1) uptake of antigen from the extracellular environment and degradation of antigen within endosomal compartments into peptides, (2) synthesis of MHC II molecules, (3) peptide-MHC II binding to form pMHC complexes, and (4) display of pMHC complexes on the APC surface. We review these briefly below, but for a full treatment we refer the reader to a recent review [Bryant & Ploegh 2004].

Exogenous antigens, constituting the bulk source of peptides for MHC II-mediated antigen presentation, generally have three routes of entry to the APC: fluid-phase pinocytosis, receptor-mediated endocytosis, and phagocytosis [Lanzavecchia 1996]. Pinocytosis is a common mode of entry and is our focus. Once taken up, antigens move through a series of increasingly acidified endosomal compartments and are either processed into peptides capable of binding MHC II molecules or degraded. Low pH-activated proteases degrade antigen as it traffics through the endocytic pathway, yielding peptides suitable for binding MHC II [Honey & Rudensky 2003].

MHC II expression is normally low in resident populations of APC that have not been exposed to antigen. However, a number of environmental cues can alter MHC II expression including chemical signals (cytokines) secreted by neighboring cells and direct contact with certain molecules native to pathogens. Such signals trigger a signal transduction cascade in the APC resulting in the up-regulation (or, in a few cases, down-regulation) of MHC II expression. For example, macrophages are often incubated with IFN-γ for *in vitro* studies; in the *in vivo* situation, this would come from T cells or natural killer cells. IFN-γ binds to receptors on the macrophage surface, increasing the expression of class II transactivator (CIITA), a master regulator of MHC II transcription, over a period of hours, leading after a time delay to increased MHC II expression and presumably increased ability to present antigen. Describing the effects of IFN-γ requires consideration of the degradation of IFN-γ in solution and the uptake of IFN-γ by macrophages [Celada & Schreiber

1997]. Shortly after appearing in the endoplasmic reticulum, a nascent MHC II molecule is coupled to invariant chain (Ii) which possesses a cytosolic domain capable of directing the molecule to the endosomal pathway and an extracytosolic domain capable of binding and protecting the MHC II binding groove.

The MHC II molecule arrives in the endosomal pathway with its binding groove still loaded with a remnant of Ii, the class II invariant chain-derived peptide (CLIP). Removal of CLIP occurs in an endosomal compartment, the MIIC, that also contains antigenic peptides and is catalyzed by the MHC-related enzyme HLA-DM [Denzin & Cresswell 195]. Self peptides derived from the body's own proteins are also present within the MIIC and compete with antigenic peptides for binding to MHC II [Adorini *et al.* 1988]. Indeed, in the absence of exogenous antigen self peptides may bind 80% or more of the available MHC II molecules [Chicz *et al.* 1993]. Once a pMHC complex is formed, whether it involves antigenic or self peptide, it is transported to the cell surface where it can be recognized by $CD4^+$ T cells for a period of time until it is either degraded or internalized. These processes appear largely unaffected by IFN-γ in contrast to MHC II expression [Boehm *et al.* 1997].

DCs and macrophages represent two types of so-called professional APCs, i.e. APC that express not only MHC II molecules but also co-stimulatory and adhesion molecules necessary to engage T cells.While DCs take up antigen at the site of infection and migrate to LNs to present antigen, macrophages primarily perform their function as APC at the infection site [Reinhardt *et al.* 2001] Thus, in examining the lung in *M. tuberculosis* infection, we focus our attention on the macrophage.

13.3.3 Many Pathogens Regularly Interfere with the Antigen Presentation Process.

Not surprisingly, since pathogens meet APCs continually as a first line of defense, many have evolved ways in which to inhibit antigen presentation, including both viral and bacterial pathogens. Cytomegalovirus is a viral pathogen that has been shown to inhibit antigen presentation, interrupting the MHC II expression pathway [Miller *et al.* 1998]. An example of one such bacterial pathogen is *M. tuberculosis*. Upon entering the lungs, *M. tuberculosis* is taken up by resident macrophages or DCs, adapts to the intraphagosomal environment, and survives or slowly replicates [Fenton 1998]. To evade immune surveillance, *M. tuberculosis* is known to inhibit antigen presentation via both class I and class II pathways in chronically infected macrophages [Grotzke & Lewinsohn 2005, Brookes 2003, Chang *et al.* 2005]. The mechanisms by which *M. tuberculosis* achieves inhibition of presentation via the class II pathway have not been completely elucidated, though several hypotheses have been proposed [Moreno *et al.* 1998, Hmama *et al.* 1998, Noss *et al.* 2000]. Without a detailed model of the molecular and cellular events of antigen presentation, it is difficult to assess the impact of various mechanisms of inhibition on the display of antigen and ultimately on the immune response. Early models by Linderman *et al* presented a first look at the dynamics of antigen presentation at the cellular level and demonstrated that the rates of endocytosis could be related to the display of antigen

[Singer & Linderman 1990, Singer & Linderman 1991, Petrovsky & Brusic 2004]. However, these models did not account for the more recently understood dynamics of antigen presentation and the role of IFN-γ in increasing MHC II expression. We developed a next-generation model of the molecular and cellular events required for display of antigen on the surface of the APC and describe how it might be used to elucidate the mechanisms pathogens use to interfere with the process [Chang et al. 2005]. We use the number of pMHC complexes on the APC surface with respect to time as our output variable and our measure of antigen presentation unless otherwise stated.

Our model uses ordinary differential equations (ODEs) to describe the time-dependent processes essential to antigen processing and presentation [Chang et al. 2005]. A previous model of the class I presentation pathway applied a method known as nueral networks [Petrovsky & Brusic 2004]. As detailed earlier, these processes include uptake of protein antigen from the extracellular environment, degradation of antigen within endosomal compartments into peptides, synthesis of MHC II molecules, peptide-MHC II binding to form pMHC complexes, and display of pMHC complexes on the APC surface. ODEs are well suited for modeling dynamical systems when species are well mixed and present in numbers large enough that they can be considered continuous. Both of these conditions are met in the case of MHC II-mediated antigen presentation by macrophages. We represent MHC II molecules using six variables to distinguish between intracellular and surface localizations as well as free, self peptide-bound, and exogenous peptide-bound forms. The portions of our model dealing with exogenous antigen and MHC class II peptide loading will be similar to the simpler model developed by [Singer & Linderman 1990].

Key assumptions made in our model development included the following: (1) Both antigen uptake and processing can be represented as single-step reactions. (2) Events leading up to MHC II expression require long periods of time relative to other events, e.g. peptide-MHC binding, and therefore should be included in our model. Long-lived intermediates of these events, mainly mRNA and protein species, will be represented explicitly, while shorter-lived intermediates such as second messengers will not. (3) Events bridging the appearance of MHC II molecules in the ER and removal of CLIP occur constitutively and therefore can be represented as one event. (4) All forms of MHC class II molecules are capable of being transported to and from the plasma membrane, including peptide-free ("empty") MHC II [Germain & Hendrix 1991, Santambrogio et al. 1999]. (5) The reaction scheme MHC + peptide $\Leftrightarrow$ pMHC is sufficiently accurate on the timescales of the experimental conditions we wish to simulate to allow us to forego more complicated models of this process (e.g. in [Beeson & McConnell 1995]). Indeed, our calculations with peptides for which we have pMHC association and dissociation rate constants indicate that we can assume equilibrium binding in the endosome in the presence of the enzyme HLA-DM. (6) Different self peptides bind to MHC II molecules with similar kinetics, despite being derived from various endogenous proteins, and can be represented as a single population. These self peptides will be available for MHC II binding or will be transported to lysosomes and degraded.

Parameters for the model were estimated from published experimental data; many parameters are similar to earlier models [Singer & Linderman 1990, Singer & Linderman 1991, Agrawal & Linderman 1996]. The model was validated under a number of control scenarios. For example, macrophage CIITA, MHC II mRNA, and MHC II protein levels have been reported at various time points by [Pai 2002] and [Cullell-Young et al. 2001]; these data were used to verify the MHC expression portions of our model. Other simulations were compared to time courses of antigen presentation in the presence and absence of IFN-γ from the data of [Delvig et al. 2002]. In each case we matched both qualitative and quantitatively to the known experimental data (see [Chang et al. 2005] for full details of the negative and positive control simulations).

Simulations were run using several ODE solvers to ensure consistency, including the NDSolve feature of Mathematica v4.2 (Wolfram Research, Inc.) and our own solver coded in C and run on Sun UNIX machines. We also performed a detailed sensitivity analysis integrated into the numerical solver.

Using the model described above, we simulated several time courses of antigen presentation. As net pMHC binding affinity was increased in the model (base +/- 25% is shown), the average number of pMHC complexes appearing on the surface over the first six hours of antigen exposure also increased (Figure 13.1). Depending on other conditions in the model, such as extracellular antigen level and level of MHC II expression, this number sometimes dipped below a threshold required to elicit T cell responses, approximately 200 pMHC complexes [Kimachi et al. 1997]. These results suggest that some variants of MHC II may hinder the development of adaptive immunity, and that binding affinity is a key parameter a successful immune response.

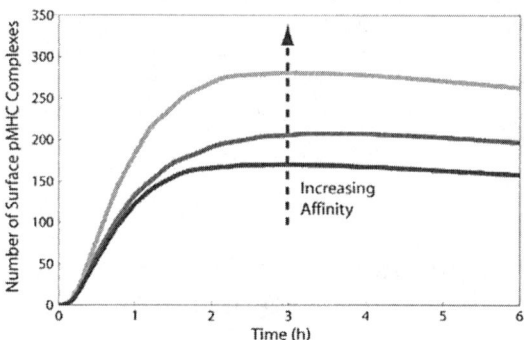

Fig. 13.1. Simulated time courses of surface pMHC levels following exposure to antigen as net pMHC affinity is increased.

13.3.4 *M. tuberculosis* Inhibits Antigen Presentation at Multiple Times using Multiple Mechanisms

Inhibiting antigen presentation at some level is a strategy that many pathogens need to employ to evade immune killing. Because the many processes that constitute antigen presentation are complex and difficult to study individually, many mechanisms have been proposed to explain how pathogens may interrupt one or more of these processes. That *M. tuberculosis* inhibits antigen presentation in macrophages is now well established. Multiple studies have provided a number of hypotheses regarding the mechanism used by *M. tuberculosis* to inhibit antigen presentation, reviewed in [Harding *et al.* 2003], including (H_1) inhibition of antigen processing [Hmama *et al.* 1998, Singer & Linderman 1990] (H_2) of MHC II protein maturation (including delivery of MHC II proteins to the MIIC and Ii processing), (H_3) of MHC II peptide loading [Hmama *et al.* 1998] or (H_4) of transcription of MHC II genes [Noss *et al.* 2000]. Our model addresses why multiple mechanisms have been observed, whether previous experimental protocols favored the detection of some mechanisms over others, and whether alternative mechanisms may exist.

We included into our model of antigen presentation those processes hypothesized to be inhibited by *M. tuberculosis*: antigen processing, MHC class II maturation, MHC class II peptide-loading, and MHC class II transcription. Parameter values were estimated from the literature, mostly *in vitro* studies on mouse cells, and major features of the output, typically surface peptide-MHC levels, were compared to other experimental data. We then used the model to simulate experimental protocols from studies proposing hypotheses and found that some were biased to detecting mechanisms targeting MHC class II expression (Figure 13.2). We also found that mechanisms differed by the timescales on which they were effective (either less than or greater than 10 hours) and therefore might be used in combination by *M. tuberculosis* to ensure continuous inhibition of antigen presentation. Finally, by analyzing the sensitivity of the model to variations in parameter values, we also identified other intracellular processes that may significantly affect antigen presentation (such as self-peptide synthesis) and be targeted by *M. tuberculosis* or other pathogens as a result.

13.4 Genetic Epidemiology of TB- a further look at the impact of antigen presentation in a broader context

One important application of a mathematical modeling approach can be to bridge gaps between biological scales of interest. Clearly, what manifests at the epidemiological level is a result of events that occur at many host-level scales. To illustrate one approach, we explore a link between effects occurring at the level of antigen presentation to effects manifesting at the population level during tuberculosis epidemics.

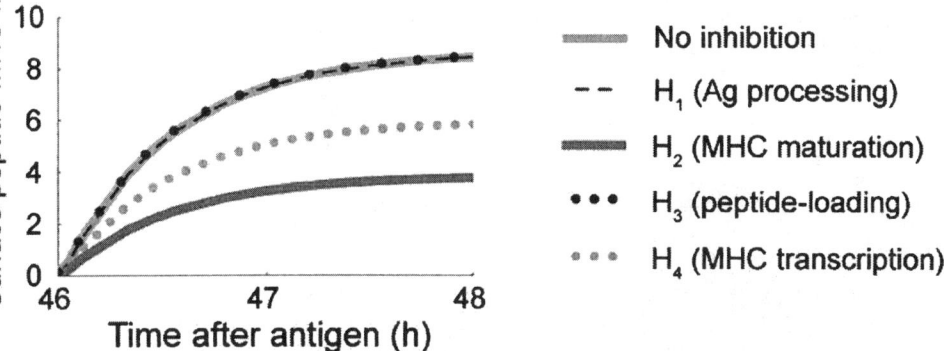

Fig. 13.2. Simulation of one experimental protocol showing that detection of MHC class II expression-targeting mechanisms is favored

Several studies have found that genetic factors influence susceptibility and resistance to *M. tuberculosis* infection [Kramnik *et al.* 2000, Bothamley *et al.* 1993, Goldfeld *et al.* 1998, Selvaraj *et al.* 1998, Bellamy & Hill 1998, Bellamy *et al.* 1998, Wilkinson *et al.* 1999, Hill 1998]. These studies employ a variety of methods including large-scale association-based population case/control studies of candidate genes, family-based linkage analysis, investigation of rare individuals with exceptional mycobacteria susceptibility, and comparison with murine models of disease. Such studies enable identification of particular host genes that influence susceptibility to TB disease.

The major components of susceptibility and resistance to TB appear to be linked directly to the immune response, and in particular to MHC class II molecules. Human MHC molecules are termed human leukocyte antigen (HLA) molecules (but the terms tend to be used interchangeably). Increased susceptibility and resistance to more than 500 diseases has been shown to be associated with various HLA antigens, alleles, or haplotypes (sets of genes that are typically inherited as a unit) [Zachary *et al.* 1996]. In some diseases, HLA expression may influence the balance and strength of the immune response [Pile 1999]. The level and type of immune response to a particular pathogen may vary among populations that have different distributions of HLA molecules.

Many HLA genotypes are implicated in susceptibility to *M. tuberculosis* infection [Bothamley *et al.* 1993, Goldfeld *et al.* 1998, Selvaraj *et al.* 1998, Meyer *et al.* 1998]. Variable binding of mycobacterial antigens to the various HLA molecules may affect the intensity of the adaptive immune response and thus influence susceptibility to TB [Lim 2000, Vordermeier 1995]. Expression of HLA-DR2 is strongly and consistently linked to pulmonary TB and the severe multibacillary form of TB in India [Selvaraj *et al.* 1998, Singh *et al.* 1983, Bothamley *et al.* 1989, Brahmajothi 1991, Rajalingam *et al.* 1996]. HLA-DR2 correlates with increased levels of serum antibody levels [Bothamley *et al.* 1993, Bellamy & Hill 1998, Bothamley *et al.* 1989], indicating an elevated humoral immune response, associated with active disease. The presence of the HLA-DR2 allele may induce tolerance to *M. tuberculosis*, leading to uncontrolled growth of the bacilli [Rajalingam *et al.* 1996]. In addition, HLA-DR2 correlates with decreased production of key proteins that play crucial roles in granuloma formation and subsequent containment of bacteria [Tracey 1997, Flynn & Chan 2001a, Flynn & Chan 2001b].

13.5 Modeling Epidemic TB

Our goal was to develop a mathematical model of epidemic TB that allowed us to investigate different demographic populations with inherent susceptibility to infection by *M. tuberculosis*. To illustrate our approach, we highlight results related to India where the frequency of the HLA-DR2 allele is high and prevalence and incidence levels of TB are significantly higher as compared with the rest of the world. We were motivated by previous work from our group which presented a first model of HIV infection within a genetically heterogeneous population, [Sullivan *et al.* 2001].

We have developed a model of epidemic TB using a modified Susceptible-Infected-Removed (SIR) model with mutually-exclusive groups of individuals who are uninfected, latently infected (those infected with *M. tuberculosis* but not infectious), or actively infected with *M. tuberculosis* (those infected AND infectious) [Murphy *et al.* 2002, Murphy *et al.* 2003]. As our goal was to study the effects of a genetically susceptible subpopulation on the dynamics of epidemic TB at the population level, we further subdivide each of these three groups to include individuals carrying a susceptibility allele for MHC II (DR2 in this case), resulting in the six mutually-exclusive populations. Due to extensive diversity in the HLA genetic system, we examine disease relationships based upon the presence of susceptibility with no distinction between homozygotes and heterozygotes. For full details of the model equations and assumptions, [Murphy *et al.* 2002, Murphy *et al.* 2003].

13.5.1 How to include the effects of genetic susceptibility

Two things are important to consider regarding including effects of a susceptibility gene into this model. First, we divided individuals entering uninfected classes into a cohort that was neutral with respect to effects of a gene and a cohort that was susceptible because of the gene. To allow for births into the population, we defined a parameter that represents the fraction of the general population exhibiting a susceptible phenotype. If we consider a specific genotype underlying this phenotype, then this value must be derived from the allelic frequency according to dominance patterns for that allele. In the model implementation, we considered this value to be constant. This could certainly be extended to include a time varying allelic frequency, as we did in [Sullivan *et al.* 2001], to examine selection processes.

Second, based on the observed significant correlations of HLA-DR2 with active TB, we proposed three possible ways that the HLA-DR2 susceptibility allele may affect the susceptible cohort:

1. HLA-DR2+ individuals have an increased probability of direct progression to active TB upon initial infection
2. HLA-DR2+ individuals exhibit an increased reactivation rate from latent to active TB
3. HLA-DR2+ individuals are more likely to transmit and/or receive *M. tuberculosis*.

To account for these potential processes within the modeling framework, we introduced a parameter to describe the possible influence(s) of genetic susceptibility from our 3 hypotheses on baseline (i.e. genetically neutral) parameters. We do not predict specific values for this parameter as none have been identified; rather we use this parameter to indicate where we included influences from hypotheses of genetic susceptibility and studied a wide range of effects. To observe the effects of this variation, we predict 95% confidence intervals on our output measures (prevalence and incidence) based on large variations in this parameter.

Parameter values and initial conditions reflect demographics of India (derived from the WHO and other data [World Health Organization 2001], as this is the population with the highest frequency of the HLA-DR2 allele. For this simple model we also assumed no treatment or therapy, as may be the case for many of the developing countries with the highest burden of TB. Worldwide, the average (baseline) prevalence of TB is approximately 33%, and the average incidence is 135/100K/yr [Bleed *et al.* 2001, Chakraborty 1993]. Figure 13.2 (dashed curves) shows baseline simulations (worldwide) prevalence and incidence simulations together with a 95% confidence interval on the mean derived from an uncertainty and sensitivity analysis, see [Murphy *et al.* 2002, Murphy *et al.* 2003] for all details).

Our goal was to determine what effects to the epidemiological system would likely have to occur to bring prevalence and incidence in line with the significantly higher level known to exist in India (where prevalence of TB is almost 50% and incidence is between 200-400/100K/yr) [World Health Organization 2001]. The model predicted that the scenario when HLA-DR2 affected all 3 hypotheses (listed above) simultaneously yields results most closely in line with current outcomes for India (Figure 13.3, solid curves). The combined effects yield increased values for incidence and prevalence closer to levels that are observed in India where HLA-DR2 is most prevalent. While the combined effects are more representative of current TB burden, they may be too high in some cases. One explanation is that the presence of known resistance alleles may balance these effects.

While the role of genetic susceptibility is not well defined, it is clearly important to understanding the dynamics of infectious diseases. This is a first attempt to show how effects occurring at the immune system scale can impact dynamics in a significant way at the population scale. Further detailed studies along these lines can likely lead to suggested strategies for intervention and control.

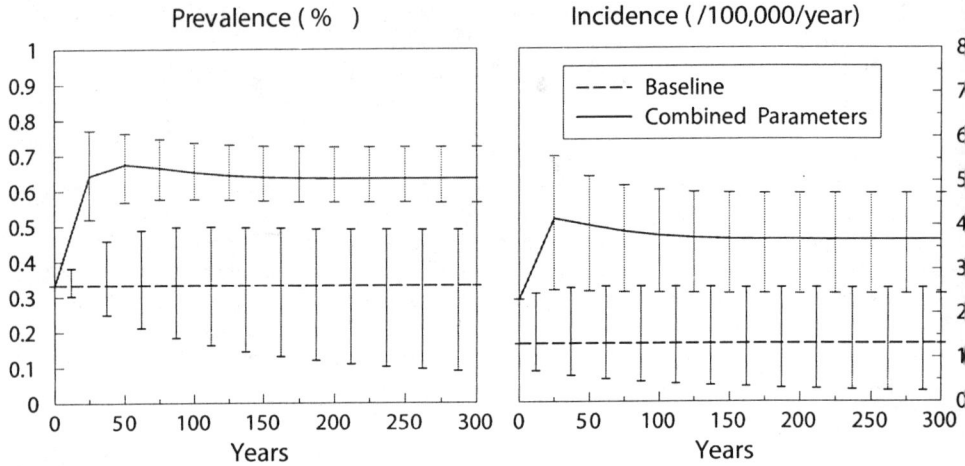

Fig. 13.3. Shown are simulations of the epidemic model for susceptibility to TB over a 300 year period. Panel A indicates the Prevalence, and Panel B shows incidence cases per 100,000/year. The horizontal dashed curves indicate the worldwide (baseline) prevalence and incidence levels with 95% confidence intervals, while the horizontal solid curves indicate the simulated outcomes when all 3 hypotheses are altered indicating the effects of the susceptibility allele (also shown with 95% confidence intervals for variations of parameter values)

13.6 A Temporal Model Tracking the Immune Response to *M. tuberculosis* in the Lung

When a CD4$^+$ T cell encounters an APC, and its T cell receptor (TCR) recognizes the specific pMHC being displayed on the surface of the APC, a series of events follows leading to T cell activation. This interaction between cells bridges to the next biological scale – that of cellular level events. As a first attempt to understand the cellular immune response to infection with *M. tuberculosis*, we have developed a temporal model that qualitatively and quantitatively characterizes the cellular and cytokine control network operational during TB infection in the whole lung [Wigginton & Kirschner 2001]. Using this model we made a first attempt at identifying key regulatory elements in the host response.

This first model was developed to capture infection with *M. tuberculosis* at the site of infection in the lung. Our 'reference space' is the entire lung tissue; however since no data are available in humans, we consider that the simulations take place in bronchoalveolar lavage (BAL) fluid, and we measure all cells and cytokines in units per ml of BAL, as data is available in humans and non-human primates.

While it is likely that the quantitative response differs between the airspace and the interstitium, we relied on the acceptance of BAL as a qualitative predictor of lung environment [Ainslie *et al.* 1992, Moodley *et al.* 2000].

We developed a mathematical system based on the interactions of a number of key cells and cytokines known to be important in TB infection. We tracked both extracellular and intracelluar mycobacteria, the cell populations: Th0, Th1 and Th2 cells, resting, activated and infected macrophages, and four cytokines: IFN-γ, IL-12, IL-10, and IL-4. Our first goal was to develop a model that represents the basic processes of the immune response to Mtb. This model serves as a template on which to add other cells, cytokines, chemokines and interactions as new data warrants to determine how their presence augments or abrogates the system dynamics.

Mathematical expressions were developed representing the interactions between the 8 cell populations and 4 cytokines and parameter values for the rates and rate constants governing each of the interactions were determined (for complete details, see [Wigginton & Kirschner 2001]. Values for most rate parameters were estimated from published experimental data, with weight given to results obtained from humans or human cells and Mtb-specific data over results based on BCG or other mycobacterial species. We outline below how we incorporate these data into the model. Estimates obtained from multiple studies are presented as a range of values. On those parameters for which we have a range, or those for which no experimental data are available, we performed uncertainty and sensitivity analyses to obtain order of magnitude estimates (see the methods outlined above). As an example, we indicate how we estimate the decay rate of IL-10. When IL-10 was administered intravenously to human volunteers, one study estimated its half-life to be 2.3-3.7 hours [Huhn *et al.* 1996]. A similar study estimated this quantity to be 2.7-4.5 hours [Huhn *et al.* 1997]. Therefore, we estimate a range for the half-life from 2.3 to 4.5 hours. The decay rate can be estimated from half-life given by the standard formula $r = ln2/half\text{-}life$. Thus, the decay rate of IL-10 lies in the range [3.69, 7.23] /day. Once the parameters values are estimated, we then simulate the model by solving the differential equations using an appropriate numerical method. Our lab utilizes both packaged software (such as Mathematica and MATLAB) as well as algorithms we coded in C/C^{++} to directly compare results of these different platforms for accuracy.

13.6.1 Simulating Infection Outcomes with *M. tuberculosis*

The negative control, if there are no Mtb present in the system, yields a results with resting macrophages at equilibrium ($3 \ 10^5$ ml of BAL) and all other populations and cytokines at zero (which agrees with estimates for resting macrophage populations in the lung in healthy individuals). The model also indicates that it is possible to be exposed to an initial bacterial inoculum and then clear infection with no memory of that response (i.e. PPD negative). This outcome is plausible, as it is thought that only 30% of individuals exposed to Mtb become infected (i.e. PPD positive) [Comstock 1982]. The other outcomes for the model are: latency and primary disease. Figure 13.4 presents representative simulations for two given sets of parameter values - one leading to latency and the other leading to active disease. The different outcomes predicted by the model begs the question: "Which elements of the dynamical system that describes the host response to *M. tuberculosis* govern the different disease outcomes observed?".

Parameter values that govern the rates and behavior of interactions in the model may change from individual to individual and over time within an individual. The virtual experiments reveal that changes in only certain parameters lead to the different disease outcomes – either latency or active disease. Our primary finding is that the rate of T cell killing (via cytotoxic or apoptotic mechanisms) of chronically infected macrophages governs infection outcome. High efficiency of T-cell killing of infected cells, and consequently bacteria, acts to maintain latency, while lower efficiencies lead to active disease. Further, a trade-off exists between the rate of activated macrophage killing of bacteria and T cell cytotoxicity; if macrophage function is compromised, the T cell response must be more potent in order to control infection. However, when the rate of activated macrophage killing of bacteria is considerably increased (beyond values estimated from experimental data), latency is consistently achieved, even for severely compromised T cell function.

13.6.2 Virtual Deletion and Depletion Experiments

The power of the models we develop is that they can be manipulated in a variety of ways to ask questions about interactions and rates within the system. By doing so, we can explore experimental outcomes on a scale that would be difficult, if not presently impossible, to analyze with other approaches. For example, we can perform both virtual deletion and depletion experiments in this virtual human model for comparison with known experimental results in mice as well as to perform new experiments. Deletion experiments mimic knockout (disruption) experiments whereby we remove an element from the system at day 0, before any infection is imposed into the system. This type of analysis allows us to elaborate which system elements control the establishment of latency. Second, we can simulate depletion experiments by setting the relevant parameters to zero after the system has already achieved latency. These depletion experiments mimic, for example, the addition of antibody that can, to a significant level, neutralize most of a cytokine of one type. This analysis allowed us to determine what elements control maintenance of latency (data not shown- see [Wigginton & Kirschner 2001] for details).

A limitation of this model is that it only tracks temporal dynamics while any spatial aspects are considered homogenous. Moving from a temporal-only model to a spatio-temporal model allows us to elaborate the immune response seen in tissues- that of granuloma formation.

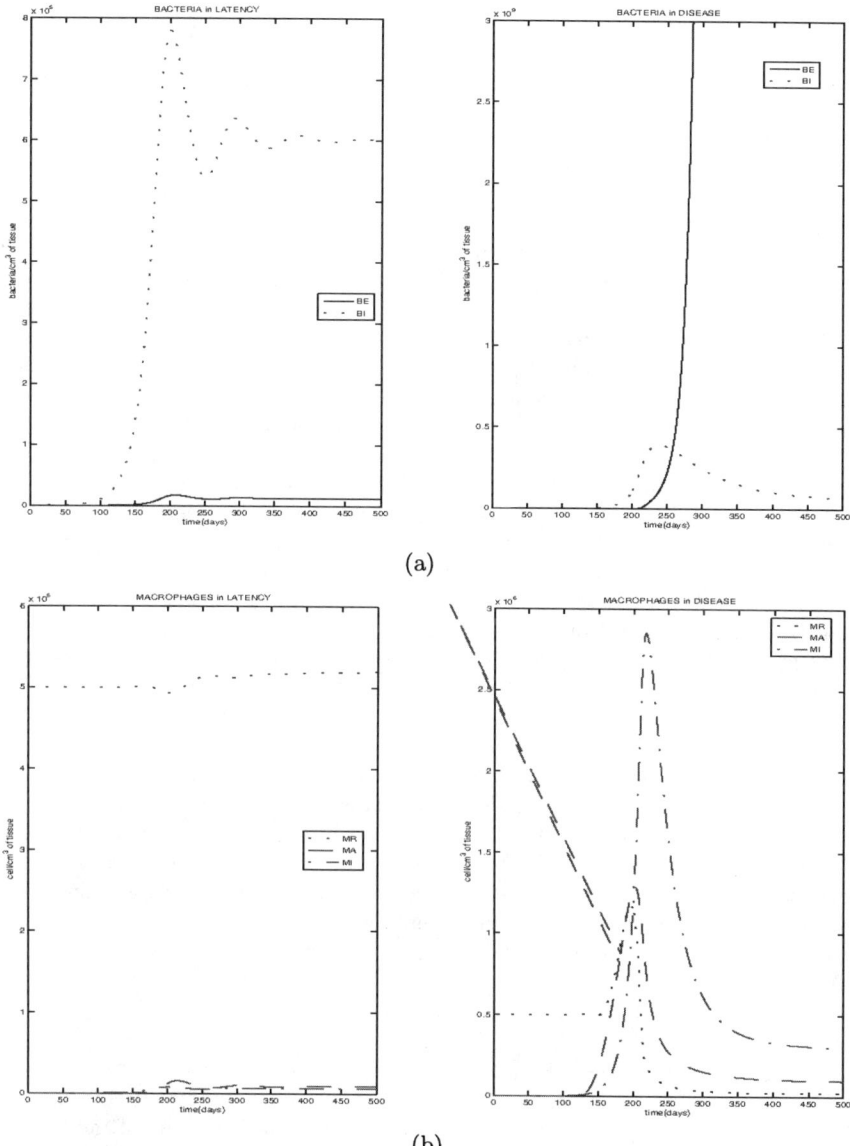

(a)

(b)

Fig. 13.4. Simulations of latency and active disease. The top two panels indicate the bacteria load during latency (left) and disease (right). Shown are the distinct intracellular bacteria (BI) and extracellular (BE) levels over a 500 day time-course. The bottom two panels indicate the macrophage populations over 500 days during latency (left) and disease (right). Shown are resting (MR), infected (MI) and activated (MA) macrophages.

13.7 A Model of Granuloma Formation- the Localized Immune Response to *M. tuberculosis*

The process of granuloma formation leads to a core of dead and infected macrophages together with a centralized necrotic region. These are encircled by activated and resting macrophages as well as CD4+ and CD8+ T cells. Infected macrophages that have not been activated have bacteria growing within them can be killed by activiated CD4+ and CD8+ T cells, which both can act by cytotoxic and apoptotic pathways [Kaleab *et al.* 1990, Kaufmann 1988, Kaufmann 1993, Lewinsohn *et al.* 1998]. Bacteria released are ingested and killed by other activated macrophages. These processes are mediated by a host of elements that must operate in concert to achieve successful granuloma formation. Cells are the key players, but their roles are orchestrated by a number of factors, including chemokines, cytokines, adhesion molecules and their corresponding receptors. Therefore, understanding the dynamic interplay between these immune elements during the time course of granuloma formation and maintenance will provide insight into the mechanisms that control this process. This should distinguish differences between proper functioning granulomas (leading to latency) from those that are unable to contain the bacteria (active disease). A clinical study by [Emile *et al.* 1997] examined granulomas from 14 patients with BCG-induced infection (from receiving the TB vaccine!) . In these cases, it is likely some immune defect (potentially genetically linked) contributed to susceptibility to BCG-induced disease. However, some children suppressed infection while others suffered acute disease. Interestingly, granulomas formed by these two groups of patients were distinct and uniform throughout a given patient. Patients with well-circumscribed, well-differentiated, solid granulomas with activated macrophages and infected macrophages surrounded by lymphocytes containing few bacteria, suppressed infection. Patients with ill-defined, poorly differentiated granulomas with few giant cells and lymphocytes containing a plethora of macrophages filled with bacteria, suffered disseminated disease. Thus, the structure of the granuloma likely determines function which in turn determines whether the host suppresses infection or progresses to active disease. Therefore, understanding granuloma formation will aid in our understanding of the elements that contribute to success or failure of the immune response towards achieving latency in TB.

The importance of the spatial aspect of the immune response to *M. tuberculosis* via granuloma formation has not yet been determined. Likely, the structure plays at least two important roles [Saunders & Cooper 2000]: first is to wall off the bacteria not allowing spread of an infection which cannot be cleared, but second is to facilitate communication between the immune cells affording an optimal, quorum sensing-like interaction [Bonecini-Almeida 1998]. The temporal model developed above is not able to capture this spatial behavior, so new models had to be developed.

To determine the appropriate mathematical tool with which to study the formation and function of granuloma, we developed a series of mathematical models each using a different application, and then performed a formal comparison of each method (see [Gammack *et al.* 2005] for details). Here, we will focus solely on the approach where we used a computational system known as an agent-based model. This allows us to capture the most discrete and stochastic representation of the forming granuloma. This approach also allows for heterogeneity in space and time.

13.8 The Agent-based Model

We have developed the first model of this type applied in the context of the immune response to a pathogen [Segovia-Juarez et al. 2004]. To develop an agent-based model 4 things are necessary: a description on the agents, the rules that govern their behavior, the environment on which they reside and the parameters that govern their interactions. The environment is a key feature of ABMs; important details about modeling environments in general can be found in Chapter 12 by Stepney in this book. The environment is a 2-dimensional lattice representing 2mm x 2mm of lung tissue. The lattice is comprised of grids where the size of each grid can hold the largest cell-type, the macrophage. A single macrophage can reside is a grid with other smaller cell types (such as T cells) and large amounts of effector molecules, such as cytokines and chemokines. The agents are a mix of discrete and continuous entities: immune cells such as macrophages and T cells are discretely tracked, while the bacterial populations and effectors such as cytokine and chemokine are continuously tracked variables. Cells can take on one of several states. A macrophage can be resting, infected or activated, while T cells can take on resting or activated status. There are a complex set of rules that govern the individual behavior of each agent, as well as rules that govern their interactions. These are based on well-documented data. For example, if a macrophage takes up mycobacteria, there is a window of opportunity where a T cell can move into the same grid space occupied by the infected macrophage and activate it via direct cell signaling together with secretion of the cytokine IFN-γ, allowing macrophages to clear the load of intracellular bacteria [Nathan et al. 1983, Flesch & Kaufmann 1990, Armstrong & Hart 1971, Sturgill-Koszycki et al. 1994]. This is one of the many rules coded into the model (see Figure 13.5).

Many of the parameter values are not known in this setting as they are probabilities and these are difficult to estimate in a wetlab. This makes the use of a detailed uncertainty and sensitivity analyses important in this context. We were the first to apply this analysis to study agent-based models [Segovia-Juarez et al. 2004]. For many of the other parameters, we could borrow from what we had estimated previously. For full details please see [Segovia-Juarez et al. 2004].

13.8.1 Simulating Granuloma Formation

The behaviors that emerge from this model are complex and of three consistent types. First, a small solid granuloma forms showing containment of bacteria with little to no necrosis forming (Figure 13.6, Panels A, C). Second, we can also generate a larger, more necrotic granuloma that is consistent with dissemination (Figure 13.6,

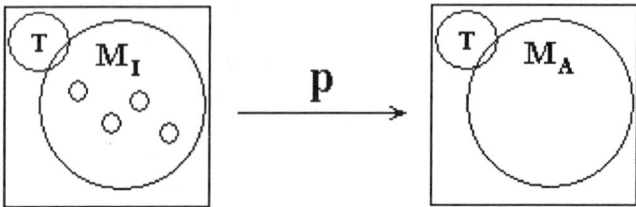

Fig. 13.5. An example of a rule for the agent-based model. If an infected macrophage (M_I) has taken up bacteria (small circles) a T cell can activate it with some probability p, which allows the macrophage to become activated (M_A) and also to clear its intracellular bacterial load.

Panels B, D). Third, we can simulate clearance of all bacteria with no trace of a granuloma (not shown). This last outcome is interesting as it predicts that under certain circumstances the immune response is efficient at clearance. This is suspected as only 30% of individuals exposed to *M. tuberculosis* become infected, however it has not been strictly documented.

The top panels of Figure 13.6 show early time points (2 weeks) in the development of the granuloma under two sets of parameter choices: on the left T cells arrive to the site of infection on day 2 as compared with the right panel where they arrive on day 14. Also, the initial number of macrophages is higher on the right panel than on the left. Within 14 days, it is clear that already the granuloma on the left is more solid and contained than the one on the right which shows more diffusivity. By 6 months (bottom panels) the amount of necrotic tissue (shown in brown) is much greater and the granuloma on the right is much larger as compared with the granuloma forming on the left. Based on the study of [Emile *et al.* 1997] this would indicate that granulomas forming similar to those in the left panels would be able to contain infection, while those on the right would lead to disseminated infection.

The benefit of mathematical modeling here lies in predicting what mechanisms determine these different granuloma outcomes. The sensitivity analysis we employ is based on a partial rank correlation and can identify (with statistical significance) the parameters in the model that when varied correlate to different outcomes. In the simulations shown in Figure 13.6, the timing of effector T cell entry onto the grid (from lymph node homing) is what was determinative. Interestingly, all of the parameters that relate to early numbers of resting macrophages present on the lattice positively correlate with bacteria load. This likely follows since they serve as the primary host for mycobacteria and their presence serves to propagate infection.

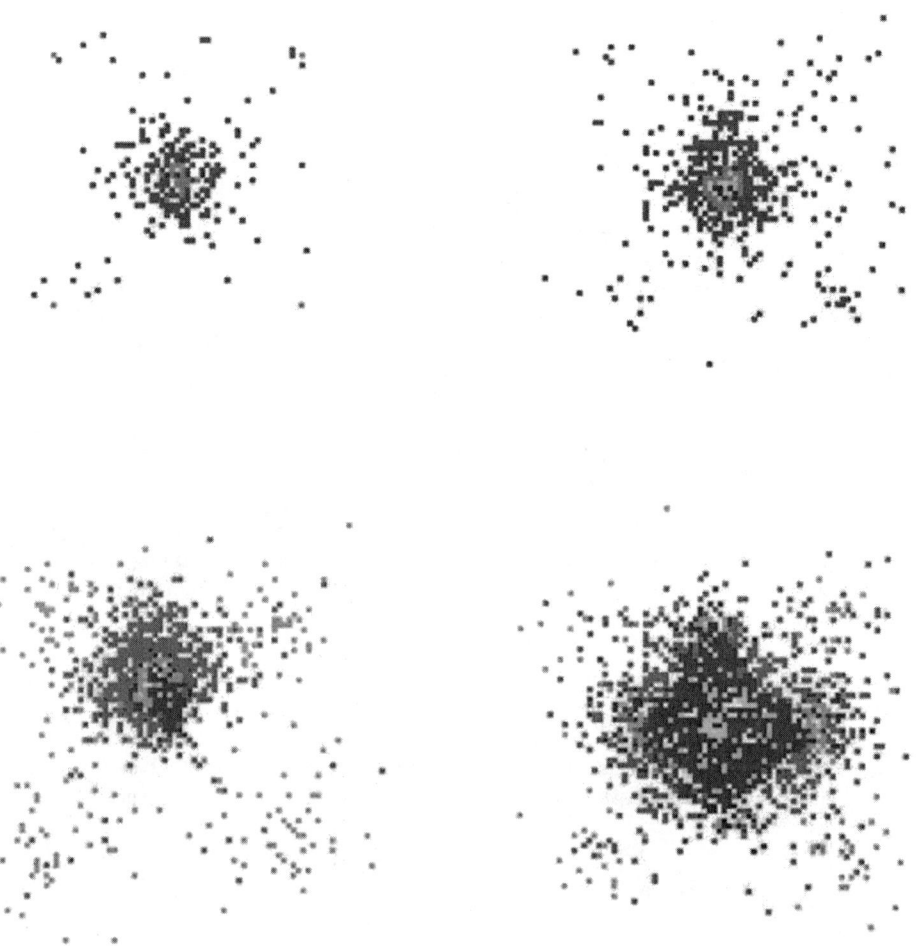

Fig. 13.6. Agent-based model simulations for containment and dissemination at early (2 week time point-top panels) and late (6 month time point bottom panels). Left panels show the containment simulations and right panels show dissemination. T cells are shown in pink, macrophages in green, activated macrophages in blue, infected macrophages in orange (and chronically infected in red). Necrosis is shown in brown with yellow indicating extracellular bacteria.

Parameter	30 days	60 Days	500 days
Chemokine diffusion rate	0.18	0.13	0.13
Prob. T cell recruitment	-0.36	-0.27	-0.31
Prob. T cell movement	-0.65	-0.54	-0.57
Prob. T cell activates a macrophage	-0.24	-0.16	-0.15
Initial number of macrophages	0.40	0.54	n.s.
Prob. a macrophage is recruited	0.56	0.61	0.75
Speed of activated macrophage	0.31	0.61	n.s.

Table 13.1. Time-dependent partial rank correlations for the 7 parameters in the model (out of 27) that behave as bifurcation parameters driving the system toward containment or dissemination as they are varied. Correlations are shown for total bacteria load as the outcome variable of interest. Similarly, the size of the granuloma or amount of necrosis could be used as outcomes ($p < .001$ in all cases, unless not significant (n.s.)).

Thus, reducing early inflammation (less than 60 days post infection via the influx of too many cells) could be beneficial towards halting infection or tipping the scales in favor of containment. Table 13.8.1 shows all 7 key host parameters with their correlation coefficients over time.

The agent-based approach has its strengths and weaknesses. The strength here is that individual cells can be tracked and at any moment in time all interactions and cell levels can be observed. Weaknesses include an inability for complete mathematical analysis. Regardless, this method uncovers some important features of the host pathogen interaction that we were unable to identify previously with any other approach.

13.9 Discussion

Despite a wealth of information in the biological literature regarding elements of the immune response over genetic, molecular, tissue and system levels, no single representation synthesizing this information into a model of the overall immune response currents exists. In this paper we present approaches for capturing each of these levels to address one specific case: the immune response to *M. tuberculosis*. The next goal is to combine information over the relevant biological and temporal scales to generate a single, integrated multi-scale representation. Such multi-scale models should be developed so that they are sufficiently general that they can be applied to answer a wide range of questions regarding immunity but adaptable enough to answer specific questions regarding, for example, pathogen invasion, tumors, vaccines or auto-immunity. One step towards achieve this goal will be to develop hybrid models (such as multiple compartment, agent-based models) that include various biological scales. Here we have presented a number of models that each include representations of multiple biological scales, but none are a complete picture of the entire immune response to *M. tuberculosis* and its manifestations at the epidemic level.

Once we can develop multi-scale models, we can apply them towards the generation of hypotheses regarding features of the roles of specific processes in immunity, such as antigen presentation. It is crucial to work under a hypothesis that events occurring at each level (genetic, molecular, cellular, and tissue) of the immune system affect the development of the overall immune response.

For example, the efficacy of vaccines are in part determined by activation of CD4+ T cells. A multi-scale model should enable testing the roles that various factors play in that activation. What is the relationship between antigen dose in the vaccine and the number of mature DCs appearing in a lymph node? Further, what aspects of the antigen presentation process should be targeted to optimize vaccine efficacy? Can our insights help to explain why BCG, the vaccine against TB used for the last 80 years, has failed to control the TB scourge? As theoretical immunologists we are poised to make a strong contribution in this area through hypothesis generation and testing using multi-scale models.

Go Dutch: Exploit Interactions and Environments with Artificial Immune Systems

Mark Neal[1] and B.C. Trapnell, Jr.[2]

[1] Department of Computer Science, University of Wales, Aberystwyth.
mjn@aber.ac.uk
[2] University of Maryland, USA. btrapnel@gmail.com

Summary. The natural immune system is composed of a diverse array of cells and proteins which cooperate to attack infections in the body. These cells interact in space and over time in two principal ways: through direct physical contact, and via intermediate signaling molecules. The network of these interactions is extremely complex and difficult to analyze. However, an attempt to understand how this network is organized is critical to further development of artificial immune systems (AIS). This chapter attempts to characterize some of the major interaction mechanisms found in the immune system from a computational perspective. We also explore how AIS might exploit the properties of such interactions in practical applications.

It is not intended or claimed that this chapter will be a complete or comprehensive discussion of immune system signals and interactions; for more detail the immunology literature should be addressed. It *is* intended that this chapter provides a reasonable summary of some of the major and widely accepted interaction mechanisms found in the immune system, as well as some insightful examination of their properties and potential uses in AIS. The final section of the chapter points out some artificial mechanisms and characteristics of applications that might benefit from particular aspects of the interactions identified.

14.1 Characterizing some Immune System Actors and Interactions

The immune system is complex and highly integrated with other organs and functional systems in the body. The approach taken here draws heavily on standard immunological views, divisions and theories. At the outset it is relatively easy to identify a large set of interaction mechanisms and components. We can separate actor interactions into two broad categories: cell to cell and cytokine-mediated. By identifying specific cellular and cytokine interactions and examining how they facilitate communication between the various components of the immune system (and other parts of the body) we can understand some relatively traditional and useful divisions and sub-components.

Cell-to-cell interactions rely on very close proximity of cell membranes allowing interaction of receptors and ligands bound to their membranes. Phagocytosis falls into this category of interaction. More generally, cell-to-cell interactions require the participants to occupy the same small region of tissue at the same time. Cytokine interactions occur when cells detect small molecules in their environment which affect their function, movement or state in some way. Cells can interact by secreting intermediary cytokines, thus relaxing the constraint that they be in the same place at the same time. It will rapidly become clear that the number of interactions identified by immunologists is large in both of these categories, and the two types of interaction are often interdependent.

Before delving into cell to cell interactions in detail it will be useful to provide names and descriptions of the actors and their roles. A list of cell types and the complement system and a brief caricature of their functions and interactions will prove useful in the discussion that follows. At this point it is worth noting that these cells are specialized "on the fly" by the immune system and that immunologists group the cell types in a number of ways. In order to retain the focus of the chapter on interactions the reader is directed to other sources for detailed descriptions of differentiation and development of different cell types and roles[Leslie 2000, Bullock et al. 2003, Luther & Cyster 2001, Janeway 2001, Sompayrac 2002]. The emphasis of the following descriptions is on interaction mechanisms.

The complement system is a set of (about 30) proteins that has a range of functions essential to the priming and triggering of many immune responses[Medzhitov & Janeway 2002, Ochsenbein & Zinkernagel 2000]. The action of these proteins involves an astonishing cascade of chemical interactions. The complement system is usually seen as the lowest layer of the mechanisms that stimulate the innate immune response and ultimately triggers the adaptive immune response as well. Triggering of the complement system occurs when particular pathogen associated molecular patterns (PAMPs) are bound by circulating proteins in the blood. Mannose on the surface of many prokaryotes is such a PAMP and triggers cleavage of complement components that begin the cascade. The most important effects of this cascade are the *opsonization* of pathogens (coating them with proteins that encourage phagocytosis), the attraction of leukocytes (in particular neutrophils and other phagocytic cells) through the release of powerful chemotactic cytokines, puncturing of bacterial membranes, and stimulating antibody production. To the best of the authors' knowledge computer scientists looking for immune system inspiration have so far entirely ignored the complement system's essential role.

Mast cells are present in large numbers in a range of tissues that are common entry points of infectious agents. They are involved in the elimination of helminthic parasites as well as some common bacteria such as *E.coli*. These responses are stimulated by the activity of the complement system, and opsonization of some pathogens is necessary for effective binding and response by mast cells. They are also apparently stimulated by early responding leukocytes such as eosinophils and neutrophils. Mast cells also produce TNFα, an inflammatory cytokine that induces endothelial cells to capture circulating leukocytes. The production of TNFα and the close proximity to endothelial cells suggests that mast cells closely interact with endothelial tissue[Abraham & Malaviya 1997]. Interestingly mast cells are also often neuroconnected and thus interact *directly* with the neural system[Dines & Powell 1997]. This is a clear example of a strong relationship between two systems often considered as separate.

Neutrophils (a type of granulocyte) are the most numerous of all white blood cells and are involved in a very wide range of innate immune responses. These include phagocytosis of foreign cells (including yeasts, fungi and bacteria), toxins and viruses[Bonnett 2005]. They (like other leukocytes) are in constant flux in the blood and are recruited and activated locally when inflammation occurs. Adherence to and passage through the endothelium is promoted by a cascade of reactions beginning with the complement system. Neutrophils respond to some of the same inflammatory cytokines that activate endothelial cells. These cytokines cause neutrophils to express the reciprocal ligands to endothelial surface proteins, causing the neutrophils to become "stuck" to the endothelium. Once attached, the direct cell-to-cell interaction with endothelial cells effects a structural change in the neutrophils: they become flattened and pliable, allowing them to squeeze between endothelial cells, crossing the vascular boundary into the local tissue. They use concentration gradients of inflammatory cytokines such as TNFα to migrate through the tissue towards the center of inflammation. Upon arrival at the seat of infection or trauma, the neutrophils will lyse pathogens to which they are sensitive via a number of different mechanisms depending on the particular pathogen in question. Neutrophils increase the inflammatory response by the release of cytokines; which particular cytokines that are released is dependent on the type of infection or trauma that is present. Neutrophils detect pathogens through receptors on the surface of the cells known as "Toll like receptors" (TLRs) which are capable of recognizing a range of PAMPs[Kurt-Jones et al. 2002]. A selection of about ten of these receptors is capable of triggering responses tailored to the elimination of the likely culprits associated with each pattern.

Epithelial cells constitute continuous physical barriers between the host and the external environment (including the lung, gut etc.). They secrete natural antimicrobial peptides called *defensins* in response to exposure to several inflammatory cytokines such as interleukin-1 (IL-1) and TNFα [Stolzenberg et al. 1997]. Epithelial cells also secrete several inflammatory cytokines when exposed to antigen, which they are known to recognize via TLRs [MacRedmond et al. 2005].

Endothelial cells have numerous physiological roles, several of which are key to host defense. They make up the majority of the surface of blood vessel walls (lung, gut etc...). They form the boundary between the blood (and thus the mobile immune cell populations) and the body's tissue. Endothelial cells are a cornerstone of the inflammatory response because they express surface proteins when activated that bind circulating leukocytes. They are the principal component of dynamic dispatch of circulating leukocytes to infected tissue. As a result, they interact with a wide variety of cells (notably macrophages and neutrophils), and are activated in response to an array of inflammatory cytokines [Ebnet et al. 1996, McIntyre et al. 2003].

Basophils are a rather enigmatic type of granulocyte, and for a number of reasons including their relatively low abundance in circulating blood have largely been ignored in the study of the immune system. They are generally accepted to have some properties similar to mast cells and to react on similar timescales to eosinophils. They have been studied mostly in the context of allergic reactions and asthma in particular, but little is known about their role in normal immune function. Discoveries about their ability to produce large quantities of some cytokines (such as IL-4 and IL-13) which have important roles in defining B cell and T cell activity do however point to a role in the stimulation and suppression of adaptive immune responses[Falcone et al. 2000].

Eosinophils are granulocytes that are generally accepted to be effective in dealing with multicellular parasites such as invasive worms. They also appear to perform signalling roles much like other innate response leukocytes. In particular they are known to produce signals which affect mast cell and basophil function. Once again much research into their role in allergic disease has been undertaken somewhat at the expense of detailed examination of their function under "normal" conditions[Flood-Page et al. 2003].

Monocytes circulate in the blood and have the potential to migrate into tissues and differentiate into macrophages or dendritic antigen presenting cells. Monocytes interact with the endothelium similarly to neutrophils. They are preferentially recruited to the site of inflammation by expressing ligands reciprocal to endothelial surface proteins upon exposure to inflammatory cytokines[Lichtman & Abbas 1997, Luscinskas et al. 1994].

Macrophages are effector cells of innate immunity. Their primary role is to phagocytose microbes, killing them and presenting microbial components to helper T cells. Thus, they are central to the effector response of the innate immune system, and are a bridge to the adaptive immune system. They are strategically placed in tissues where contact with pathogens is likely, such as subepithelial connective tissue and lymph nodes. They secrete TNFα upon phagocytosis of microbes, amplifying inflammation. Monocytes differentiate into tissue macrophages upon recruitment to the site of inflammation, so macrophages are, in effect, recruitable [Abbas et al. 2000].

Dendritic cells can be derived from a number of sources (including monocytes as described above), and migrate from the site where they engulfed the pathogen to lymph nodes where they present the pathogenic material (suitably preprocessed). The way in which the material is presented and the cytokines that are produced during this process define the precise effect that presentation has on the B cells and T cells in the lymph node[McKenna et al. 2005]. The type of antigenic material (and the way that it interacts with TLRs on the monocyte) selects the outcome of this process.

T cells that are capable of recognizing the fragments of molecule presented by dendritic cells are then activated and proliferate in order to attack pathogens which display similar molecular patterns and also stimulate macrophage activity. These reactions involve the production of transcription factors which in turn switch on cytokine production genes causing the production of TNFα, IL-1 and chemokines which attract other leukocytes (neutrophils, etc.). The end result of these actions is localized inflammation of the infection site. T cells which differentiate in this way are known as Th1. Dendritic cells can also cause T cells to interact with B cells and cause them to produce large numbers of antibodies. Some of the mechanisms which distinguish this mechanism from the previous mechanism are unclear, but it seems likely that the cytokines produced by the dendritic cell in the interaction with the T cell are different and thereby evoke different behaviour. This is an example of a key paracrine interaction which mediates cell to cell interaction. T cells which differentiate in this way are known as Th2[Abbas *et al.* 2000, Sompayrac 2002].

B cells are workhorses of antibody production by the immune system and like T cells occur in more than one subtype. B cells are produced in the bone marrow with the ability to recognize particular random antigens[Abbas *et al.* 2000, Sompayrac 2002]. B cells circulating through lymph nodes become activated when in contact with antigen that they recognize as well as a T cell that stimulates it simultaneously. Once stimulated the B cell will proliferate and generate plasma cells which migrate into the bloodstream where they produce very large numbers of antibodies very rapidly. There is also evidence that B cells can stimulate each other in an *idiotypic* fashion and thereby remain active even in the absence of antigen[Jerne 1974]. The precise mechanisms of this interaction and its role in immune memory is both unclear and somewhat controversial.

NK cells are effector cells of the innate immune system that specialize in lysing (killing) virus-infected host cells. They interact with macrophages via the cytokines IL-12 and interferon gamma (IFNγ). Their proliferation and effector functions are increased in response to several inflammatory cytokines [Abbas *et al.* 2000].

Figure 14.1 contains a number of the actors and interactions that are commonly reported in the immunology literature. Diffusible mediator interactions (signified by dashed lines in the diagram) are commonly classified into three types: paracrine, autocrine and endocrine. Autocrine interactions involve the production and detection of cytokines by cells of the same type. Paracrine interactions involve the production and detection of cytokines by cells of different types, for example the activation of macrophages by the production of IFNγ in T cells. Endocrine interactions are generally indiscriminate in nature and affect large numbers of cell types throughout the body with common signalling molecules. In general the cytokines involved in these long-range diffusible media interactions fall into three functional categories: inflammation mediators, specific immune response regulators and growth and differentiation factors. The complexity of this network of cytokines is horrifying to most computer scientists; accurate simulation or analysis of such systems is inevitably extremely difficult, computationally expensive and prone to cumulative errors.

The diagram is not intended to be a comprehensive description of the elements shown, nor to contain all components commonly attributed to the immune system. It is however intended to highlight the highly connected nature of the graph of interactions. By including in the diagram even an incomplete set of the movements and interactions of the immune system components it is clear that the system is

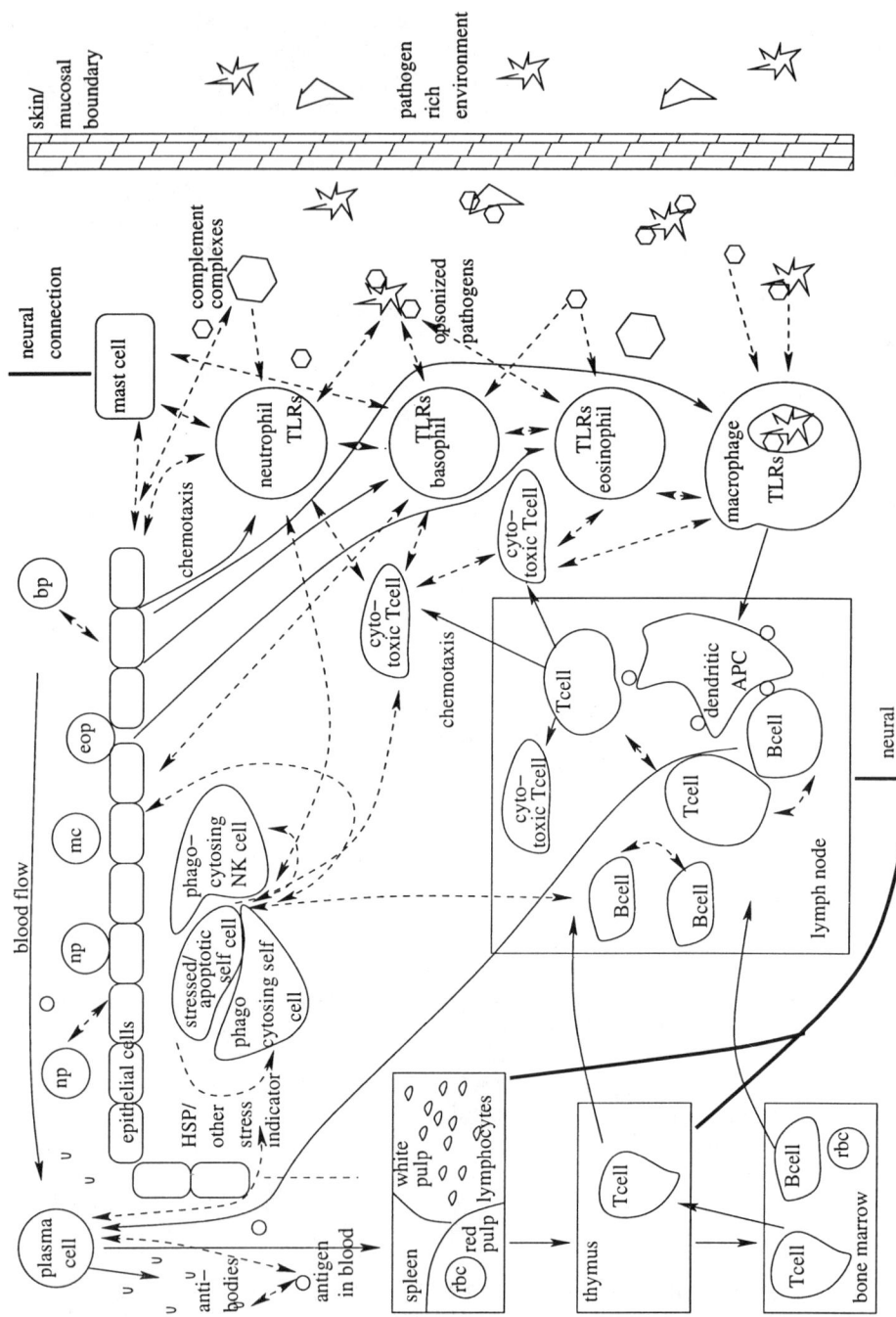

Fig. 14.1. A condensed and incomplete view of immune system actors and interactions. Solid arrows indicate cell movements, dashed lines indicate cytokine/protein/diffusible mediator signalling.

close to being a heterogeneous fully connected graph of interactions: a very large number of components affect and are affected by a very large number of other components. Such networks are by definition extremely difficult to analyse and characterize, they also appear to be quite common in naturally evolved systems. Brains, multi-cellular organisms, genetic expression networks, economic and cultural systems, and ecosystems exhibit similar levels of connectivity and have so far proved impossible to understand, characterize and control despite enormous scientific effort in their study.

In general the study of such systems as a whole relies on the use of complex system analysis techniques which neglect local detail in favour of global properties and behaviour. Isolation of small sections of such networks and their study may well reveal detailed information about local mechanisms, but in general is not guaranteed to provide useful information about the full *in vivo* operation of the components under examination. A closed environment containing a dozen people for a few weeks provides entertainment for poor television shows, but not necessarily useful information about the performance of the national economy or the functioning of normal social systems.

14.2 Properties of Interactions

The network of actors and their interactions described above has functional features that are of great interest to computer scientists[Cohen 2000b]. The immune network manages to achieve a level of coverage, reliability, and decentralized control that for many systems, is the "holy grail".

14.2.1 Heterogeneity

The wide range of actors and interactors within the immune system is the most obvious marker of heterogeneity within the system. This indicates that environments as varied as those presented to organisms such as humans may *require* a wide range of defenders, and that evolution has found no single actor, mechanism, chemical or cell type capable of dealing with all of the pathogens posing a threat. There is an obvious lesson to be learnt from this in the realm of computational systems: diverse problems require a range of specialist solutions[Wolpert & G. 1997]. Within the innate immune system we see a range of mechanisms and cell types for dealing with different circumstances ranging from heat shock proteins, the complement system and cytokines to eosinophils, neutrophils and macrophages. Each class of actor and interaction deals with different threats and stimulates different responses from other parts of the immune system and organism in general. This may in part be a result of the way in which evolution and co-evolution work: co-option and adjustment of existing mechanisms is commonplace which leads to initial diversity; and subsequent evolution of pathogens in response forces further diversity of mechanisms and interactions to be produced. It may also be in part due to the simple observation that a wide range of potential mechanisms (and by implication interactions) is more likely to contain *something* for dealing with any individual pathogen: the shotgun approach. More realistically it is likely to be a combination of the two, with the latter effect taking over once diversity had been introduced by the former.

14.2.2 Redundancy

A consequence of the variety of mechanisms that evolution has retained within the immune system is that many of the mechanisms will be effective to some degree in many different situations and with many different pathogens and classes of pathogen. Examples of interactive redundancy occur throughout the immune system and its mechanisms. These include multiple receptors capable of recognizing (often different parts of) the same pathogen. This effect occurs within the innate immune system with TLRs and PAMPs as well as with antibodies produced by the adaptive immune system. A second layer of redundancy is within the signalling network of the immune system (and body) itself. Many functions are promoted by a range of different cytokines and many cytokines promote the same functions, sometimes in the same cell types and sometimes in others. Redundancy seems to increase the likelihood of the development of perceptual and functional degeneracy (see below) and to decrease the probability of a particular function being completely arrested by pathogenic or chemical activity. The existence of viruses that bind to, exploit, block or mimic particular cytokines also indicates that such redundancy and the co-evolutionary pressure to exploit it have significantly sculpted the human immune system in its present form. For further treatment on the topic of degeneracy, see chapter 7 and 6 in this book.

14.2.3 Perceptual degeneracy

Perceptions of proteins and other molecules via receptors on cell membranes are degenerate in a number of ways. Imprecise recognition of proteins whether by T cells of MHC or by TLRs of PAMPs in a neutrophil ensures that slight mutations and alterations do not result in complete failure to recognize a possibly important feature. Binding of multiple sites on a molecule provides a further safeguard and both of these features are a natural consequence of the chemistry of the large molecules present in living systems. The concept of *recognition balls* as regions of the space of shapes of antigens is a direct consequence of this degeneracy and carries with it the possibility of somatic hyper-mutation of B cells. This allows fine tuning of the adaptive immune system to very precisely adjusted and specific antibody production. If perception by antigen receptors was *not* degenerate and was instead an "all or nothing" affair then the ability to optimize antibodies in this way would be removed. Thus perceptual degeneracy fulfils at least two vital roles: one in the innate and one in the adaptive immune system.

14.2.4 Functional degeneracy/pleiotropy

Most of the actors and molecules active in the immune system elicit a wide range of responses in a wide range of actors. This can be extremely confusing when working with *in vivo* systems, and can result in surprising effects such as suppression of a function when a molecule is at very low concentration, excitation at a slightly higher concentration, and suppression at a slightly higher concentration again. Of course each gene may result in a number of other functions, some of which may be contradictory. Equally the functions may be completely unrelated and merely the result of serendipitous reuse during the evolutionary process. To add a little interest into this mix the location, internal state and external environment can modify the behaviour of actors further.

14.2.5 Co-respondence

One feature common to a number of immune system interactions is that a number of molecules/interactions must be present simultaneously for particular functions to be activated. Examples include the activation of B cells in the lymph nodes which requires both antigen presentation and T cell interaction to arise. The MHC is a key part of the immune system's chemistry and is used in a number of ways including the detection of "dangerous" self cells that do not display it on their surfaces, as well as in T cell binding. The use of multiple simultaneous interactions to avoid erroneous damage and to control responses more precisely is already well known for systems such as opening bank vaults and firing nuclear missiles; it provides security and flexibility especially when combined with the ability to combine signal presence and signal absence in controlling responses. Little has been made of such interactions computationally but their importance in the immune system implies that perhaps we ought to look more carefully.

14.2.6 Gross parallelism: incomplete temporal ordering

The gross parallelism of the immune system undoubtedly enhances its attractiveness to computer scientists. It is also one of the first features to be discarded as computationally intractable. There is however a more subtle aspect to the gross parallelism of the immune system, and that is the impossibility of perfectly ordering the responses of such huge numbers of actors and components. The temporal evolution of an immune response *is* marked by more or less defined stages, but this is by no means reflected at the level of the individual cells involved. The shift from an initial inflammatory response to tissue damage through to the later stages of wound healing follows a clear temporal path, but it is not possible to precisely predict the states of individual cells: cells of a particular type will tend to be in a range of states clustered about some "average" state. Thus there is a loosely coupled temporal organization of actor states which evolves (presumably) controlled by the properties outlined in the previous few sections. This ability to maintain reasonable temporal coordination of a highly parallel system without perfect or complete communication channels is truly enviable for computer scientists. In general, highly parallel systems in computer science rely on well-defined, fixed and precise global coordination. The immune system achieves an appropriate level of approximate temporal coordination in a far larger system without using a central clock or control mechanism, instead the diffusion of cytokines and hormones locally defines the responses of individual actors.

14.2.7 What about the rest of the body?

Up to this point in the discussion the involvement of the parts of the body not considered to be parts of the immune system has been ignored. The immune system does however operate in this much wider context and much of its activity is stimulated by interaction with the body at large and its interaction with the environment. The overlap in functionality between the immune system and other parts of the body not usually considered to be parts of the immune system is also considerable. For instance removal of the spleen (generally considered to be a part of the immune system) is usually not a particularly serious problem: its functions are partially replaced by the liver and bone marrow. Far more widespread than this is the ability of any cell in the body to generate signals that invoke immune system responses. Molecules released when cells experience stress (heat shock proteins etc...) call into action NK cells and macrophages in order to clean up the mess. It is also now generally accepted that microglial cells in particular are capable of phagocytosis, and indeed that almost any cell in the body is capable of phagocytosis should the correct circumstances arise. This raises the question of whether the immune system is truly separable in any meaningful sense from the other parts of the body[Ottaviani & Franceschi 1996]. Perhaps parts of it can be meaningfully plucked from their nests, but it does not seem at all certain that alot is not lost in doing so.

14.2.8 What about the rest of the world?

A rather more fundamental question is about the validity of considering the immune system or indeed a complete organism in isolation from its environment. The trend of the last few decades in considering cognitive systems in a "situated" manner has led to a number of new ways of thinking about such matters[Clancey 1997]. Arguably, immune system researchers were at the head of the field in this respect[Varela 1981]. If one takes this stance seriously then the consideration of software and robotic systems in conjunction with their environments is essential. Thus we are left to consider how such systems (and the immune system) coordinate with their environments, and how their environments coordinate with them. The notion of mutual coordination of any system and its environment is clearly attractive in the light of co-evolution and the development of homeostatic organisms, but it is open to debate as to whether this really adds anything new to the design and development possibilities for artificial systems. We will return to discussion of how the environment that a system is to be deployed in might affect a particular system toward the end of the chapter.

14.3 Identifying "sub-systems"

In this section we identify a few of the sub-systems that are widely accepted to be present in the immune system. We would however like to clarify our stance and point out a few potential pitfalls with the activities of subdivision, simplification and in particular biologically inspired computing.

The nature of human thinking, and especially of reductionism as a scientific paradigm encourages the sub-division of complicated structures and systems into intellectually manageable pieces. The study of the immune system is no exception to this and the identification of pathways and sub-systems within the immune system forms a large part of immunological research. The types of system that can be identified are defined to a significant degree by the research interests (often medical and biased towards pathological cases), experimental methods and existing knowledge exploited by the researchers. As such they are unlikely to be the only or "best" way of splitting up the complex web that makes up the immune system. That mast cells are not well characterized and are most frequently studied in the context of allergy and that basophils are not a popular topic of research due to the difficulty of working with them provides reasonable cause to believe that such effects are likely to be commonplace. Indeed most biologists would (perfectly sensibly) see such effects as necessary consequences of working with complex *in-vivo* systems.

Such arguments apply to most biological systems, but especially to the immune system due to its pervasive, diverse and spatially distributed nature. There *are* a few fixed structures that provide definite boundaries to be exploited in this analysis process such as the bone marrow and thymus, but in general the components and actors that make up the immune system are mobile, variably concentrated, and dramatically multi-functional. Whilst this line of thought may call in to doubt the genuine biological validity of particular descriptions of immune agent functions and interactions, we do not deny their usefulness in furthering the understanding of the immune system and its functions. The fact that *in vivo* the immune system is very complicated does not necessarily imply that all of the complexity must be considered at every step. The fact that the *in vivo* system is closer to being a fully connected graph of interactions than to a set of disconnected pathways implies that any abstraction (and arguably any understanding of cause and effect) will require the removal of large amounts of important detail. That these abstractions and divisions can be, and are, made by biologists is the main reason for the attractiveness of biological systems to computer scientists. The computer scientist's wish to exploit the biological models is a natural consequence of the belief that it is actually the arrangement, properties and interactions of immune agents *as identified by the biologists* that lead to the interesting, useful and unique behaviour of the organism as a whole. The particular attraction of the immune system is that a number of its actors (or at least the biological properties *as identified by the biologists* appear to map neatly to existing computational techniques (see for example [Farmer *et al.* 1986, Kelsey & Timmis 2003]). A set of assumptions, one of which afflicts most computer scientists (including the authors) to some degree can be identified:

i. the sub-systems and components as described by biologists are circumscribed by "natural" boundaries and by implication have some degree of functional separability over and above that likely to be found for similar sized arbitrary sets of linked components selected at random.

ii. that sub-systems and components *as identified by the biologists* will be of interest and utility to computer scientists seeking inspiration.

iii. that biologists really believe that they have identified the important features of the immune system when describing these sub components.

iv. that it is valid to make computational simplifications and caricatures by using "off the shelf" components when building computational analogues of the sub-systems and their components.

Clearly if none of these assumptions were ever valid then the exercise of biology and biologically inspired computing would be essentially meaningless, and there are those who believe that this is so. If however one is prepared to accept (i) and (ii) most of the time, to talk to biologists in order to ascertain when (iii) holds and to guard against (iv) by maintaining sufficient complexity and constantly re-examine the biology to avoid over-simplification then both activities maintain value. To a certain extent (i) is beyond the capabilities of current biological techniques to prove beyond doubt for many systems, and the authors would argue that large parts of the immune system probably fall into this category, so computer scientists are essentially acting in good faith and hoping that the biologists are getting it right. The blind acceptance of (ii) is rather more dangerous. Features of immense biological importance such as the way that the complement system can puncture bacterial membranes without damaging host cells are perceived from a computer science perspective as analogous to "clever" heuristic solutions. Careful examination of the biology is required to ensure that a useful and valuable analogy can be made before embarking on detailed analysis and development. Guarding against (iii) and (iv) is however much more manageable and it is to this end that work such as that presented in [Stepney *et al.* 2005b, Stepney *et al.* 2004] is directed.

14.4 Immunologically Inspired Organization of Complexity

Having cleared the air of a few fundamental limitations of biologically inspired computation in general and immune inspired computation in particular we can consider the immune system in this context. It seems clear that the connectivity of the network of interactions within the immune system and its connectivity to the rest of the functions of the body preclude any attempts at full emulation of such a system at present. It is also interesting to consider the implications of such a highly connected system for identification of potentially useful subsystems. Interestingly it is the potentially valuable aspects of the immune system that make it so difficult to tease apart functionally. The heterogeneity of functionally interesting components and interactions is in stark contrast to other systems used for biological inspiration in computation. The electrical properties of the neural system are relatively easy to study in isolation from other aspects of the central nervous system, and despite enormous connectivity have proved valuable as a computational tool. Likewise, the representation and manipulation of computational problems in ways similar to natural genetic systems has resulted in some useful computational tools and systems. It is by no means clear what components of the immune system are best suited to this type of subdivision and emulation, although the clonal selection theory and idiotypic network theory have proved popular targets[de Castro & Von Zuben 2002, Neal 2003]. These systems have relied on gross caricatures of their natural counterparts and despite meeting with some success have not resulted in systems that genuinely reflect the maintenance, cognitive and homeostatic roles of the complete immune system. This may well be because they fail to reflect the heterogeneity, redundancy and pleiotropy of the natural immune system. If this is so then computer scientists and mathematicians are in serious trouble: such systems are generally intractable with current computational and analytic techniques.

A less depressing possibility is that researchers attempting to draw inspiration from the immune system have yet to identify the most suitable subsystems and features for emulation. Perhaps closer adherence to the natural systems is the route to success. A list of immune system components will generally include items such as the following:

The complement system detects classes of problems and evokes specific responses in a decentralized and locally efficient manner.

The innate immune system deals with common intruders and problems locally without resorting to complex computation and matching.

The adaptive immune system is able to recognize and adapt to a wide range of potentially dangerous patterns.

The bone marrow produces a range of leukocytes capable of completely covering the range of search space required for defence of the body.

The thymus hones the T cell population to avoid damage to the self.

The spleen removes infective agents and debris from circulation.

Antibodies can be produced in a staggering range of shapes in order to eliminate pathogens with very high specificity.

Phagocytes clean up messy and dangerous debris, and destroy foreign bodies with low specificity recognition.

Toll-like receptors are effective at recognizing common patterns representative of *classes* of invaders.

B cells can home in through hyper-mutation on a particular solution.

T cells facilitate highly specific removal of pathogens. They are the main bridge between the innate and adaptive immune systems, and also play a central role in the effector response to systemic viral infection.

Dendritic cells preprocess and transport information in order to alert more flexible systems of potential problems.

Clearly these components do not *only* contain immune system components, and equally some of them overlap with each other. For instance the bone marrow produces red blood cells as well as white blood cells, and white blood cells encompass both the innate and adaptive sections of the immune system. Categorization at this level appears to be a difficult activity when considering components for potential emulation in artificial immune system research. It is also apparent that at their various levels each of these components possesses some interesting and potentially useful properties. This overwhelming set of individual capabilities represents a range of modes of computation, any one of which may prove useful. For example it is easy to imagine a "virtual spleen" installed on a network router which is populated with virtual splenocytes that filter dangerous network traffic, or a population of phagocytes which move around the memory space of a Java virtual machine removing unreferenced structures in the memory space. Producing analogies is not difficult, and justifying them is often a matter of careful wording and good journalism. Producing analogies and systems built around them that genuinely capture the complexity and essential properties of the immune system is a different matter. Indeed it can be argued that systems such as the virtual spleen or memory phagocyte as suggested above are actually missing the point and are unlikely to capture anything of the essence of the immune system and may well be better implemented in a range of conventional ways. It is also arguable that the components listed above and their individual properties as described above are nothing new in computing: they rely on pattern recognition, pattern generation and communication. Pattern recognition is a relatively well developed field in computing with several journals and hundreds of books, algorithms, proofs, and conferences to its name. Pattern generation through the types of manipulation inherent in biological systems (mutation and other genetic operations) have also been very keenly examined in the evolutionary computing literature. Transmission of signals is also well studied and extremely well understood. The main feature of the immune system which is really of interest over and above those mentioned is *how it is organized*.

The immune system is dramatically heterogeneous, redundant and many of its elements are pleiotropic: and yet it is extremely robust to major perturbations such as removal of the spleen or amputation of a leg (and thereby half of its bone marrow). No computer system displays anything approaching this ability for reorganization, especially when the range of functions and the level of their integration into the rest of the body/system is considered. It seems clear that it is very unlikely to be the individual mechanisms and components that are important, but instead their interrelationships, interactions, redundancies and functional overlaps. It also seems very likely that the close coordination of the immune system with the environment assists with both maintaining stability directly, and with developing structures within the immune system that are inherently stable within that environment. An example of a direct contribution to maintaining stability might be characterized by the flow of nutrients and hormones into the bone marrow which permits appropriate production rates to be maintained. Examples of the second type of contribution to stability might be the exposure to particular antigens which result in the acquisition of immune memories capable of preventing serious infection by common pathogens. Thus, perhaps consideration of how the mechanisms and features described earlier might be exploited in order to result in coordination with the environment might result in more immune-like computer systems than are currently under consideration.

Interaction within the immune system and with its environment is both the problem for researchers in immunology and the reason for the immune system's successful operation. It is doubtful that many biologists would argue with this statement, and yet as computer scientists we are constantly attempting to dissect and break up the system in order to capture its properties. This is not to suggest that the entire system must be emulated in order to capture any of the interesting properties, but instead to suggest that in order to capture the most interesting behaviour we must concentrate on the organization and properties of the interactions as well as the functions of the components that we wish to emulate. Thus we must ensure that the properties of the interactions are retained in addition to the functional properties of the components. Some of these key properties are likely to include those mentioned above: heterogeneity, redundancy, perceptual degeneracy, pleiotropy and co-respondence. By ignoring most or all of these properties in our artificial systems we are likely to be dramatically reducing the probability of reproducing the useful properties that we seek.

14.5 A Computational View of Immune Complexity

By now it should be quite clear that analyzing the immune network is no mean feat. Each individual immune actor is connected to many others, and those connections only vaguely describe functional interactions that are often incompletely understood. Further, those interactions occur on widely varying timescales and within the context of a volatile, interactive intercellular environment. Interactions between immune actors are as much a function of the cytokines in the local environment as of the actors themselves. Many AIS that feature "recruitment" of actors do so at the population or concentration level, and essentially place all actors in the system in the same local environment. In other words, "recruitment" is limited to the fluctuation of concentration levels of species of antibodies within a common area of "tissue". These systems do not address environmental context, they focus entirely on classification by affinity. This is tantamount to ignoring any structure that is inherent in the environment in which the AIS resides. A good example of this is the traditional clonal selection AIS [de Castro & Von Zuben 2002, Watkins *et al.* 2004] which fits within the framework of a general population-selection algorithm[Newborough & Stepney 2005]. Such algorithms evaluate the fitness (affinity) of the current population at each iteration. In stark contrast the natural immune system evaluates affinity through biochemical interactions among actors that are non-uniformly distributed throughout the body. The body also has a topology that is organized with host-defence in mind, and the distribution of immune actors complements that organization. For example, CD4+ T cell activation occurs after exposure to peptide fragments embedded within class II MHCs on professional antigen presenting cells (APCs). However, exposure to such peptides occurs in the peripheral lymphoid organs (notably draining lymph nodes). The peptides were probably carried to those organs by dendritic cells, which matured from Langerhans cells in the skin. The activated CD4+ T cells then enter circulation to conduct their effector response [Abbas *et al.* 2000]. The immune system's evaluation of affinity is thus conducted on a highly restricted subset of the total adaptive actor population which implies that a large proportion of the actors will not be prone to selection and therefore a very large number of highly diverse untested actors will always be present.

The immune system is ultimately tasked with calculating affinity between all possible antibody-antigen pairs. We use "antigen", because the immune system must of course also discriminate between self and non-self along the way to selecting a response (a high affinity match). Moreover, the immune system must make these comparisons all over the body, all the time. That is, in order to provide adequate host defence, a high-affinity antibody must be available for all possible antigens in all tissues, all the time. Of course, the body achieves coverage not with an omnipresent, large, and perfectly diverse antibody population, but rather with a network of actors that dramatically reduce the set of antigenic patterns that must be examined.

The first, and possibly most effective actor in the immune network is the epithelium. This barrier provides an immeasurable reduction in the number of antigenic patterns that the rest of the immune system must later evaluate. Moreover, since it is totally non-specific, epithelial action could be considered to be "constant-time" in terms of computational complexity.

An antigen that has overcome epithelial barriers and invaded tissue presents a major challenge to the immune system. Because no assumptions about the location of the invasion can be made, the body must be prepared to fight infection in any of its tissues. However, that fight will involve a multi-pronged assault, only part of which is tailored specifically to the invader. Thus, part of the fight may be begun as soon as the self/non-self determination has been made. That determination is typically performed by a set of actors from the innate and adaptive immune systems. Professional APCs make the first half of the determination: they break down protein that they encounter and then attempt to install the resulting protein fragments into MHC molecules. Those fragments that can be successfully bound by MHC are transported from within the APC to its surface for expression or presentation. This processing represents a low complexity computation step performed by the immune system as the first phase of self/non-self determination.

The second phase is performed by a key controller of peptide antigen immune response: the CD4+ helper T cell. Helper T cells are specific to peptide fragments displayed by APCs. However, helper T cells only bind peptide-MHC complexes; they will not bind free peptides or MHC by itself. Further, the interaction between helper T cells and peptide-MHC complexes is of much lower affinity than the interaction between the equivalent protein antigen and its specific antibody species. So this interaction should be interpreted as one of intermediate complexity: lower than full antibody recognition, but higher than APC uptake/processing.

Finally the adaptive layers of the immune system result in the production of highly specific cells and antibodies that eliminate the carriers of particular antigens. This process can be viewed as a higher computational complexity search of the space of possible patterns and is less localized to sites of damage and infection. The systemic nature of antibody production by B cells has led to the types of algorithm which are currently prevalent in AIS[de Castro & Timmis 2002a] which essentially ignore environmental structure and localization of response, but if systems that exploit the full richness of the natural immune system are ever to be achieved, then inclusion of environmental context within which to situate analogues of the innate actors seems to be indispensable. A corollary of the inclusion of such components is that the computational complexity of such systems is likely to be reduced significantly as a large portion of the computation carried out by such a system is likely to be in the lower computational cost (innate) layers of the system as described above, with only occasional breaches of these defences resulting in deployment of full clonal selection based responses. The price to be paid is of course in the *structural* complexity of the computational system which implements such multi-layer systems. It is encouraging that AIS researchers are now turning to 'alternative' immune components, such as Dendritic Cells (DCs), as the inspiration for their work. In one such case, Greensmith et al have described an algorithm for anomaly detection inspired by DCs. The algorithm models DCs that classify antigen as malignant or benign based on environmental signals, and features regulatory and inflammatory cytokines. [Greensmith *et al.* 2005] We refer you to their work for the mechanics, but would like to point out their stated hope that such a system would find its way into a working, distributed artificial immune system. The system exemplifies the benefit to be gained by offloading classification load to low-specificity immune components.

14.6 Implications

If the importance of the properties and complexity of the interactions in the immune system is accepted as central to the generation of many of the key features that the immune system displays, then a number of conclusions about the types of system and computational analogies that should be approached in the future can be drawn.

Systems requiring multiple representations of actors, information and interactions should become tractable and appropriate if suitable analogies can be constructed. This is merely the recognition that the heterogeneity that the immune system supports and exploits can also be supported and exploited in computational analogies. A fundamental problem for many types of artificial intelligence solution is the selection of a suitable representation which captures the essence of the problem at hand whilst permitting suitable operations to be performed on the data in all cases. Selection of representations is rarely simple and often excludes some cases or situations from inclusion in the system. For example most systems that depend upon continuous values usually deal in very unsatisfactory ways with binary valued fields and missing values. The use of an immune inspired solution in such circumstances may allow the use of a number of different representations and/or processing regimes in different circumstances and/or different parts of the system and thereby allow a principled approach to tackling such problems.

Systems requiring distributed detection of correlations of events and/or patterns can be considered, and exploration of immune system interaction analogues that are capable of temporally and spatially organizing themselves can be expected to be fruitful. Thus the straightforward use of artificial immune systems for detection and classification can be rapidly extended to cases requiring consideration of timing and locality of events and patterns. For example the use of computational analogues of cytokine diffusion, decay and detection could contribute to the development of such systems.

Systems requiring close integration and distribution of pattern recognition mechanisms with existing complex engineered systems can be considered. The ability of the immune system to both maintain the integrity of the normally operating body and to detect and repel invasive agents ought to be replicable in engineered systems. This is a reflection of a possibly bigger challenge for biologically inspired computing: that of considering systems as complete organisms situated within their environments and attempting to reproduce homeostasis at the organism level. Systems such as robots are complex heterogeneous systems in themselves and have a number of different components and subsystems that may lend themselves to maintenance and monitoring in layered, immunologically analogous ways.

Mechanisms inspired by the layers of immune response found in both the innate and adaptive parts of the immune system described above could be included and exploited by using both hand coded remediation routines and adaptive components similar to those currently prevalent in AIS and by integrating them in immunologically plausible ways.

The system engineer would be wise to look for embodiment opportunities. Essentially, embodied systems compute in cooperation with their environment. In brief, an embodied agent perceives its environment and alters it, which alters the agent's subsequent perception. This simple-sounding feedback process produces astonishing complexity in the system. Luckily, many applications that welcome an immune approach also offer opportunities to use environmental dynamics to parallelize computation. For a detailed explanation of embodiment, please see Chapter 12.

¿From the preceding descriptions of potential application types it seems that the acceptance of the importance of the nature of the interactions in the immune system leads naturally to the expectation that far more complex and ambitious immune inspired computation than is currently attempted is required and should be possible. The careful exploitation of multiple signals, mechanisms and representations coupled with redundancy and pleiotropy ought to lead to successful construction and control of quite complex systems using immune inspired techniques. Managing this additional complexity ought naturally to emerge from the systems if the analogies are correctly drawn. It is by no means obvious that this will actually occur in practice without considerable engineering effort and experimentation. Indeed it would be truly miraculous should little additional effort be required at least initially. There is not going to be a free lunch. Even if it seems unlikely that artificial immune systems will be able to dine for free, perhaps we can hope that they will be the first to "go Dutch" by exploiting rather than suffering from the complexity in the environments that they are addressing.

Immune Inspired Learning in a Distributed Environment

Andrew Watkins[1]

Department of Computer Science and Engineering
Mississippi State University
Box 9637, Mississippi State, MS 39762 USA andrew@cse.msstate.edu

Summary. This chapter explores the distributed nature of the immune system as a source of inspiration for a distributed learning algorithm. It discusses modifications to the AIRS immune-inspired learning system which offers an initial examination of using distributed computing techniques for immune-inspired systems. A variety of results from a computational efficiency and classification accuracy standpoint are presented. It is argued that this basic step toward a distributed computing algorithm may be fruitful for future explorations of the distributed nature of the immune system and how it might inspire computational solutions.

15.1 Introduction

One of the many reasons for exploring verterbrate immune systems as a source of inspiration for computational problem solving include the observations that the immune system is inherently parallel and distributed with many diverse components working simultaneously and in cooperation to provide all of the services that the immune system provides [de Castro & Timmis 2002a, Dasgupta 1999]. Within the AIS community, there has been some exploration of the distributed nature of the immune system as evidenced in algorithms for network intrusion detection (e.g., [Hofmeyr & Forrest 2000, Kim 2002]) as well as some ideas for distributed robot control (e.g., [Lee *et al.* 1999, Lau & Wong 2003]), to name a number of examples.

As Stepney discusses elsewhere in this book, the environment in which our computational system are embodied can play a key role in the overall behavior of the system. While the basic ideas discussed in this chapter undoubtedly suffer from a fairly impoverished sense of environmental richness, the exploration of a more distributed environment presented here offers an initial direction for a larger discussion on the impact of the role the environment of the computational system plays in the overall results generated.

In distributed computing, the impetus tends to be an exploration of the use of computational resources for increased diversity of reaction or on problem solving in a highly decentralized manner where each computational resource requires independent decision making facilities with little to no input from a centralized mechanism. Beyond just the development of learning algorithms inspired from observations of nature, this chapter explores the ability to harness greater computational power for these tasks. With the recent proliferation of clusters of computers due to the decreasing costs of commodity computing components, it has now become extremely feasible to dedicate multiple computers to a given problem solving task. This leads to several possible consequences. The most obvious one is the ability to speed up the overall processing time in order to arrive at more timely solutions. This is extremely appealing from a machine learning/data mining perspective. One of the motivations for developing machine learning algorithms is in their abilities to save time for humans in complex tasks. By utilizing multiple computers or large parallel processors, machine learning algorithms, which can take advantage of this increased power, will also increase their benefits to the user.

This use of multiple processes in our learning algorithm also provides the ability for the development of more robust, or in depth, solutions. In other words, we can take the known computational technology of a parallel system and look for ways to incorporate this power in our developing biologically-inspired system. With many bio-inspired algorithms taking on an evolutionary component, the ability to evolve and explore separate niches and species of solutions on individual processors and then bring these individual populations to bear on the problem as a whole is potentially invaluable. The use of these distributed processing techniques allows us to further enhance our immune model. The immune system is inherently distributed and decentralized; therefore, it is important to the evolution of the field of artificial immune systems that we explore the use of distributed and parallel computing in order to more fully explore this biological potentials. While there has been some work on machine learning and evolutionary algorithms using large multi-computers or clusters of computers (see [Cantú-Paz 1998, Chattratichat et al. 1997] for some basic examples), very little in the application of parallel and distributed computing techniques for immune-inspired learning has been explored.

This chapter examines an initial approach to applying distributed computing techniques to the AIRS learning algorithm [Watkins et al. 2004]. AIRS is an immune-inspired learning algorithm which employs the concepts of B-Cell interactions and clonal selection to develop a set of memory cells capable of classifying previously unseen data. In previous work, our initial approaches to utilizing multiple processors on the AIRS learning algorithm provided mixed results [Watkins & Timmis 2004, Watkins 2005]. While the methods explored allowed us to maintain classification accuracy and provided modest computational gains, the parallel efficiency of those modifications were less than desirable. In keeping with our theme of the developmental process of a biologically-inspired algorithm, this chapter examines the decentralized nature of the immune system as a source of inspiration for learning.

While we still are exploiting standard computational techniques, we introduce a distributed processing version of AIRS that allows us to explore this biological concept in more detail.

In our previous parallelization schemes we were extremely focused on the gathering and merging of the memory cells at the root process. We felt this memory cell merging was necessary in order to preserve the final predictive model that was so attractive in the serial version of AIRS. However, if we remove this self-imposed goal of maintaining a single pool of memory cells, we can explore more interesting options.

This chapter examines a different slant on parallelizing the AIRS algorithm. Rather than viewing the goal of our parallelization as developing a memory representation identical to the serial version, we now choose to exploit the behavior of the entire parallel system. If we treat the entire parallel system as a learning model, rather than just a memory cell generation factory, we can offer much more solid computational gains. This approach also has the attractive feature of pointing us to a more distributed AIS algorithm. Since biological immune systems do not have one central antigen identification location, we would like to explore algorithms that also are divorced of this idea. This decentralized view would also allow us to explore more fully the ability of the system to develop a localized response. This would be achieved through localized learning, as well. So beyond just a global evaluation, the local evaluation can provide interesting information about the nature of the learning system, also.

We begin this chapter by briefly discussing the salient aspects of the AIRS algorithm and providing an overview of this new distributed version of AIRS. We highlight here the changes made to our parallelization of AIRS and the implications these have in terms of classification. We follow this with a series of experiments. We find that our computational gains are much greater through this distributed approach and that our new version of AIRS is much more scalable. Yet these changes present several question with regards to the purpose of the AIRS algorithm. We conclude this chapter with a discussion of how this new version of AIRS provides for the potential for more biologically plausible immune algorithms.

15.2 AIRS and Distributed AIRS

This section provides a high-level overview of the AIRS algorithm and then discusses the modifications to this algorithm for the creation of a distributed learning system.

15.2.1 AIRS

The Artificial Immune Recognition System (AIRS) is a supervised learning algorithm inspired by the immune system. Initially proposed by Watkins [Watkins 2001], AIRS has undergone a series of refinements and augmentations over the years [Watkins *et al.* 2004, Watkins & Timmis 2004, Goodman *et al.* 2002, Goodman *et al.* 2003]. AIRS is a specifically designed one-shot, supervised learning algorithm which appears robust to the parameter space, and performs well on a number of test problems [Watkins *et al.* 2004]. Immunologically speaking, AIRS is inspired by the clonal selection theory of the immune system. AIRS capitalises on this immunological mechanism, and through a process of matching, cloning and mutation, evolves a set of memory detectors that are capable of being used as classifiers for unseen data items. Unlike other immune inspired approaches such as negative selection, AIRS is specifically designed for use in classification, more specifically one-shot supervised learning.

Essentially, AIRS evolves with two populations, a memory pool (**M**) and an ARB pool (**C**). It has a separate training and test phase, with the test phase being akin to a k-nearest neighbour classifier. During the training phase, a training data item is presented to **M**. This set can be seeded randomly, and experimental evidence would suggest that AIRS is insensitive to the initial starting point. The training item is matched against all memory cells in the set **M**, and a single cell is identified as the highest match, MC_{match}. This MC_{match} is then cloned and mutated. Cloning is performed in proportion to stimulation (the higher the stimulation, the higher the clonal rate), and mutation is inversely proportional (the higher the stimulation, the lower the mutation rate). These clones are inserted into the ARB pool, **C**. The training item is then presented to the members of the ARB pool, where an iterative procedure is adopted which allows for the cloning and mutation of new candidate memory cells. Through a process of population control, where survival is dictated by the number of resources an ARB can claim, a new candidate memory cell is created. This new candidate is compared against the MC_{match}, with the training item. If the affinity between the candidate cell and MC_{match} is higher, then the memory cells is replaced with the candidate cell. This process is performed for each training item, where upon the memory set will contain a number of cells, capable of being used for classification. Classification of an unseen data item is performed in a k-nearest neighbour fashion.

15.2.2 A Distributed Approach

The reaction of the biological immune system to incoming antigens takes place in numerous places throughout the body. Unlike the nervous system which is centrally located, there is no one site that we can point to as the source of all immune responses. Immunological components circulate throughout the system and react in place as needed. We would like to be able to capture this idea in our artificial immune algorithms as well. Our new approach to exploiting increased computational resources for AIRS begins this process.

We start with a simple model of distribution. For this method, we begin by distributing a random subset of the training data to each process. We continue, by allowing each process to react to this subset of data with the usual AIRS training routine. However, unlike in our previous approach [Watkins & Timmis 2004], we no longer gather the developed memory cells back to the root process. Instead, we now view the entire parallel system as a distributed classifier with multiple classification sites. Figure 15.1 presents a graphical view of this concept. Some of these

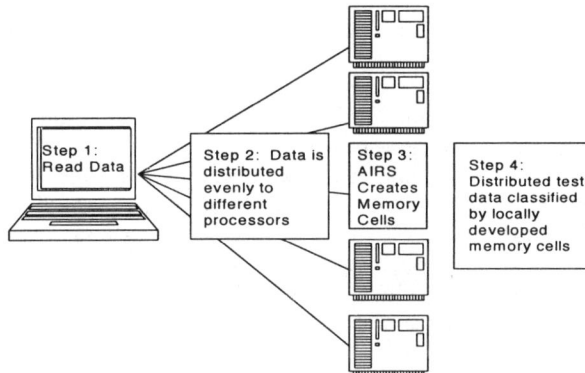

Fig. 15.1. Distributed AIRS

sites may be better trained to handle certain classes of the data than others. This is in keeping with our biological systems where there is a need for cell recruitment to certain areas that are under attack if the cells currently at that location are not trained to handle the given invader. We do not tackle the issue of recruitment and the communication of developed memory cells at this time. Rather, we focus only on allowing each processing site to develop its own miniature model of the data set based on its training subset.

For classification, we again distribute the test data throughout the parallel system. This may lead to uneven classification of some data items if the data items are assigned to sites that are not as equipped to recognize them as others sites may be. Nevertheless, we can then evaluate the performance of the system globally on the test data. We can also investigate the local reactions made by each processing site to its assigned pieces of data.

This introduces the need for multiple ways of assessing the performance of the system. One of our chief concerns is the runtime of the system. We want to examine how utilizing more processing power increases our computational gains. We can also still grade the performance of the entire system based on the same classification criterion used earlier. However, the introduction of this distributed memory and reaction model means that we also want to assess the performance at the local level as well. To this extent, we examine the classification capabilities of individual processors on their randomly assigned data. Eventually, this could lead to a network or recruitment based model in which individual populations of memory cells can pass on their classifications and their degrees of confidence in that classification. For now, however, we present a model that is divorced of any interaction other than the initial scattering of the data.

We find that the core characterization of AIRS as a learning algorithm does not change. However, the final memory model and subsequent decisions based on this model do need to be reevaluated. Each processing site continues to develop its own set of memory cells in the same way as the serial version of AIRS. However, the decision that the system makes concerning a given input is now completely decentralized. While the mechanics of this decision remain the same (i.e., an affinity based approach involving the closest memory cells), the memory structure itself has been altered due to less information available at any one given site. For the current formulation, each site remains limited in this way; however, this is only an initial prototype step. A next step in the evolution of this algorithm could be to incorporate meta-learning strategies or other distributed learning approaches to the disparate memory models [Brazdil et al. 1991, Chan & Stolfo 1993], or it could be to examine communication strategies available in the immune system, such as cytokine networks or immune network models, to more fully integrate the localized reactions of the system.

15.3 Verification Experiments

We begin by performing experiments three machine learning benchmark data sets which have been previously used to test the serial version of AIRS [Watkins et al. 2004]. We want to examine both the global and local performance of our new classifier. Additionally, we are also concerned with computational gains, and this is presented here as well.

15.3.1 Experimental Design

For the experiments presented in this section, we used the Iris, Pima Diabetes, and Sonar data sets that were used in previous studies of AIRS [Watkins et al. 2004]. For all of these we took an average over 30 cross-validated runs and tested the parallel version on an increasing number of processors. In keeping with previous experiments on these data sets, we used a 5-fold cross-validation for the Iris data set, a 10-fold cross-validation for the Pima Diabetes data set, and a 13-fold cross validation for the Sonar data set. A cluster of dual-processor 2.4Ghz Xeons were used. The Message Passing Interface (MPI)[Gropp et al. 1999, Snir & Otto 1998] was used as the communication library and communication took place over a Gigabit Ethernet network.

In order to assess the performance of this new formulation, we offer several metrics. We begin with the global accuracy of the system. This is measured by counting the number of correctly classified test items at each processor and dividing it by the total number of test items distributed to the system. While this global accuracy is not identical to that achieved through a single merged memory cell pool, it does provide a quick overview of how, on average, the system would react to a random data item presented to a random processing site. We then examine the local accuracy. For this, we report the average minimum local accuracy and the average maximum local accuracy. That is, for each run we record which site did the poorest on its assigned test data and which did the best. We also explore the size of the memory model developed. We look globally at the number of memory cells developed throughout the system, and we also look at the minimum and maximum number of memory cells developed at individual sites. Finally, we look at the parallel performance characteristics. In keeping with our goals of achieving faster and more efficient processing, we measure the average run times, speedup, and parallel efficiency of the system.

15.3.2 Results

Tables 15.1, 15.2, and 15.3 give the global accuracy and memory cells developed from this distributed version of AIRS on the Iris, Pima Diabetes, and Sonar data sets.

np	Accuracy	MCs
1	95.16%(3.06)	63.11(4.70)
2	94.56%(3.98)	74.15(4.53)
4	94.38%(4.59)	84.17(4.12)
8	93.53%(4.04)	95.49(3.56)
16	88.33%(4.88)	104.49(2.91)

Table 15.1. Distributed Iris: Global Accuracy

For all of these results we see a drop-off in this measure of global accuracy as we increase the number of processing sites. However, what this actually means as far as the classification performance of our new algorithm is less clear. As we mentioned in section 15.3.1, this metric is not exactly equivalent to the accuracy measures for AIRS when using a global memory cell set. This is merely a measurement of the sum of the number of correctly classified items at each processing site divided by the total number of test items distributed throughout the system. Section 15.3.3 provides more discussion into this matter.

np	Accuracy	MCs
1	73.00%(4.40)	279.04(10.11)
2	72.29%(5.00)	317.44(11.00)
4	71.80%(4.75)	358.16(11.39)
8	71.67%(5.06)	400.03(11.63)
16	69.24%(5.15)	445.96(11.06)

Table 15.2. Distributed Pima Diabetes: Global Accuracy

np	Accuracy	MCs
1	84.79%(8.15)	173.11(3.62)
2	78.81%(9.00)	179.89(3.04)
4	72.61%(11.24)	184.65(2.41)
8	66.97%(11.64)	187.80(1.90)
16	61.94%(12.28)	190.05(1.25)

Table 15.3. Distributed Sonar: Global Accuracy

Tables 15.4, 15.5, and 15.6 provide the average minimum and maximum test set accuracies and minimum and maximum number of memory cells developed at individual processing sites for distributed AIRS on our three bench mark learning problems.

np	Min. Acc.	Max. Acc.	Min MCs	Max MCs
1	95.16%(3.06)	95.16%(3.06)	63.11(4.70)	63.11(4.70)
2	90.58%(7.25)	98.53%(2.88)	35.11(2.81)	39.03(2.48)
4	87.36%(8.88)	99.42%(2.65)	18.83(1.43)	23.45(1.52)
8	64.94%(20.24)	100.00%(0.00)	9.56(1.00)	14.03(0.70)
16	19.00%(29.92)	100.00%(0.00)	4.80(0.60)	7.92(0.27)

Table 15.4. Distributed Iris: Local Accuracy

What this metric provides is a glimpse of the range of reactions by individual processors to their assigned data sets. Again, we see the widest range of values for the maximum number of processors. We discuss this further in section 15.3.3. Figures

np	Min. Acc.	Max. Acc.	Min MCs	Max MCs
1	73.00%(4.40)	73.00%(4.40)	279.04(10.11)	279.04(10.11)
2	67.83%(6.20)	76.75%(5.39)	154.71(6.51)	162.73(6.20)
4	60.62%(7.69)	82.40%(5.78)	83.33(3.94)	95.62(3.89)
8	48.87%(9.87)	90.42%(7.05)	43.76(2.68)	55.85(2.39)
16	30.33%(12.71)	99.40%(3.42)	22.73(1.53)	32.98(1.48)

Table 15.5. Distributed Pima Diabetes: Local Accuracy

np	Min. Acc.	Max. Acc.	Min MCs	Max MCs
1	84.79%(8.15)	84.79%(8.15)	173.11(3.62)	173.11(3.62)
2	70.64%(12.10)	86.99%(9.60)	88.66(2.02)	91.23(1.58)
4	48.40%(18.25)	94.55%(10.64)	44.86(1.04)	47.39(0.63)
8	19.62%(24.44)	100.00%(0.00)	22.39(0.65)	24.00(0.05)
16	0.00%(0.00)	100.00%(0.00)	11.06(0.38)	12.00(0.00)

Table 15.6. Distributed Sonar: Local Accuracy

15.2, 15.3, and 15.4 present these accuracy measurements as log-linear graphs with 3σ error bars, and figures 15.5, 15.6, and 15.7 present the memory cells results.

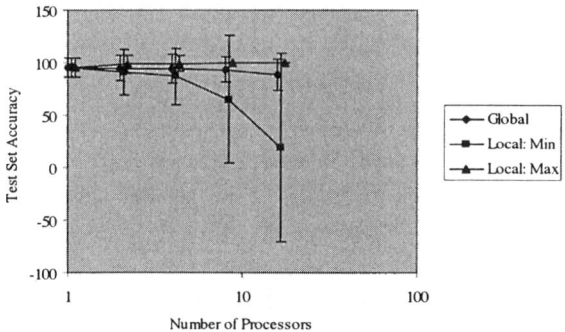

Fig. 15.2. Distributed AIRS: Iris: Accuracies (x-axis offset applied for visual clarity)

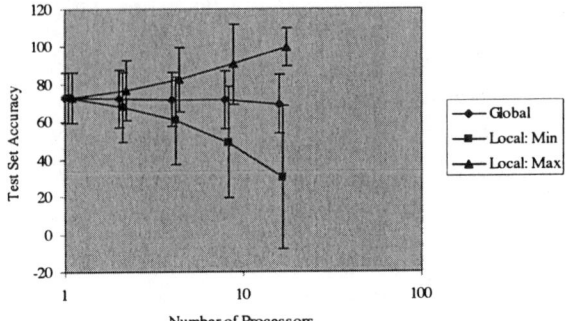

Fig. 15.3. Distributed AIRS: Pima Diabetes: Accuracies (x-axis offset applied for visual clarity)

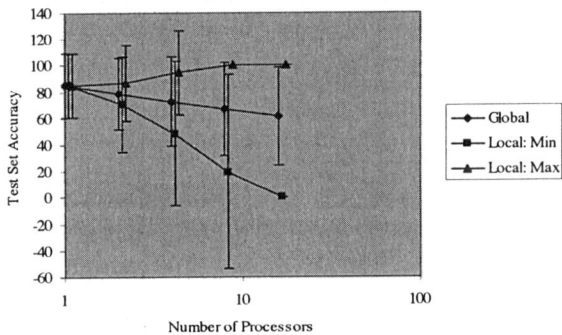

Fig. 15.4. Distributed AIRS: Sonar: Accuracies (x-axis offset applied for visual clarity)

Finally, figures 15.8, 15.9, and 15.10 provide the parallel performance of the distributed version of AIRS on our three learning problems.

Here we find solid parallel gains. We see inconclusive results Iris data set: it is simply too small and too easy of a classification task to gain much from the use of multiple processors. That is, the time to setup the communication fabric and distribute the data items to the individual processors is greater than the actual time to classify the data. However, for our other two data sets we continue to see runtime improvement through the use of our parallelization schemes.

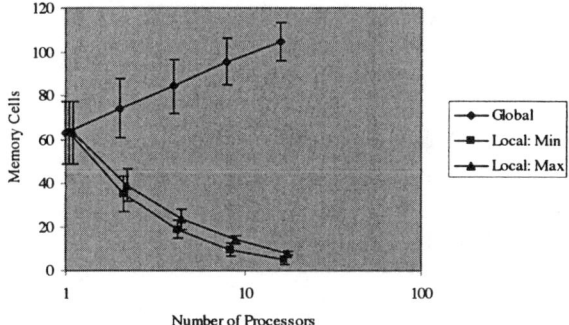

Fig. 15.5. Distributed AIRS: Iris: Memory Cells (x-axis offset applied for visual clarity)

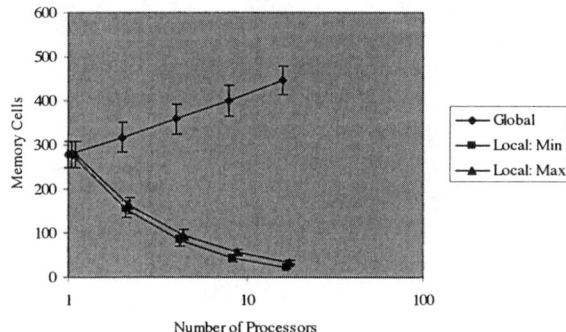

Fig. 15.6. Distributed AIRS: Pima Diabetes: Memory Cells (x-axis offset applied for visual clarity)

15.3.3 Discussion

The accuracy results presented above raise some potentially troubling questions about this distributed version of AIRS. The behavior exhibited is not surprising. An increase in the number of processing sites decreases the amount of data each site has available for learning. With fewer examples to learn from, the generalizations possible at the individual site are much more limited. That is, with a more incomplete picture of the world, each site's model of the world is also less complete. So, how then, should we view these results? Do they indicate that this approach is useless?

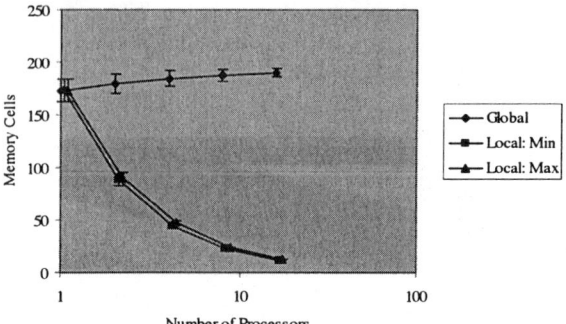

Fig. 15.7. Distributed AIRS: Sonar: Memory Cells (x-axis offset applied for visual clarity)

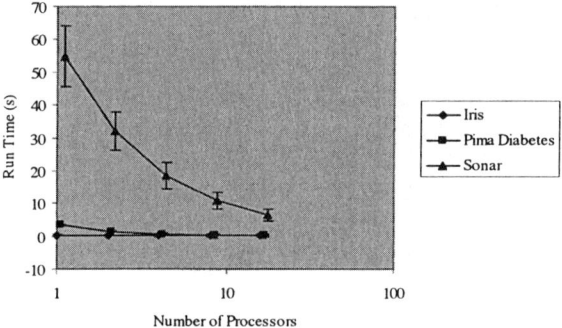

Fig. 15.8. Distributed AIRS: Run Times (x-axis offset applied for visual clarity)

One way of addressing these issues would be through a re-formulation of the approach. Looking at tables 15.7, 15.8, and 15.9 and figures 15.11, 15.12, and 15.11 we find that the distributed version of AIRS is capable of classifying the training data, both at a global and at a local level.

Since AIRS is a supervised learning algorithm, we could embody this knowledge somehow at each processing site. That is, each site can keep track of what type of data (i.e., what classes of data) it has been trained on and its individual performance on that data. This knowledge could then be used in a more global reaction sense. That is, when an individual site is asked to classify some piece of data, it could do so

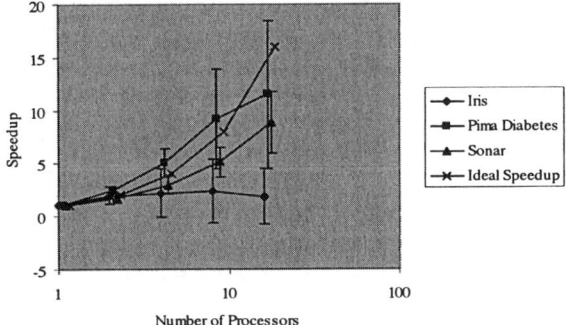

Fig. 15.9. Distributed AIRS: Speedup (x-axis offset applied for visual clarity)

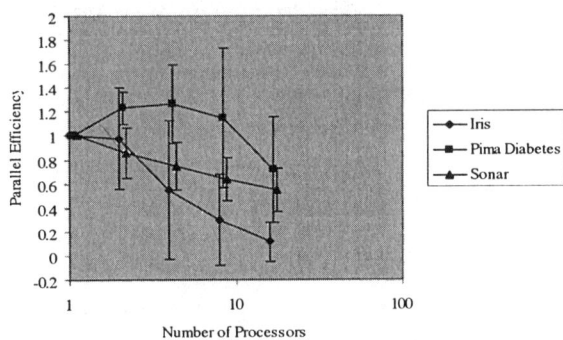

Fig. 15.10. Distributed AIRS: Parallel Efficiency (x-axis offset applied for visual clarity)

while attaching a degree of confidence to that classification. This degree of confidence could be based on the data's similarity to the training data seen at that particular site as well as the individual site's ability to classify that training data. The site could then pass on this confidence to other sites within the system. Basically, what we are proposing is that there could be more interaction among the processing sites, rather than the simple limited isolation in the current model. This approach would not be an attempt to reformulate a global memory cell pool presented in [Watkins & Timmis 2004]. Rather, we could begin to model local interactions, define areas of communication, and introduce concepts of a topology of reaction to a given test data item. This would allow certain sites to share information with other sites, which is

np	Global Acc.	Min. Acc.	Max. Acc.
1	97.94%(0.99)	97.94%(0.99)	97.94%(0.99)
2	98.47%(0.96)	97.59%(1.36)	99.36%(0.96)
4	97.94%(1.20)	95.00%(2.72)	99.91%(0.54)
8	97.74%(1.03)	92.71%(2.60)	100.00%(0.00)
16	97.07%(1.00)	80.52%(6.99)	100.00%(0.00)

Table 15.7. Distributed Iris: Training Set Accuracy

np	Global Acc.	Min. Acc.	Max. Acc.
1	76.47%(1.25)	76.47%(1.25)	76.47%(1.25)
2	76.17%(1.21)	74.79%(1.63)	77.55%(1.44)
4	76.01%(1.12)	73.41%(1.70)	78.65%(1.69)
8	75.84%(1.30)	69.90%(2.67)	82.00%(2.67)
16	75.22%(1.40)	63.07%(4.22)	85.68%(2.52)

Table 15.8. Distributed Pima Diabetes: Training Set Accuracy

np	Global Acc.	Min. Acc.	Max. Acc.
1	97.79%(1.12)	97.79%(1.12)	97.79%(1.12)
2	96.72%(1.29)	95.67%(1.73)	97.77%(1.32)
4	95.06%(1.50)	91.49%(2.74)	98.13%(1.59)
8	93.68%(1.69)	86.05%(4.10)	99.37%(1.49)
16	93.00%(1.74)	77.91%(6.25)	100.00%(0.00)

Table 15.9. Distributed Sonar: Training Set Accuracy

more akin to the biological model, while limiting the need for global communication, which is not as biologically plausible.

15.4 Scalability

This section presents scalability tests of distributed AIRS. Our chief concern for this section is parallel performance.

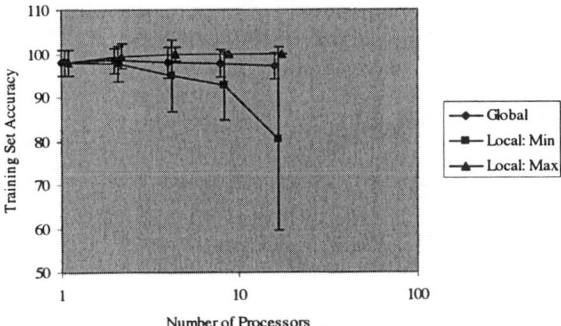

Fig. 15.11. Distributed AIRS: Iris: Training Set Accuracies (x-axis offset applied for visual clarity)

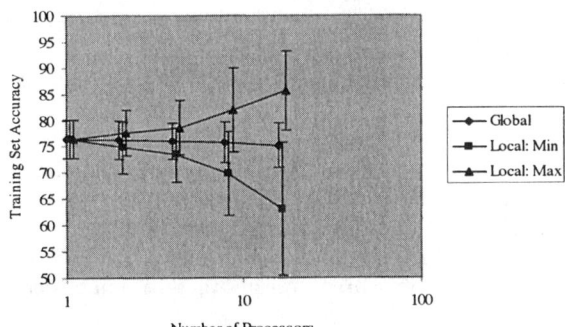

Fig. 15.12. Distributed AIRS: Pima Diabetes: Training Set Accuracies (x-axis offset applied for visual clarity)

15.4.1 Experimental Design

To generate our datasets for the experiments varying the number and length of the input vectors, we utilized Powell Bendict's DGP-2 data generation program for inductive learning tasks [Benedict 1990]. This program is designed to produce synthetic data for testing learning algorithms. It allows the user to specify the number of features in each data instance, the amount of data, and the number of "peaks" (or centroids) to be used for the positive data examples. The program then generates synthetic data with both positive and negative examples based on these user

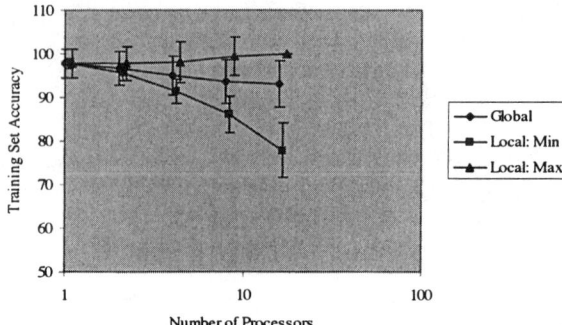

Fig. 15.13. Distributed AIRS: Sonar: Training Set Accuracies (x-axis offset applied for visual clarity)

parameters. For the experiments varying the number of training vectors, we kept the number of features constant at 64. For the experiments varying the number of features, we maintained the number of training vectors at 256. We did an average of 30 runs of 5-fold cross validation; so $n_1 = n_2 = 150$. The number of test data items was relative to the number of training vectors such that $T = \frac{N}{4}$, where T is the number of test data items and N is the number of training data items. All experiments utilized the processor dependent, affinity-based merging scheme with a dampener value of 0.1. Again we performed 30 runs of 5-fold crossvalidation. We examined the parallel performance of distributed AIRS as we varied the number of training vectors and the number of features in these training vectors.

15.4.2 Results

Varying the Number of Training Items

Figure 15.14 give the runtimes and Figures 15.15 and 15.16 provide the speedup and parallel efficiency measures on the simulated data for distributed AIRS when varying the number of training instances. The number of test data items is $\frac{N}{4}$. These experiments show that the distributed version of AIRS is scalable in terms of the number of training items used. We see that an increase in the number of training vectors increases the runtime, but this increase can be counteracted by an increase in the number of processors used. Interestingly, while we see very small runtimes for these experiments, we also find a high degree of variability as noted by the large error bars. In fact, for some of the runtimes there is no distinguishable difference between running the same dataset on more processors. Still, the general trend with

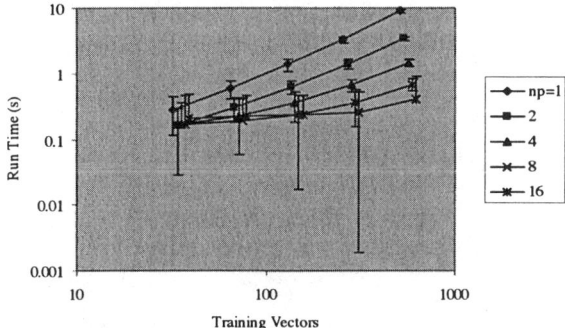

Fig. 15.14. Distributed AIRS: Run Times when Varying the Number of Training Vectors (x-axis offset applied for visual clarity)

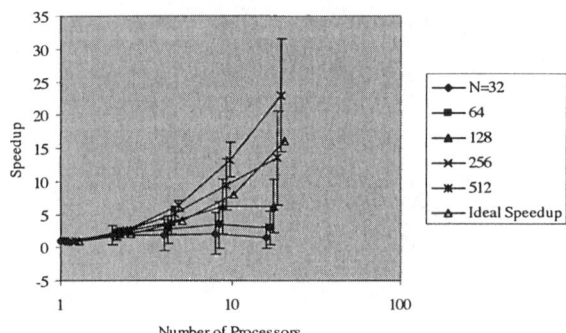

Fig. 15.15. Distributed AIRS: Speedup when Varying the Number of Training Vectors (x-axis offset applied for visual clarity)

these results is that an increase in processing power coupled with an increase in data size leads to scalable efficiency

Varying the Number of Features

Figure 15.17 give the run times and figures 15.18 and 15.19 provide the speedup and parallel efficiency measures on the simulated data for distributed AIRS when

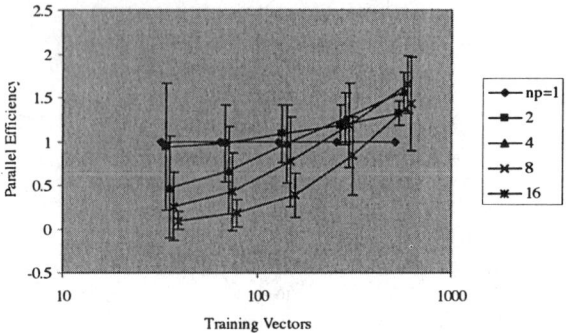

Fig. 15.16. Distributed AIRS: Parallel Efficiency when Varying the Number of Training Vectors (x-axis offset applied for visual clarity)

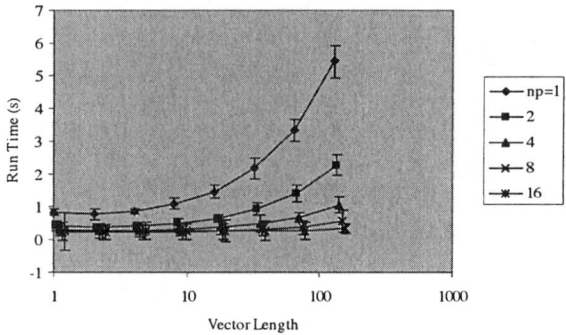

Fig. 15.17. Distributed AIRS: Run Times when Varying the Length of the Input Vectors (x-axis offset applied for visual clarity)

varying the length of the input vector. As with the results seen when varying the number of input items, we find that our distributed version of AIRS appears to scale well when increasing the number of features in the data set. The fundamental relationship between the runtime and the number of features has not changed. That is, with an increase in the number of features in the data set there is a corresponding increase in the runtime of the system.

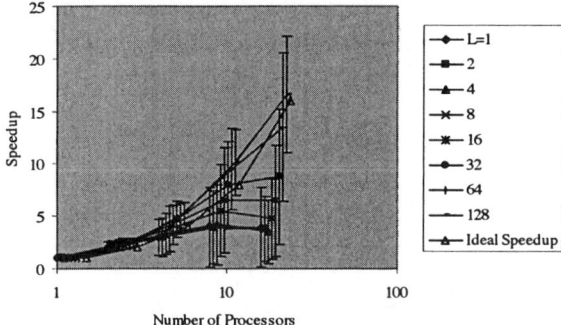

Fig. 15.18. Distributed AIRS: Speedup when Varying the Length of the Input Vector (x-axis offset applied for visual clarity)

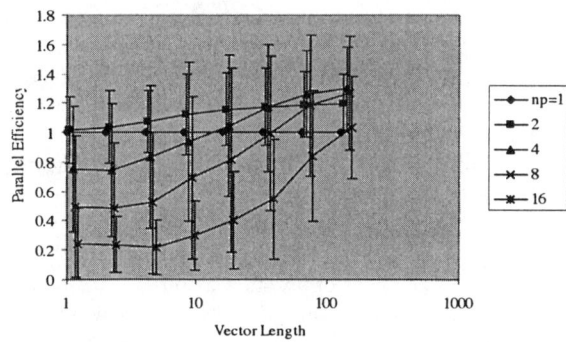

Fig. 15.19. Distributed AIRS: Parallel Efficiency when Varying the Length of the Input Vector (x-axis offset applied for visual clarity)

15.4.3 Discussion

This section has demonstrated that our distributed version of AIRS is much more scalable in terms of the number of training items and number of features in the data set than was parallel AIRS. Results presented in [Watkins 2005] provide the global and local accuracy measures for these experiments along with the number of memory cells developed. As discussed in section 15.3.3, the interpretation of these "accuracy" numbers is somewhat problematic. However, what they do reveal is this continued sense of a local reaction. This lack of global interaction together with the development of locally learned models and local reactions is much more biologically appealing. Couple this with the stable performance gains, and the distributed version of AIRS offers interesting areas of exploration. Still, one of the key issues that must be addressed is how to utilize these local reactions for solving real-world problems.

15.5 Discussion and Conclusions

This distributed version of AIRS presented in this chapter offers a basic alternative model to using multiple processors when compared to the parallel version [Watkins & Timmis 2004, Watkins 2005]. While this model is undeniably faster, its usefulness as a classifier is much more difficult to assess. One of the goals from using this approach was to remove the need for global interaction that was present in previous parallel versions. By doing this, we begin to explore the distributed concepts exhibited in biological immune systems. However, this also possibly limits the predictive capabilities of the system. This distributed design presented here is, admittedly, limited in scope. It removes all interactions among processing sites in the development of individual miniature world views at each site. One way of extending this work and recapturing some of the predictive capabilities of AIRS would be to allow for communication among the sites. While all of the interaction in our parallel model occurred on a global level, we could develop more local interactions. In this way we could begin to simulate cell recruitment and the diversity of reactions seen in the biological system. While this again may impact our overall runtime, it may offer us more insight into our learning task.

Another key aspect that this distributed approach could allow us to investigate is that of emergence. Much of the field of immune-inspired learning has focused on the engineering of desired behavior into the given system. If what is needed is a classification algorithm, then the biological metaphors are manipulated or engineered to provide this behavior. However, this is counter-intuitive to the development of the biological system itself. Within the immune system the properties that computer scientists find so attractive are in fact emergent properties of the system as a whole. It is through the distributed, diverse reactions of the system that the cognitive capabilities of learning emerge. A distributed approach to learning with local reactions leading to global interaction can provide a truer path to exploring ways that these attractive characteristics evolve and emerge apart from the *a priori* intent engineered into such a system. Or, as Stepney discusses, it is through the interaction of our systems embodied within a particular environment that allows the richness for the emergence to appear.

Mathematical Analysis of Artificial Immune System Dynamics and Performance

Andrew Hone[1] and Hugo van den Berg[2]

[1] Institute of Mathematics, Statistics & Actuarial Science, University of Kent, UK. anwh@kent.ac.uk
[2] Warwick Systems Biology, University of Warwick, UK. hugo@maths.warwick.ac.uk

Summary. The theory of nonlinear dynamical systems is reviewed, with the aim of showing that it has a range of useful tools to offer to both immunologists and computer scientists. The theory is illustrated with a simple model for the interaction of B cell clones, namely the B-model of Perelson and Weisbuch. A simple model is provided for the dynamics of an artificial immune response, and this is analysed using optimal control theory. A new model for the interaction of signalling molecules (cytokines) with immune cells is also outlined.

16.1 Introduction

Mathematical modelling plays an essential role in theoretical immunology. Whenever one wishes to make quantitative statements about the response of the immune system to antigens, mathematics is the language of choice. Nonlinear dynamical systems [Haken & Mikhailov 1993, Ott 1993, Peitgen & Richter 1986, Scott 1999], and especially systems of nonlinear differential equations [Hone 2005, Jordan & Smith 1999], are ubiquitous in the physical sciences, and they have also been applied very successfully in population biology and ecology to describe the growth and spread of plant and animal populations [Murray 1990]. This approach, employing nonlinear differential equations, has further been used to model the interactions of populations of lymphocytes with virus or antigen populations [Perelson & Weisbuch 1997, Nowak & May 2000]. Other examples of dynamical systems applied to immunology range from mathematical models of immune receptors [Goldstein *et al.* 2004], to the response of B and T cells from the viewpoint of optimal control theory [Kepler & Perelson 1993, Perelson *et al.* 1976, van den Berg & Kiselëv 2004], to stochastic differential equations in the description of immunosenescence [Luciani *et al.* 2001].

The emerging field of artificial immune systems, as described in Chapter 3 by Timmis and Andrews, has been inspired by mathematical models of immune interactions, and in particular network models of lymphocyte populations (see for example [Perelson 1989], references in the review [Perelson & Weisbuch 1997], as well as Chapter 4 of this book by Lee and Perelson). Jerne introduced the notion of an *idiotypic network* of immune cells that are able to recognize one another as well as antigen [Jerne 1974]. In [Farmer *et al.* 1986] it was suggested that the adaptive responses of immune networks might provide useful paradigms for machine learning. Subsequently, immune network algorithms [de Castro & Timmis 2003, Timmis & Neal 2001] were based on the idea of replacing nonlinear differential equations with analogous discrete or iterative schemes, together with the inclusion of stochastic effects due to mutations. Many other artificial immune system algorithms, such as CLONALG [de Castro & Von Zuben 2002], were based on simplified mechanisms for clonal selection. In the mean time, Jerne's idiotypic network theory had already been discredited by some immunologists [Langman & Cohn 1986]. However, the biological implausibility of a theoretical model does not necessarily preclude its usefulness in a computational context.

The aim of this chapter is threefold. Firstly, we believe that mathematical analysis, and in particular the theory of dynamical systems, has a whole host of useful tools and concepts to offer to two very different communities: immunologists and computer scientists. By translating abstract processes into precise mathematical models, it should become much easier to decide exactly what the natural immune system has in common with an artificial immune system (AIS). Thus in the next section we aim to give a brief introduction to the basic notions and methods required for analysing and interpreting nonlinear systems. Secondly, mathematical models of immunological processes can be adapted to AIS in order to provide objective measures of their performance, and further to work out how they can best be controlled. In order to address the latter, we aim to provide a simple model for the dynamics of an artificial immune response, and indicate how this can be analysed using ideas from optimal control theory [Jacobs 1996, Pinch 1993, Kirk 1997]. Thirdly, AIS have been inspired by abstract models of immunological processes, so in the fourth section we aim to provide further inspiration for computer scientists by outlining a new model for the interaction of signalling molecules (cytokines) with immune cells. In turn, this cytokine network model could also provide insights for biology.

From the comparison of our second and third aims it is clear that mathematical models can play a dual role in the development of AIS algorithms: on the one hand, as suitable tools for analysing and improving their performance; and on the other, as iterative dynamical processes from which the actual algorithms can be built. The distinction between these two roles is not always clear cut. In our concluding section, we discuss whether one role might be more appropriate than the other.

16.2 Mathematical Preliminaries

Continuous dynamical systems usually take the form of a system of coupled first order ordinary differential equations (ODEs),

$$\dot{x}_1 = F_1(t, x_1, x_2, \ldots, x_N),$$

$$\dot{x}_2 = F_2(t, x_1, x_2, \ldots, x_N),\tag{16.1}$$

$$\vdots$$

$$\dot{x}_N = F_N(t, x_1, x_2, \ldots, x_N).$$

The dot above a letter denotes the derivative d/dt, so $\dot{x}_1$ is $\frac{dx_1}{dt}$, and so on. The above system describes the rate of change of each of the quantities x_1, $x_2, \ldots, x_N$ with respect to the continuously varying parameter t, which in most applications corresponds to time. Given a specific system in the form (16.1), our goal would then be to solve for each of the functions $x_j(t)$, $j = 1, \ldots, N$. (If we also want to consider how these quantities vary in space, then we need to consider systems of *partial* differential equations; these will be touched on very briefly in section 16.4.) In order to describe generic systems, it will be convenient to introduce the state vector

$$\mathbf{x} = \begin{pmatrix} x_1 \\ x_2 \\ \vdots \\ x_N \end{pmatrix},$$

which allows us to rewrite the system concisely in vector form, as

$$\dot{\mathbf{x}} = \mathbf{F}(t, \mathbf{x}),\tag{16.2}$$

with $\mathbf{F}(t, \mathbf{x})$ being the column vector given by

$$\mathbf{F} = \begin{pmatrix} F_1 \\ F_2 \\ \vdots \\ F_N \end{pmatrix}.$$

In the generic case, we have no hope of obtaining an explicit analytical solution of a nonlinear system of the form (16.1). If the state vector $\mathbf{x}(0)$ is given at time $t = 0$ then it is natural to consider the initial value problem: how does the state vector $\mathbf{x}(t)$ evolve for $t > 0$? In general, if we want to get numerical answers, then we must resort to numerical integration methods (e.g. Runge-Kutta methods [Burden & Faires 2005, Kincaid & Cheney 2002]), which replace the differential equation (16.1) with a system of finite difference equations that approximate the change in the state vector for a small time step Δt. Another alternative is to use Gillespie's algorithm [Gillespie 1977], which makes direct use of discrete stochastic dynamics in order to solve differential equations. Note that, very often in the natural sciences, the deterministic equations themselves can be regarded as an approximation to an underlying stochastic process. There is no reason why the underlying process we are trying to model should not be thought of as being fundamentally discrete rather than continuous, or ultimately stochastic rather than deterministic, and thus we need not only regard numerical schemes as being approximations: they can be valid models in their own right. However, continuous systems are more convenient from a mathematical point of view, since the assumption of continuity allows a great deal of analytical machinery and exact solution methods to be applied. In the rest of this section we will review some of the techniques that can be used to obtain qualitative and quantitative information about continuous dynamical systems.

16.2.1 Linearization and steady states

Although most systems of interest are nonlinear, it is useful to start by considering homogeneous *linear* systems, which can be written in matrix form as

$$\dot{\mathbf{x}} = \mathbf{M}\mathbf{x}, \tag{16.3}$$

where $\mathbf{M}$ is an $N \times N$ constant matrix (independent of time). This is just a special case of (16.2), obtained by choosing $\mathbf{F}(t, \mathbf{x})$ to be the linear function $\mathbf{F} = \mathbf{M}\mathbf{x}$. (More generally, we could consider the linear function $\mathbf{F} = \mathbf{M}\mathbf{x}+\mathbf{c}$ for some constant vector $\mathbf{c}$, but for nonsingular $\mathbf{M}$ this would be equivalent to (16.3) upon transforming the state vector as $\tilde{\mathbf{x}} = \mathbf{x} - \mathbf{M}^{-1}\mathbf{c}$). What characterises linear systems? From a practical point of view, the main feature of the system (16.3) is that the output is directly proportional to the input. From a theoretical point of view, the main point is that any linear combination of solutions is also a solution, so that the solutions form a vector space. Homogeneous linear systems like (16.3) that are autonomous - so that t does not appear explicitly on the right hand side - can also be solved explicitly.

The exact solution of the initial value problem for (16.3) is

$$\mathbf{x}(t) = \exp(\mathbf{M}t)\,\mathbf{x}(0), \tag{16.4}$$

and from this solution it is immediately clear that the output $\mathbf{x}(t)$ at time t is proportional to the input $\mathbf{x}(0)$ at time zero. However, as it stands the formula (16.4) does not give us much specific information about the behaviour of the system for large t. The way that the solution changes with time depends on the matrix $\mathbf{M}$, or more precisely on its eigenvalues, which govern how the system will ultimately behave. The matrix $\mathbf{M}$ is said to be diagonalizable if it can be transformed into diagonal form, with all non-zero entries appearing along the diagonal. More precisely, if $\mathbf{M}$ is diagonalizable then we can find an invertible matrix $\mathbf{U}$ such that

$$\Lambda = \mathbf{U}\mathbf{M}\mathbf{U}^{-1} = \mathrm{diag}(\lambda_1, \lambda_2, \ldots, \lambda_N)$$

is the diagonal matrix formed by the eigenvalues λ_j of $\mathbf{M}$, which are the roots of the characteristic polynomial $\det(\mathbf{M} - \lambda\mathbf{I})$ of degree N, with $\mathbf{I}$ being the $N \times N$ identity matrix. (Some matrices $\mathbf{M}$ are not diagonalizable, but they still have a

Jordan normal form which is as near as possible to being diagonal, but contains some non-zero off-diagonal entries [Cohn 1994].) If we transform to the variable $\mathbf{y} = \mathbf{U}\mathbf{x}$ then the system (16.3) becomes

$$\frac{d\mathbf{y}}{dt} = \Lambda\mathbf{y} \quad \Longrightarrow \quad \mathbf{y}(t) = \exp(\Lambda t)\mathbf{y}(0). \tag{16.5}$$

The point of this linear transformation of the state vector is that the equations for the components of

$$\mathbf{y} = \begin{pmatrix} y_1 \\ y_2 \\ \vdots \\ y_N \end{pmatrix},$$

decouple from each other, and from (16.5) these components depend on t according to

$$y_j(t) = e^{\lambda_j t} y_j(0), \qquad j = 1, \ldots, N.$$

In general the eigenvalues λ_j of matrix $\mathbf{M}$ are complex numbers, so y_j can be complex even if the original state vector $\mathbf{x}$ is real. Thus, in general, we can say that as $t \to \infty$, the components corresponding to eigenvalues with $\operatorname{Re}\lambda_j > 0$ will diverge in amplitude, while the components whose eigenvalues satisfy $\operatorname{Re}\lambda_j < 0$ will decay to zero. So it is clear that the eigenvalues of the matrix $\mathbf{M}$ determine the ultimate growth or decay properties of the state vector in a linear system. In particular, only when all components decay to zero does the system ultimately come to rest at a stationary point (the origin). This fact will be very important as we now move on to consider nonlinear systems.

As already mentioned, for the typical nonlinear system (16.2) we do not expect to have any form of exact solution at our disposal, and we must resort to numerical integration if we want to solve the associated initial value problem for the evolution of the state vector $\mathbf{x}(t)$. However, we can get a good qualitative idea of the range of possible behaviours of the system by finding the steady states and performing *linearization* around them. For the sake of simplicity let us consider an autonomous system, where the function $\mathbf{F}$ on the right hand side of (16.2) is independent of t, so we have

$$\dot{\mathbf{x}} = \mathbf{F}(\mathbf{x}). \tag{16.6}$$

The *steady states* are the values of the state vector that make the right hand side of (16.6) vanish, so to find them we must solve the system of equations

$$\mathbf{F}(\hat{\mathbf{x}}) = \mathbf{0}. \tag{16.7}$$

Given any solution vector $\hat{\mathbf{x}}$ for (16.7), it is clear from (16.6) that if we start from the initial state $\mathbf{x}(0) = \hat{\mathbf{x}}$, then we must have $\mathbf{x}(t) = \hat{\mathbf{x}}$ (constant) for all t, so if the system starts at a steady state value then it remains there in equilibrium (hence the name).

What does a nonlinear system look like in the neighbourhood of one of its steady states? Assuming that the vector function $\mathbf{F}(\mathbf{x})$ is differentiable, we can calculate the matrix of its first partial derivatives. This $N \times N$ Jacobian matrix $\hat{\mathbf{M}}$ has components

$$\hat{M}_{jk} = \left.\frac{\partial F_j}{\partial x_k}\right|_{\mathbf{x}=\hat{\mathbf{x}}}.$$

Let us write $\mathbf{x}(t) = \hat{\mathbf{x}} + \boldsymbol{\delta}(t)$ and suppose that the magnitude of the perturbation vector $\boldsymbol{\delta}$ is small so that the state $\mathbf{x}$ is near to the steady state $\hat{\mathbf{x}}$. If we look at the components $F_j(\mathbf{x})$ of the vector $\mathbf{F}(\mathbf{x})$ near any point $\hat{\mathbf{x}}$, then we can calculate their Taylor expansions in the form

$$F_j(\hat{\mathbf{x}} + \boldsymbol{\delta}) = F_j(\hat{\mathbf{x}}) + \sum_{k=1}^{N} \left.\frac{\partial F_j}{\partial x_k}\right|_{\mathbf{x}=\hat{\mathbf{x}}} \delta_k + \dots,$$

where the dots correspond to terms of quadratic or higher degree in δ_k, which are the components of $\boldsymbol{\delta}$. However, by the steady state assumption (16.7) we know that each component $F_j(\hat{\mathbf{x}}) = 0$, and so substituting for $\mathbf{x}$ in (16.6) and applying Taylor's theorem to the function $\mathbf{F}(\mathbf{x})$ implies that the perturbation away from $\hat{\mathbf{x}}$ satisfies

$$\dot{\boldsymbol{\delta}} = \hat{\mathbf{M}}\boldsymbol{\delta} + o(\boldsymbol{\delta}) \tag{16.8}$$

where $o(\boldsymbol{\delta})$ denotes a small quantity such that $|o(\boldsymbol{\delta})|/|\boldsymbol{\delta}| \to 0$ as $\boldsymbol{\delta} \to 0$. Thus the equation for small perturbations is approximately linear, i.e.

$$\dot{\boldsymbol{\delta}} \approx \hat{\mathbf{M}}\boldsymbol{\delta}.$$

It follows from the foregoing discussion of linear systems that if all the eigenvalues of the matrix $\hat{\mathbf{M}}$ associated with the point $\hat{\mathbf{x}}$ satisfy $\mathrm{Re}\,\lambda_j < 0$, then these perturbations will decay exponentially to zero; so if the state vector starts near to this steady state then

$$\mathbf{x}(t) \to \hat{\mathbf{x}} \quad \text{as} \quad t \to \infty.$$

The above condition means that the steady state $\hat{\mathbf{x}}$ is *asymptotically stable* (see e.g. [Jordan & Smith 1999], p.289 for a definition). If, on the other hand, we are in the exactly opposite situation that $\mathrm{Re}\,\lambda_j > 0$ for all j, then clearly the system will diverge away from the steady state, so the assumption that $\boldsymbol{\delta}$ is small in magnitude will cease to hold and the linear approximation will break down. In the latter case, the steady state $\hat{\mathbf{x}}$ is called an *unstable node*. Between these two extremes there are all the possible combinations of signs for the real parts of the eigenvalues, and each of these combinations corresponds to a different kind of unstable behaviour.

The N-dimensional space of possible values for the state vector $\mathbf{x}$ is known as the *phase space*. Usually the phase space is $\mathbb{R}^N$ or some subregion therein. From the above analysis of the nonlinear system (16.6) it is apparent that the approximate shape of the trajectories of $\mathbf{x}(t)$ in phase space is determined locally (near a steady state) by the eigenvalues of the associated linearized system. It is possible to patch together the whole phase space by gluing together the different neighbourhoods of each steady state, but this becomes gradually more complicated as the dimension N and the number of different steady states increases.

The simplest non-trivial situation is when there are two state variables, so $N = 2$, and in that case the system evolves in the phase plane. Yet even in the two-dimensional case there are various different types of possible behaviour near a steady state. For a real system with $N = 2$ there are two eigenvalues, λ_1 and λ_2, and

Eigenvalues:	Real λ_j	Complex λ_j
$\operatorname{Re}\lambda_1 > 0,\ \operatorname{Re}\lambda_2 > 0:$	unstable node	unstable spiral
$\operatorname{Re}\lambda_1 < 0,\ \operatorname{Re}\lambda_2 < 0:$	stable node	stable spiral
$\operatorname{Re}\lambda_1 > 0 > \operatorname{Re}\lambda_2:$	unstable saddle	—
$\operatorname{Re}\lambda_1 = 0 = \operatorname{Re}\lambda_2:$	—	centre

Table 16.1. Classification of steady states in two dimensions.

these can either be both real with all possible combinations of signs, or a complex conjugate pair with $\operatorname{Re}\lambda_1 = \operatorname{Re}\lambda_2$. In Table 1.1, we have classified the main types of steady states in two dimensions, in terms of the eigenvalues (up to permutation), indicating whether they are stable or unstable. Note that we have omitted certain special cases, namely degenerate nodes (stable/unstable), that occur when $\lambda_1 = \lambda_2$. These degenerate cases, and also centres, for which $\operatorname{Re}\lambda_1 = \operatorname{Re}\lambda_2 = 0$, lie on the boundaries between the other cases given in the table. These cases are non-generic and they do not persist if the parameters in the system are slightly altered. In fact a centre is neutrally stable in a linearized system, but for the corresponding nonlinear system it is usually either an absorbing or a repelling state. This discussion of steady states is best illustrated by the concrete example that follows, namely the B-model for interacting populations of B cells.

16.2.2 An example: the B-model

As a simple example of a nonlinear system, we consider the B-model (see [Perelson & Weisbuch 1997]), which is a system of ordinary differential equations used to describe a network of B cell clones that are able to recognize and respond to one another, as well as to the presence of antigen. The equations can be written in the form

$$\dot{\mathbf{x}} = m\mathbf{e} + (p\mathbf{f}(\mathbf{x}) - d)\mathbf{x}. \tag{16.9}$$

The constant m is a source term corresponding to continual production of new cells in the bone marrow, multiplying the constant vector

$$\mathbf{e} = \begin{pmatrix} 1 \\ 1 \\ \vdots \\ 1 \end{pmatrix},$$

while the constant parameters p and d correspond to the proliferation rate and death rate respectively. The N components of the state vector $\mathbf{x}$ correspond to the populations x_j of each of N different clone types, $j = 1, \ldots, N$, and the nonlinearity is encoded in the matrix

$$\mathbf{f}(\mathbf{x}) = \mathrm{diag}(f(h_1), f(h_2), \ldots, f(h_N))$$

which depends on the quantities

$$h_j = \sum_{k=1}^{N} J_{jk}\, x_k \qquad j = 1, \ldots, N$$

(with the constant matrix J_{jk} specifying the strength of the interaction between the clones), via the activation function f, which is given by the formula

$$f(h) = \frac{\theta_2 h}{(\theta_1 + h)(\theta_2 + h)} \tag{16.10}$$

for suitable parameters θ_1 and θ_2 with $\theta_2 \gg \theta_1 > 0$. (The notation "$\gg$" is used to mean "of a larger order of magnitude than" here and in what follows.)

For our purposes, it will be sufficient to consider the case where there are only two clone types, and following Perelson and Weisbuch (see [Perelson & Weisbuch 1997], p.1238) we choose the interaction matrix in the form $J_{12} = 1 = J_{21}$ and $J_{11} = 0 = J_{22}$, so that $h_1 = x_2$, $h_2 = x_1$, and the system has the form

$$\dot{x}_1 = m + x_1(p\, f(x_2) - d), \tag{16.11}$$

$$\dot{x}_2 = m + x_2(p\, f(x_1) - d). \tag{16.12}$$

Further analysis of the two coupled equations (16.11) and (16.12) shows that in general (for a relevant range of parameter values) there are five steady states in this two-dimensional system. Due to the permutation symmetry $x_1 \leftrightarrow x_2$, these can be divided as follows: three states for which $x_1 = x_2$, that correspond to the three numbers V, L, H with $H \gg L \gg V$ (for typical parameter values), which are the roots of the equation

$$m + x(p\, f(x) - d) = 0$$

(equivalent to a cubic), giving the steady states $(x_1, x_2) = (V, V)$, (L, L), (H, H); and a pair of states $(x_1, x_2) = (\hat{L}, \hat{H})$ and $(\hat{H}, \hat{L})$ which satisfy

$$m + \hat{L}(p\, f(\hat{H}) - d) = 0 = m + \hat{H}(p\, f(\hat{L}) - d).$$

For small values of the parameter m, we have $\hat{L} \approx L$ and $\hat{H} \approx H$, and henceforth we shall drop the hats, assuming that this approximation holds. The "H" and "L" stand for "high" and "low" respectively, since receptors of type 1 clones are supposed

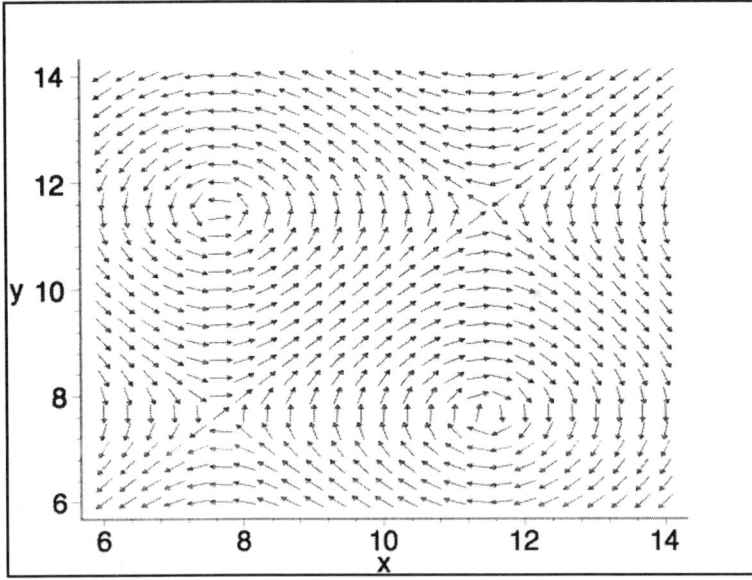

Fig. 16.1. *Part of the phase portrait for the system of two coupled equations (16.11) and (16.12), plotted on the logarithmic scale $x = \ln x_1$, $y = \ln x_2$ for the two clone types. (For parameter values, see main text.)*

have high affinity for a particular antigen, while type 2 clones have low affinity. (The interaction with antigen is not yet included; see (16.16) and (16.17) below.)

By considering the eigenvalues associated with each steady state of the B-model, it is possible to show that both (L, L) and (H, H) have one positive and one negative eigenvalue, so (from Table 1.1) we see that these states are saddle points and hence unstable. Similarly, (L, H) and (H, L) are stable spirals (having a complex conjugate pair of eigenvalues with negative real parts), and similarly (V, V) is a stable node, so these latter three steady states are attractors. Taking the terminology of [Perelson & Weisbuch 1997], the attractor (V, V) is called the *virgin* state, while (H, L) and (L, H) are called the *immune* and *tolerant* states respectively. Their attracting behaviour is best appreciated by looking at Figure 16.1, which portrays the phase paths of the system with parameter values

$$d = 0.5, \quad p = 1.0, \quad m = 1.0 \times 10^{-3}, \quad \theta_1 = 2.0 \times 10^3, \quad \theta_2 = 1.0 \times 10^5. \quad (16.13)$$

In fact Figure 16.1 shows part of the phase portrait of the system plotted on a logarithmic scale, so the x-axis corresponds to $\ln x_1$ and the y-axis corresponds to $\ln x_2$. The MAPLE package DEplot has been used here to solve the system for a range of different initial conditions in the range $6 \leq x \leq 14$, $6 \leq y \leq 14$, and the direction of the trajectory at each point is represented by an arrow whose length corresponds to the magnitude of the vector field $\dot{x} = (\dot{x}, \dot{y})^T$. This allows an immediate appreciation of where the steady states lie in the phase plane, and the directions of the arrows around them indicates whether they are stable or unstable (attracting or repelling).

Observe that Figure 16.1 contains four of the steady states - the two spirals and two saddles are clearly visible - but the (V, V) state is off the scale.

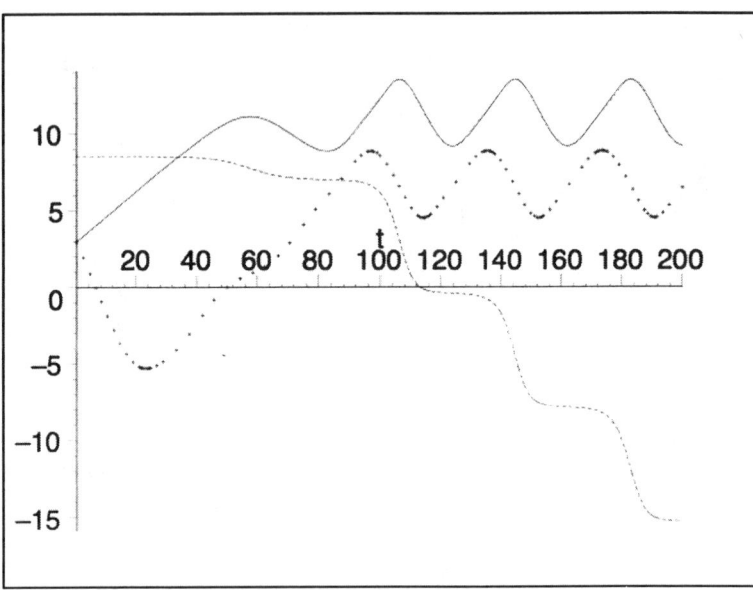

Fig. 16.2. *Time evolution of two B-cell populations and one antigen in the B-model, plotted on a logarithmic scale with* ln(*Population*) *on the vertical and time* t *on the horizontal. The upper oscillation (full curve) is clone type 1, with clone type 2 plotted with circles, while the lowermost (dotted) curve corresponds to antigen population. (See text for details of parameters.)*

The behaviour of the B-model becomes more interesting in the presence of antigen. With two clone types and one population of an antigen $A(t)$, which is eliminated by the response of cells of type 1, we can write down the system

$$\dot{x}_1 = m + x_1(p\,f(h_1) - d), \tag{16.14}$$

$$\dot{x}_2 = m + x_2(p\,f(h_2) - d), \tag{16.15}$$

$$\dot{A} = -kx_1A, \tag{16.16}$$

where

$$h_1 = Z\,x_2 + J\,A, \qquad h_2 = x_1, \tag{16.17}$$

for suitable parameters k, Z, J. The parameter k is a decay rate, while Z and J determine the strength of the stimulation of clone 1 by clone 2 and antigen respectively. It is clear from the third equation (16.16) that clone 1 responds to the presence of antigen and removes it at a rate proportional to its population.

From the form of the three-dimensional system (16.14), (16.15), (16.16) we see that to have the steady states we require $A = 0$, and then the steady state population values are the same as for the two clone system uncoupled to antigen. So the analysis of steady state values is almost identical to that for the two-component system (16.11), (16.12), with three attractors that can be referred to as the virgin, immune and tolerant states respectively, corresponding to three different stable stimulation regimes for the B cell clones that have been exposed to antigen. In Figure 16.2 we have plotted a numerical solution of the time evolution for the two clone system with antigen, where the parameters and initial conditions were chosen as

$$k = 1.0 \times 10^{-6}, \quad Z = 3.0, \quad J = 0.9, \quad x_1(0) = 20, \quad x_2(0) = 20, \quad A(0) = 5000,$$

with the other values being the same as in (16.13).

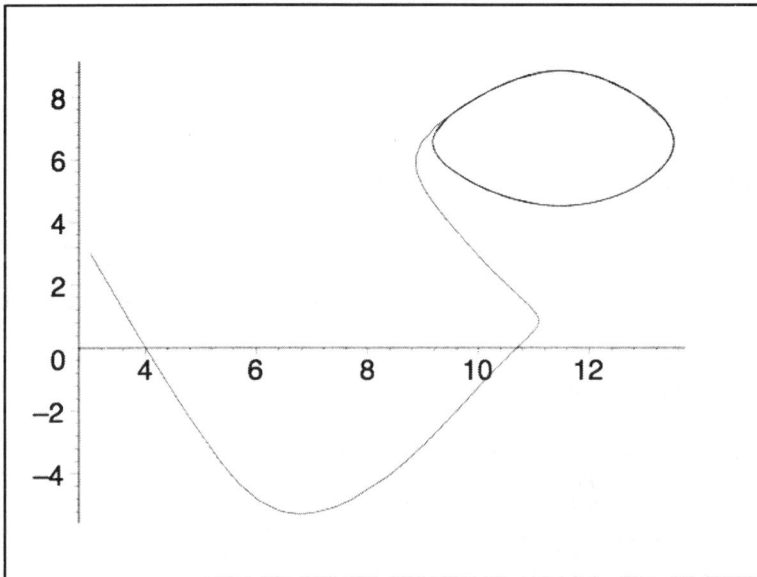

Fig. 16.3. *Phase plane dynamics of two B cell populations in the B-model with antigen, corresponding to the time evolution in Figure 16.2. The clone populations are plotted as $x = \ln x_1$, $y = \ln x_2$.*

Note that in the Figure (16.2), which is plotted on a logarithmic population scale, the population of x_1 rises very rapidly at first due to the presence of antigen, while the population of type 2 clones drops initially, before these two fall into a synchronized pattern of oscillations. The antigen population, on the other hand, decreases gradually before starting to drop very rapidly until it is effectively eliminated (when it becomes vanishingly small). Although the ultimate fate of the two B clones appears to be a periodic oscillation, in fact this oscillation is decaying in magnitude (very slowly on the response timescale) and tending towards a steady state of the form (H, L) with $H \gg L$ (an *immune state*). (To see that this state is really a stable spiral, it is necessary to calculate the eigenvalues which are found to have a small negative real part, hence a slow decay rate, cf. Table 1.1.) The state is so called because if antigen were presented to the system again then the elevated level of type 1 clones would cause a very strong and rapid immune response to eliminate the corresponding pathogen (and hence its associated antigen would be removed). The oscillation can be seen in a different way from the phase plane plot of $\ln x_1$ against $\ln x_2$ (Figure 16.3), where the loop corresponds to the decaying oscillation towards the point (H, L).

For a full discussion of idiotypic network models, in particular the response mechanism for more than two clone types and the comparison between immune, tolerant and virgin states, the reader is referred to the review [Perelson & Weisbuch 1997]. For a survey of other models of B cell and T cell receptors, we refer the reader to Lee and Perelson's contibution to this volume (Chapter 4). In the next section we change gear and describe how to model an artificial immune response from an optimal control viewpoint.

16.3 Dynamics of an Artificial Immune Response

An artificial immune response is a process (or more properly: a runnable) operating on data structures. A key concern is how well the AIS executes these operations relative to the computational resources allocated to it. This relative performance or *specific efficacy* informs higher-level decisions: for instance, should a more efficacious process become available, the artificial immune response in question might be halted, while otherwise the allocation of resources must be adjusted to the specific efficacy to ensure that the required operations are achieved to a given standard, or within a given timeframe. Such considerations of efficacy and time horizons are critical to the ability of the artificial immune response to contribute meaningfully to the proper embodiment of the system deploying this response; this is discussed in more detail by Stepney, in Chapter 12 of this book.

In this section we analyse the performance of an artificial immune response using a system of ODEs. The state variables are x, the computational resources allocated to the artificial immune response, y, the *job load*, i.e. the amount of data awaiting operation by the AIR, and z, the computational resources allocated toward improvement of the AIR's specific efficacy. The *target data* are data structures upon which the response operates. The analysis will focus mostly on the variables x and y, and on the (x, y) phase portrait.

This mathematical set-up is considerably idealized. First, computational resources allocated to the AIS or processes aimed at improving its specific efficacy may take various different forms: executor (CPU time) as well as different types of memory space. Taking this diversity into account would require x and z to be replaced by four or more state variables; but for our present purposes it will suffice to discuss only the simplest situation.

Second, the variables x and z are real numbers representing the fraction of CPU time allocated to the artificial immune response. Strictly speaking, one would only expect either $x = 1$ or $x = 0$ to be true at any point in time t, according to whether or not the CPU is devoted to the artificial immune response at time t. However, letting $\tilde{x}(t)$ denote this Boolean ($\tilde{x} \in \{0, 1\}$), and defining

$$x(t) \stackrel{\text{def}}{=} \int_{t-T}^{t+T} \tilde{x}(s)ds \tag{16.18}$$

it is reasonable to treat x as a real value in the interval $[0, 1]$ if the constant T is chosen large enough. One could think of x as the *propensity* of the scheduler to award CPU time to the artificial immune response. (More generally, one could consider a large number of parallel processors, and then $x(t)$ would denote the fraction of these devoted to the artificial immune response at time t.) We are interested in the evolution of the performance of an AIS as it develops and ultimately resolves. This dynamics generally takes place on a time scale much longer than T; we call this longer time scale the *response time scale*. The state variable z is interpreted in the same way as x.

16.3.1 Dynamics of the job load

In this subsection we concentrate on the interaction between the resources allocated and the job load, i.e. between x and y, with the possible effect of z being ignored. On the response time scale, we model the dynamics of the job load by means of an ordinary differential equation:

$$\dot{y} = G(x, y, t) - R(x, y, t) \tag{16.19}$$

where the dot indicates the derivative with respect to time t. The term $G \geq 0$ represents the increase ("growth") of the job load, the term $-R \leq 0$ its decrease due to the actions of the immune response. Now one could specify a particular differential equation for the rate of change $\dot{x}$, and then consider the dynamics of this equation coupled with (16.19), but this is *not* the approach to be adopted here. Instead, we will follow the approach of optimal control theory [Kirk 1997, Pinch 1993], and try to solve the following problem: what is the best way to vary $x(t)$ with time t in order to ensure that the job load $y(t)$ eventually decreases, subject to the differential equation (16.19)?

Of central interest is the *specific efficacy* of the response:

$$r(x, y, t) \stackrel{\text{def}}{=} \frac{R(x, y, t)}{x} \tag{16.20}$$

which tells us how well the response performs per unit computational resources allocated. Since some of the activities associated with the response involve scanning and evaluating data, r must depend on y. We define a *maximum specific efficacy* $r_{\max}$ and a *saturation factor* f, as follows:

$$r_{\max}(x, t) \stackrel{\text{def}}{=} \sup_{y \geq 0} \{r(x, y, t)\} \quad \text{and} \quad f(x, y, t) \stackrel{\text{def}}{=} \frac{r(x, y, t)}{r_{\max}(x, t)}. \tag{16.21}$$

The saturation factor f is a dimensionless quantity in the interval $[0, 1]$. Any reasonable model of this saturation factor must satisfy

$$\lim_{y \to 0+} f(x, y, t) = 0 \; ; \tag{16.22}$$

this is equivalent to $\lim_{y \to 0+} R(x, y, t) = 0$, so there is no response for a zero job load. In many cases one expects that f increases monotonically with y; a generic model that follows from quite general arguments derived from queuing theory [van den Berg & Kiselëv 2004] is as follows:

$$f(x, y, t) \equiv f(y, t) = \frac{q(t)y}{1 + q(t)y} \; . \tag{16.23}$$

The variable $q(t) > 0$ expresses the quality of the response: for large q, f will be close to 1 even for low y values. (Note that in equation (16.23), f does not depend on x.) The quality $q(t)$ usually increases in the course of an AIS response, a phenomenon analogous to *affinity maturation* in biological immune responses.

Let us make the following two assumptions about this process of response maturation:

$$r_{\max}(x, t) \sim \overline{r}_{\max} \quad \text{and} \quad f(y, t) \sim \overline{f}(y) \quad \text{as} \quad t \to \infty. \tag{16.24}$$

These assumptions mean that, in the latter stages of the response, $r_{\max}(x, t)$ no longer depends on x or t (the response is asymptotically attaining a locally optimal specific efficacy) and the saturation factor depends on y alone (e.g. in terms of equation (16.23), the quality q likewise tends to some final value $\overline{q}$). Furthermore, we consider two possible cases for the eventual behaviour of job load growth:

$$\text{either (i)} \quad G(x, y, t) \to \overline{G}, \quad \text{or (ii)} \quad \frac{G(x, y, t)}{y} \to \overline{g}. \tag{16.25}$$

Case (i) corresponds to an ultimately stationary external input, while case (ii) corresponds to cases where the target data are 'autocatalytic', that is, tend to spawn further target structures at an intrinsic growth rate $\overline{g}$. In the absence of a response, targets would grow linearly with t ($y(t) \sim \overline{G}t$) in case (i), and exponentially ($y(t) \sim c \exp(\overline{g}t)$, for c constant) in case (ii). In fact for case (i), the asymptotic form of equation (16.19) is

$$\dot{y} \sim \overline{G} - \overline{r}_{\max} \overline{f}(y)x, \tag{16.26}$$

while in case (ii) the equation becomes

$$\dot{y} \sim \overline{g}y - \overline{r}_{\max} \overline{f}(y)x. \tag{16.27}$$

The response is successful if we have $\dot{y} < 0$ in the final stage (i.e. as $t \to \infty$). This implies a condition of the form $x(t) \geq \overline{x}$ for large t; the quantity $\overline{x}$ is the *minimum surveillance level*, which is of considerable interest, since this is the allocation that the job will require in perpetuity. In case (i), we either have $\overline{x} = 0$, when $\overline{G} = 0$, or we can define a tolerance level $\hat{y}$ (i.e. a maximum allowable load), and then the minimum surveillance level is given as

$$\overline{x} = \frac{\overline{G}}{\overline{f}(\hat{y})\overline{r}_{\max}} \tag{16.28}$$

which shows that $\bar{x} > \overline{G}/\bar{r}_{\max}$. When $\overline{G} > 0$ we cannot take $\hat{y}$ as small as we please, since $\bar{x} \to \infty$ as $\hat{y} \to 0^+$, cf. equation (16.22). Thus, total "eradication" is ruled out. By contrast, we find for case (ii) that total "eradication" is possible provided the condition $\lim_{y \to 0+} y/\overline{f}(y) < +\infty$ is satisfied. For the queing model, equation (16.23) this limit equals $1/\bar{q}$, and we find that the minimal surveillance level (to ensure that $\dot{y} < 0$) is given by

$$\bar{x} = \bar{g}/(\bar{r}_{\max}\bar{q}),$$

reflecting the balance between $\bar{g}$, the target's intrinsic tendency towards growth, and the quality of the response, expressed in the values of $\bar{r}_{\max}$ and $\bar{q}$. For saturation curves described by $\overline{f}(y) = (\bar{q}y)^h/(1 + (\bar{q}y)^h)$ with a Hill coefficient $h > 1$, $y/\overline{f}(y)$ tends to infinity as y tends to zero, and "eradication" is impossible.

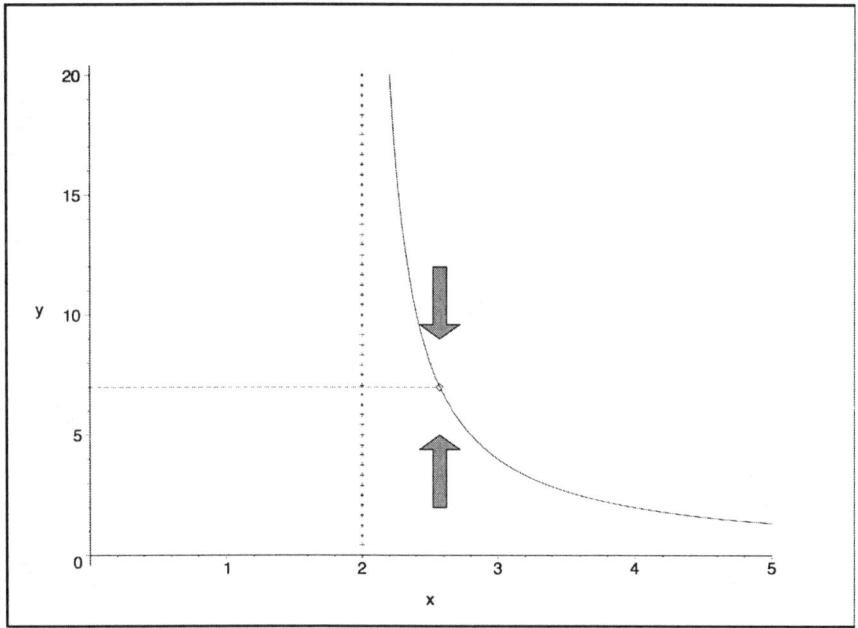

Fig. 16.4. *The (x, y) plane for case (i). The plane is divided into a "safe" region (above the curve) where $\dot{y} < 0$, and an "unsafe" region where $\dot{y} > 0$. The dotted vertical line corresponds to $x = \overline{G}/\bar{r}_{\max}$, and here we have set $\overline{G}/\bar{r}_{\max} = 2$, $\bar{q} = 2$. When x is kept at a constant level $\bar{x}$, the phase paths converge towards the point $(\bar{x}, \hat{y})$, as indicated by the arrows. The horizontal line is $y = \hat{y}$, where $\hat{y}$ is the tolerance associated with $\bar{x}$.*

These various possibilities are illustrated in Figures (16.4), (16.5) and (16.6). The first of these shows the (x, y) phase plane for case (i), with saturation model $\overline{f}(y) = \bar{q}y/(1+\bar{q}y)$. The regions where $\dot{y} > 0$ and $\dot{y} < 0$ respectively lie below/above the curved line (a hyperbola). When we fix $\bar{x}$ at any value greater than the infimum $\overline{G}/\bar{r}_{\max}$, the job load tends to the value that marks the boundary between these two regions, and the associated steady state clearly is stable.

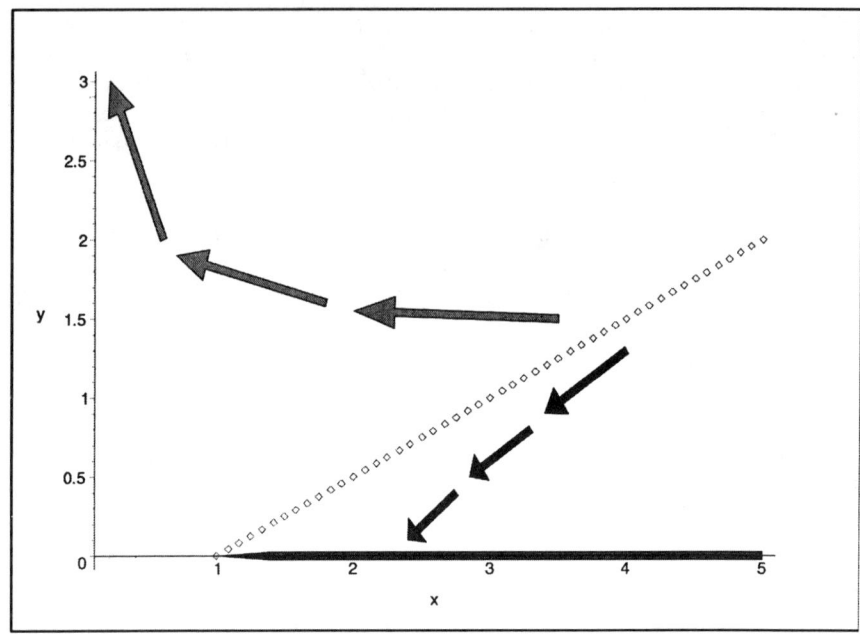

Fig. 16.5. *The (x, y) plane for case (ii). The plane is divided into a "safe" region (between the dotted line and the thickened section of the x-axis) where $\dot{y} < 0$, and an "unsafe" region where $\dot{y} > 0$. Arrows show the directions of phase paths that result when x is slowly decreased. The saturation function is $\overline{f}(y) = \overline{q}y/(1 + \overline{q}y)$, and here we have set $\overline{q} = 2$ and $\overline{g}/\overline{r}_{max} = 2$.*

The situation is different in case (ii), illustrated by Figures (16.5) and (16.6). In Figure (16.5), the saturation model is again $\overline{f}(y) = \overline{q}y/(1 + \overline{q}y)$. The boundary of the region where $\dot{y} < 0$ now has a stable and an unstable branch. The (thickened) stable branch is the part of the x-axis where $x > \overline{g}/(\overline{r}_{max}\overline{q})$, and the (dotted) unstable branch is the remainder of the boundary, a straight line in the interior of the plane, corresponding to the equation

$$x = \frac{\overline{g}}{\overline{r}_{max}} \cdot \frac{y}{\overline{f}(y)} \ . \tag{16.29}$$

Fixing x at any value greater than $\overline{g}/(\overline{r}_{max}\overline{q})$ does not guarantee a successful resolution of the response. Rather, the primary objective is to ensure that the phase point is in the safe region where $\dot{y} < 0$; x can then be gradually reduced as the phase point tends to an asymptotic value somewhere on the stable branch. For the class of saturation curves $\overline{f}(y) = (\overline{q}y)^h/(1 + (\overline{q}y)^h)$ with $h > 1$, illustrated for $h = 2$ in Figure (16.6), the unstable branch lies in the interior of the phase plane; the branches meet at an x-value greater than $\overline{g}/(\overline{r}_{max}\overline{q})$, and equation (16.29) describes a stationary steady state when $\hat{y} < (h - 1)^{1/h}/\overline{q}$.

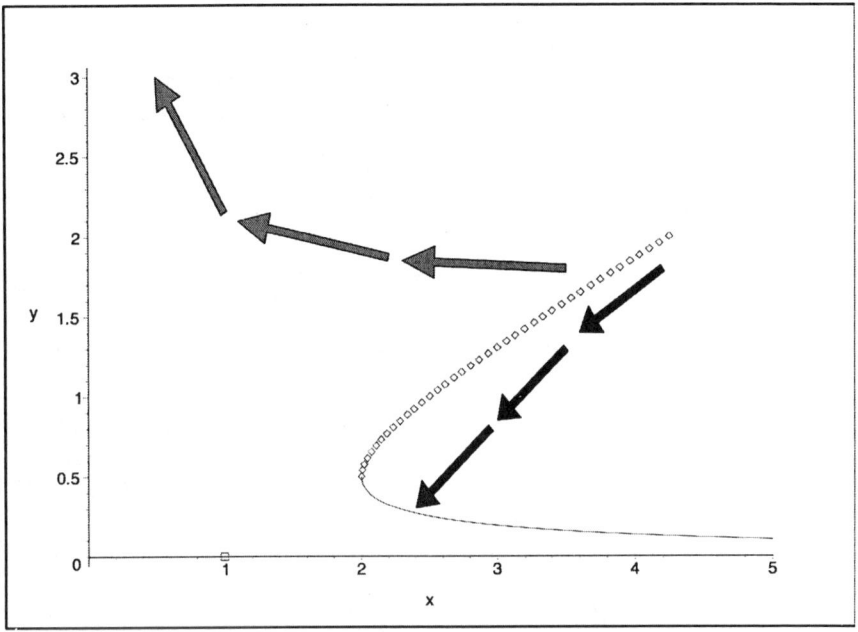

Fig. 16.6. *The (x, y) plane for case (iii). This is as in the previous figure, but with saturation function $\overline{f}(y) = (\overline{q}y)^2/(1 + (\overline{q}y)^2)$. The "safe" region, where $\dot{y} < 0$, is to the right of the dotted and undotted boundary curve. The unstable (dotted) and stable (undotted) branches of this boundary curve point meet at an x value which is larger than $\overline{g}/(\overline{r}_{max}\overline{q})$; the latter value is marked by a box on the x-axis.*

16.3.2 Dynamics of response allocation

In the biological immune system, x corresponds to the numbers of various effector lymphocytes. An increase of these numbers is due to cellular proliferation (clonal expansion), and as the rate of cell division cannot become arbitrarily large, $\frac{d}{dt} \ln x(t)$ will be bounded above. Such dynamic constraints are of great importance in the dynamics of biological immune responses [van den Berg & K"{i}selëv 2004]; in an AIS, however, they need not be nearly as prominent, and we can allow the dynamics of x to contain weighted Dirac pulses (i.e. 'jumps' of any desired magnitude).

The basic problem then is how to choose a trajectory $x(t)$. We first consider an approach that is naive, but instructive. Given $y(t)$, define

$$\mathcal{R}(t) \stackrel{\text{def}}{=} \{(x, y) \in \mathbb{R}^+ \times \mathbb{R}^+ : \dot{y} < 0\} \tag{16.30}$$

i.e. $\mathcal{R}(t)$ is the "safe region" at time t. The idea is to keep the phase path in the safe region $\mathcal{R}$. This region varies with time; it tends to expand as the quality parameters $r_{\max}$ and q improve, although variations in the job load growth rate may cause it to shrink as well. The points $(x, y(t))$ with $x \in \mathbb{R}^+$ constitute an horizontal line $L(t)$ in the phase plane. Now consider the set of points that lie in the intersection $\mathcal{R}(t) \cup L(t)$, and let $x_{\min}(t)$ denote the infimum of the x-coordinates of these points. We can assume that $\dot{y} < 0$ for all $x(t) > x_{\min}(t)$ at any time t (although this assumption may not be warranted when there are mutually inhibitory interactions between the processes that share the allocation $x(t)$). Then we propose the following dynamics:

$$x(t) = x_{\min}(t) + \varepsilon(t) \tag{16.31}$$

where $\varepsilon(t)$ denotes the *safety margin* at time t.

For case (i)-type dynamics of the work load, the safety margin may be adjusted to satisfy the condition that $y(t) \leq \hat{y}$ as $t \to \infty$, for instance:

$$\dot{\varepsilon} = \lambda(y - \hat{y})\varepsilon \tag{16.32}$$

where λ is a positive parameter and $\varepsilon(0) > 0$ (other schemes, e.g. PID-control [Kirk 1997], are possible). For case (ii)-type workload dynamics, we may take $\varepsilon(t) \equiv \bar{\varepsilon}$ or $\varepsilon(t) = \bar{\varepsilon} + (\epsilon_0 - \bar{\varepsilon}) \exp\{-\lambda t\}$ where $\bar{\varepsilon}$ and ε_0 are positive parameters. The dynamics (16.31) and (16.32) ensure that the phase path stays inside the safe region with a minimum expenditure x, while observing a safety margin to keep the state (x, y) away from the unstable boundary of $\mathcal{R}$.

Next, we sketch an analytic design approach to choosing $x(t)$. Let $\xi(x)$ denote the cost of the allocation toward responders. Accounting for the (presumably desirable) effect of the response, we have two terms: $\eta(y)$, denoting the cost of having a job load y; and $\vartheta(w)$, denoting the cost of jobs accomplished so far, where

$$w(t) \stackrel{\text{def}}{=} \int_0^t R(x(s), y(s), s)ds \tag{16.33}$$

(this latter cost is typically a negative cost i.e. a benefit). The η term dominates if the response is *supply-driven*, that is, if failure to operate on the data represented by y has adverse consequences, whereas the ϑ term dominates if the response is *demand-driven*, that is, if the accomplishment of operations yields beneficial effects to the system. The natural immune system would appear to be predominantly supply-driven, but demand-driven situations may be equally relevant for AIS. In general, both terms may be present. Finally, let $\zeta(z)$ denote the cost of the allocation toward response improvement (that is, for carrying out "affinity maturation" sensu lato). It is assumed that a commensurable cost scale can be found, applicable to all variables. Then $x(t)$ and $z(t)$ are chosen to minimize the *cost functional J*, defined by

$$J \stackrel{\text{def}}{=} \int_0^\infty \left(\xi(x(t)) + \eta(y(t)) + \vartheta(w(t)) + \zeta(z(t))\right) dt . \tag{16.34}$$

According to optimal control theory [Kirk 1997, Pinch 1993], the pair $(x(t), z(t))$ that minimizes J is found by minimizing a certain Hamiltonian H with respect to x and z at each moment t. This Hamiltonian is defined by

$$H(t) \stackrel{\text{def}}{=} Y(t) \left(G(x, y, t) - R(x, y, t)\right) + W(t)R(x, y, t)$$
$$- \left\{\xi(x(t)) + \eta(y(t)) + \vartheta(w(t)) + \zeta(z(t))\right\} \tag{16.35}$$

where the *co-state variables* $Y(t)$ and $W(t)$ obey the dynamics

$$\dot{Y} = -\partial_y H, \qquad \dot{W} = -\partial_w H.$$

Finally, the development of the response quality needs to be specified. For instance, $\dot{r}_{max}$ and $\dot{q}$ are to be given as functions of z; these ordinary differential equations will generally take the form of stochastic differential equations [Oksendal 2003].

In summary, the key control decisions in mounting an artificial immune response are (i) the allocation of computational resources $(x + z)$ to the AIS at any point in time; and (ii) how the resources are to be distributed between the "frontline" agents of the response itself (x) and the improvement of the efficacy of these agents (z). We have outlined here how this analysis might proceed for the simplest possible (three-dimensional) dynamic model of the artificial immune response. The second choice, between development and attack, is crucial: while resources are being spent on improving the agents, the job load might grow beyond tolerable levels; on the other hand, if too little has been invested in agent efficacy, the job might ultimately prove to be impossible to accomplish even at very high levels of expenditure of computational resources.

It may be difficult to formulate cost functions $(\xi, \eta, \vartheta, \zeta)$ that accurately model the situation. This motivates an interest in fairly robust, qualitative relationships between various plausible choices for these functions and the optimal control regime that they entail, since such relationships indicate how critically the results depend on the particular choice of cost functions. In particular, an interesting hypothesis is that for a wide range of reasonable choices of cost functions one will generically find an optimal pattern of investment in development in the early stages of the artificial immune response, followed by a terminal phase of expenditure (almost) exclusively on frontline agents (cf. [Perelson *et al.* 1976]). In this section we have tried to model an artificial immune response in complete generality, without restricting ourselves to a particular type of algorithm or choice of problem to be solved. We expect that it would be worthwhile to apply this optimal control model to analyse the performance of some existing AIS algorithms such as CLONALG [de Castro & Von Zuben 2002], and this will give a better indication of what sort of cost functions are appropriate.

16.4 Cytokine Networks

The allocation of computational resources to the processes of an artificial immune response is critical to AIS performance. The urgency of the job load, the tasks at hand, the specific efficacies of the available artificial immune response mechanisms, the likelihood of discovering more efficacious responses (e.g. a higher-affinity receptor), demands on CPU time by processes outside the AIS: these are all factors that determine how much should be allocated to the available AIS and developing better responses.

The natural immune system regulates this allocation by means of a system of various immune cells that mutally influence each other's activities via hormone-like intercellular messenger molecules called cytokines [Balkwill 2000, Roitt 1997]. Cytokines stimulate proliferation of various immune cells, with different immune cell types responding to different cytokines; these immune cells include the effector cells that carry out the response as well as the cells that produce cytokines themselves; indeed, immune cells often have a dual role, producing cytokines in addition to an effector function. The activities modulated by the cytokines produced by one cell include the production and secretion of cytokines by other immune cells. These interactions form a lymphoid endocrine system called the *cytokine network*. Each cell type in this network is characterised by its own subset of the cytokines that it can secrete, as well as its own subset of cytokine receptors that govern its activity. The cytokine network integrates stimuli from a variety of sources (e.g. distressed cells, effector cells, naive response-precursor cells), and the cytokines produced by the network regulate the development and growth of the responding immune cells.

From a computational point of view, the cytokine network has an input of information about the state (extent and severity) of the disease, the state (extent and efficacy) of the ongoing responses, and an output that governs proliferation of selected effector cells as well as the organization of new responses (e.g. antigen presentation, germinal centre reaction). Both input and output are encoded by the concentrations of the various cytokines. From a modelling point of view, the complexity of cytokine networks poses considerable challenges [Callard *et al.* 1999]. Moreover, the cytokine network operates both locally and more globally, and is thus intermediate between a paracrine system and an endocrine system; however, here we ignore these spatial aspects for the sake of simplicity.

Below we give a mathematical specification of the cytokine network which highlights it as a computational paradigm. To emphasise this intent, we call it an *artificial cytokine network* (ACN). The ACN is one example of a computational system inspired by biological para-/endocrine systems. As will become apparent, the ACN has much in common with the associative memory models studied by neural network theory [Meade & Sonneborn 1996]. This is not too surprising since the ACN is likewise a system that matches a vector representing a given situation to a vector representing a (hopefully suitable) response. However, there are a few interesting points of contrast. The analogues of "synaptic weights" in the ACN are non-changing. However, the ACN is in some sense a superposition of a number of associative memory structures, with the relative contributions of these structures changing on a second, slower time scale.

16.4.1 General model

To model the cytokine network, we consider an intercellular medium in which n distinct chemical species of cytokines diffuse and are well mixed. We can then define cytokine concentrations $u_1, \ldots, u_n$. The cytokines are produced by cytokine-producing cells, of which there are m types. Cytokine production by a cell of any one of these types depends on external stimuli $s_1, \ldots, s_r$ (i.e. signals arising outside of the cytokine network) as well as the cytokines themselves. The density of cell type ℓ in the medium is denoted as v_ℓ. We thus have the following kinetics:

$$\dot{u}_k = \sum_{\ell=1}^{m} \psi_{\ell k}(u_1, \ldots, u_n, s_1, \ldots, s_r)v_\ell - \nu_k u_k, \quad k = 1, \ldots, n, \tag{16.36}$$

$$\dot{v}_\ell = \Big(\varphi_\ell(u_1, \ldots, u_n, s_1, \ldots, s_r) - \mu_\ell\Big)v_\ell, \quad \ell = 1, \ldots, m; \tag{16.37}$$

the function $\psi_{\ell k} > 0$ expresses the effect of the cytokines and external stimuli on the production of cytokine k by a cell of type ℓ; $\nu_k > 0$ is the rate of degradation of the kth cytokine; $\varphi_\ell > 0$ expresses the effect of the cytokines and external stimuli on the proliferation rate of a cell of type ℓ; and $\mu_\ell > 0$ is the death rate of cells of type ℓ. There is often a separation of time scales between the dynamics of the cytokines, equation (16.36), and the dynamics of the cytokine-producing cells, equation (16.37). The system as a whole is a functional, mapping the external stimuli into a cytokine profile,

$$s_1(t), \ldots, s_r(t) \mapsto u_1(t), \ldots, u_n(t)$$

where the latter directs the immune response.

When the stimuli evolve slowly (much slower than the typical timescale $1/\lambda_\ell$), and when there is just one cell type ($m = 1$), this mapping is quasi-static, with the stimuli at time t uniquely determining the cytokine profile at that moment in time (up to transient behaviour). The cytokine network then behaves essentially like a look-up table. However, when $m \geq 2$, this look-up table will itself evolve over time, depending on the *history* of the stimuli. Moreover, certain rapid changes in the stimuli (on the fast $1/\lambda_\ell$ timescale) may precipitate sudden transitions to a different look-up table.

To illustrate these points more concretely, consider the following specification:

$$\psi_{\ell k}(u_1, \ldots, u_n, s_1, \ldots, s_r) = \overline{\psi}_{\ell k} S(\textstyle\sum_{i=1}^{n} w_{\ell k i} u_i - \tilde{\theta}_{\ell k}) \tag{16.38}$$

with

$$\tilde{\theta}_{\ell k} \stackrel{\text{def}}{=} \theta_{\ell k} - \sum_{j=1}^{r} \tilde{w}_{\ell k j} s_j \tag{16.39}$$

where $\theta_{\ell k} > 0$ represents the *stimulation threshold* of cell type ℓ as regards production of cytokine k, and $\overline{\psi}_{\ell k} > 0$ represents the maximum cell-specific secretion rate of cytokine k by a cell of type ℓ. The function S is monotonically increasing, with $S(x) \in [0, 1]$ for all $x \in \mathbb{R}$, with $\lim_{x \to -\infty} S(x) = 0$ and $\lim_{x \to +\infty} S(x) = 1$; an example is $S(x) = 1/(1 + \exp\{-x\})$. The parameters $w_{\ell k i}$ and $\tilde{w}_{\ell k j}$, which may be negative, zero, or positive, characterize how the production of cytokine k by cell type ℓ is affected by stimulation by cytokine i or external stimulus j.

To gain an insight into the dynamics of this model, consider first the case $m = 1$ (just one cell type), $v_1(t) \equiv \overline{v}$ (timescale separation) and with the $\tilde{\theta}_{1k}$ fixed for all cytokines k. Also, let us take S to be the Heaviside step function, i.e.

$$S(x) = 0, \quad x < 0, \tag{16.40}$$
$$= 1, \quad x \geq 1; \tag{16.41}$$

the $\tilde{\theta}_{1k}$s behave as 'crisp' thresholds for this choice. Let $(\overline{u}_1, \ldots, \overline{u}_n)$ be a (quasi) stationary point which solves $\dot{u}_k = 0$ for all k (there are *at most* 2^n such points), and consider the region around this stationary point bounded by the hyperplanes that are the locus of $\sum_{i=1}^{n} w_{\ell k i} u_i = \tilde{\theta}_{\ell k}$; stationary points do not generically lie on such a hyperplane. This region is the basin of attraction of the asymptotically stable point $(\overline{u}_1, \ldots, \overline{u}_n)$, since we have either $S = 1$ or $S = 0$ throughout this region for all k.

When S is a smooth sigmoid function (such as $S(x) = 1/(1 + e^{-x})$), the situation becomes more complicated. However, we will in general be able to define regions around the stationary points in which $\sum_{k=1}^{n} (\dot{u}_k)^2$ satisfies the properties of a Lyapunov function (see chapter 10 in [Jordan & Smith 1999]). The weight parameters characterize cell types; thus "learning" in the classical neural network sense only takes place over the much slower evolutionary timescale on which novel cell types arise.

16.4.2 Shape space model

It is particularly useful to make a sharp distinction between different cell types in certain contexts, e.g. in virology [Nowak & May 2000], when one is mainly interested in the population of T lymphocytes that respond to the presence of a particular virus. However, in practice the immune system consists of a vast number of cells that display a broad range of genetic diversity in terms of the different receptors that are expressed on their cell surfaces. If we are interested in modelling a fairly broad range of responses within a population of cells, then in that case rather than thinking of a discrete set of cell types, it is convenient to model this diversity with a continuum. Since the specifities of the receptors (e.g. cytokine receptors) on the surface of a lymphocyte are determined by their shapes, it is useful to consider the cells as belonging to a shape space Σ of dimension M, and represent a cell type (that is, a cell with a particular type of receptor) by a vector $\mathbf{l} \in \Sigma$. A review of shape space models in immunology can be found in [Perelson & Weisbuch 1997] for instance, while possible applications to AIS are discussed in [Hart & Ross 2004].

In order to model a population of cells with different receptors, we introduce a shape space density $v(\mathbf{l}, t)$, such that

$$n_S(t) = \int_S v(\mathbf{l}, t) \, d^M \ell \tag{16.42}$$

is the number of cells in the subregion S at time t, where this subregion of Σ will correspond to cells of similar specificity. The notation $d^M \ell$ indicates that we are integrating over a subregion of the M-dimensional shape space. For example, in the cytokine network model we could say that all points near to some fixed point $\mathbf{l}_0$ in shape space correspond to cells having receptors with high affinity for a particular cytokine, e.g. IL-12, and take S to be a ball of fixed radius R around $\mathbf{l}_0 \in \Sigma$. Thus, at each time t, $n_S(t)$ would count the number of cells that are strongly stimulated by IL-12, for instance.

Taking this shape space approach we can reformulate the cytokine network dynamics in terms of an integral partial differential system, as follows:

$$\dot{\mathbf{u}} = \int_{\Sigma} \Psi(\mathbf{l}, \mathbf{u}, \mathbf{s}) \, v(\mathbf{l}) \, d^M \ell - \mathbf{N} \, \mathbf{u}, \tag{16.43}$$

$$\frac{\partial v}{\partial t} = \Big(\varphi(\mathbf{l}, \mathbf{u}, \mathbf{s}) - \mu(\mathbf{l}) \Big) v(\mathbf{l}). \tag{16.44}$$

In the above, $\mathbf{u} = (u_1, \ldots, u_n)$ is the cytokine state vector (a function of t); $v = v(\mathbf{l}, t)$ is the shape space density (with dependence on t suppressed above); $\mathbf{N} = \mathrm{diag}(\nu_1, \ldots, \nu_n)$ is the death rate matrix for cytokines; and $\mu(\mathbf{l})$ is the local cell death rate in shape space. The network functions Ψ and φ must then be specified over the whole of shape space.

From the point of view of computer simulation, or designing ACN algorithms, the continuum version of the cytokine model, given by the coupled ordinary and partial differential equations (16.43) and (16.44), appears to be more awkward than the ODEs (16.36) and (16.37), since for computer modelling one requires discrete dynamics. However, we expect that for modelling a large number of cytokine receptor types, the shape space version of the cytokine model might be more tractable for doing analytical calculations.

16.5 Outlook

We have outlined some of the basic notions in the theory of dynamical systems, and reviewed the B-model as one example of a receptor repertoire-based model of the adaptive immune system; alternative approaches to the dynamics of the adaptive T cell repertoire can be found in [van den Berg & Kiselëv 2004, van den Berg 2004, Luciani *et al.* 2001], for instance. We have also outlined an optimal control approach to the allocation of computational resources in an artificial immune response. Furthermore, we have sketched a model of an artificial cytokine network (ACN) based on a decision-making subsystem of the natural immune system. Viewed as computational systems, the adaptive repertoire as well as the ACN are characterized by a few common properties: (i) they consist of a vast number of fairly complex computational units (agents) with a fairly stereotypical, limited set of possible behaviours; (ii) the execution of tasks by the system arises out of a myriad of mutual interactions between these agents; (iii) reliable, robust behaviour of the system as a whole emerges as a statistical effect from the vastness of the number of agents.

The natural computational medium for systems of the kind described above is massively parallel computing which, unfortunately, remains relatively underdeveloped. Thus the question arises as to whether these immune mechanisms can in fact be effectively exploited: if we attempt to simulate the massively parallel dynamics of these vast populations, we quickly find ourselves constrained by scalability issues [Stibor *et al.* 2005a], and we are at risk of losing the useful properties that arise by virtue of the "thermodynamic limit" effect. We can think of various ways out of this quandary: (a) accept the various limitations on scalability, and attempt to obtain useful applications within these constraints; (b) perform the thermodynamic limit through theoretical calculations, and develop mean field models with correlations to simulate the dynamics directly at the emergent system level; (c) adopt AIS as a programming paradigm for massively parallel hardware. Each of these avenues deserves to be explored in the future.

Conceptualizing the Self-Nonself Discrimination by the Vertebrate Immune System

Melvin Cohn

Conceptual Immunology Group, The Salk Institute for Biological Studies, 10010 N. Torrey Pines Rd., La Jolla, CA 92037
and
Instituto Gulbenkian de Ciência, 6 Rua da Quinta Grande, 2780-156 Oeiras, Portugal. cohn@salk.edu

Summary. A Preamble: In the end, biological phenomena are best understood in terms of evolution, which implies among other things, interactive selection on variants such that the selector and the selectee evolve as lineages [Hull *et al.* 2001]. The formulation of the concept should precede any attempt at modeling or simulation of any family of observations. The construction of a computer web around a random collection of facts is of marginal interest. Abstractions must be heuristic implying a testable output accompanied by a sharpening of the concept. This essay is an effort to conceptualize the most divisively discussed segment of immunology, namely, the Self-Nonself discrimination.

17.1 Introduction

The property referred to as a "Self-Nonself discrimination" is a requirement of any mechanism that has a bio-destructive and ridding effector output. All free living organisms defend themselves against parasitism by using such mechanisms. Bio-destructive and ridding protective mechanisms require a recognitive site that is of sufficient specificity to distinguish the Not-To-Be-Ridded (NTBR) components of the host from the To-Be-Ridded (TBR) components of the parasite/pathogen.

The recognitive sites (referred to as *paratopes*) that guide these effector mechanisms to their targets are of two origins. There are those paratopes that are germline-encoded and germline-selected (Type I). These are present in all free living organisms. Then, there are those paratopes that are somatically-derived and somatically-selected (Type II). Non-vertebrates express only Type I paratopes, whereas vertebrates express both Types I and II.

As the interactive selection between pathogen and host progressed, there came a point when germline-selection on the non-vertebrate Type I paratopes became too slow to match the ability of the various members of the pathogenic load to escape recognition. This necessitated a unique solution that appeared at the time when vertebrates emerged, namely to generate somatically a large, random (with respect to the recognition of Self and Nonself) paratopic repertoire that divided the antigenic universe into combinatorials of determinants (ligands) referred to as *epitopes*. An antigen is defined as a combinatorial of linked epitopes [Cohn & Langman 1990, Cohn 1997, Cohn 2005a].

While this Type II paratopic repertoire effectively solved the problem of the escape of pathogens from recognition, it created two new problems that had to be solved in parallel.

First, this random paratopic repertoire had to be sorted into those specificities (anti-Self) which, if expressed, would debilitate the host by autoimmunity and those specificities (anti-Nonself) which, if not expressed, would result in the death of the host by infection. I will refer to the sorting of the Type II paratopic repertoire as **Decision 1.**

Second, the sorted repertoire had to be coupled to the biodestructive and ridding effector mechanisms such that the response to each antigen is both coherent and independent. I will refer to the appropriate coupling of the sorted repertoire to the effector mechanisms as **Decision 2**, the regulation of effector class.

The effector mechanisms were, in large measure, invented by the non-vertebrates. However, when the Type II somatically-derived repertoire emerged, these effector mechanisms with their Type I germline-selected paratopes co-evolved with it, such that the effector output of the vertebrates could be controlled by either paratopic input.

There is an asymmetry to consider. The Type I paratopic repertoire cannot recognize a large portion of the epitopic universe that the Type II paratopic repertoire can see. The somatically-derived Type II repertoire recognizes everything that the germline-selected Type I repertoire sees, not *vice versa*. This has many consequences the most important being that Decision 1 cannot be mediated by germline-selected paratopes.

17.2 Decision 1. The Sorting of the Paratopic Repertoire

This essay is essentially about Decision 1, the sorting of the Type II paratopic repertoire. The term "Self-Nonself discrimination" has invited all manners of tergiversation and semantic debate that has been largely unproductive. Unfortunately, it is very difficult to avoid using the terms Self and Nonself and still communicate, because they have been cemented into a vast literature. I have tried referring to a more precise Not-To-Be-Ridded (NTBR)—To-Be-Ridded (TBR) discrimination but this has not taken hold. Consequently, I will continue to use Self (S) and Nonself (NS), constantly warning the reader of various ambiguities and misuses of the concept.

To begin, I will define the Self-Nonself (S-NS) discrimination as the mechanism by which the Type II repertoire is sorted (Decision 1). This is the only valid meaning of the term because an adequate Decision 1 is both necessary and sufficient to enable the ridding of the pathogen without debilitating the host. An attack on Self (NTBR) must have a debilitating consequence if it is to be a selectable factor in the evolution of the mechanism of Decision 1.

Stated differently, the function of Decision 2, the regulation of effector class, is to optimize the biodestruction and ridding of the target, which is defined by the paratopes put under its control by Decision 1. Decision 2 is not concerned with whether the paratope is anti-Self (NTBR) or anti-Nonself (TBR). It rids any target defined by the paratopes of the activated cells that leave Decision 1.

The function of Decision 1, the sorting of the repertoire, is to subtract the anti-Self (anti-NTBR) specificities from the repertoire and leave the residue as anti-Nonself (anti-TBR), which protects the host. The consequence of this dichotomy into Decision 1, the sorting of the repertoire, and Decision 2, the regulation of effector class, is that Decision 1, not Decision 2, can be properly described as the Self-Nonself or the NTBR-TBR discrimination. Decision 2 does not make this discrimination; it optimizes the ridding independent of the specificity of the paratope, anti-S or anti-NS, that Decision 1 puts under its control.

17.2.1 What does it take to sort the Type II repertoire?

In the case of the germline-selected paratopic repertoire (Type I), the cells expressing it are born as effectors. This is only possible because the Type I repertoire is germline-selected to purge anti-Self and, therefore, only anti-Nonself paratopes survive to be encoded in the germline.

In the case of the somatically-generated paratopic repertoire (Type II), if the cells expressing it were born as effectors, the individual would die of autoimmunity. The somatically-encoded repertoire cannot be sorted at the level of effectors (i.e., at the level of Decision 2).

Consequently, in order to sort this repertoire, we must envisage an immune system arising as initial state cells (i-cells) expressing the somatically generated repertoire on a one cell-one paratope basis and devoid of effector function. The antigen-responsive or initial state cells (i-cells), must have two pathways open to them, inactivation and activation, because they cannot know what is the specificity of their expressed paratopes, anti-S or anti-NS. Those i-cells expressing anti-Self must be inactivated; those expressing anti-Nonself must have the potential to be activated to enter the Decision 2 pathway.

17.2.2 Decision 1 requires that the antigenic universe be sorted into Self and Nonself

The sorting of the paratopes into anti-S and anti-NS depends upon the ability of the individual to sort the antigenic universe into Self and Nonself. It is the somatic sorting of the antigenic universe that permits the sorting of the somatically generated repertoire. This is model-independent; no signal via the antigen-receptor, T-cell receptor (TCR) or B-cell receptor (BCR) can tell the i-cell if it is anti-S or anti-NS.

17.2.3 What option is not open to the individual in order to sort the antigen universe into Self and Nonself?

We might begin with the assumption that all Self is red and all Nonself is blue; or all Self is harmless and all Nonself is dangerous [Matzinger 2002], or pathogenic [Janeway 1992] or cytopathic [Zinkernagel & Hengartner 2004]; or all Self is included in the thymus whereas all Nonself is excluded from it [Kyewski & Derbinski 2004, Gotter & Kyewski 1994]; or conversely, all Self is excluded from lymph nodes whereas all Nonself enters them [Zinkernagel & Hengartner 2004]; or all Self arises inside the individual whereas all Nonself comes from the outside (often declared); or all Self differs from Nonself by emergent properties [Cohen et al. 2004], or tuning [Grossman & Paul 2001] or context [Cohen 2000b] or integrity [Dembic 2000] or morphostasis [Cunliffe 1997], and so on.

If any of these properties were to divide the antigenic universe into Self and Nonself, then they would have to be recognized by germline-encoded recognitive elements. However, the germline-selected repertoire is *per force* blind to a large portion of the antigenic universe that the somatically-derived repertoire recognizes. An antigenic universe sorted by a germline-selected repertoire (Type I) would leave most of the somatically selected random repertoire (Type II) unsorted. A somatically generated repertoire can only be sorted by an antigenic universe that itself has been sorted into Self and Nonself by a somatic learning or historical process. In other words, the mechanism used to sort the antigenic universe must be independent of the "innate" or germline-encoded (Type I) recognitive system. Decision 1 must be dependent solely on interactions of the presorted antigenic universe with the somatically generated repertoire (Type II) itself.

This argument can be formulated in another way. There is no physical or chemical property of antigens as classes that can be used by an individual's immune system to sort the antigenic universe into Self and Nonself. This follows because *what is Self for one individual of a species is Nonself for another.* A random paratopic repertoire cannot be sorted into anti-S and anti-NS by germline encoding of the recognition of Self. Further, the immune system has no way to determine if an antigen is or is not encoded in the host genome or whether the antigen originates from the inside or the outside of the host.

It might be stressed that an immune attack on only one Self-component would debilitate the individual. Therefore, any model that allows for the sorting of a portion of the Self-antigenic family is lacking; all must be sorted. Any partial solution only invites, how does this system deal with the remainder?

17.2.4 What solution is open?

Essentially only one formulation has withstood repeated critical analyses although it now requires tweaking. I will refer to it as the "Developmental Time Model."

Under this model, as a first approximation to initiate the discussion, all Self and no Nonself must be expressed during a developmental time window when i-cells arise under conditions such that they are inactivatable-only. Interaction with Self-epitopes, the only ones present, deletes them. When the window closes, meaning that the system becomes responsive, the persistence of Self maintains the state of unresponsiveness to Self-antigen. This process operates throughout the entire individual viewed as a single space.

In all vertebrates with somatically derived paratopic repertoires, the fetus is protected by maternal immunity until its own immune system becomes responsive. During this period, the fetal immune system arises in the presence of Self and in the absence of Nonself.

It is important to appreciate that all somatic processes are built on germline-encoded components, and, in the case of a Decision 1 sorting mechanism that is built on developmental time, the germline-encoded component is the duration of the time window. It is the time that it takes before the system becomes responsive that is germline-selected.

Consider the following scenario (Figure 17.1):

If there were no overlap between the *de novo* appearance of Self and the entrance of Nonself into the system then germline-selection could set the length of the developmental time window such that the window remained open (i.e., the system remained unresponsive or inactivatable-only) until the last Self-antigen had appeared, at which point the window would close (i.e., the system would become responsive to the appearance of Nonself). No new Self is allowed to appear after the window closes. The Self that was expressed while the window was open, must persist. Clearly the individual must learn (a somatic process) what is Self (Not-To-Be-Ridded) at a time when Nonself (To-Be-Ridded) is absent. Another way to state this is that Self is

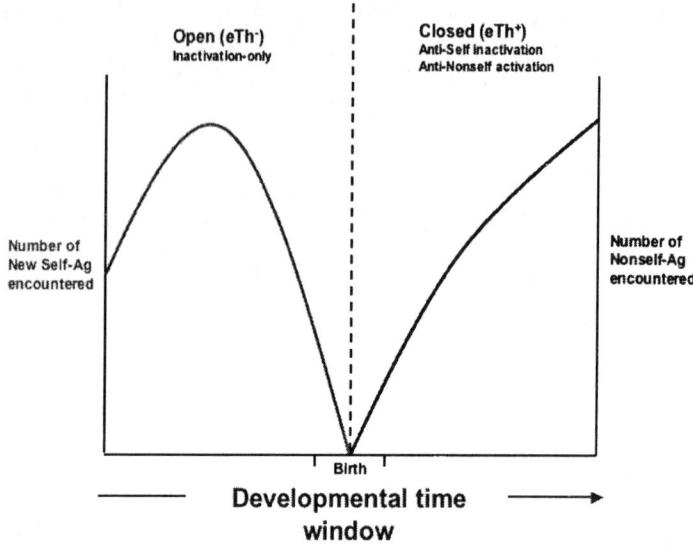

Fig. 17.1. The idealized Developmental Time Model

defined by the immune system as **prior and persistent**. Nonself is defined by the immune system as **posterior and transient**.

However, if there is an overlap between the *de novo* appearance of Self- and Nonself-antigens then evolution would have no way to set the length of the time window. That this is a likely scenario derives from the fact that Self-components are evolutionarily selected to function in the physiology of the host, not to escape the immune system. The immune system is selected upon not to attack these Not-To-Be-Ridded or Self-components. It is not surprising, then, that some Self, for reasons of function, would be expressed after the developmental time window closes and the system is responsive.

This problem can be illustrated as follows (Figure 17.2);

If evolution closed the window when Nonself first appears, the individual would die of autoimmunity to the late appearing Self-antigen (NTBR). If evolution closed the window when the last Self-antigen had appeared, then the individual would die of infection by Nonself pathogens to which the system would be unresponsive. Given that there is an overlap, the Developmental Time Model needs to be tweaked.

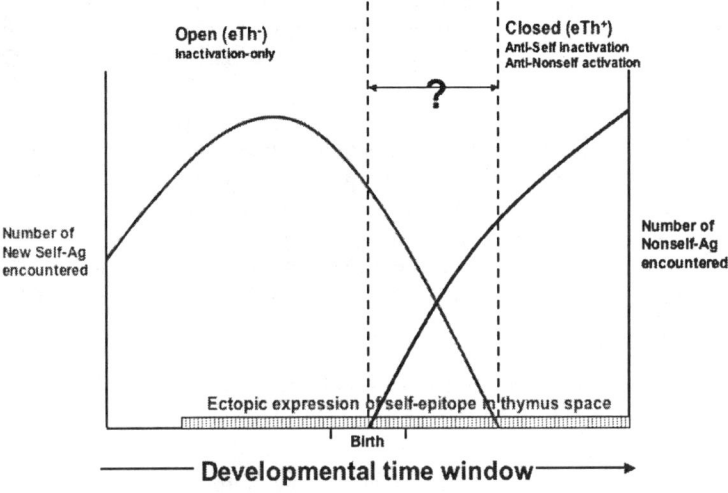

Fig. 17.2. The generalized (tweaked) Developmental Time Model

17.2.5 Tweaking the Developmental Time Model

In order to deal with late appearing Self-antigens (NTBR), I envisage the following scenario. The developmental time window must close (i.e., the immune system becomes responsive) just before the first Nonself enters the system (in all likelihood around birth). This protects the neonate against infection but requires that the late appearing Self be ectopically expressed initially in an anatomical space where crucial regulatory i-cells arise under inactivatable-only conditions. This requires some detailing.

First, i-cells are of two categories, iT-cells that arise in the thymus and iB-cells that arise in the bone marrow (or bursa). Evolution had to decide where to ectopically express the late appearing Self. Fortunately, this choice was simplified because there is an asymmetry between the two categories. The T-helpers (Th) which arise in the thymus are required, when in the effector state (eTh), for the activation of all categories of i-cell, iT and iB, including the iTh itself. In the absence of effector T-helpers (eTh), the i-cell is inactivatable-only upon interaction of its antigen-receptor, TCR or BCR, with ligand. The appropriate ectopic expression in thymus when the window is open permits deletion of iTh anti-Self leaving all other i-cells anti-Self inactivable-only. The crucial role of the effector T-helper (eTh) as a regulatory cell will be detailed later. At this point it is only necessary to understand that it is the insufficiency or sufficiency of eTh that determines responsiveness to an antigen.

Second, the iTh-cell can only recognize as ligand peptide (P) derived from the late appearing Self presented by an MHC-encoded Class II restricting element (RII). The ectopic expression in thymus then must present the late appearing Self as an appropriate [Ps-RII] ligand for the iTh. There are new endogenously generated antigens that arise by mutation or other during the life of an individual but they are often non-immunogenic because they are not expressed functionally by presentation on RII. Many *de novo* arising tumor antigens as well as idiotypic determinants fall into this category.

The generalized Developmental Time Model then has two germline-selected elements:

1. The length of the developmental time window, and
2. The appropriate ectopic expression in thymus as [Ps-RII] of all peripheral Self-components that appear after the window closes.

The assumption here is that ectopic expression in thymus was evolutionarily selected to cope with Self (Not-To-Be-Ridded)-antigens that appear after the developmental window closes and the system is responsive. The ectopic expression in thymus must occur before the developmental time window closes and the system is responsive. If it occurred after the window closed, iTh specific for the given late expression Self-antigen would accumulate in the periphery and initiate a response before deletion of iTh by ectopic expression in thymus could have an effect. The timing is important (selectable) here.

Thus far I have analyzed how the individual sorts the antigenic universe into Self and Nonself. Sorting by an individual implies a somatic learning process. The Self components of the host are categorized by this process as shown in Table 17.1.

17.3 How Does the Paratopic Repertoire Respond to the Antigenic Univerise?

The initial state or i-cells arise when the developmental time window is open; they encounter Self-only and are inactivated. Those that do not encounter antigen are defined as anti-Nonself and accumulate to a steady state, homeostatically determined level. This process can be symbolized as:

$$
\begin{array}{ccc}
 & \text{S-epitope} & \\
\text{i-cell} & \longrightarrow & \text{INACTIVATION} \\
 & \text{Signal[1]} & \\
\end{array}
$$

| Category of self-antigens | Expression in | | | Observation |
| | Thymus | periphery when developmental window is | | |
		open	closed	
I	+	+	+	unresponsive to SI
II Aire$^+$(wt)	+(ectopic)	-	+	unresponsive to SII
Aire$^-$(mutant)	-	-	+	autoimmunity to SII (eTh-dependent)
III	-	+	+	unresponsive to SIII

Table 17.1. Categories of host-encoded Self-antigens defined by the Developmental Time Model. *Aire* is a transactivating transcription factor required for the ectopic expression of some peripheral antigens as [P-RII] in thymus. These are presumably late expression Self-antigens. A mutation of *Aire, Aire$^-$*, results in failure to ectopically express the peripheral antigens in thymus with resultant autoimmunity [Su & Anderson 2004, Anderson *et al.* 2002].

As long as Self persists, the inactivation of i-cells anti-Self maintains the state of unresponsiveness to Self even when the window closes.

No signal via the antigen-receptors, TCR or BCR, can tell the i-cell whether its specificity is anti-S or anti-NS. Signal[1] upon interaction with ligand is the same and eventually inactivating whether the antigen is Self or Nonself. In order to distinguish Self from Nonself a second signal must be delivered to the i-cell receiving Signal[1] from an antigen-specific source that itself has independently undergone a sorting process. This pathway would be:

$$
\text{i-cell} \quad \xrightarrow[\text{Signal}([1]+[2])]{\text{NS-antigen}} \quad \text{ACTIVATION} \longrightarrow\longrightarrow \text{e-cell (effector)}
$$

The origin of this Signal[2] is at the crux of the S-NS discrimination. In order to clarify this we must review the overall pathway that i-cells take in becoming effectors anti-NS [Cohn 2005b].

17.3.1 The response pathway (Figure 17.3)

There is a steady state production of i-cells anti-S and anti-NS [Cohn & Langman 1990, Cohn 1997, Cohn 1998, Cohn 2002]. Upon interaction with their ligand, Signal[1] converts the i-cell to an anticipatory cell or a-cell that has two pathways open to it, inactivation or activation.

The continued delivery of Signal[1] results in inactivation.

384   Melvin Cohn

ASSOCIATIVE RECOGNITION OF ANTIGEN OR "TWO SIGNAL" MODEL

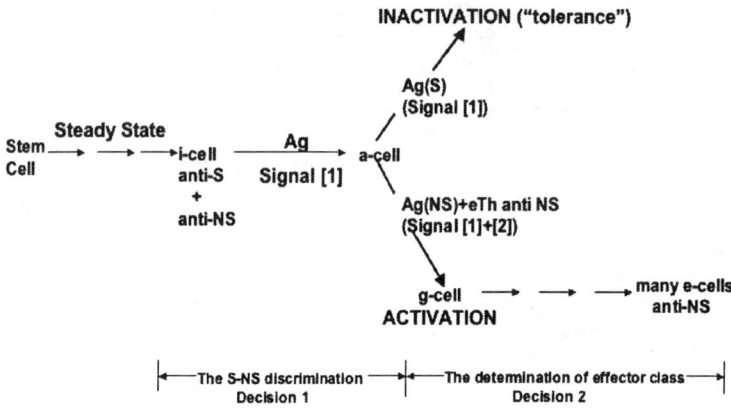

Fig. 17.3. The pathway of the sorting of the repertoire (Decision 1.)

However, if an a-cell on the pathway to inactivation receives Signal[2] then it is activated to what is symbolized as the g-cell; this is the first step of the Decision 2 pathway to effectors. The anticipatory or a-cell intermediate insures that no i-cell can be activated that, in principle, could not have been inactivated. Signals ([1]+[2]) are required for activation.

17.3.2 What is the source of Signal[2]?

As I pointed out earlier, Signal[2] is delivered by an effector T-helper (eTh) to an a-cell.

The requirement of Signal[2] is that it be epitope-specific and Nonself-antigen specific. Nonself-antigen specificity depends on the sorting mechanism used by the cells delivering Signal[2]. Thus, there are two points to develop.

First, the pathway of Decision 2 is different for T- and B-cells because T-cells are restricted and have their effector classes, helper (Th), suppressor (Tsu), cytotoxic, (Tc), predetermined intrathymically. Further, as T-cells only see peptide derived by processing, the Antigen-Presenting Cell (APC) becomes an obligatory player and an eTh-APC-aT interaction must be envisaged. For B-cells, because they themselves present the antigen only an eTh-aB interaction is required for activation.

The crux of the interaction is that Signal[2] must only be delivered by an eTh recognizing one epitope of the NS-antigen to an a-cell recognizing another epitope from that same antigen. This is referred to as Associative Recognition of Antigen (ARA). In the absence of ARA, coherent and independent regulation of the response would be impossible. The mechanism that assures ARA in the eTh-APC-aT interaction has been ignored by immunologists but that doesn't make it less crucial [Cohn 1992, Bretscher 1999].

It should be noted in this regard that inactivation (Signal[1]) occurs epitope-by-epitope whereas activation (Signal([1]+[2]) occurs antigen-by-antigen (ARA).

Second, the iTh undergoes the same pathway of activation as all other i-cells, meaning that it requires an eTh delivered Signal[2], thus raising the question, "Where does this primer eTh come from?"

This question was raised for the first time by the ARA model [Cohn 1969] and solutions to it have been proposed [Bretscher 1972]. The most likely [Cohn et al. 2002, Langman et al. 2003, Cohn 1983] is that there is an NS antigen-independent pathway to primer eTh anti-NS (Figure 17.4).

Essentially, if the rate of inactivation of the iTh on encountering prior and persistent Self is rapid compared to an Nonself-independent conversion to effectors then the primer eTh will be essentially anti-NS and function as the source of eTh that initiate an "autocatalytic" induction of iTh anti-NS to eTh anti-NS. This, in turn, determines the response of all other i-cells specific for that antigen [Cohn 1992].

Returning now to an earlier discussion, in referring to the developmental time window as open or closed, I meant that when open, the system is unresponsive and when it closed it is responsive. Now this needs precision.

When the developmental time window is open the immune system is unresponsive because of an insufficiency of eTh. The i-cells are not inherently inactivatable-only. They always have two pathways open to them; it is the system, because it is lacking in eTh, that renders the i-cell inactivatable-only upon interaction with an epitope (Signal[1] only).

When the developmental time window closes and the system is responsive, it is because a priming level of eTh anti-NS has been reached.

Most immunologists treat Signal[2] as a "costimulatory" event determined by the APC. This cannot be substituted for the eTh in the activation of aT-cells without providing a whole new framework accounting for APC function, and at present this is lacking. As it is viewed today, "costimulation" as a source of Signal[2], is antigen-unspecific; "costimulation" is essentially innocent bystander activation, the

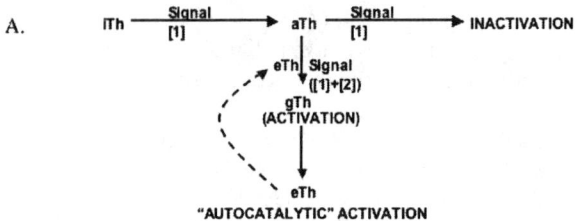

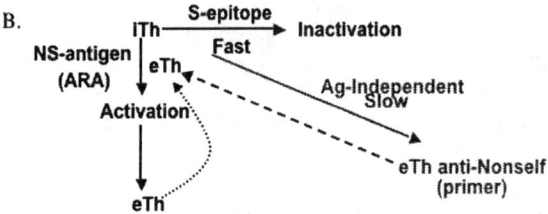

Fig. 17.4. A. Induction iTh to effectors (eTh). B.The antigen-independent pathway to eTh

requirement for ARA being totally ignored. Further, the APC is treated as functioning independently of whether the developmental time window is open or closed.

As the APC presents both S and NS peptides, it has no way to selectively activate anti-NS and delete anti-S. Therefore, it could not be a factor in Decision 1, the sorting of the repertoire. The repertoire must be presorted before antigen-unspecific signals like costimulation can play a role in directing responsiveness (i.e., initiating Decision 2).

Further, if the APC can present both S and NS, then it can present two NS-antigens. As each different antigen requires coherent and independent regulation of effector class [Cohn 2005a], processing multiple antigens to peptides that are displayed mixed and scrambled on the APC surface, would make impossible coherent and independent regulation of class for each antigen. Associative Recognition of antigen (ARA) is obligatory to a regulated response. This requires that the processed peptides from each antigen be kept together or linked for presentation in an eTh-APC-aTh interaction. I have discussed these problems elsewhere [Cohn 2005a, Cohn 2005b, Cohn 1992].

17.3.3 An aside on T-suppression

One cannot leave the question of the sorting of the Type II repertoire without commenting on the immensely popular subject of T-suppression usually referred to as regulatory T-cells (Tregs) without distinguishing T-help (an equally poor term) from Tsuppression. As would be true for any subject that barrels along without any conceptualization and with no rationalized role for its place in the regulation of responsiveness, the interpretations are made on an "I see-I believe" basis replete with optimistic emphasis on "the cure." I have discussed the role of feedback suppression [Cohn 2004], so that here I would simply like to point out that T-suppression on *a priori* grounds has no role to play in Decision 1, the sorting of the repertoire but it does have a key role in Decision 2 to control the magnitude of the response. As feedback is essential to a regulated response, a role for T-suppression is predictable. However, this implies that the iT-suppressor must be sorted like every other i-cell to be anti-Nonself, a point sharply disputed by the workers in this field [Picca & Caton 2005, Hsieh *et al.* 2004]. Suffice it to point out here that, if the T-suppressor is sorted to be anti-S, then it cannot play a role in Decision 2, the regulation of class. If it is sorted to be anti-NS, it cannot play a role in Decision 1, the sorting of the repertoire. The T-suppressor repertoire cannot be left unsorted and be functional. I have yet to see a heuristic model that either leaves the repertoire unsorted (i.e., no Decision 1) and determines the S-NS discrimination uniquely at the level of regulation of effector output, or that sorts the repertoire by feedback from the effector output, without the latter debilitating the host. Further, no model of the mechanism of sorting has been able to avoid a step of negative selection, not even models of so-called "dominant tolerance" because specificities directed against the cells of the immune system itself must be purged by deletion (apoptosis) for suppression to be functional.

A straightforward argument ruling out T-supression (Tregs) in making Decision 1, is that if it did have such a role, it would have to play that role via associative recognition of antigen. This would make the individual unresponsive ("tolerant") to the Self-of-the-species. If there is one fact all immunologists accept, it is that "what is Self for one individual is Nonself for another!"

17.3.4 A detail of Self-antigen presentation

The three categories of cell, T-helper (Th), T-cytotoxic (Tc) and B-cell respond to different ligands (epitopes). The T-helper recognizes [P-RII]; the cytotoxic T-cell recognizes [P-RI]; and the B-cell recognizes a "shape-patch" epitope. The MHC-encoded Class I restricting element (RI) is expressed by every cell, which means that most intracellular proteins are presented to cytotoxic T-cells (Tc). However, only a limited number of cells express MHC-encoded Class II restricting element (RII), which means that many host proteins are not seen directly as Self by the T-helper (Th), although some may be seen indirectly as "housekeeping" antigens taken up from effete and necrosing cells by APCs. The B-cell recognizes epitopes encountered on soluble components and cell surfaces. The three Self-ligand or epitopic repertoires

Presentation of a Self-component as:			Consequence of a challenge by Self
P-RII	P-RI	Shape-patch	
+	+	+	Unresponsive in all categories, Th, Tc and B
-	+	+	Unresponsive in Tc, B, but responsive in Th
+	-	-	Unresponsive in Th, but potentially responsive in Tc and B (see text)

Table 17.2. The effect of differential presentation of Self-antigens

of the Th-, Tc- and B-cell are different. While inactivation only requires interaction with epitope (Signal[1]), activation (Signal[1]+[2]) introduces an asymmetry because eTh is required to activate iTc and iB not vice versa. Any host antigens not presented on RII while the developmental window is open would be treated as Nonself by Th (Table 17.2).

As the response to all Self-ligands is eTh-dependent, a Self-component not presented on RII would not be inactivating for Th, which would treat the epitope as Nonself were it later presented as [P-RII]. This would occur if the individual were confronted with a Nonself-antigen that shared an epitope with the Self-antigen in question. Whether the state of unresponsiveness to that Self-component would be broken depends on several factors but most often the NS-antigen is ridded before autoimmunity becomes self-generating and debilitating. In any case, autoimmunity via this route occurs rarely, but does occur. This is the limit to evolutionary selection which never achieves perfection.

17.3.5 An important characteristic of Signal[1] inactivation

Inactivation upon receiving Signal[1] cannot be instantaneous. There must be an interim after interaction with an epitope during which the a-cell on the pathway to inactivation can be diverted to activation. If the interim (half-life) were too short, activation would be impossible because delivery of Signal[2] involving cell-cell interactions (eTh-APC-aT) is slow compared to epitope-cell interactions. The half-life for inactivation determines the steady state level of a-cells anti-Self on the pathway to inactivation. I refer to this as the autoimmune boundary [Cohn & Langman 1990]. The half life of inactivation must be short enough to make autoimmunity acceptably rare, yet long enough to permit activation.

17.3.6 In order to give the discussion balance............

Many immunologists [Zinkernagel & Hengartner 2004, Silverstein & Rose 1997, Miller & Basten 1996, Miller 2004] do not consider developmental time as a factor in determining the Self-Nonself discrimination (i.e., Decision 1). Their arguments have been analyzed elsewhere [Cohn 2001]. Suffice it to point out here, the experimental demonstration that one can induce a response while the developmental time window is open using procedures that provide a source of Signal[2] (e.g., LPS) does not challenge the Time Model; it supports it. Secondly, epitopes detectable by an immunologist that are generated after the window closes (e.g., idiotypic determinants) do not test the Time Model as they are non-immunogenic, in most cases obligatorily "tolerogenic" due to a lack of eTh that recognize them [Parks et al. 1978, Parks & Weigle 1980]. Third, the assumption [Kyewski & Derbinski 2004, Gotter & Kyewski 1994] that the sorting of the repertoire can be adequately accomplished by postulating a space (e.g., thymus) where **all** Self-antigens are expressed as [Ps-RII], does not obviate the Developmental Time Model. If any of these Self-antigens were expressed both in thymus and periphery after the window closed, the iTh anti-Self that accumulated extra-thymically while the window was open would initiate an autoimmune response when the delayed Self-antigen appeared in the periphery. Pure space Models are also constrained by developmental time. Lastly, it is **not** *"implicit in the"* Developmental time Model that there *"is a requirement for prenatal generation of the entire immune repertoire"* [Miller 2004]. This is a conceptual error, the history of which has been analyzed [Cohn 2001]. The state of unresponsiveness is maintained as a steady state process discussed both here and elsewhere [Cohn 2005b, Cohn 1992, Bretscher 1999, Bretscher 1972, Cohn et al. 2002, Bretscher & Cohn 1970].

17.4 A Summary of what has been Argued

First, the response of the immune system can be reduced to two processes:

- the sorting of the repertoire (Decision 1)
- the regulation of effector class (Decision 2)

Second, in order to sort a somatically-derived, random paratopic repertoire, a prior somatic sorting of the antigenic universe into Self and Nonself is required.

These two points are matters of principle. In order to extrapolate principle to mechanism, I generalized the Developmental Time Model to include those Self-antigens that appear after the time window closes and the system becomes responsive. For these delayed expression Self-antigens early ectopic expression in the thymus when the system is unresponsive, is required. The generalized Developmental Time Model defines:

- Self as prior and persistent
- Nonself as posterior and transient.

This sorting mechanism results in an individual that expresses a unique Self and permits us to understand why "What is Self for one individual is Nonself for another."

Given this, I then considered how the somatic repertoire might respond to the sorted antigenic universe.

The purging of anti-Self from the repertoire leaving the residue as anti-Nonself requires that the initial state or i-cell be able to enter one of two pathways, inactivation or activation. Two pathways require two signals and I assigned Signal[1] via the antigen-receptor to the inactivation pathway and Signals ([1]+[2]) to the activation pathway. The source of Signal[2] is the effector T-helper (eTh), thus making this cell the key regulator of responsiveness. For any given antigen, the response depends on the sufficiency or insufficiency of eTh specific for that antigen. All regulation of responsiveness in one way or the other passes via the eTh.

This introduced the problem of the origin of eTh (the primer question) and I proposed a pathway unique to it namely an antigen-independent conversion of iTh to eTh anti-NS.

Now I would like to illustrate these principles by referring to several classic experiments and reinterpreting them in the light of the present framework.

17.5 A Reinterpretation of Illustrative Experiments

17.5.1 Experiment 1. [Ohki *et al.* 1987, Coutinho *et al.* 1993, Le Douarin *et al.* 1996]

A quail limb bud was grafted onto a chicken embryo before any i-cells appeared in the system (i.e., when the developmental time window was open). My expectation would have been that the chicken would be born with a quail wing that would be accepted throughout life as a Self or NTBR host component. This is not what happens. The chicken is born with a healthy integrated quail wing but shortly after birth an acute rejection reaction occurs. In the framework of the generalized Time Model, the only explanation would be that the quail limb bud expresses a new NTBR- or Self-component after the developmental window has closed and the immune system is responsive.

If this reinterpretation has any validity then we must ask, why isn't the quail limb rejected in quail? We have two possible answers. The postulated quail component that is delayed in expression in chick either is expressed early in quail while the window is open or is equally delayed in quail but ectopically expressed in the quail thymus while the window is open (the generalized Developmental Time Model).

What experiment distinguishes the two possibilities? A graft of the embryonic quail thymic epithelium onto chick results in a chimeric quail-chick thymus. In such an animal, if the quail limb were rejected, then abnormal delayed expression of the quail component in chick would be validated. If it is accepted then ectopic expression in thymus of the delayed component is the explanation. It turns out that the quail limb is accepted. Therefore, ectopic expression in thymus is the favored interpretation.

It should be stressed that delayed expression for T-helpers means delayed functional presentation of peptide (P) on Class II MHC-encoded restricting elements (RII), not the mere presence of the component.

Further, and as an aside, the ectopic expression of the delayed antigen on quail thymic epithelium resulting in graft acceptance means that chick can see a quail [P-RII] complex. Other examples of allele-specific recognition of Xeno-MHC by a given TCR locus are known [Swain et al. 1983]. This has important implications for the pathway of speciation [Cohn & Mata 2006].

One might justifiably in the framework of the generalized Time Model cite these experiments as the first demonstration of the ectopic expression of peripheral Self-antigens in the thymus. The discovery of Aire controlled ectopic expression [Su & Anderson 2004, Anderson et al. 2002] confirmed these studies in that it gave us one among several mechanisms for ectopic thymic expression [Gillard & Farr 2005]. Under the generalized Time Model, all peripheral Self-antigens that appear after the window closes would be expected to be ectopically expressed in thymus while the window was open.

17.5.2 Experiment 2 [Thomas-Vaslin et al. 1987]

This experiment deals with a mutant mouse (NOD, non-obese diabetic) that spontaneously develops after birth an autoimmune disease due to an attack on a specific antigen expressed on the β-cells of the pancreas resulting in diabetes. Under the generalized Time Model, the mutant NOD mouse is expected to express a delayed component of β-cells that fails to be appropriately ectopically expressed in the thymus (i.e., as a [P-RII] complex). Given a failure to ectopically and normally express the β-cell component in thymus, it is predictable that a graft of mutant thymus onto a wild type syngeneic normal mice would result in diabetes. The failure of NOD thymus to ectopically express the peripheral target that appears after the time window closes would make a peripheral Self-component that is delayed in expression indistinguishable from a Nonself-component.

This is what happens. The NOD thymic epithelium grafted onto an athymic syngeneic mouse results in detectable autoimmunity. If NOD is a mutation that deletes ectopic expression in thymus of a delayed expression β-cell antigen, then the otherwise normal animal would suffer autoimmunity.

These two experiments are complementary. In the quail-chick experiment [Ohki *et al.* 1987, Coutinho *et al.* 1993, Le Douarin *et al.* 1996], the graft of thymic epithelium added a Self-component to the Self of the chick, thereby permitting the acceptance of a xenograft. In the NOD mouse experiment [Thomas-Vaslin *et al.* 1987] the graft of mutant thymic epithelium subtracted a Self-component from the Self of the wild-type mouse triggering an autoimmune attack. In the framework of the Developmental Time Model, these two experiments established the important role of appropriate ectopic thymic expression of delayed appearing peripheral Self.

17.5.3 Experiment 3 [Adams *et al.* 1987]

Two transgenic murine lines were studied that express the T-antigen of the SV40 virus in the β-cells of the pancreas. One line expresses the T-antigen early in embryonic life while the developmental time window is open (eTh$^-$). The other line expresses the T-antigen delayed until after the window is closed (eTh$^+$) and the immune system is responsive. The early expressor treats the T-antigen as Self (NTBR) and there is no immune response to it. The late expressor treats the T-antigen as Nonself (TBR) and the response to it results in destruction of the β-cells and diabetes.

If the T-antigen provokes a response when expressed delayed peripherally, it could not have been appropriately expressed in thymus as a negative selector of iTh or as a positive selector of iT-suppressors (iTsu). Given that it is not functionally expressed in thymus, then the unresponsive state in the early expressor cannot be due to suppression; it must be due to deletion of iTh anti-T-antigen in the periphery when the window was open (i.e., an insufficiency of eTh). Of course, it is possible to argue that the early expressor is functionally expressed in thymus, whereas the late expressor, in addition to being late, is in some way also defective in thymic expression. Although unlikely, this experiment is admittedly incomplete. Nevertheless, a straightforward interpretation of it is that the transgene is not **functionally** expressed in thymus and, therefore, a deletional mechanism of peripheral unresponsiveness exists.

17.5.4 Experiment 4 [Lafaille *et al.* 1994]

Two murine lines expressing as a transgene a Class II MHC-restricted TCR anti-'a-peptide-from-myelin-basic-protein (Pmbp)' were established. One line referred to as T/R$^+$ expresses the endogenous TCR and BCR loci in addition to the transgenic TCR (T). The other line referred to as T/R$^-$, due to a Rag$^-$ mutation, cannot express the endogenous TCR and BCR loci. It only expresses the transgene (T).

The T/R$^-$ line at 5-7 weeks after birth begins to develop EAE (experimental allergic encephalitis) due to an attack by eTh expressing the transgene on the myelin basic protein (MBP) in the central nervous system. By 40 weeks, 100% of the T/R$^-$ mice have succumbed to EAE.

The T/R$^+$ line expresses a much lower incidence of EAE, which reaches 14% of mice in 40 weeks and is initiated somewhat later, around 10-12 weeks.

In order to discuss this experiment the role of MBP must be addressed. Given the existence of EAE, the MBP is in one sense obviously accessible to the immune system. However, because it is behind the blood-brain barrier, the effector T-helpers (eTh) must breach the barrier and this contributes to the delay in onset after birth of 5-7 weeks.

As negative selection is not detectable, functional ectopic expression of MBP in thymus is unlikely. Given this, if MBP were viewed as a Self-antigen, then the induction of regulatory T-suppressors anti-MBP would also be ruled out. Further, deletion in the periphery while the window was open, is also unlikely. MBP is sequestered behind the blood-brain barrier and does not act normally as a deletional peripheral Self-component. MBP behaves as Nonself to the immune system. For this reason, iT- and iB-cells are present in the periphery at the same level as any other i-cell anti-NS. There is no response normally because MBP is not presented to the immune system. There are host components that are sequestered behind barriers that prevent interaction with the immune system (e.g., the eye, parts of the nervous system, the fetus and most intracellular constituents) (Table 17.3). When the developmental time window is open, these components do not interact with the immune system to negatively select (or to induce suppression) and, therefore, are indistinguishable from Nonself, and, if rendered available, are treated as such.

Consequently, Self is defined by the immune system as a Not-To-Be-Ridded host component that was encountered when the developmental window was open and that persists. In short, once again, Self is prior and persistent; Nonself is posterior and transient.

Now let's face several questions that have not been addressed regarding this experiment.

1 – The TCR transgene is expressed by i-cells with no effector function. How does the induction of the iTh to eTh get started? How are the iTh anti-[Pmbp-RII] induced to effectors (eTh anti-[Pmbp-RII]) in T/R$^-$ (Rag$^-$)? Here we have a clear hint of an antigen-independent pathway to primer eTh.

2 – Why does it take close to 7 weeks after birth for T/R$^-$ mice to show any signs of EAE?

There are several possible additive reasons:

Barrier	Nonself Component
Blood-brain	parts of nervous system [Yan *et al.* 2003]
Blood-eye	parts of the eye [Streilein 1999, McKenna & Kapp 2004]
	[Streilein 2003]
Blood-placenta	fetus of placental mammals [Mellor *et al.* 2002]
	[Poole & Claman 2004, Koch & Platt 2003]
Blood-mammary gland	milk of the lactating mammal
Cell membrane	- intracellular constituents that are isolated
	from interacting with iB-cells and antibody
	- host constituents not presented as
	[P-RII] are ignored by iTh but if presented
	as [P-RII] to that individual, they would
	be responded to as Nonself

Table 17.3. Categories of host-encoded Nonself-antigens because they are isolated by a barrier

First, it takes time to reach an effective eTh primer level, but because the entire iTh population is anti-MBP, the primer eTh anti-MBP level would be abnormally high and might itself (without "autocatalytic" induction) be capable of initiating autoimmunity.

Second, the blood-brain barrier must be breached [Yan *et al.* 2003]. To do this, endothelial cells must express [Pmbp-RII] from MBP scavenged from necrosing glia. The primer eTh might be at a sufficient level to breach this barrier.

Third, it takes time to do revealable nervous system damage once the blood-brain barrier is reached.

3 – Why is EAE less frequent in T/R^+ than in T/R^-?

There are many possible reasons, among them:

— dilution of the transgenic TCR by endogenous TCR expression. Therefore, the effective level of primer eTh from the transgene will be much lower. The result is a less effective breaching of the blood-brain barrier and EAE is initiated poorly.

— induction of T-suppressors (iTsu) in T/R^+, not T/R^-. It does not appear that the thymus expresses MBP in T/R^- and is unlikely to do so in T/R^+. However, if like all other T-cells (iTh, iTc), the iTsu leave the thymus as anti-Nonself, and if MBP behaves as Nonself, then the subliminal autoimmunity in T/R^+ might provide a source of MBP for presentation by APC and the subsequent induction of iTsu to eTsu specific for MBP in the periphery. The role of eTsu anti-MBP would be to reduce the magnitude of the eTh1 response to an ineffective level and maintain it low. This would be an effect at the level of Decision 2, not Decision 1.

— A switch from eTh1 to eTh2. I haven't discussed Decision 2 regulation but for completeness, eTh2 is less effective (not zero) than eTh1 in provoking EAE [Lafaille *et al.* 1997]. The response in T/R^- might be eTh1, whereas in T/R^+ it might be eTh2.

4 - Why doesn't T/R^- develop suppression?

Under the standard assumption that T-suppressors (Tsu) are a special lineage selected in thymus to be anti-Self, suppression is not expected either in T/R^- or T/R^+ because MBP does not appear to be functionally ectopically expressed in thymus. The generalization from this assumption would be that there are non-thymically expressed peripheral Self-antigens, the response to which cannot be regulated by suppression. MBP treated as a "Self-antigen" would have to be viewed as an example. The transgenic line of early expressed SV40T-antigen, discussed as Experiment 3, would also have to be viewed as another example. Consequently, I would expect that this popular assumption that Tsu are thymically selected to be anti-Self will meet demise.

Under the theory suggested here, namely that Tsu are a special lineage positively selected in thymus to be anti-Nonself like all other T-cells, then MBP would be viewed as Nonself by the immune system. It would not be attacked because it is sequestered and not normally expressed as a functional [Pmbp-RII] ligand outside of the blood-brain barrier.

If Tsu are sorted to be anti-NS, they cannot contribute to Decision 1, the sorting of the repertoire. They must function at the level of Decision 2 as a feedback mechanism regulating the magnitude of the effector response.

17.5.5 Experiment 5 [Avrameas 1991]

A normal individual is under a steady state antigenic load that engages roughly 10% of the total paratopic repertoire. Analyses of the specificities engaged shows that most of them are directed at host epitopes. This has engendered a flurry of views that range from a putative disproof of the Developmental Time Model, to a denial of a protective role of the immune system accompanied by a semantic argument as to the meaning of Self. Some of the debate has been productive; most of it has been derouting and sterile.

During fetal development, when the time window is open, cells die by apoptosis. The resultant apoptotic granules that encapsulate the intracellular contents are phagocytosed without their contents being exposed to the newly arising i-cells. When the window closes around birth and pathogens enter the system, cells die by necrosis releasing their intracellular contents to immune attack. The immune system treats as Nonself all autogenously generated host waste (e.g., necrosing or senescing cells, denatured or effete protein) and rids it. I have referred to this as a housekeeping function [Cohn 1986]. Because it is autogenously generated does not make it Self (NTBR) to the immune system. The biodestructive and ridding response to it is salutary as it would be to any Nonself-antigen (TBR). No principle of the Developmental Time Model is violated by this finding. It is expected.

17.6 Concept should guide Computer Analyses

It is an unfortunate but obvious fact that theoretical studies in immunology have had little impact on the experimentalist. The field progresses by formulating mini-models based on inductive extrapolation from "I believe what I see," lacking as a rule, any consideration of a generalizing principle. As expected of a subject driven by crass empiricism, it is replete with dead-ends, wasteful experimentation and questionable assumptions. The theoretical immunologists should ask themselves, why, unlike the theoretical physicist, is there an absence of influence of their output on the direction and development of the field.

For some 15 years now we have been writing small computer programs available on our website (www.cig.salk.edu) that allow analysis of the output of a conceptualized system. The operator can choose the values of the parameters defined by the theory to test their validity. For example:

- The humoral output of a unit immune system (the B-Protecton) [Cohn & Langman 1990, Cohn 1997, Langman & Cohn 1993b, Langman 2000]
- The configuration of rearranged TCR and BCR loci (haplotype exclusion) [Langman & Cohn 1993b, Langman & Cohn 2002a, Langman & Cohn 1993a]
- The cell-mediated output of a unit immune system based on the Tritope Model (TUNI I [Mata & Cohn 2006b]).
- The antigen-independent pathway of iTh to eTh [Bretscher 1972, Cohn *et al.* 2002]

These have been programs to analyze small segments of the immune system. They have been theory-dependent.

Recently we have attempted to link the programs (Synthetic Immune system, SIS) so that they can deal with a larger segment of the response (SIS I) [Mata & Cohn 2006a]. This led us to consider the ideal.

The ideal would be a program that is theory-independent. This means that the program should allow the operator to state the theory and use it to find optimal or acceptable values of the parameters consistent with the known behavior of the output of the system. A first approach to this goal is SIS II [Mata & Cohn 2006a]. Unfortunately, the more one hardwires (makes theory-dependent), the simpler the program; the less one hardwires (makes theory-independent), the more complex and cumbersome the program and the less user-friendly it becomes. Nevertheless, SIS III with theory independence as its goal, is slowly emerging.

At the moment, most generalizations about immune responsiveness are stymied by the lack of an adequate theory of the regulation of class (Decision 2). Decision 1, the sorting of the repertoire has a solid conceptual basis. Decision 2 is in need of a set of principles to guide experiment. This is where the real vacuum exists and where more effort should go. Here computer modeling will be precious.

Lastly, a good theory challenges the observation with the same validity that an observation challenges the theory. Not all observations are what they seem to be. Reinterpreting them is an important function of theory. Similarly and conversely, as theoreticians, we must be prepared to change our minds. In fact we should lead the way.

Acknowledgements

This work was supported by a grant (RR07716) from the National Center for Research Resources at the National Institutes of Health. This paper was written while Melvin Cohn was a visiting scholar at the Gulbenkian Science Institute, Oeiras, Portugal, April-July 2005. The fellowship support from the Ministério Da Ciência E Da Tecnologia is gratefully acknowledged as is the encouragement and criticism of the Director, Professor Antonio Coutinho.

A List of Abbreviations

ARA   associative recognition of antigen
APC   antigen presenting cell
iTh   initial state T-helper cell
aTh   anticipatory T-helper cell
eTh   effector T-helper cell
S   Self
NS   Nonself
NTBR   not-to-be-ridded
TBR   to-be-ridded
MHC   major histocompatibility complex
RI   class I restricting element
RII   class II restricting element
iTsu   initial state T-suppressor cell
eTsu   effector T-suppressor cell
MBP   myelin basic protein
EAE   experimental allergic encephalitis
TCR   T-cell antigen-receptor
BCR   B-cell antigen-receptor

References

[Abbas *et al.* 2000]   A. K. Abbas, A. H. Lichtman, and J. S. Pober. *Cellular and Molecular Immunology*. W. B. Saunders Company, 2000.

[Abraham & Malaviya 1997]   S. Abraham and R. Malaviya. Mast cells in infection and immunity. *Infection and Immunity*, 65(9):3501–3508, 1997.

[Adams & Koziol 1995]   H. P. Adams and J. A. Koziol. Prediction of binding to MHC class I molecules. *J. Immunol. Methods*, 185:181–190, 1995.

[Adams *et al.* 1987]   T. E. Adams, S. Alpert, and D. Hanahan. Non-tolerance and autoantibodies to a transgenic self antigen expressed in pancreatic β cells. *Nature*, 325:223–228, 1987.

[Adams *et al.* 1995]   L. B. Adams et al. Exacerbation of acute and chronic murine tuberculosis by administration of a tumor-necrosis-factor receptor-expressing adenovirus. *J. Infect. Dis.*, 171(2):400–405, 1995.

[Adorini *et al.* 1988]   L. Adorini, S. Muller, F. Cardinaux, P. V. Lehmann, F. Falcioni, and Z. A. Nagy. In vivo competition between self peptides and foreign antigens in T-cell activation. *Nature*, 344(6183):623–625, 1988.

[Agrawal & Linderman 1996]   N. G. Agrawal and J. J. Linderman. Mathematical modeling of helper T lymphocyte/antigen-presenting cell interactions: analysis of methods for modifying antigen processing and presentation. *J. Theor. Biol.*, 182(4):487–504, 1996.

[Aguilar *et al.* 2004]   A. Aguilar, G. Roemer, S. Debenham, M. Binns, D. Garcelon, and R. K. Wayne. High MHC diversity maintained by balancing selection in an otherwise genetically monomorphic mammal. *Proc. Natl. Acad. Sci. USA*, 101(10):3490–3494, 2004.

[Ahmed & Hashish 2003]   E. Ahmed and A. H. Hashish. On modelling of immune memory mechanisms. *Theory Biosci.*, 122:339–342, 2003.

[Aickelin *et al.* 2003]   U. Aickelin, P. Bentley, S. Cayzer, J. Kim, and J. McLeod. Danger theory: The link between AIS and IDS? In [Timmis *et al.* 2003], pages 147–155.

[Ainslie *et al.* 1992]   G. M. Ainslie, J. A. Solomon, and E. D. Bateman. Lymphocyte and lymphocyte subset numbers in blood and in bronchoalveolar lavage and pleural fluid in various forms of human pulmonary tuberculosis at presentation and during recovery. *Thorax*, 47(7):513–518, 1992.

[Altuvia & Margalit 2000]   Y. Altuvia and H. Margalit. Sequence signals for generation of antigenic peptides by the proteasome: implications for proteasomal cleavage mechanism. *J. Molec. Biol.*, 295:879–890, 2000.

[Alves *et al.* 2004]   R. T. Alves, M. R. Delgado, H. S. Lopes, and A. A. Freitas. An artificial immune system for fuzzy-rule induction in data mining. volume 3242, pages 1011–1020, 2004.

[Amaro *et al.* 1995]   J. D. Amaro, J. G. Houbiers, J. W. Drijfhout, R. M. Brandt, R. Schipper, J. N. Bavinck, C. J. Melief, and W. M. Kast. A computer program for predicting possible cytotoxic T lymphocyte epitopes based on HLA class I peptide-binding motifs. *Hum. Immunol.*, 43:13–18, 1995.

[Anderson & Matzinger 2000a]   C. C. Anderson and P. Matzinger. Anderson and Matzinger: Round 2. *Seminars in Immunology*, 12(3):277–292, 2000.

[Anderson & Matzinger 2000b]   C. C. Anderson and P. Matzinger. Danger: the view from the bottom of the cliff. *Seminars in Immunology*, 12(3):231–238, 2000.

[Anderson *et al.* 2002]   M. S. Anderson, E. Venanzi, L. Klein, Z. Chen, S. Berzins, S. Turley, H. von Boehmer, R. Bronson, A. Dierich, C. Benoist, and D. Mathis. Projection of an immunological self shadow within the thymus by the aire protein. *Science*, 298(5597):1395–1401, 2002.

[Andrews & Timmis 2005]   P. S. Andrews and J. Timmis. On diversity and artificial immune systems: Incorporating a diversity operator into aiNet. In *Proceedings of the International Conference on Natural and Artificial Immune Systems (NAIS05)*, volume 391 of *LNCS*, pages 293–306. Springer, 2005.

[Andrews & Timmis 2006]   P. S. Andrews and J. Timmis. A computational model of degeneracy in a lymph node. In *To appear in 5th International Conference on Artificial Immune Systems (ICARIS)*, 2006.

[Anfinsen 1973]   C. B. Anfinsen. Principles that govern the folding of protein chains. *Science*, 181:223–230, 1973.

[Antia *et al.* 1998]   R. Antia, S. S. Pilyugin, and R. Ahmed. Models of immune memory: On the role of cross-reactive stimulation, competition, and homeostasis in maintaining immune memory. *Proc. Natl. Acad. Sci. USA*, 95:14926–14931, 1998.

[Antia *et al.* 2005]   R. Antia, V. V. Ganusov, and R. Ahmed. The role of models in understanding $CD8^+$ T-cell memory. *Nature Review of Immunology*, 5:101–111, 2005.

[Aoki 1980]   K. Aoki. A criterion for the establishment of a stable polymorphism of higher order with an application to the evolution of polymorphism. *J. Math. Biol.*, 9:133–146, 1980.

[Apanius *et al.* 1997]   V. Apanius, D. Penn, P. R. Slev, L. R. Ruff, and W. K. Potts. The nature of selection on the major histocompatibility complex. *Crit. Rev. Immunol.*, 17:179–224, 1997.

[Apostolopoulos *et al.* 2001]   V. Apostolopoulos, I. F. McKenzie, and I. A. Wilson. Getting into the groove: unusual features of peptide binding to MHC class I molecules and implications in vaccine design. *Front. Biosci.*, 6(D):1311–1320, 2001.

[Armstrong & Hart 1971]   J. A. Armstrong and P. D. Hart. Response of cultured macrophages to mycobacterium tuberculosis, with observations on fusion of lysosomes and phagosomes. *J. Exp. Med.*, 134:713–740, 1971.

[Arnold *et al.* 2002]   P. Y. Arnold, N. L. La Gruta, T. Miller, K. M. Vignali, P. S. Adams, D. L. Woodland, and D. A. Vignali. The majority of immunogenic epitopes generate CD4+ T cells that are dependent on MHC class II bound peptide-flanking residues. *J. Immunol.*, 169:739–749, 2002.

[Asachenkov 1994]   A. Asachenkov. *Disease Dynamics.* Birkhauser, 1994.

[Atlan & Cohen 1998]   H. Atlan and I. R. Cohen. Immune information, self-organization and meaning. *International Immunology*, 10(6):711–717, 1998.

[Avrameas 1991]   S. Avrameas. Natural autoantibodies: From 'horror autotoxicus' to 'gnothi seauton'. *Immunology Today*, 12:154–159, 1991.

[Ayara 2005]   M. Ayara. *An immune inspired solution for adaptable error detection in embedded systems.* PhD thesis, University of Kent, 2005.

[Baas & Gao 1999]   A. Baas and X. Gao. Peptide binding motifs and specificities for HLA-DQ molecules. *Immunogenetics*, 50(1-2):8–15, 1999.

[Bachmaier *et al.* 1999]   K. Bachmaier, N. Neu, L. M. De la Maza, S. Pal, A. Hessel, and J. M. Penninger. *Chlamydia* infections and heart disease linked through antigenic mimicry. *Science*, 283:1335–1339, 1999.

[Bader & Hogue 2000]   G. D. Bader and C. W. Hogue. BIND–a data specification for storing and describing biomolecular interactions, molecular complexes and pathways. *Bioinformatics*, 16:465–477, 2000.

[Baggiolini 1998]   M. Baggiolini. Chemokines and leukocyte traffic. *Nature*, 392(6676):565–568, 1998.

[Bak 1997]   P. Bak. *How Nature Works: the science of self-organized criticality.* Oxford University Press, 1997.

[Balasubramanian *et al.* 2005]   S. Balasubramanian, Y. Xia, E. Freinkman, and M. Gerstein. Sequence variation in G-protein-coupled receptors: analysis of single nucleotide polymorphisms. *Nucleic. Acids. Res.*, 33(5):1710–1721, 2005.

[Baldwin *et al.* 1999]   K. K. Baldwin, B. P. Trenchak, J. D. Altman, and M. M. Davis. Negative selection of T cells occurs throughout thymic development. *J. Immunol.*, 163:689–698, 1999.

[Balkwill 2000]   F. Balkwill, editor. *The Cytokine Network.* Oxford University Press, 2000.

[Bankovich *et al.* 2004]   A. J. Bankovich, A. T. Girvin, A. K. Moesta, and K. C. Garcia. Peptide register shifting within the MHC groove: theory becomes reality. *Mol Immunol*, 40:1033–1039, 2004.

[Banzhaf *et al.* 1998]   W. Banzhaf, P. Nordin, R. Keller, and F. Francome. *Genetic Programming: An Introduction.* Morgan Kauffman, 1998.

[Barouch *et al.* 1995]   D. Barouch, T. Friede, S. Stevanovic, L. Tussey, K. Smith, S. Rowland-Jones, V. Braud, A. McMichael, and H. G. Rammensee. HLA-A2 subtypes are functionally distinct in peptide binding and presentation. *J. Exp. Med.*, 182:1847–1856, 1995.

[Baum 1986]   E. B. Baum. Intractable computations without local minima. *Phys. Rev. Lett.*, 57:2764–2767, 1986.

[Bazzani *et al.* 2003]   A. Bazzani, D. Remondini, N. Intrator, and G.C. Castellani. The effect of noise on a class of energy-based learning rules. *Neural Computing*, 15:1621–1640, 2003.

[Bean *et al.* 1999]   A. G. D. Bean et al. Structural deficiencies in granuloma formation in TNF gene-targeted mice underlie the heightened susceptibility to aerosol mycobacterium tuberculosis infection, which is not compensated for by lymphotoxin. *J. Immunol.*, 162(6):3504–3511, 1999.

[Beck 1984]   K. Beck. Coevolution: mathematical analysis of host–parasite interactions. *J. Math. Biol.*, 19:63–77, 1984.

[Beer 1995]   R. D. Beer. A dynamical systems perspective on agent-environment interaction. *Artificial Intelligence*, 72:173–215, 1995.

[Beeson & McConnell 1995]   C. Beeson and H. M. McConnell. Reactions of peptides with class-II proteins of the major histocompatibility complex. *Journal of the American Chemical Society*, 117(42):10429–10433, 1995.

[Bellamy & Hill 1998]   R. Bellamy and A. Hill. Genetic susceptibility to mycobacteria and other infectious pathogens in humans. *Curr. Op. Immunol.*, 10:483–487, 1998.

[Bellamy et al. 1998]   R. Bellamy, C. Ruwende, T. Corrah, K. P. McAdam, H. C. Whittle, and A. V. Hill. Variations in the NRAMP1 gene and susceptibility to tuberculosis in West Africans. *N. Engl. J. Med.*, 338(10):640–4., 1998.

[Beltman et al. 2002]   J. B. Beltman, J. A. M. Borghans, and R. J. De Boer. Major Histocompatibility Complex: Polymorphism from coevolution. In U. Dieckmann, J. A. J. Metz, M. W. Sabelis, and K. Sigmund, editors, *Adaptive Dynamics of Infectious Diseases. In Pursuit of Virulence Management*, pages 210–221. Cambridge University Press, 2002.

[Benedict 1990]   P. Benedict.   Data generation program/2 v1.0. http://ftp.ics.uci.edu/pub/machine-learning-databases/dgp-2/, 1990. Inductive Learning Group, Beckman Institute for Advanced Technology and Sciences, University of Illinois at Urbana.

[Bentley & Timmis 2004]   P. J. Bentley and J. Timmis. A fractal immune network. In [Nicosia et al. 2004], pages 133–145.

[Bentley et al. 2005]   P. J. Bentley, J. Greensmith, and S. Ujjin. Two ways to grow tissue for Artificial Immune Systems. In [Jacob et al. 2005], pages 139–152.

[Bentley 2004]   P. J. Bentley. Fractal proteins. *Genetic Programming and Evolvable Machines*, 5(1):71–101, 2004.

[Berek & Ziegner 1993]   C. Berek and M. Ziegner. The maturation of the immune response. *Immunology Today*, 14:200–402, 1993.

[Bernasconi et al. 2002]   N. L. Bernasconi, E. Traggiai, and A. Lanzavecchia. Maintenance of serological memory by polyclonal activation of human memory B-cells. *Science*, 298:2199–2202, 2002.

[Bersini & Varela 1994]   H. Bersini and F. Varela. *The Immune Learning Mechanisms: Recruitment, Reinforcement and their Applications.* Chapman Hall, 1994.

[Bersini 1991]   H. Bersini. Immune network and adaptive control. In *Proceedings of the 1st European Conference on Artificial Life (ECAL)*, pages 217–226. MIT Press, 1991.

[Bersini 1992]   H. Bersini. Reinforcement and recruitment learning for adaptive process control. In *Proc. Int. Fuzzy Association Conference (IFAC/IFIP/IMACS) on Artificial Intelligence in Real Time Control*, pages 331–337, 1992.

[Bersini 2002]   H. Bersini. Self-assertion versus self-recognition. In [Timmis & Bentley 2002], pages 107–112.

[Bertoni & Sidney 1997]   R. Bertoni and J. Sidney. Human histocompatibility leukocyte antigen-binding supermotifs predict broadly cross-reactive cytotoxic T lymphocyte responses in patients with acute hepatitis. *J. Clin. Invest.*, 100(3):503–515, 1997.

[Bezerra et al. 2005]   G. Bezerra, T. Barra, L. N. de Castro, and F. Von Zuben. Adaptive Radius Immune Algorithm for Data Clustering. In [Jacob et al. 2005], pages 290–303.

[Bhasin & Singh 2003]   M. Bhasin and H. Singh.   MHCBN: a comprehensive database of MHC binding and non-binding peptides. *Bioinformatics*, 19(5):665–666, 2003.

[Bienenstock *et al.* 1982]   E. L. Bienenstock, L. N. Cooper, and P. W. Munro. Theory for the development of neuron selectivity: Orientation specificity and binocular interaction in visual cortex. *J. Neurosci*, 2:32–48, 1982.

[Bindewald *et al.* 1998]   E. Bindewald, J. Hesser, and R. Männer. Implementing genetic algorithms with sterical constrains for protein structure prediction. In *Proc. International Conference on Parallel Problem Solving from Nature*, pages 959–967, Amsterdam, Netherland, 1998.

[Binley *et al.* 2004]   J. M. Binley et al. Comprehensive cross-clade neutralization analysis of a panel of anti-human immunodeficiency virus type 1 monoclonal antibodies. *J. Virol.*, 78:13232–13252, 2004.

[Bisset & Fierz 1993]   L. R. Bisset and W. Fierz. Using a neural network to identify potential HLA-DR1 binding sites within proteins. *J. Mol. Rec.*, 6:41–48, 1993.

[Bjorkman *et al.* 1987]   P. J. Bjorkman, M. A. Saper, B. Samraoui, W. S. Bennett, J. L. Strominger, and D. C. Wiley. Structure of the human class I histocompatibility antigen, HLA-A2. *Nature*, 329:506–512, 1987.

[Bleed *et al.* 2001]   D. Bleed et al. World health report 2001: Global tuberculosis control, 2001.

[Blythe & Flower 2005]   M. J. Blythe and D. R. Flower. Benchmarking B cell epitope prediction: Underperformance of existing methods. *Protein Sci.*, 14:246–248, 2005.

[Blythe *et al.* 2002]   M. J. Blythe, I. A. Doytchinova, and D. R. Flower. Jenpep, a database of quantitative functional peptide data for immunology. *Bioinformatics*, 18:434–439, 2002.

[Bodmer 1972]   W. F. Bodmer. Evolutionary significance of the HL-A system. *Nature*, 237:139–145, 1972.

[Boehm *et al.* 1997]   U. Boehm, T. Klamp, M. Groot, and J. C. Howard. Cellular responses to interferon-gamma. *Ann. Rev. Immunol.*, 15:749–95, 1997.

[Bogarad & Deem 1999]   L. D. Bogarad and M. W. Deem. A hierarchical approach to protein molecular evolution. *Proc. Natl. Acad. Sci. USA*, 96:2591–2595, 1999.

[Bohm *et al.* 1999]   M. Bohm, J. Sturzebecher, and G. Klebe. Three-dimensional quantitative structure-activity relationship analyses using comparative molecular field analysis and comparative molecular similarity indices analysis to elucidate selectivity differences of inhibitors binding to trypsin, thrombin, and factor Xa. *J. Med. Chem.*, 42:458–477, 1999.

[Bonecini-Almeida 1998]   M. G. Bonecini-Almeida. Induction of in vitro human macrophage anti-mycobacterium tuberculosis activity: requirement for IFN-gamma and primed lymphocytes. *J. Immunol.*, 160(9):4490–9, 1998.

[Bonnett 2005]   C. Bonnett. The role of the recruited neutrophil in the innate response to aspergillus fumigatus. Master's thesis, Montana State University, USA, 2005.

[Borghans & De Boer 2001]   J. A. M. Borghans and R. J. De Boer. Diversity in the immune system. [Cohen & Segal 2001], pages 161–183.

[Borghans & De Boer 2002]   J. A. M. Borghans and R. J. De Boer. Memorizing innate instructions requires a sufficiently specific adaptive immune system. *Int. Immunol.*, 14:525–532, 2002.

[Borghans *et al.* 1999]   J. A. M. Borghans, A. J. Noest, and R. J. De Boer. How specific should immunological memory be? *J. Immunol.*, 163:569–575, 1999.

[Borghans et al. 2003] J. A. M. Borghans, A. J. Noest, and R. J. De Boer. Thymic selection does not limit the individual MHC diversity. *Eur. J. Immunol.*, 33:3353–3358, 2003.

[Borghans et al. 2004] J. A. M. Borghans, J. B. Beltman, and R. J. De Boer. MHC polymorphism under host-pathogen coevolution. *Immunogenetics.*, 55:732–739, 2004.

[Botha & Ryffel 2003] T. Botha and B. Ryffel. Reactivation of latent tuberculosis infection in tnf-deficient mice. *J. Immunol.*, 171(6):3110–3118, 2003.

[Bothamley et al. 1989] G. H. Bothamley et al. Association of tuberculosis and M. tuberculosis-specific antibody levels with HLA. *J. Infect. Dis.*, 159(3):549–555, 1989.

[Bothamley et al. 1993] G. H. Bothamley et al. Association of antibody responses to the 19-kDa antigen of Mycobacterium tuberculosis and the HLA-DQ locus. *J. Infect. Dis.*, 167(4):992–993, 1993.

[Bowie et al. 1991] J. U. Bowie, R. Luthy, and D. Eisenberg. A method to identify protein sequences that fold into a known three-dimensional structure. *Science*, 253:164–170, 1991.

[Brahmajothi 1991] V. Brahmajothi. Association of pulmonary tuberculosis and HLA in south india. *Tubercle*, 72(2):123–32, 1991.

[Brazdil et al. 1991] P. Brazdil, M. Gams, S. Sian, L. Torgo, and W. van de Velde. Learning in distributed systems and multi-agent environments. volume 482 of *LNAI*, pages 412–423. Springer, 1991.

[Bretscher & Cohn 1970] P. Bretscher and M. Cohn. A theory of self-nonself discrimination. *Science*, 169:1042–1049, 1970.

[Bretscher 1972] P. Bretscher. The control of humoral and associative antibody synthesis. *Transplant. Rev.*, 11:217–267, 1972.

[Bretscher 1999] P. Bretscher. A two-step, two-signal model for the primary activation of precursor helper T cells. *Proc. Natl. Acad. Sci. USA*, 96:185–190, 1999.

[Bretscher 2000] P. Bretscher. Contemporary models for peripheral tolerance and the classical 'historical postulate'. *Seminars in Immunology*, 12(3):221–229, 2000.

[Brookes 2003] R. H. Brookes. CD8+ T cell-mediated suppression of intracellular Mycobacterium tuberculosis growth in activated human macrophages. *European J. Immunol.*, 33(12):3293–302, 2003.

[Brooks et al. 1983] B. R. Brooks, R. E. Bruccoleri, B. D. Olafson, D. J. States, S. Swaminathan, and M. Karplus. CHARMM: A program for macromolecular energy, minimization, and dynamics calculations. *J. Comp. Chem.*, 4:187–217, 1983.

[Brooks 1991a] R. A. Brooks. Intelligence without reason. In *Proc. 12th IJCAI, Sydney, Australia*, pages 569–595. Morgan Kaufmann, 1991.

[Brooks 1991b] R. A. Brooks. Intelligence without representation. *Artificial Intelligence*, 47:139–159, 1991.

[Brown et al. 1996] M. Brown, M. B. Rittenberg, C. Chen, and V. A. Roberts. Tolerance to single, but not multiple, amino acid replacements in antibody V-H CDR2: A means of minimizing B cell wastage from somatic hypermutation? *J. Immuonol.*, 156:3285–3291, 1996.

[Brusic & Flower 2004] V. Brusic and D. R. Flower. Bioinformatics tools for identifying T-cell epitopes. *Drug Discovery Today: BioSilico*, 2(1):18–23, 2004.

[Brusic *et al.* 1998]   V. Brusic, G. Rudy, and L. C. Harrison. MHCPEP, a database of MHC-binding peptides: update 1997. *Nucleic Acids Res*, 26:368–371, 1998.

[Brusic *et al.* 2002]   V. Brusic, J. Zeleznikow, and N. Petrovsky. Molecular immunology databases and data repositories. *J. Immunol. Methods*, 238:17–28, 2002.

[Bryant & Ploegh 2004]   P. Bryant and H. Ploegh. class II MHC peptide loading by the professionals. *Curr. Op. Immunol.*, 16(1):96–102, 2004.

[Bryngelson & Wolynes 1987]   J. D. Bryngelson and P. G. Wolynes. Spin glasses and the statistical mechanics of protein folding. *Proc. Natl. Acad. Sci. USA*, 84:7524–7528, 1987.

[Bullock *et al.* 2003]   T. N. J. Bullock, D. W. Mullins, and V. H. Engelhard. Antigen density presented by dendritic cells in vivo differentially affects the number and avidity of primary, memory, and recall CD8(+) T cells. *J. Immunol.*, 170:1822–1829, 2003.

[Burden & Faires 2005]   R. L. Burden and J. D. Faires. *Numerical Analysis*. Thomson-Brooks/Cole Publishing, 2005.

[Burnet 1959]   F. M. Burnet. *The Clonal Selection Theory of Acquired Immunity*. Cambridge University Press, 1959.

[Burrows & Elkington 2003]   S. R. Burrows and R. A. Elkington. Promiscuous CTL recognition of viral epitopes on multiple human leukocyte antigens: biological validation of the proposed HLA A24 supertype. *J. Immunol.*, 171(3):1407–1412, 2003.

[Callard *et al.* 1999]   R. Callard, A. J. T. George, and J. Stark. Cytokines, chaos, and complexity. *Immunity*, 11:507–513, 1999.

[Canetti 1955]   G. Canetti. *The tubercle bacillus in the pulmonary lesion in man*. Springer, 1955.

[Canham & Tyrrell 2002]   R. O. Canham and A. M. Tyrrell. A multilayered immune system for hardware fault tolerance within an embryonic array. In [Timmis & Bentley 2002], pages 3–11.

[Cano & Fan 1998]   P. Cano and B. Fan. A geometric study of the amino acid sequence of class I HLA molecules. *Immunogenetics*, 45(1):15–26, 1998.

[Cantú-Paz *et al.* 2003]   E. Cantú-Paz et al., editors. volume 2724 of *LNCS*. Springer, 2003.

[Cantú-Paz 1998]   E. Cantú-Paz. A survey of parallel genetic algorithms. *Calculateurs Parallèles*, 10(2):141–171, 1998.

[Carlson & Doyle 2001]   J. Carlson and J. Doyle. Robustness and complexity. Technical Report RS-2001-017, SFI, 2001.

[Carrington & Martin 2006]   M. Carrington and M. P. Martin. The impact of variation at the KIR gene cluster on human disease. *Curr. Top. Microbiol. Immunol.*, 298:225–257, 2006.

[Carrington *et al.* 1999]   M. Carrington, G. W. Nelson, M. P. Martin, T. Kissner, D. Vlahov, J. J. Goedert, R. Kaslow, S. Buchbinder, K. Hoots, and S. J. O'Brien. HLA and HIV-1: Heterozygote advantage and B*35-Cw*04 disadvantage. *Science*, 283:1748–1752, 1999.

[Caruso *et al.* 1999]   A. M. Caruso et al. Mice deficient in CD4 T cells have only transiently diminished levels of IFN-gamma, yet succumb to tuberculosis. *J. Immunol.*, 162(9):5407–5416, 1999.

[Casson & Manser 1995]   L. P. Casson and T. Manser. Random mutagenesis of 2 complementarity-determining region amino-acids yields an unexpectedly high-frequency of antibodies with increased affinity for both cognate antigen and autoantigen. *J. Exp. Med.*, 182:743–750, 1995.

[Castellani *et al.* 1998]   G. C. Castellani, C. Giberti, C. Franceschi, and F. Bersani. Stable state analysis of an immune network model. *Int. J. Bifurcation and Chaos*, 8:1285–1301, 1998.

[Castellani *et al.* 1999]   G. C. Castellani, N. Intrator, H. Shouval, and L. N. Cooper. Solutions of the BCM learning rule in a network of lateral interacting non-linear neurons networks. *Comput. Neur. Syst*, 10:111–121, 1999.

[Castellani *et al.* 2001]   G. C. Castellani, E. M. Quinlan, L. N. Cooper, and H. Z. Shouval. A biophysical model of bidirectional synaptic plasticity: dependence on AMPA and NMDA receptors. *Proc. Natl. Acad. Sci. USA*, 98:12772–12777, 2001.

[Castellani *et al.* 2005]   G. C. Castellani, E. M. Quinlan, F. Bersani, L. N. Cooper, and H. Z. Shouval. A model of bidirectional synaptic plasticity: from signaling network to channel conductance. *Learn. Mem.*, 12:423–432, 2005.

[Castelli & Buhot 202]   F. A. Castelli and C. Buhot. HLA-DP4, the most frequent HLA II molecule, defines a new supertype of peptide-binding specificity. *J. Immunol.*, 169(12):6928–6934, 202.

[Castiglione *et al.* 2003]   F. Castiglione, V. Selitser, and Z. Agur. The effect of drug schedule on hypersensitive reactions: a study with a cellular automata model of the immune system. In Luigi Preziosi, editor, *Cancer Modelling and Simulation*, chapter 12. CRC Press, 2003.

[Castro & Timmis 2002]   L. N. de Castro and J. Timmis. An Artificial Immune Network for Multi Modal Optimisation. In *Proceedings of the World Congress on Computational Intelligence WCCI*, pages 699–704, Honolulu, HI., 2002.

[Celada & Schreiber 1997]   A. Celada and R. D. Schreiber. Internalization and degradation of receptor-bound interferon-gamma by murine macrophages. demonstration of receptor recycling. *J. Immunol.*, 139(1):147–53, 1997.

[Celada & Seiden 1992]   F. Celada and P. E. Seiden. A computer model of cellular interactions in the immune system. *Immunology Today*, 13:56–62, 1992.

[Cerottini & Luescher 1991]   J. C. Cerottini and I. F. Luescher. Direct analysis of peptide binding to cell-associated MHC class I molecules. *Immunol. Lett.*, 30(2):171–5, 1991.

[Chakraborty 1993]   A. Chakraborty. Tuberculosis situation in India: Measuring it through time. *Indian Journal of Tuberculosis*, 40:215–225, 1993.

[Chan & Kaufmann 1995]   J. Chan and H. Kaufmann. Immune mechanisms of protection. *Tuberculosis. Pathogenesis, Protection and Control*, pages 389–415, 1995.

[Chan & Stolfo 1993]   P. K. Chan and S. J. Stolfo. Toward parallel and distributed learning by meta-learning. In *Working Notes AAAI Work. Knowledge Discovery in Databases*, pages 227–240, 1993.

[Chan 1993]   V. Chan. Secondary lymphoid-tissue chemokine (SLC) is chemotactic for mature dendritic cells. *Blood*, 93:3610–3616, 1993.

[Chang *et al.* 2005]   S. T. Chang, J. J. Linderman, and D. E. Kirschner. Multiple mechanisms allow mycobacterium tuberculosis to continuously inhibit MHC class II -mediated antigen presentation by macrophages. *Proc. Natl. Acad. Sci. USA*, 102(12):4530–4535, 2005.

[Chao *et al.* 2004]   D. L. Chao, M. P. Davenport, S. Forrest, and A. S. Perelson. A stochastic model of cytotoxic T cell responses. *J. Theor. Biol.*, 228:227–240, 2004.

[Chao *et al.* 2005]   D. L. Chao, M. P. Davenport, S. Forrest, and A. S. Perelson. The effects of thymic selection on the range of T cell cross-reactivity. *Eur. J. Immunol.*, 35:3452–3459, 2005.

[Chapman 1998]   H. A. Chapman. Endosomal proteolysis and MHC class II function. *Curr. Op. Immunol.*, 10:93–102, 1998.

[Chattratichat *et al.* 1997]   J. Chattratichat, J. Darlington, M. Ghanem, Y. Guo, H. Hunning, M. Kohler, J. Sutiwaraphun, H. Wing To, and D. Yang. Large scale data mining: Challenges and responses. In *KDD-97*, pages 143–146, 1997.

[Chelvanayagam 1997]   G. Chelvanayagam. A roadmap for HLA-DR peptide binding specificities. *Hum. Immunol.*, 58(2):61–69, 1997.

[Chen & Parham 1989]   B. P. Chen and P. Parham. Direct binding of influenza peptides to class I HLA molecules. *Nature*, 337(6209):743–745, 1989.

[Chen *et al.* 1992]   C. Chen, V. A. Roberts, and M. B. Rittenberg. Generation and analysis of random point mutations in an antibody CDR2 sequencemany mutated antibodies lose their ability to bind antigen. *J. Exp. Med.*, 176:855–866, 1992.

[Chen *et al.* 1994]   Y. Chen, J. Sidney, S. Southwood, A. L. Cox, K. Sakaguchi, R. A. Henderson, E. Appella, D. F. Hunt, A. Sette, and V. H. Engelhard. Naturally processed peptides longer than nine amino acid residues bind to the class I MHC molecule HLA-A2.1 with high affinity and in different conformations. *J. Immunol.*, 152:2874–2881, 1994.

[Chen *et al.* 2002]   X. Chen, Y. Lin, M. Liu, and M. K. Gilson. The binding database: data management and interface design. *Bioinformatics*, 18:130–139, 2002.

[Chicz *et al.* 1992]   R. M. Chicz, R. G. Urban, W. S. Lane, J. C. Gorga, L. J. Stern, D. A. A. Vignali, and J. L. Strominger. Predominant naturally processed peptides bound to HLA-DR1 are derived from MHC-related molecules and are heterogeneous in size. *Nature*, 358:764–768, 1992.

[Chicz *et al.* 1993]   R. M. Chicz et al. Specificity and promiscuity among naturally processed peptides bound to HLA-DR alleles. *J. Exp. Med.*, 178(1):27–47, 1993.

[Chiel & Beer 1997]   H. J. Chiel and R. D. Beer. The brain has a body: adaptive behavior emerges from interactions of nervous system, body and environment. *Trends in Neurosciences*, 20(12):553–557, 1997.

[Christinck & Luscher 1991]   E. R. Christinck and M. A. Luscher. Peptide binding to class I MHC on living cells and quantitation of complexes required for CTL lysis. *Nature*, 352(6330):67–70, 1991.

[Churchill *et al.* 2000]   H. R. O. Churchill, P. S. Anderson, E. A. Parke, R. A. Mariuzza, and D. M. Kranz. Mapping the energy of superantigen staphylococcus enterotoxin C3 recognition of an alpha/beta T cell receptor using alanine scanning mutagenesis. *J. Exp. Med.*, 191:835–846, 2000.

[Clancey 1997]   W. Clancey. *Situated Cognition: On Human Knowledge and Computer Representations*. Cambridge University Press, 1997.

[Clark & Forman 1984]   S. S. Clark and J. Forman. Functional aspects of class I MHC molecule domains. *Surv. Immunol. Res.*, 3((2-3)):179–83, 1984.

[Clark *et al.* 2005a]   E. Clark, A. Hone, and J. Timmis. A Markov Chain Model of the B-cell Algorithm. In [Jacob *et al.* 2005], pages 318–330.

[Clark *et al.* 2005b]   J. A. Clark, S. Stepney, and H. Chivers. Breaking the model: finalisation and a taxonomy of security attacks. In *REFINE 2005*, volume 137 of *ENTCS*, pages 225–242, 2005.

[Clark 1997]   A. Clark. *Being There : putting brain, body and world together again.* Oxford University Press, 1997.

[Cohen & Segal 2001]   I. R. Cohen and L. A. Segal. *Design Principles for the Immune System and Other Distributed Autonomous Systems.* SFI, 2001.

[Cohen et al. 2004]   I. R. Cohen, Uri Hershberg, and Sorin Solomon. Antigen-receptor degeneracy and immunological paradigms. *Molecular Immunology,* 40:993–996, 2004.

[Cohen 2000a]   I. R. Cohen. Discrimination and dialogue in the immune system. *Seminars in Immunology,* 12(3):215–219, 2000.

[Cohen 2000b]   I. R. Cohen. *Tending Adam's Garden: Evolving the Cognitive Immune Self.* Elsevier Academic Press, 2000.

[Cohen 2001]   I. R. Cohen. The creation of immune specificity. [Cohen & Segal 2001], pages 151–159.

[Cohn & Langman 1990]   M. Cohn and R. E. Langman. The protecton: the evolutionarily selected unit of humoral immunity. *Immunol. Reviews,* 15:1–131, 1990.

[Cohn & Mata 2006]   M. Cohn and J. Mata. The tritope model for restrictive recognition of antigen by T-cells: II. implications for ontogeny, evolution and physiology. *Molecular Immunology,* in press, 2006.

[Cohn et al. 2002]   M. Cohn, R. E. Langman, and J. J. Mata. A computerized model for the self-nonself discrimination at the level of the T-helper (Th-genesis). I. the origin of 'primer' effector T-helpers. *Int. Immunol.,* 14:1105–1112, 2002.

[Cohn 1969]   M. Cohn. Immunological tolerance. pages 281–338. Academic Press, 1969.

[Cohn 1983]   M. Cohn. Progress in immunology. pages 839–851. Academic Press, 1983.

[Cohn 1985]   M. Cohn. Diversity in the immune system: "Preconceived ideas" or ideas preconceived? *Biochimie,* 67:9–27, 1985.

[Cohn 1986]   M. Cohn. In P. Matzinger et al., editors, *Proceedings of the EMBO Workshop on Tolerance,* pages 3–35, Basle, 1986. Editiones (Roche).

[Cohn 1992]   M. Cohn. The self-nonself discrimination: Reconstructing a cabbage from sauerkraut. *Res. Immunol.,* 143:323–334, 1992.

[Cohn 1994]   P. M. Cohn. *Elements of Linear Algebra.* Chapman and Hall, 1994.

[Cohn 1997]   M. Cohn. A new concept of immune specificity emerges from a consideration of the self-nonself discrimination. *Cell. Immunol.,* 181:103–108, 1997.

[Cohn 1998]   M. Cohn. At the feet of the master: The search for universalities. Divining the evolutionary selection pressures that resulted in an immune system. *Cytogenet. Cell Genet.,* 80:54–60, 1998.

[Cohn 2001]   M. Cohn. *Historical Issues and Contemporary Debates in Immunology,* chapter Dialogues with Selves, pages 53–85. Elsevier, 2001.

[Cohn 2002]   M. Cohn. The immune system: a weapon of mass destruction invented by evolution to even the odds during the war of the DNAs. *Immunol. Rev.,* 185:24–38, 2002.

[Cohn 2003]   M. Cohn. Does complexity belie a simple decision—on the Efroni and Cohen critique of the minimal model for a self–nonself discrimination. *Cellular Immunology,* 221:138–142, 2003.

[Cohn 2004]   M. Cohn. Whither T-suppressors: If they didn't exist would we have to invent them? *Cell. Immunol.,* 27:81–92, 2004.

[Cohn 2005a]   M. Cohn. A biological context for the self-nonself discrimination and the regulation of effector class by the immune system. *Immunologic Research*, 31:133–150, 2005.

[Cohn 2005b]   M. Cohn. The common sense of the self-nonself discrimination. *Springer Seminars in Immunopathology*, 27:3–17, 2005.

[Comstock 1982]   G. Comstock. Epidemiology of tuberculosis. *American Review of Respiratory Disease*, 125:8–16, 1982.

[Cooke & Hunt 1995]   D. Cooke and J. Hunt. Recognising promoter sequences using an Artificial Immune System. In *Proceedings of Intelligent Systems in Molecular Biology*, pages 89–97. AAAI Press, 1995.

[Cooper *et al.* 1993]   A. M. Cooper, D. K. Dalton, T. A. Stewart, J. P. Griffin, D. G. Russell, and I. M. Orme. Disseminated tuberculosis in interferon gamma gene-disrupted mice. *J. Exp. Med.*, 178(6):2243–2247, 1993.

[Cooper *et al.* 1997a]   A. M. Cooper, J. E. Callahan, M. Keen, J. T. Belisle, and I. M. Orme. Expression of memory immunity in the lung following re-exposure to mycobacterium tuberculosis. *Tuber. Lung Dis.*, 78(1):67–73, 1997.

[Cooper *et al.* 1997b]   A. M. Cooper, J. Magram, J. Ferrante, and I. M. Orme. Interleukin 12 (il-12) is crucial to the development of protective immunity in mice intravenously infected with mycobacterium tuberculosis. *J. Exp. Med.*, 186(1):39–45, 1997.

[Cooper *et al.* 2003]   L. R. Cooper, D. W. Corne, and M. J. Crabbe. Use of a novel hill-climbing genetic algorithm in protein folding simulations. *Comp. Bio. Chem.*, 27:575–580, 2003.

[Copeland 2002]   B. J. Copeland. The Church-Turing Thesis. In E. N. Zalta, editor, *The Stanford Encyclopedia of Philosophy* (Fall 2002 edition), 2002. http://plato.stanford.edu/archives/fall2002/entries/church-turing/.

[Corper *et al.* 2000]   A. L. Corper, T. Stratmann, V. Apostolopoulos, C. A. Scott, K. C. Garcia, A. S. Kang, I. A. Wilson, and L. Teyton. A structural framework for deciphering the link between I-Ag7 and autoimmune diabetes. *Science*, 288:505–511, 2000.

[Coutinho *et al.* 1993]   A. Coutinho, J. Salaun, C. Corbel, A. Bandeira, and N. Le Douarin. The role of thymic epithelium in the establishment of transplantation tolerance. *Immunol. Rev.*, 133:225–240, 1993.

[Crippen *et al.* 1998]   T. L. Crippen, D. W. H. Riches, and D. M. Hyde. Differential regulation of the expression of cytokine-induced neutrophil chemoattractant by mouse macrophages. *Pathobiology*, 66(1):24–32, 1998.

[Crotty *et al.* 2003]   S. Crotty, P. Felgner, H. Davies, J. Glidewell, L. Villarreal, and R. Ahmed. SAP is required for generating long-term humoral immunity. *Nature*, 421:282–287, 2003.

[Cruciani & Watson 1994]   G. Cruciani and K. A. Watson. Comparative molecular field analysis using GRID force-field and GOLPE variable selection methods in a study of inhibitors of glycogen phosphorylase b. *J. Med. Chem.*, 37(16):2589–2601, 1994.

[Crutchfield 1994]   J. P. Crutchfield. The calculi of emergence: computation, dynamics, and induction. *Physica D*, 75:11–54, 1994.

[Cui *et al.* 1998]   Y. Cui, R. S. Chen, and W. H. Wong. Protein folding simulation using genetic algorithm and supersecondary structure constraints. *Proteins: Structure, Function and Genetics*, 31(3):247–257, 1998.

[Cullell-Young *et al.* 2001]   M. Cullell-Young et al. From transcription to cell surface expression, the induction of MHC class II I-A alpha by interferon-gamma in macrophages is regulated at different levels. *Immunogenetics*, 53(2):136–44, 2001.

[Cunliffe 1997]   J. Cunliffe. Morphostasis: an evolving perspective. *Medical Hypotheses*, 49:449–459, 1997.

[Cutello & Nicosia 2004]   V. Cutello and G. Nicosia. The clonal selection principle for in silico and in vitro computing. In L. N. de Castro and F. J. von Zuben, editors, *Recent Developments in Biologically Inspired Computing*. Idea Group Publishing, Hershey, PA, 2004.

[Cutello *et al.* 2004]   V. Cutello, G. Nicosia, and M. Parvone. Exploring the capability of immune algorithms: A characterisation of hypermutation operators. In [Nicosia *et al.* 2004], pages 263–276.

[Cutello *et al.* 2006]   V. Cutello, G. Narzisi, and G. Nicosia. A multi-objective evolutionary approach to the protein structure prediction problem. *J. Royal Soc. Inter.*, 3:139–151, 2006.

[Czermak *et al.* 1999]   B. J. Czermak et al. In vitro and in vivo dependency of chemokine generation on C5a and TNF-alpha. *J. Immunol.*, 162(4):2321–2325, 1999.

[Daida & Hilss 2003]   J. M. Daida and A. M. Hilss. Identifying structural mechanisms in standard genetic programming. In [Cantú-Paz *et al.* 2003], pages 1639–1651.

[Daida *et al.* 2003a]   J. M. Daida, A. M. Hilss, D. J. Ward, and S. L. Long. Visualizing tree structures in genetic programming. In [Cantú-Paz *et al.* 2003], pages 1652–1664.

[Daida *et al.* 2003b]   J. M. Daida, H. Li, R. Tang, and A. M. Hilss. What makes a problem GP-hard? Validating a hypothesis of structural causes. In [Cantú-Paz *et al.* 2003], pages 1665–1677.

[Dalton *et al.* 1993]   D. Dalton et al. Multiple defects of immune cell function in mice with disrupted inteferon gamma genes. *Science*, 259:1739–1742, 1993.

[Dandekar & Argos 1996]   T. Dandekar and P. Argos. Identifying the tertiary fold of small proteins with different topologies from sequence and secondary structure using the genetic algorithm and extended criteria specific for strand regions. *J. Molec. Biol.*, 256:645–660, 1996.

[Dasgupta 1999]   D. Dasgupta. *Artificial Immune Systems and their Applications*. Springer, 1999.

[Davenport *et al.* 1995]   M. P. Davenport, C. L. Quinn, R. M. Chicz, B. N. Green, A. C. Willis, W. S. Lane, J. I. Bell, and A. V. Hill. Naturally processed peptides from two disease-resistance-associated HLA-DR13 alleles show related sequence motifs and the effects of the dimorphism at position 86 of the HLA-DR β chain. *Proc. Natl. Acad. Sci. USA*, 92:6567–6571, 1995.

[Day *et al.* 2001]   R. O. Day, J. B. Zydallis, and G. B. Lamont. Solving the protein structure prediction problem through a multiobjective genetic algorithm. *ICNN*, 2:31–35, 2001.

[De Boer & Noest 1998]   R. J. De Boer and A. J. Noest. T cell renewal rates, telomerase, and telomere length shortening. *J. Immunol.*, 22:5832–5837, 1998.

[De Boer *et al.* 1992]   R. J. De Boer, L. A. Segal, and A. S. Perelson. Pattern formulation in one-and two-dimensional shape-space models of the immune system. *J. Theor. Biol.*, 155(3):295–333, 1992.

[De Boer *et al.* 2004]   R. J. De Boer, J. A. Borghans, M. van Boven, C. Kesmir, and F. J. Weissing. Heterozygote advantage fails to explain the high degree of polymorphism of the MHC. *Immunogenetics.*, 55:725–731, 2004.

[de Castro & Timmis 2002a]   L. N. de Castro and J. Timmis. *Artificial Immune Systems: A New Computational Intelligence Approach.* Springer, 2002.

[de Castro & Timmis 2002b]   L. N. de Castro and J. Timmis. Hierarchy and convergence of immune networks: Basic ideas and preliminary results. In [Timmis & Bentley 2002], pages 231–240.

[de Castro & Timmis 2003]   L. N. de Castro and J. Timmis. Artificial immune systems as a novel soft computing paradigm. *Soft Computing*, 7:526–544, 2003.

[de Castro & Von Zuben 2000]   L. N. de Castro and F. J. Von Zuben. The clonal selection algorithm with engineering applications. In *GECCO Workshop on Artificial Immune Systems and Their Applications*, pages 36–37, 2000.

[de Castro & Von Zuben 2001]   L. N. de Castro and F. J. Von Zuben. *aiNet: An Artificial Immune Network for Data Analysis*, pages 231–259. Idea Group Publishing, USA, 2001.

[de Castro & Von Zuben 2002]   L. N. de Castro and F. J. Von Zuben. Learning and optimization using the clonal selection principle. *IEEE Transactions on Evol. Comp.*, 6(3):239–251, 2002.

[de Groot *et al.* 2001]   A. S. de Groot, A. Bosma, N. Chinai, J. Frost, B. M. Jesdale, M. A. Gonzalez, W. Martin, and C. Saint-Aubin. From genome to vaccine: in silico predictions, ex vivo verification. *Vaccine*, 19:4385–4395, 2001.

[de Groot *et al.* 2002]   N. G. de Groot, N. Otting, G. G. Doxiadis, S. S. Balla-Jhagjhoorsingh, J. L. Heeney, J. J. van Rood, P. Gagneux, and R. E. Bontrop. Evidence for an ancient selective sweep in the MHC class I gene repertoire of chimpanzees. *Proc. Natl. Acad. Sci. USA*, 99:11748–11753, 2002.

[Deavin *et al.* 1996]   A. J. Deavin, T. R. Auton, and P. J. Greaney. Statistical comparison of established T-cell epitope predictors against a large database of human and murine antigens. *Mol. Immunol.*, 33(145-155), 1996.

[Deb *et al.* 2002]   K. Deb, A. Pratap, S. Agarwal, and T. Meyarivan. A fast and elitist multiobjective genetic algorithm: Nsga-2. *IEEE Trans. Evol. Comp.*, 6:182–197, 2002.

[De Boer & Perelson 1993]   R. J. De Boer and A. S. Perelson. How diverse should the immune system be? *Proc. R. Soc. Lond. B Biol. Sci.*, 252:171–175, 1993.

[Deem & Lee 2003]   M.W. Deem and H.Y. Lee. Sequence space localization in the immune system response to vaccination and disease. *Phys. Rev. Lett.*, 91:068101:1–4, 2003.

[Degli-Eposti & Smyth 2005]   M. A. Degli-Eposti and M. J. Smyth. Close encounters of a different kind: dendritic cells and NK cells take centre stage. *Nat. Rev. Imm.*, 5(2):112–124, 2005.

[del Guercio & Sidney 1995]   M. F. del Guercio and J. Sidney. Binding of a peptide antigen to multiple HLA alleles allows definition of an a2-like supertype. *J. Immunol.*, 154(2):685–693, 1995.

[Delvig *et al.* 2002]   A. A. Delvig et al. TGF-beta1 and IFN-gamma cross-regulate antigen presentation to CD4 T cells by macrophages. *J. Leukoc. Biol.*, 72(1):163–166, 2002.

[Dembic 2000]   Z. Dembic. Immune system protects integrity of tissues. *Molec. Immunol.*, 37:563–569, 2000.

412   References

[Dempsey et al. 1996]   P. W. Dempsey, M. E. D. Allison, S. Akkaraju, C. C. Good-now, and D. T. Fearon. C3D of complement as a molecular adjuvant: Bridging innate and acquired immunity. *Science*, 271(5247):348–350, 1996.

[Denzin & Cresswell 195]   L. K. Denzin and P. Cresswell. HLA-DM induces CLIP dissociation from MHC class II alpha beta dimers and facilitates peptide loading. *Cell*, 82(1):155–65, 195.

[Derrida 1980]   B. Derrida. Random-energy model: limit of a family of disordered models. *Phys. Rev. Lett.*, 45:79–82, 1980.

[Dessen et al. 1997]   A. Dessen, C. M. Lawrence, S. Cupo, D. M. Zaller, and D. C. Wiley. X-ray crystal structure of HLA-DR4 (DRA*0101, DRB*0401) complexed with a peptide from human collagen II. *Immunity*, 7:473–481, 1997.

[Detours & Perelson 1999]   V. Detours and A. S. Perelson. Explaining high allore-activity as a quantitative consequence of affinity-driven thymocyte selection. *Proc. Natl. Acad. Sci. USA*, 96:5153–5158, 1999.

[Detours & Perelson 2000]   V. Detours and A. S. Perelson. The paradox of allore-activity and self MHC restriction: Quantitative analysis and statistics. *Proc. Natl. Acad. Sci. USA*, 97:8479–8483, 2000.

[Detours et al. 1999]   V. Detours, R. Mehr, and A. S. Perelson. A quantitative theory of affinity-driven T cell repertoire selection. *J. Theor. Biol.*, 200:389–403, 1999.

[Dewar et al. 1985]   M. J. S. Dewar, E. G. Zoebisch, E. F. Healy, and J. J. P. Stewart. A new general purpose quantum mechanical molecular model. *J. Am. Chem. Soc*, 107:3902–3909, 1985.

[Dines & Powell 1997]   K. Dines and H. Powell. Mast cell interactions with the ner-vous system: relationship to mechanisms of disease. *Journal of neuropathology and experimental neurology*, 56(6):627–640, 1997.

[Doherty & Zinkernagel 1975]   P. C. Doherty and R. M. Zinkernagel. Enhanced im-munological surveillance in mice heterozygous at the H-2 gene complex. *Nature*, 256:50–52, 1975.

[Doherty et al. 2000]   P. C. Doherty, J. M. Riberdy, and G. T Belz. *Philos. Trans. R. Soc. Lond. B. Biol. Sc.*, 355:1093, 2000.

[Dolin et al. 1994]   P. J. Dolin, M. C. Raviglione, and A. Kochi. Global tubercu-losis incidence and mortality during 1990-2000. *Bulletin of the World Health Organization*, 72(2):213–220, 1994.

[Doolan & Hoffman 1997]   D. L. Doolan and S. L. Hoffman. Degenerate cytotoxic T cell epitopes from P. falciparum restricted by multiple HLA-A and HLA-B supertype alleles. *Immunity*, 7(1):97–112, 1997.

[Doytchinova & Flower 2002a]   I. A. Doytchinova and D. R. Flower. A compara-tive molecular similarity index analysis (CoMSIA) study identifies an HLA-A2 binding supermotif. *J. Comput.-Aid. Molec. Des.*, 16:535–544, 2002.

[Doytchinova & Flower 2002b]   I. A. Doytchinova and D. R. Flower. Physicochem-ical explanation of peptide binding to HLA-A*0201 major histocompatibility complex: a three-dimensional quantitative structure-activity relationship study. *Proteins*, 48:505–518, 2002.

[Doytchinova & Flower 2002c]   I. A. Doytchinova and D. R. Flower. Quantitative approaches to computational vaccinology. *Immunol Cell Biol*, 80:270–279, 2002.

[Doytchinova & Flower 2003]   I. A. Doytchinova and D. R. Flower. Towards the in silico identification of class II restricted T-cell epitopes: a partial least squares it-erative self-consistent algorithm for affinity prediction. *Bioinformatics*, 19:2263–2270, 2003.

[Doytchinova et al. 2002] I. A. Doytchinova, M. J. Blythe, and D. R. Flower. Additive method for the prediction of protein-peptide binding affinity. application to the MHC class I molecule HLA-A*0201. J. Proteome Res, 1:263–272, 2002.

[Doytchinova et al. 2004a] I. Doytchinova, S. Hemsley, and D. R. Flower. Transporter associated with antigen processing preselection of peptides binding to the MHC: A bioinformatic evaluation. J. Immunol., 173:6813–6819, 2004.

[Doytchinova et al. 2004b] I. A. Doytchinova, P. Guan, and D. R Flower. Identifying human MHC supertypes using bioinformatic methods. J. Immunol., 172(7):4314–4123, 2004.

[Doytchinova et al. 2004c] I. A. Doytchinova, V. A. Walshe, N. A. Jones, S. E. Gloster, P. Borrow, and D. R. Flower. Coupling in silico and in vitro analysis of peptide-MHC binding: A bioinformatic approach enabling prediction of superbinding peptides and anchorless epitopes. J. Immunol., 172:7495–7502, 2004.

[Dunbrack & Cohen. 1997] R. L. Dunbrack and F. E. Cohen. Bayesian statistical analysis of protein sidechain rotamer preferences. Protein Science, 6:1661–1681, 1997.

[Dutton et al. 1998] R. W. Dutton, L. M. Bradley, and S. L. Swain. T cell memory. Ann. Rev. Immunol., 16:201–223, 1998.

[Dybdahl & Lively 1998] M. F. Dybdahl and C. M. Lively. Host-parasite co-evolution: evidence for rare advantage and time-lagged selection in a natural population. Evolution, 52:1057–1066, 1998.

[Ebnet et al. 1996] K. Ebnet, E. P. Kaldjian, A. O. Anderson, and S. Shaw. Orchestrated information transfer underlying leukocyte endothelial interactions. Ann. Rev. Immunol., 14:155–177, 1996.

[Edelman & Gally 2001] G. M. Edelman and J. A. Gally. Degeneracy and complexity in biological systems. Proc. Natl. Acad. Sci. USA, 98(24):13763–13768, 2001.

[Efroni & Cohen 2002] S. Efroni and I. R. Cohen. Simplicity belies a complex system: a response to the minimal model of immunity of Langman and Cohn. Cellular Immunology, 216:23–30, 2002.

[Efroni & Cohen 2003] S. Efroni and I. R. Cohen. The heuristics of biologic theory: the case of self–nonself discrimination. Cellular Immunology, 223:87–89, 2003.

[Efroni et al. 2003] S. Efroni, D. Harel, and I. R. Cohen. Towards rigorous comprehension of biological complexity: Modeling, execution, and visualization of thymic T-cell maturation. Genome Research, 13:2485–2497, 2003.

[Egerton et al. 1990] M. Egerton, R. Scollay, and K. Shortman. Kinetics of mature T-cell development in the thymus. Proc. Natl. Acad. Sci. USA, 87:2579–2582, 1990.

[Ehlers 2003] S. Ehlers. Role of tumour necrosis factor (TNF) in host defence against tuberculosis: implications for immunotherapies targeting TNF. Annals of the Rheumatic Diseases, 62:37–42, 2003.

[Ellegren et al. 1993] H. Ellegren, G. Hartman, M. Johansson, and L. Andersson. Major histocompatibility complex monomorphism and low levels of DNA fingerprinting variability in a reintroduced and rapidly expanding population of beavers. Proc. Natl. Acad. Sci. USA, 90(17):8150–8153, 1993.

[Emile et al. 1997] J. F. Emile et al. Correlation of granuloma structure with clinical outcome defines two types of idiopathic disseminated BCG infection. Journal of Pathology, 181:25–30, 1997.

[Falcone *et al.* 2000]   F. Falcone, H. Haas, and B. Gibbs. The human basophil: a new appreciation of its role in immune responses. *Blood*, 96(13):4028–4038, 2000.

[Fan & Long 2001]   Q. R. Fan and E. O. Long. Crystal structure of the human natural killer cell inhibitory receptor KIR2DL1-HLA-CW4 complex. *Nat. Immunol.*, 2(5):452–460, 2001.

[Farmer *et al.* 1986]   J. D. Farmer, N. H. Packard, and A. S. Perelson. The immune system, adaptation, and machine learning. *Physica D*, 22:187–204, 1986.

[Farmer *et al.* 1987]   J. D. Farmer, S. A. Kauffmann, N. H. Packard, and A. S. Perelson. Adaptive dynamic networks as models for the immune-system and autocatalytic sets. *Ann. New York Acad. Sci.*, 504:118–131, 1987.

[Fauchere & Pliska 1983]   J. Fauchere and V. Pliska. Hydrophobic parameters π of amino acid side chain from the partitioning of n-acetyl-amino-acid amides,. *Eur. J. Med. Chem.*, 18:369–375, 1983.

[Fell 2005]   D. A. Fell. Enzymes, metabolites and fluxes. *J. Exp. Bot.*, 56:267–272, 2005.

[Fenton 1998]   M. J. Fenton. Macrophages and tuberculosis. *Curr. Op. Immunol.*, 5(1):72–78, 1998.

[Fink & Bevan 1995]   P. J. Fink and M. J. Bevan. Positive selection of thymocytes. *Adv. Immunol.*, 59:99–133, 1995.

[Flesch & Kaufmann 1990]   I. Flesch and S. Kaufmann. Activation of tuberculostatic macrophage functions by gamma interferon, interleukin-4, and tumor necrosis factor. *Infect. Immun.*, 58:2675–2677, 1990.

[Flood-Page *et al.* 2003]   P. Flood-Page, A. Menzies-Gow, A. B. Kay, and D. Robinson. Eosinophilas role remains uncertain as anti–interleukin-5 only partially depletes numbers in asthmatic airway. *American Journal of Respiratory Critical Care Medicine*, 167:199–204, 2003.

[Flower *et al.* 2002]   D. R. Flower, I. A. Doytchinova, K. Paine, P. Taylor, M. J. Blythe, D. Lamponi, C. Zygouri, P. Guan, H. McSparron, and H. Kirkbride. Computational vaccine design. In *Drug Design: Cutting Edge Approaches*. RSC, 2002.

[Flower 2003]   D. R. Flower. Towards in silico prediction of immunogenic epitopes. *Trends Immunology*, 24:667–674, 2003.

[Flynn & Chan 2001a]   J. Flynn and J. Chan. Immunology of tuberculosis. *Annual Reviews in Immunology*, 19:93–129, 2001.

[Flynn & Chan 2001b]   J. L. Flynn and J. Chan. Tuberculosis: latency and reactivation. *Infect Immun*, 69(7):4195–201., 2001.

[Flynn & Ernst 2000]   J. L. Flynn and J. D. Ernst. Immune responses in tuberculosis. *Curr. Op. Immunol.*, 12(4):432–6., 2000.

[Flynn *et al.* 1992]   J. L. Flynn et al. Major histocompatibility complex class-I-restricted T-cells are required for resistance to Mycobacterium-tuberculosis infection. *Proc. Natl. Acad. Sci. USA*, 82(24):12013–12017, 1992.

[Flynn *et al.* 1993]   J. L. Flynn, J. Chan, K. J. Triebold, D. K. Dalton, T. A. Stewart, and B. R. Bloom. An essential role for interferon gamma in resistance to mycobacterium tuberculosis infection. *J. Exp. Med.*, 178(6):2249–2254, 1993.

[Flynn 1995]   J. L. Flynn. IL-12 increases resistance of BALB/c mice to Mycobacterium tuberculosis infection. *J. Immunol.*, 155:2515, 1995.

[Fogel 2000]   D. B. Fogel. *Evolutionary Computation: Toward a New Philosophy of Machine Intelligence*. IEEE Press, 2000.

[Fontenot & Rudensky 2005]   J. D. Fontenot and A. Y. Rudensky. A well adapted regulatory contrivance: regulatory T cell development and the forkhead family transcription factor Foxp3. *Nat. Immunol.*, 6(4):331–337, 2005.

[Forrest *et al.* 1994]   S. Forrest, A. S. Perelson, L. Allen, and R. Cherukuri. Self–nonself discrimination in a computer. In *Proc. IEEE Symposium on Research Security and Privacy*, pages 202–212, 1994.

[Forrest *et al.* 1997]   S. Forrest, S. Hofmeyr, and A. Somayaji. Computer Immunology. *Comm. ACM*, 40(10):88–96, 1997.

[Foxman *et al.* 1997]   E. Foxman, J. Campbell, and E. Butcher. Multistep navigation and the combinatorial control of leukocyte chemotaxis. *J. Cell Biol.*, 139:1349–1360, 1997.

[Freitas & Timmis 2003]   A. Freitas and J. Timmis. Revisiting the foundations of artificial immune systems: A problem oriented perspective. In [Timmis *et al.* 2003], pages 229–241.

[Fremont *et al.* 1998]   D. H. Fremont, D. Monnaie, C. A. Nelson, W. A. Hendrickson, and E. R. Unanue. Crystal structure of I-Ak in complex with a dominant epitope of lysozyme. *Immunity*, 8:305–317, 1998.

[Freter *et al.* 1983]   R. Freter, R. R. Freter, and H. Brickner. Experimental and mathematical-models of escherichia-coli plasmid transfer in-vitro and in-vivo. *Infection and Immunity*, 39(1):60–84, 1983.

[Gallucci *et al.* 1999]   S. Gallucci, M. Lolkema, and P. Matzinger. Natural adjuvants: endogenous activators of dendritic cells. *Nature Medicine*, 5:1249–1255, 1999.

[Gammack *et al.* 2005]   D. Gammack et al. Understanding granuloma formation using different mathematical and biological scales. *SIAM Journal of Multiscale Modeling and Simulation*, 3:312–345, 2005.

[Gao *et al.* 2001]   X. Gao, G. W. Nelson, P. Karacki, M. P. Martin, J. Phair, R. Kaslow, J. J. Goedert, S. Buchbinder, K. Hoots, D. Vlahov, S. J. O'Brien, and M. Carrington. Effect of a single amino acid change in MHC class I molecules on the rate of progression to AIDS. *N. Engl. J. Med.*, 344(22):1668–1675, May 2001.

[Gardner 1970]   M. Gardner. Mathematical games: the fantastic combinations of John Conway's new solitaire game "life". *Scientific American*, 223:120–123, 1970.

[Garrett 2003]   S. M. Garrett. A paratope is not a epitope: implications for immune network models and clonal selection. In [Timmis *et al.* 2003], pages 217–228.

[Garrett 2005]   S. M. Garrett. How do we evaluate artificial immune systems? *Evolutionary Computation*, 13(2):145–177, 2005.

[Gaspar & Hirsbrunner 2002]   A. Gaspar and B. Hirsbrunner. From optimization to learning in learning in changing environments: The pittsburgh immune classifier system. In [Timmis & Bentley 2002], pages 190–199.

[George *et al.* 2005]   A. J. T. George, J. Stark, and C. Chan. Understanding specificity and sensitivity of T cell recognition. *Trends in Immunology*, 26(12):653–659, 2005.

[Germain & Hendrix 1991]   R. N. Germain and L. R. Hendrix. MHC class II structure, occupancy and surface expression determined by post-endoplasmic reticulum antigen binding. *Nature*, 353(6340):134–139, 1991.

[Germain 2004]   R. N. Germain. An innately interesting decade of research in immunology. *Nature Medicine*, 10:1307–1320, 2004.

[Gillanders *et al.* 1997]   W. E. Gillanders, H. L. Hanson, R. J. Rubocki, T. H. Hansen, and J. M. Connolly. Class I-restricted cytotoxic T cell recognition of split peptide ligands. *Int. Immunol.*, 9:81–89, 1997.

[Gillard & Farr 2005]   G. O. Gillard and A. G. Farr. Contrasting models of promiscuous gene expression by thymic epithelium. *J. Exp. Med.*, 202:15–19, 2005.

[Gillespie 1977]   G. T. Gillespie. Exact stochastic simulation of coupled chemical reactions. *J. Phys. Chem.*, 81:2340 – 2361, 1977.

[Godkin *et al.* 2001]   A. J. Godkin, K. J. Smith, A. Willis, M. V. Tejada-Simon, J. Zhang, T. Elliott, and A. V. Hill. Naturally processed HLA class II peptides reveal highly conserved immunogenic flanking region sequence preferences that reflect antigen processing rather than peptide-MHC interactions. *J. Immunol.*, 166:6720–6727, 2001.

[Golbraikh & Tropsha 2002]   A. Golbraikh and A. Tropsha. Beware of q2! *J. Molec. Graph Model*, 20:269–76, 2002.

[Goldfeld *et al.* 1998]   A. E. Goldfeld et al. Association of an HLA-DQ allele with clinical tuberculosis. *J. Am. Med. Assoc.*, 279(3):226–228, 1998.

[Goldstein *et al.* 2004]   B. Goldstein, J. R. Faeder, and W. Hlavacek. Mathematical models of immune receptor signalling. *Nat. Rev. Imm.*, 4:445–456, 2004.

[Gomez *et al.* 2003]   J. Gomez, F. Gonzalez, and D. Dasgupta. An immuno-fuzzy approach to anomaly detection. In *Proceedings of the 12th IEEE International Conference on Fuzzy Systems*, pages 137–142. IEEE Press, 2003.

[Goodman *et al.* 2002]   D. Goodman, L. Boggess, and A. Watkins. Artificial immune system classification of multiple-class problems. In *Proc. of Intelligent Engineering Systems*, pages 179–184. ASME, 2002.

[Goodman *et al.* 2003]   D. Goodman, L. Boggess, and A. Watkins. An investigation into the source of power for AIRS, an artificial immune classification system. In *Proc. Int. Joint Conf. Neural Networks*, pages 1678–1683. IEEE, 2003.

[Goodwin 1994]   B. C. Goodwin. *How the Leopard Changed Its Spots: the evolution of complexity*. Phoenix, 1994.

[Gopalakrishnan & Roques 1992]   B. Gopalakrishnan and B. P. Roques. Do antigenic peptides have a unique sense of direction inside the MHC binding groove? a molecular modelling study. *FEBS Lett*, 303((2-3)):224–228, 1992.

[Gordon & Riley 1992]   D. M. Gordon and M. A. Riley. A theoretical and experimental-analysis of bacterial-growth in the bladder. *Molecular Microbiology*, 6(4):555–562, 1992.

[Gotter & Kyewski 1994]   J. Gotter and B. Kyewski. Regulating self-tolerance by deregulating gene expression. *Curr. Op. Immunol.*, 16:741–745, 1994.

[Goust 1993]   J. M. Goust. Major histocompatibility complex. *Immunol. Ser.*, 58:29–48, 1993.

[Govindarajan *et al.* 2003]   K. R. Govindarajan, P. Kangueane, T. W. Tan, and S. Ranganathan. MPID: MHC-peptide interaction database for sequence-structure-function information on peptides binding to MHC molecules. *Bioinformatics*, 19:309–310, 2003.

[Grayson *et al.* 2002]   J. Grayson, L. E. Harrington, J. G. Lanier, E. J. Wherry, and R. Ahmed. Differential sensitivity to naïve and memory CD8+ T cells to apoptosis in vivo. *J. Immunol.*, 169:3760–3770, 2002.

[Greensmith *et al.* 2005]   J. Greensmith, U. Aickelin, and S. Cayzer. Introducing dendritic cells as a novel immune-inspired algorithm for anomaly detection. In [Jacob *et al.* 2005], pages 153–167.

[Gropp et al. 1999]   W. Gropp, E. Lusk, and A. Skjellum. *Using MPI: Portable Parallel Programming with the Message Passing Interface.* MIT Press, 2nd edition, 1999.

[Grossman & Paul 2000]   Z. Grossman and W. E. Paul. Self-tolerance: context dependent tuning of T cell antigen recognition. *Seminars in Immunology,* 12(3):197–203, 2000.

[Grossman & Paul 2001]   Z. Grossman and W. E. Paul. Autoreactivity, dynamic tuning and selectivity. *Curr. Op. Immunol.,* 13(687-698), 2001.

[Grotzke & Lewinsohn 2005]   J. E. Grotzke and D. M. Lewinsohn. Role of CD8+ T lymphocytes in control of Mycobacterium tuberculosis infection. *Microbes Infect,* 7(4):776–788, 2005.

[Guan et al. 2003a]   P. Guan, I. A. Doytchinova, and D. R. Flower. A comparative molecular similarity indices (CoMSIA) study of peptide binding to the HLA-A3 superfamily. *Bioorg. Med. Chem.,* 11:2307–2311, 2003.

[Guan et al. 2003b]   P. Guan, I. A. Doytchinova, and D. R. Flower. HLA-A3 super-motif defined by quantitative structure-activity relationship analysis. *Protein Eng.,* 16:11–18, 2003.

[Gulukota & DeLisi 2001]   K. Gulukota and C. DeLisi. Neural network method for predicting peptides that bind major histocompatibility complex molecules. *Methods Mol. Biol.,* 156:201–209, 2001.

[Gulukota et al. 1997]   K. Gulukota, J. Sidney, A. Sette, and C. DeLisi. Two complementary methods for predicting peptides binding major histocompatibility complex molecules. *J. Molec. Biol.,* 267:1258–1267, 1997.

[Haken & Mikhailov 1993]   H. Haken and A. Mikhailov, editors. *Interdisciplinary Approaches to Nonlinear Complex Systems,* volume 62 of *Synergetics.* Springer, 1993.

[Hamilton et al. 1990]   W. D. Hamilton, R. Axelrod, and R. Tanese. Sexual reproduction as an adaptation to resist parasites (a review). *Proc. Natl. Acad. Sci. USA,* 87:3566–3573, 1990.

[Harding et al. 2003]   C. V. Harding, L. Ramachandra, and M. J. Wick. Interaction of bacteria with antigen presenting cells: influences on antigen presentation and antibacterial immunity. *Curr. Op. Immunol.,* 15(1):112–9, 2003.

[Hart & Ross 2002]   E. Hart and P. Ross. Exploiting the analogy between immunology and sparse distributed memories: A system for clustering non-stationary data. In [Timmis & Bentley 2002], pages 49–58.

[Hart & Ross 2004]   E. Hart and P. Ross. Studies on the implications of shape-space models for idiotypic networks. In [Nicosia et al. 2004], pages 413–426.

[Hart & Ross 2005]   E. Hart and P. Ross. The impact of the shape of antibody recognition regions on the emergence of idiotypic networks. *Intl. J. Unconventional Computing,* 1(3):281–314, 2005.

[Hart & Timmis 2005]   E. Hart and J. Timmis. Application areas of AIS: The past, the present and the future. In [Jacob et al. 2005], pages 483–497.

[Hart 2005]   E. Hart. Not all balls are round: An investigation of alternative recognition-region shapes. In [Jacob et al. 2005], pages 29–42.

[Hattotuwagama et al. 2004]   C. K. Hattotuwagama, P. Guan, I. A. Doytchinova, C. Zygouri, and D. R. Flower. Quantitative online prediction of peptide binding to the major histocompatibility complex. *J. Molec. Graph. Model.,* 22:195–207, 2004.

[Haykin 1999]   S. Haykin. *Neural Networks-A Comprehensive Foundation.* Prentice-Hall, 1999.

[Hennecke & Wiley 2002]   J. Hennecke and D. C. Wiley. Structure of a complex of the human T-cell receptor (TCR) HA1.7, influenza hemaglutinin peptide, and major histocompatibility complex class II molecule, HLA-DR4 (DRA*0101 and DRB1*0401): insight into TCR cross-restriction and alloreactivity. *J. Exp. Med*, 195:571–581, 2002.

[Hiemstra et al. 2000]   H. S. Hiemstra, J. W. Drijfhout, and B. O. Roep. Antigen arrays in T cell immunology. *Curr. Op. Immunol.*, 12:80–84, 2000.

[Hightower et al. 1995]   R. R. Hightower, S. A. Forrest, and A. S. Perelson. The evolution of emergent organization in immune system gene libraries. In *Proceedings of the 6th International Conference on Genetic Algorithms*, pages 344–350. Morgan Kaufmann, 1995.

[Hill 1998]   A. Hill. The immunogenetics of human infectious diseases. *Ann. Rev. Immunol.*, 16:593–617, 1998.

[Hmama et al. 1998]   Z. Hmama et al. Attenuation of HLA-DR expression by mononuclear phagocytes infected with mycobacterium tuberculosis is related to intracellular sequestration of immature class II heterodimers. *J. Immunol.*, 161(9):4882–93, 1998.

[Hoebe et al. 2004]   K. Hoebe, E. Janssen, and B. Beutler. The interface between innate and adaptive immunity. *Nat. Immunol.*, 5(10):971–974, 2004.

[Hofmeyr & Forrest 2000]   S. Hofmeyr and S. Forrest. Architecture for an artificial immune system. *Evolutionary Computation*, 7(1):1289–1296, 2000.

[Holland 1975]   J. H. Holland. Genetic algorithms and classifer systems: Foundations and future directions. In *Proc. of International Conference on Genetic Algorithms*, pages 82–89, 1975.

[Hone 2005]   A. Hone. Ordinary differential equations. In *Encyclopedia of Nonlinear Science*. Routledge, 2005.

[Honey & Rudensky 2003]   K. Honey and A. Y. Rudensky. Lysosomal cysteine proteases regulate antigen presentation. *Nat. Rev. Imm.*, 3(6):472–82, 2003.

[Horig et al. 1999]   H. Horig, A. C. Young, N. J. Papadopoulos, T. P. DiLorenzo, and S. G. Nathenson. Binding of longer peptides to the H-2Kb heterodimer is restricted to peptides extended at their C terminus: refinement of the inherent MHC class I peptide binding criteria. *J. Immunol*, 163:4434–4441, 1999.

[Hsieh et al. 2004]   C. S. Hsieh, Y. Linng, J. Tyznik, S. G. Self, D. Liggit, and A. Y. Rudensky. Recognition of the peripheral self by naturally arising CD4+ CD25+ T cell receptors. *Immunity*, 21(2):267–277, 2004.

[Huang et al. 1996]   C. C. Huang, G. S. Couch, E. F. Pettersen, and T. E. Ferrin. Chimera: An extensible molecular modeling application constructed using standard components. *Pacific Symposium on Biocomputing*, 1:724, 1996.

[Huang et al. 1999]   S. E. Huang, R. Samudrala, and J. W. Ponder. Ab initio folding prediction of small helical proteins using distance geometry and knowledge-based scoring functions. *J. Molec. Biol.*, 290:267–281, 1999.

[Hughes & Nei 1988]   A. L. Hughes and M. Nei. Pattern of nucleotide substitution at major histocompatibility complex class I loci reveals overdominant selection. *Nature*, 335:167–170, 1988.

[Hughes & Nei 1989]   A. L. Hughes and M. Nei. Nucleotide substitution at major histocompatibility complex class II loci: Evidence for overdominant selection. *Proc. Natl. Acad. Sci. USA*, 86:958–962, 1989.

[Hughes & Nei 1992]   A. L. Hughes and M. Nei. Models of host–parasite interaction and MHC polymorphism. *Genetics*, 132:863–864, 1992.

[Hughes & Yeager 1998]   A. L. Hughes and M. Yeager. Natural selection at major histocompatibility complex loci of vertebrates. *Annu. Rev. Genet.*, 32:415–435, 1998.

[Huhn *et al.* 1996]   R. D. Huhn, E. Radwanski, S. M. O'Connell, M. G. Sturgill, L. Clarke, R. P. Cody, M. B. Affrime, and D. L. Cutler. Pharmacokinetics and immunomodulatory properties of intravenously administered recombinant human interleukin-10 in healthy volunteers. *Blood*, 87(2):699–705., 1996.

[Huhn *et al.* 1997]   R. D. Huhn et al. Pharmacodynamics of subcutaneous recombinant human IL-10 in healthy volunteers. *Clinical Pharmacology and Therapeutics*, 62:171–180, 1997.

[Hull *et al.* 2001]   D. L. Hull, R. E. Langman, and S. S. Glenn. A general account of selection: Biology, immunology and behavior. *Behavioral and Brain Sciences*, 245:11–28, 2001.

[Hunt & Cooke 1996]   J. Hunt and D. Cooke. Learning using an artificial immune system. *Journal of Network and Computer Applications*, 19:189–212, 1996.

[Hunt *et al.* 1992]   D. F. Hunt, H. Michel, T. A. Dickinson, J. Shabanowitz, A. L. Cox, K. Sakaguchi, and E. Appella. Peptides presented to the immune system by the murine class II major histocompatibility complex molecule I-Ad. *Science*, 256:1817–1820, 1992.

[Hunt *et al.* 1998]   J. Hunt, J. Timmis, D. Cooke, M. Neal, and C. King. JISYS: Development of an artificial immune system for real-world applications. In D. Dasgupta, editor, *Artificial Immune Systems and their Applications*, pages 157–186. Springer, 1998.

[Huseby *et al.* 2005]   E. S. Huseby, J. White, F. Crawford, T. Vass, D. Becker, C. Pinilla, P. Marrack, and J. W. Kappler. How the T cell repertoire becomes peptide and MHC specific. *Cell*, 122(2):247–260, July 2005.

[Jacob *et al.* 2004]   C. Jacob, J. Litorco, and L. Lee. Immunity through swarms: Agent-based simulations of the human immune system. In [Nicosia *et al.* 2004], pages 477–489.

[Jacob *et al.* 2005]   C. Jacob, M. Pilat, P. Bentley, and J. Timmis, editors. *Proc. of the 4th International Conference on Artificial Immune Systems(ICARIS)*, volume 3627 of *LNCS*. Springer, 2005.

[Jacobs 1996]   O. L. R Jacobs. *Introduction to Control Theory*. Oxford University Press, corrected 2nd edition, 1996.

[Jakobsen & Gao 1998]   I. B. Jakobsen and X. Gao. Correlating sequence variation with HLA-A allelic families: implications for T cell receptor binding specificities. *Immunol. Cell Biol.*, 76(2):135–142, 1998.

[James & Tawfik 2003]   L. C. James and D. S. Tawfik. The specificity of cross-reactivity: promiscuous antibody binding involves specific hydrogen bonds rather than nonspecific hydrophobic stickiness. *Protein Sci.*, 12:2183–2193, 2003.

[Janeway & Katz 1984]   C. A. Janeway and M. E. Katz. Self Ia-recognizing T cells undergo an ordered series of interactions with Ia-bearing substrate cells of defined function during their development: a model. *Surv. Immunol. Res.*, 3:45–54, 1984.

[Janeway & Travers 1997]   C. A. Janeway and P. Travers. *Immunobiology: The Immune System in Health and Disease*. Garland Publications, 1997.

[Janeway 1992]   C. A. Janeway. The immune system evolved to discriminate infectious nonself from noninfectious self. *Immunology Today*, 13:11–16, 1992.

[Janeway 2001]   C. A. Janeway. Inaugural article: How the immune system works to protect the host from infection: A personal view. *Proc. Natl. Acad. Sci. USA*, 98(13):7461–7468, 2001.

[Jardetzky *et al.* 1991]   T. S. Jardetzky, W. S. Lane, R. A. Robinson, D. R. Madden, and D. C. Wiley. Identification of self peptides bound to purified HLA-B27. *Nature*, 353:326–329, 1991.

[Jeffery & Bangham 2000]   K. J. Jeffery and C. R. Bangham. Do infectious diseases drive MHC diversity? *Microbes. Infect.*, 2:1335–1341, 2000.

[Jeffery *et al.* 2000]   K. J. Jeffery, A. A. Siddiqui, M. Bunce, A. L. Lloyd, A. M. Vine, A. D. Witkover, S. Izumo, K. Usuku, K. I. Welsh, M. Osame, and C. R. Bangham.   The influence of HLA class I alleles and heterozygosity on the outcome of human T cell lymphotropic virus type I infection.   *J. Immunol.*, 165:7278–7284, 2000.

[Jen 2001]   E. Jen. Stable or robust? What's the difference? *Robustness in Natural, Engineering and Social Systems*, (RS-2001-024), 2001.

[Jerne 1974]   N. K. Jerne. Towards a network theory of the immune system. *Ann. Immunol. (Inst. Pasteur)*, 125C:373–389, 1974.

[Jiang & Chess 2006]   H. Jiang and L. Chess. Mechanisms of disease - regulation of immune responses by T cell. *New England Journal Of Medicine*, 354(11):1166–1176, 2006.

[Jiang *et al.* 2002]   S. Jiang, N. J. Borthwick, P. Morrison, G. F. Gao, and M. W. Steward. Virus-specific CTL responses induced by an H-2K(d)-restricted, motif-negative 15-mer peptide from the fusion protein of respiratory syncytial virus. *J. Gen. Virol.*, 83:429–438, 2002.

[Johnson *et al.* 2004]   C. G. Johnson, J. P. Goldman, and W. J. Gullick. Simulating complex intracellular processes using object-oriented computational modelling. *Progress in Biophysics and Molecular Biology*, 86(3):379–406, 2004.

[Jones *et al.* 1993]   D. R. Jones, C. D. Perttunen, and B. E. Stuckman. Lipschitzian optimization without the Lipschitz constant. *J. Optimization Theory and Application*, 79:157–181, 1993.

[Jones 2005]   S. A. Jones. Directing transition from innate to acquired immunity: Defining a role for IL-6. *J. Immunol.*, 175(6):3463–3468, 2005.

[Jordan & Smith 1999]   D. W. Jordan and P. Smith. *Nonlinear Ordinary Differential Equations*. Oxford University Press, 3rd edition, 1999.

[Kaiser *et al.* 1997]   C. E. Kaiser, G. B. Lamont, L. D. Merkle, G. H. Gates Jr., and R. Patcher. Polypeptide structure prediction: Real-valued versus binary hybrid genetic algorithms. In *Proc. ACM Symp. on Applied Computing (SAC)*, pages 279–286, San Jose, CA, 1997.

[Kaleab *et al.* 1990]   B. Kaleab et al.   Mycobacterial-induced cytotoxic T cells as well as nonspecific killer cells derived from healthy individuals and leprosy patients. *European J. Immunol.*, 20:2651–2659, 1990.

[Kamradt & Volkmer-Engert 2004]   T. Kamradt and R. Volkmer-Engert. Cross-reactivity of T lymphocytes in infection and autoimmunity. *Mol. Divers.*, 8:271–280, 2004.

[Kapsenberg 2003]   M. L. Kapsenberg. Dendritic cell control of pathogen driven T cell polarization. *Nat. Rev. Imm.*, 3(12):984–993, 2003.

[Karlin & Mrazek 2000]   S. Karlin and J. Mrazek. Predicted highly expressed genes of diverse prokaryotic genomes. *J. Bacteriol.*, 182:5238–5250, 2000.

[Karre *et al.* 1986]   K. Karre, H. G. Ljunggren, G. Piontek, and R. Kiessling. Selective rejection of H-2-deficient lymphoma variants suggests alternative immune defense strategy. *Nature*, 319(6055):675–678, 1986.

[Karre 1995]   K. Karre. Express yourself or die: peptides, MHC molecules, and NK cells. *Science*, 267:978–979, 1995.

[Kast & Melief 1991]   W. M. Kast and C. J. Melief. Fine peptide specificity of cytotoxic T lymphocytes directed against adenovirus-induced tumours and peptide-MHC binding. *Int. J. Cancer Suppl.*, 6:90–94, 1991.

[Kast *et al.* 1994]   W. M. Kast, R. M. Brandt, J. Sidney, J. W. Drijfhout, R. T. Kubo, H. M. Grey, C. J. Melief, and A. Sette. Role of HLA-A motifs in identification of potential CTL epitopes in human papillomavirus type 16 E6 and E7 proteins. *J. Immunol.*, 152:3904–3912, 1994.

[Kastrup *et al.* 2000]   I. B. Kastrup, S. Stevanovic, G. Arsequell, G. Valencia, J. Zeuthen, H. G. Rammensee, T. Elliott, and J. S. Haurum. Lectin purified human class I MHC-derived peptides: evidence for presentation of glycopeptides in vivo. *Tissue Antigens*, 56:129–135, 2000.

[Kauffman & Weinberger 1989]   S. A. Kauffman and E. D. Weinberger. The NK model of rugged fitness landscapes and its application to maturation of the immune response. *J. Theor. Biol.*, 141:211–246, 1989.

[Kauffman *et al.* 1988]   S. A. Kauffman, E. D. Weinberger, and A. S. Perelson. Theoretical Immunology, Part one. In *Maturation of the immune response via adaptive walks on affinity landscapes*. 1988.

[Kauffman 1995]   S. A. Kauffman. *At Home in the Universe: the search for laws of complexity*. Viking, 1995.

[Kauffman 2000]   S. A. Kauffman. *Investigations*. Oxford University Press, 2000.

[Kaufmann 1988]   S. Kaufmann. CD8 + T-lymphocytes in intracellular microbial infections. *Immunol. Today*, 9:168–174, 1988.

[Kaufmann 1993]   S. Kaufmann. Immunity to intracellular bacteria. *Annu. Rev. Immunol.*, 11:129–163, 1993.

[Kelsey & Timmis 2003]   J. Kelsey and J. Timmis. Immune inspired somatic contiguous hypermutation for function optimisation. In *Proc. of Genetic and Evolutionary Computation Conference (GECCO)*, volume 2723 of *LNCS*, pages 207–218. Springer, 2003.

[Kelsey *et al.* 2003]   J. Kelsey, J. Timmis, and A. Hone. Chasing chaos. In *Proc. of Congress on Evolutionary Computation (CEC)*, pages 89–98, Canberra, Australia., 2003. IEEE.

[Kelso 1995]   J. A. S. Kelso. *Dynamic Patterns: the self-organization of brain and behavior*. MIT Press, 1995.

[Kennedy & Eberhart 2001]   J. Kennedy and R. Eberhart. *Swarm Intelligence*. Morgan Kaufmann, 2001.

[Kepler & Perelson 1993]   T. B. Kepler and A. S. Perelson. Somatic hypermutation in B cells: an optimal control treatment. *J. Theor. Biol.*, 164:37 – 64, 1993.

[Kesmir *et al.* 2002]   C. Kesmir, A. K. Nussbaum, H. Schild, V. Detours, and S. Brunak. Prediction of proteasome cleavage motifs by neural networks. *Protein Eng.*, 15:287–296, 2002.

[Khanna & Burrows 1997]   R. Khanna and S. R. Burrows. Hierarchy of epstein-barr virus-specific cytotoxic T-cell responses in individuals carrying different subtypes of an HLA allele: implications for epitope-based antiviral vaccines. *J Virol*, 71(10):7429–7439, 1997.

[Kim & Bentley 2002a]   J. Kim and P. J. Bentley. Immune memory in the dynamic clonal selection algorithm. In [Timmis & Bentley 2002], pages 59–67.

[Kim & Bentley 2002b]   J. Kim and P. J. Bentley. A model of gene library evolution in the dynamic clonal selection algorithm. In [Timmis & Bentley 2002], pages 182–189.

[Kim et al. 2005]   S. Kim et al. Licencing of natural killer cells by host major histocompatibility complex class I molecules. Nature, 436:709–713, 2005.

[Kim 2002]   J. Kim. Integrating Artificial Immune Algorithms for Intrusion Detection. PhD thesis, UCL, 2002.

[Kimachi et al. 1997]   K. Kimachi, M. Croft, and H. M. Grey. The minimal number of antigen-major histocompatibility complex class II complexes required for activation of naive and primed T cells. European J. Immunol., 27(12):3310–7, 1997.

[Kincaid & Cheney 2002]   D. Kincaid and W. Cheney. Numerical Analysis: Mathematics of Scientific Computing. Brooks/Cole, 3rd edition, 2002.

[Kindler 1989]   V. Kindler. The inducing role of tumor necrosis factor in the development of bactericidal granulomas during BCG development. Cell, 56:731–740, 1989.

[King et al. 2005]   R. D. King, S. M. Garrett, and G. M. Coghill. On the use of qualitative reasoning to simulate and identify metabolic pathways. Bioinformatics, 21:2017–2026, 2005.

[Kirk 1997]   D. E. Kirk. Optimal Control Theory-An Introduction. Prentice Hall Inc., 1997.

[Kirschner & Blaser 1995]   D. Kirschner and M. Blaser. The dynamics of H. pylori infection of the human stomach. J. Theor. Biol., 176:281–290, 1995.

[Kisielow et al. 1988]   P. Kisielow, H. Bluthmann, U. D. Staerz, M. Steinmetz, and H. Von Boehmer. Tolerance in T-cell-receptor transgenic mice involves deletion of nonmature $CD4^+8^+$ thymocytes. Nature, 333:742–746, 1988.

[Klebe & Abraham 1999]   G. Klebe and U. Abraham. Comparative molecular similarity index analysis (comsia) to study hydrogen-bonding properties and to score combinatorial libraries. J. Comput. Aid. Molec. Des., 13:1–10, 1999.

[Klebe et al. 1994]   G. Klebe, U. Abraham, and T. Mietzner. Molecular similarity indices in a comparative analysis (comsia) of drug molecules to correlate and predict their biological activity. J. Med. Chem., 37:4130–4146, 1994.

[Kleinstein & Seiden 2000]   S. H. Kleinstein and Philip E. Seiden. Simulating the immune system. Computing in Science and Engineering, 2(4):69–77, 2000.

[Kleinstein et al. 2003]   S. H. Kleinstein, Y. Louzoun, and M. J. Shlomchik. Estimating hypermutation rates from clonal tree data. J. Immunol., 171(9):4639–4649, 2003.

[Klyubin et al. 2005]   A. S. Klyubin, D. Polani, and C. L. Nehaniv. All else being equal be empowered. In ECAL 2005, volume 3630 of LNAI, pages 744–753. Springer, 2005.

[Knight & Timmis 2001]   T. Knight and J. Timmis. AINE: An immunological approach to data mining. In Proc. IEEE Int. Conf. Data Mining, pages 297–304. IEEE Press, 2001.

[Knight & Timmis 2003]   T. Knight and J. Timmis. A multi-layered immune inspired machine learning algorithm. In A. Lotfi and M. Garibaldi, editors, Applications and Science in Soft Computing, pages 195–202. Springer, 2003.

[Knowles & Corne 1999]   J. D. Knowles and D. W. Corne. The Pareto archived evolution strategy : A new baseline algorithm for Pareto multiobjective optimisation. *Proc. of Congress on Evolutionary Computation (CEC)*, 10:98–105, 1999.

[Knowles & Corne 2000]   J. D. Knowles and D. W. Corne. Approximating the nondominated front using the pareto archived evolution strategy. *Evol. Comp.*, 8(2):149–172, 2000.

[Koch & Platt 2003]   C. A. Koch and J. L. Platt. Natural mechanisms for evading graft rejection: the fetus as an allograft. *Springer Seminars Immunopathology*, 25:95–117, 2003.

[Kocher et al. 1999]   A. Kocher, J. Jaffe, and B. Jun. Differential power analysis. In *Crypto '99*, volume 1666 of *LNCS*, pages 388–397. Springer, 1999.

[Kocher 1996]   A. Kocher. Timing attacks on implementations of Diffie-Hellman, RSA, DSS, and other systems. In *Crypto '96*, volume 1109 of *LNCS*, pages 104–113. Springer, 1996.

[Korber et al. 2001a]   B. Korber, B. Gaschen, K. Yusim, R. Thakallapally, C. Kesmir, and V. Detours. Evolutionary and immunological implications of contemporary HIV-1 variation. *Br. Med. Bull.*, 58:19–42, 2001.

[Korber et al. 2001b]   B. T. M. Korber, C. Brander, B. F. Haynes, R. Koup, C. Kuiken, J. P. Moore, B. D. Walker, and D. Watkins. HIV molecular immunology. Technical report, Los Alamos National Laboratory, 2001.

[Koyama et al. 1999]   S. Koyama et al. Monocyte chemotactic factors released from type II pneumocyte-like cells in response to TNF-alpha and IL-1 alpha. *European Respiratory Journal*, 13(4):820–828, 1999.

[Krakauer 2001]   D. Krakauer. Robustness and overdesign. Technical Report RS-2001-050, SFI, 2001.

[Kramnik et al. 2000]   I. Kramnik et al. Genetic control of resistance to experimental infection with virulent mycobacterium tuberculosis. *Proc. Natl. Acad. Sci. USA*, 97(15):8560–8565, 2000.

[Krohling et al. 2002]   R. Krohling, Y. Zhou, and A. Tyrrell. Evolving FPGA-based robot controllers using an evolutionary algorithm. In [Timmis & Bentley 2002], pages 41–46.

[Kubinyi & Kehrhahn 1976]   H. Kubinyi and O. H. Kehrhahn. Quantitative structure-activity relationships. 3.1 a comparison of different Free-Wilson models. *J. Med. Chem.*, 19:1040–1049, 1976.

[Kumar & Bentley 2003]   S. Kumar and P. J. Bentley, editors. *On Growth, Form and Computers*. Elsevier, 2003.

[Kundu & Faulkes 2004]   S. Kundu and C. G. Faulkes. Patterns of MHC selection in African mole-rats, family Bathyergidae: the effects of sociality and habitat. *Proc. Biol. Sci.*, 271(1536):273–278, 2004.

[Kurata & Berzofsky 1990]   A. Kurata and J. A. Berzofsky. Analysis of peptide residues interacting with MHC molecule or T cell receptor. Can a peptide bind in more than one way to the same MHC molecule? *J. Immunol.*, 144(12):4526–4535, 1990.

[Kurt-Jones et al. 2002]   E. Kurt-Jones, L. Mandell, C. Whitney, A. Padgett, K. Gosselin, P. Newburger, and R. Finberg. Role of toll-like receptor 2 (TLR2) in neutrophil activation: GM-CSF enhances TLR2 expression and TLR2-mediated interleukin 8 responses in neutrophils. *Blood*, 100(5):1860–1868, 9 2002.

[Kushmerick 1997]   N. Kushmerick. Software agents and their bodies. *Minds and Machines*, 7(2):227–247, 1997.

[Kyewski & Derbinski 2004]   B. Kyewski and J. Derbinski. Self-representation in the thymus: An extended view. *Nat. Rev. Imm.*, 4(9):688–698, 2004.

[Lafaille *et al.* 1994]   J. J. Lafaille, K. Nagashima, M. Katsuki, and S. Tonegawa. High incidence of spontaneous autoimmune encephalomyelitis in immunodeficient anti-myelin basic protein T cell receptor transgenic mice. *Cell*, 78:399–408, 1994.

[Lafaille *et al.* 1997]   J. J. Lafaille, E. van de Keere, A. Hsu, J. Baron, W. Haas, C. Raine, and S. Tonegawa. Myelin basic protein-specific T helper 2 (Th2) cells cause experimental autoimmune encephalomyelitis in immunodeficient hosts rather than protect them from the disease. *J. Exp. Med.*, 186(2):307–312, 1997.

[Lakoff & Johnson 1980]   G. Lakoff and M. Johnson. *Metaphors We Live By.* University of Chicago Press, 1980.

[Lakoff & Núñez 2000]   G. Lakoff and R. E. Núñez. *Where Mathematics Comes From.* Basic Books, 2000.

[Lakoff 1987]   G. Lakoff. *Women, Fire, and Dangerous Things: what categories reveal about the mind.* University of Chicago Press, 1987.

[Lamas *et al.* 1998]   J. R. Lamas, J. M. Brooks, B. Galocha, A. B. Rickinson, and J. A. Lopez de Castro. Relationship between peptide binding and T cell epitope selection: a study with subtypes of HLA-B27. *Int. Immunol.*, 10(3):259–266, 1998.

[Lane *et al.* 1999]   B. R. Lane et al. TNF-alpha inhibits HIV-1 replication in peripheral blood monocytes and alveolar macrophages by inducing the production of RANTES and decreasing C-C chemokine receptor 5 (CCR5) expression. *J. Immunol.*, 63(7):3653–3661, 1999.

[Langman & Cohn 1986]   R. E. Langman and M. Cohn. The 'complete' idiotype network is an absurd immune system. *Immunology Today*, 7(4):100 – 101, 1986.

[Langman & Cohn 1993a]   R. E. Langman and M. Cohn. The challenges of chickens and rabbits to immunology. In *52nd Forum in Immunology*, volume 144, pages 421–495, 1993.

[Langman & Cohn 1993b]   R. E. Langman and M. Cohn. A theory of the ontogeny of the chicken humoral immune system: The consequences of diversification by gene hyperconversion and its extension to rabbit. *Res. Immunol*, 144:421–446, 1993.

[Langman & Cohn 2000a]   R. E. Langman and M. Cohn. Editorial introduction. *Seminars in Immunology*, 12(3):159–162, 2000.

[Langman & Cohn 2000b]   R. E. Langman and M. Cohn. Editorial summary. *Seminars in Immunology*, 12(3):343–344, 2000.

[Langman & Cohn 2000c]   R. E. Langman and M. Cohn. A minimal model for the self–nonself discrimination: a return to basics. *Seminars in Immunology*, 12(3):189–195, 2000.

[Langman & Cohn 2002a]   R. E. Langman and M Cohn. Haplotype exclusion: the solution to a problem in natural selection. *Seminars in Immunology*, 14:153–162, 2002.

[Langman & Cohn 2002b]   R. E. Langman and M. Cohn. If the immune repertoire evolved to be large, random, and somatically generated, then... *Cellular Immunology*, 216:15–22, 2002.

[Langman *et al.* 2003]   R. E. Langman, J. J. Mata, and M. Cohn. A computerized model for the self-nonself discrimination at the level of the T-helper (Th genesis) II. The behavior of the system upon encounter with nonself antigens. *Int. Immunol.*, 15:593–609, 2003.

[Langman 2000]   R. E. Langman. The specificity of immunological reactions. *Molecular Immunology*, 37:555–561, 2000.

[Langton 1991]   C. G. Langton. *Computation at the Edge of Chaos: phase-transitions and emergent computation*. PhD thesis, University of Michigan, USA, 1991.

[Lanzavecchia 1996]   A. Lanzavecchia. Mechanisms of antigen uptake for presentation. *Curr. Op. Immunol.*, 8(3):348–54, 1996.

[Lapedes & Farber 2001]   A. Lapedes and R. Farber. The geometry of shape space: application to influenza. *J. Theor. Biol.*, 212:57–69, 2001.

[Lau & Wong 2003]   H. Y. K. Lau and V. W. K. Wong. Immunologic control framework for automated material handling. In [Timmis *et al.* 2003], pages 57–68.

[Lau *et al.* 1994]   L. Lau, B. Jamieson, T. Somasundraram, and R. Ahmed. Cytotoxic T-cell memory without antigen. *Nature*, 369:648–652, 1994.

[Lawlor & Warren 1991]   D. A. Lawlor and E. Warren. Gorilla class I major histocompatibility complex alleles: comparison to human and chimpanzee class I. *J. Exp. Med.*, 174:6, 1491-1509 1991.

[Lawlor *et al.* 1990]   D. A. Lawlor, J. Zemmour, P. D. Ennis, and P. Parham. Evolution of class-I MHC genes and proteins: From natural selection to thymic selection. *Annu. Rev. Immunol.*, 8:23–63, 1990.

[Le Douarin *et al.* 1996]   N. Le Douarin, C. Corbel, A. Bandeira, V. Thomas-Vaslin, Y. Modigliani, A. Coutinho, and J. Salaun. Evidence for a thymus-dependent form of tolerance that is not based on elimination or anergy of reactive T cells. *Immunol. Rev.*, 149(1):35–53, 1996.

[Lee *et al.* 1999]   D. W. Lee, H. B. Jun, and K. B. Sim. Artificial immune system for realisation of co-operative strategies and group behaviour in collective autonomous mobile robots. In *Proceedings of Fourth International Symposium on Artificial Life and Robotics*, pages 232–235. AAAI, 1999.

[Lee *et al.* 2000]   P. U. Y. Lee, H. R. O. Churchill, and M. Daniels. Role of 2C T cell receptor residues in the binding of self- and allo-major histocompatibility complexes. *J. Exp. Med.*, 191:1355–1364, 2000.

[Leng & Bentwich 2002]   Q. Leng and Z. Bentwich. Beyond self and nonself: Fuzzy recognition of the immune system. *Scandinavian J. Immunol.*, 56:224–232, 2002.

[Leslie *et al.* 2004]   A. J. Leslie, K. J. Pfafferott, P. Chetty, R. Draenert, M. M. Addo, M. Feeney, Y. Tang, E. C. Holmes, T. Allen, J. G. Prado, M. Altfeld, C. Brander, C. Dixon, D. Ramduth, P. Jeena, S. A. Thomas, A. St John, T. A. Roach, B. Kupfer, G. Luzzi, A. Edwards, G. Taylor, H. Lyall, G. Tudor-Williams, V. Novelli, J. Martinez-Picado, P. Kiepiela, B. D. Walker, and P. J. Goulder. HIV evolution: CTL escape mutation and reversion after transmission. *Nat. Med.*, 10(3):282–289, 2004.

[Leslie *et al.* 2005]   A. Leslie, D. Kavanagh, I. Honeyborne, K. Pfafferott, C. Edwards, T. Pillay, L. Hilton, C. Thobakgale, D. Ramduth, R. Draenert, S. Le Gall, G. Luzzi, A. Edwards, C. Brander, A. K. Sewell, S. Moore, J. Mullins, C. Moore, S. Mallal, N. Bhardwaj, K. Yusim, R. Phillips, P. Klenerman, B. Korber, P. Kiepiela, B. Walker, and P. Goulder. Transmission and accumulation of CTL escape variants drive negative associations between HIV polymorphisms and HLA. *J. Exp. Med.*, 201(6):891–902, Mar 2005.

[Leslie 2000]   R. G. Q. Leslie. The role of complement in the acquired immune response. *Immunology*, 100(1):4–12, 2000.

[Leveton et al. 1989]   C. Leveton et al. T-cell-mediated protection of mice against virulent mycobacterium-tuberculosis. *Infection and Immunity*, 57(2):390–395, 1989.

[Levinthal 1969]   C. Levinthal. How to fold graciously. In P. De Brunner, J. Tsibris, and E. Munck, editors, *Mossbauer Spectroscopy in Biological Systems*. University of Illinois Press, 1969.

[Levitsky & Liu 2000]   V. Levitsky and D. Liu. Supermotif peptide binding and degeneracy of MHC: peptide recognition in an EBV peptide-specific CTL response with highly restricted TCR usage. *Hum. Immunol.*, 61(10):972–984, 2000.

[Lewinsohn et al. 1998]   D. M. Lewinsohn et al. Human purified protein derivative specific CD4+ T cells use both CD95-dependent and CD95-independent cytolytic mechanisms. *J. Immunol.*, 160(5):2374–2379, 1998.

[Lewontin et al. 1978]   R. C. Lewontin, L. R. Ginzburg, and S. D. Tuljapurkar. Heterosis as an explanation for large amounts of genic polymorphism. *Genetics*, 88:149–170, 1978.

[Li & Scheraga 1988]   Z. Li and H. A. Scheraga. Structure and free energy of complex thermodynamics systems. *J. Molec. Stru.*, 179:333–352, 1988.

[Li et al. 2000]   Y. Li, H. Li, R. Martin, and R. A. Mariuzza. Structural basis for the binding of an immunodominant peptide from myelin basic protein in different registers by two HLA-DR2 proteins. *J. Molec. Biol.*, 304:177–188, 2000.

[Lichtman & Abbas 1997]   A. Lichtman and A. Abbas. T-cell subsets: recruiting the right kind of help. *Current Biology*, 7(4):242–244, 4 1997.

[Lim 2000]   T. K. Lim. Human genetic susceptibility to tuberculosis. *Ann. Acad. Med.*, 29(3):298–304, 2000.

[Lipsitch & Levin 1997]   M. Lipsitch and B. R. Levin. The population dynamics of antimicrobial chemotherapy. *Antimicrobial Agents and Chemotherapy*, 41(2):363–373, 1997.

[Liu et al. 2002]   X. Liu, S. Dai, F. Crawford, R. Fruge, P. Marrack, and J. Kappler. Alternate interactions define the binding of peptides to the MHC molecule iab. *Proc. Natl. Acad. Sci. USA*, 99:8820–8825, 2002.

[Lively & Dybdahl 2000]   C. M. Lively and M. F. Dybdahl. Parasite adaptation to locally common host genotypes. *Nature*, 405:679–681, 2000.

[Ljung 1999]   L. Ljung. *System Identification: Theory for the User*. PRT Prentice Hall, 2nd edition, 1999.

[Lorenz 1963]   E. N. Lorenz. Deterministic nonperiodic flow. *J. Atmos. Sci.*, 20:130–141, 1963.

[Lucchiari-Hartz et al. 2003]   M. Lucchiari-Hartz, V. Lindo, N. Hitziger, S. Gaedicke, L. Saveanu, P. M. Van Endert, F. Greer, K. Eichmann, and G. Niedermann. Differential proteasomal processing of hydrophobic and hydrophilic protein regions: contribution to cytotoxic T lymphocyte epitope clustering in HIV-1-Nef. *Proc. Natl. Acad. Sci. USA*, 100(13):7755–7760, 2003.

[Luciani et al. 2001]   F. Luciani, S. Valensin, R. Vescovin, P. Sansoni, F. Fagnoni, C. Bonafe M. Franceschi, and G. Turchetti. A stochastic model for CD8+ T cell dynamics in human immunosenescence: Implications for survival and longevity. *J. Theor. Biol.*, 213:587–597, 2001.

[Lund et al. 2004]   O. Lund, M. Nielsen, C. Kesmir, A. G. Petersen, C. Lundegaard, P. Worning, C. Sylvester-Hvid, K. Lamberth, G. Roder, S. Justesen, S. Buus, and S. Brunak. Definition of supertypes for HLA molecules using clustering of specificity matrices. *Immunogenetics*, 55(12):797–810, 2004.

[Luscinskas *et al.* 1994]   F. Luscinskas, G. S Kansas, H. Ding, P. Pizcueta, B. E. Schleiffenbaum, T. F. Tedder, and M. A. Gimbrone. Monocyte rolling, arrest and spreading on IL-4-activated vascular endothelium under flow is mediated via sequential action of L-selectin, β_1-integrins, and β_2-integrins. *J. Cell Biol.*, 125(6):1417–1427, 6 1994.

[Luther & Cyster 2001]   Sanjiv A. Luther and Jason G. Cyster. Chemokines as regulators of T cell differentiation. *Nat. Immunol.*, (2), 2001. 10.1038/84205.

[Macken & Perelson 1989]   C. A. Macken and A. S. Perelson. Protein evolution on rugged landscapes. *Proc. Natl. Acad. Sci. USA*, 86:6191–6195, 1989.

[MacKerell Jr. *et al.* 1998]   A. D. MacKerell Jr., B. Brooks, III C. L. Brooks, L. Nilsson, B. Roux, Y. Won, and M. Karplus. CHARMM: The energy function and its parameterization with an overview of the program. In P. v. R. Schleyer et al., editors, *The Encyclopedia of Computational Chemistry*, volume 1. John Wiley and Sons, 1998.

[MacRedmond *et al.* 2005]   R. MacRedmond, C. Greene, C. Taggart, N. McElvaney, and S. O'Neill. Respiratory epithelial cells require toll-like receptor 4 for induction of human β-defensin 2 by lipopolysaccharide. *Respiratory Research*, 6:116, 2005.

[Madden *et al.* 1991]   D. R. Madden, J. C. Gorga, J. L. Strominger, and D. C. Wiley. The structure of HLA-B27 reveals nonamer self-peptides bound in an extended conformation. *Nature*, 353:321–325, 1991.

[Mallios 2001]   R. R. Mallios. Predicting class II MHC/peptide multi-level binding with an iterative stepwise discriminant analysis meta-algorithm. *Bioinformatics*, 17:942–948, 2001.

[Mamitsuka 1998]   H. Mamitsuka. Predicting peptides that bind to MHC molecules using supervised learning of hidden markov models. *Proteins*, 33:460–474, 1998.

[Manz *et al.* 2002]   R. A. Manz, S. Arce, G. Cassese, A. E. Hauser, F. Hiepe, and A. Radbruch. Humoral immunity and long-lived plasma cells. *Curr. Opin. Immunol.*, 14:517, 2002.

[Maruyama & Nei 1981]   T. Maruyama and M. Nei. Genetic variability maintained by mutation and overdominant selection in finite populations. *Genetics*, 98:441–459, 1981.

[Mason 1998]   D. Mason. A very high level of crossreactivity is an essential feature of the T-cell receptor. *Immunol. Today*, 19:395–404, 1998.

[Mason 2001]   D. Mason. Some quantitative aspects of T-cell repertoire selection: the requirement for regulatory T cells. *Immunol. Rev.*, 182:80–88, 2001.

[Mata & Cohn 2006a]   J. Mata and M. Cohn. A cellular automata based synthetic immune system (SIS) that can be used to evaluate hypotheses describing responsiveness. *Immunol. Rev.*, in press, 2006.

[Mata & Cohn 2006b]   J. Mata and M. Cohn. The tritope model for restrictive recognition of antigen by T-cells: III. a computer program (TUNI I) to analyze the parameters of the model. *Molecular Immunology*, in press, 2006.

[Maturana & Varela 1980]   H. R. Maturana and F. J. Varela. *Autopoiesis and Cognition*. Reidel, 1980.

[Matzinger *et al.* 1984]   P. Matzinger, R. Zamoyska, and H. Waldmann. Self tolerance is H-2-restricted. *Nature*, 308:738–741, 1984.

[Matzinger 1994a]   P. Matzinger. Memories are made of this? *Nature*, 369:605–606, 1994.

[Matzinger 1994b]   P. Matzinger. Tolerance, danger and the extended family. *Ann. Rev. Immunol.*, 12:991–1045, 1994.

[Matzinger 2002]   P. Matzinger. The danger model: A renewed sense of self. *Science*, 296:301–305, 2002.

[McDonough *et al.* 1993]   K. McDonough, Y. Kress, and B. Bloom. Pathogenesis of tuberculosis: interaction of mycobacterium tuberculosis with macrophages. *Infect. Immun.*, 61:2763–2773, 1993.

[McFarland *et al.* 1999]   B. J. McFarland, A. J. Sant, T. P. Lybrand, and C. Beeson. Ovalbumin (323-339) peptide bind to the major histocompatibility complex class II , I-A(d) protein using two functionally distinct registers. *Biochemistry*, 38:16663–16670, 1999.

[McHeyzer-Williams & McHeyzer-Williams 2005]   L. J. McHeyzer-Williams and M. G. McHeyzer-Williams. Antigen-specific memory B cell development. *Ann. Rev. Immunol.*, 23:487–513, 2005.

[McIntyre *et al.* 2003]   T. McIntyre, S. Prescott, A. Weyrich, and G. Zimmerman. Cell-cell interactions: leukocyte-endothelial interactions. *Curr. Op. Hematology*, 10(2):150–158, 3 2003.

[McKenna & Kapp 2004]   K. C. McKenna and J. A. Kapp. Ocular immune privilege and CTL tolerance. *Immunol Res*, 29:103–112, 2004.

[McKenna *et al.* 2005]   K. McKenna, A. S. Beignon, and N. Bhardwaj. Plasmacytoid dendritic cells: Linking innate and adaptive immunity. *J. Virology*, 79(1):17–27, 1 2005.

[McKenzie & Pecon-Slattery 1992]   L. M. McKenzie and J. Pecon-Slattery. Taxonomic hierarchy of HLA class I allele sequences. *Genes Immunol.*, 1(2):120–129, 1992.

[McKenzie *et al.* 2006]   B. S. McKenzie, R. A. Kastelein, and D. J. Cua. Understanding the IL-23-IL-17 immune pathway. *Trends in Immunology*, 27(1):17–23, 2006.

[McLachlan 1982]   A. D. McLachlan. Rapid comparison of protein structures. *Acta Cryst.*, A38:871–873, 1982.

[McSparron *et al.* 2003]   H. McSparron, M. J. Blythe, C. Zygouri, I. A. Doytchinova, and D. R. Flower. Jenpep: A novel computational information resource for immunobiology and vaccinology. *J. Chem. Inf. Comp. Sc.*, 43:1276–1287, 2003.

[Meade & Sonneborn 1996]   A. J. Meade and H. C. Sonneborn. Numerical solution of a calculus of variations problem using the feedforward neural network architecture. *Advances in Engineering Software*, 27:213–225, 1996.

[Medzhitov & Janeway 2000]   R. Medzhitov and C. A. Janeway. How does the immune system distinguish self from nonself? *Seminars in Immunology*, 12(3):185–188, 2000.

[Medzhitov & Janeway 2002]   R. Medzhitov and C. A. Janeway. Decoding the patterns of self and nonself by the innate immune system. *Science*, 296(5566):298–300, 2002.

[Meerwijk *et al.* 1997]   J. P. Van Meerwijk, S. Marguerat, R. K. Lees, R. N. Germain, B. J. Fowlkes, and H. R. MacDonald. Quantitative impact of thymic clonal deletion on the T cell repertoire. *J. Exp. Med.*, 185:377–383, 1997.

[Meier-Schellersheim & Mack 1999]   M. Meier-Schellersheim and G. Mack. SIMMUNE, a tool for simulating and analyzing immune system behavior, 1999. http://www-library.desy.de/.

[Mellor *et al.* 2002]   A. L. Mellor, P. Chandler, G. K. Lee, T. Johnson, D. B. Keskin, and D. H. Munn. Indoleamine 2,3-dioxygenase, immunosuppression and pregnancy. *J. Reprod. Immunol.*, 57(1-2):143–150, 2002.

[Mendez-Samperio & Jimenez-Zamudio 1991]   P.   Mendez-Samperio   and
L. Jimenez-Zamudio. Peptide competition at the level of MHC-binding
sites using T cell clones from a rheumatoid arthritis patient. *J. Autoimmunity*,
4(5):795–806, 1991.

[Merkenschlager et al. 1997]   M. Merkenschlager, D. Graf, M. Lovatt,
U. Bommhardt, R. Zamoyska, and A. G. Fisher. How many thymocytes
audition for selection? *J. Exp. Med.*, 186:1149–1158, 1997.

[Meyer et al. 1998]   C. G. Meyer, J. May, and K. Stark. Human leukocyte antigens
in tuberculosis and leprosy. *Trends in Microbiology*, 6(4):148–154, 1998.

[Middleton et al. 2003]   D. Middleton, L. Menchaca, H. Rood, and R. Komerof-
sky. New allele frequency database: http://www.allelefrequencies.net. *Tissue
Antigens*, 61(5):403–407, 2003.

[Mikko & Andersson 1995]   S. Mikko and L. Andersson. Low major histocompatibility complex class II diversity in European and North American moose. *Proc.
Natl. Acad. Sci. USA*, 92(10):4259–4263, 1995.

[Milik et al. 1998]   M. Milik, D. Sauer, L. Brunmark, A. P.and Yuan, A. Vitiello,
MR. Jackson, P. A. Peterson, J. Skolnick, and C. A. Glass. Application of
an artificial neural network to predict specific class I MHC binding peptide
sequences. *Nature Biotechnology*, 16:753–756, 1998.

[Miller & Basten 1996]   J. Miller and A. Basten. Mechanism of tolerance to self.
Curr. Op. Immunol, 8:815–821, 1996.

[Miller et al. 1998]   D. M. Miller et al. Human cytomegalovirus inhibits major
histocompatibility complex class II expression by disruption of the Jak/Stat
pathway. *J. Exp. Med.*, 187(5):675–83, 1998.

[Miller 2004]   J. Miller. Self-nonself discrimination by T lymphocytes. *C.R. Biolo-
gies*, 327:399–408, 2004.

[Mitchell 1997]   T. Mitchell. *Machine Learning*. McGraw-Hill, 1997.

[Mohan et al. 2001]   V. Mohan et al. Effects of tumor necrosis factor alpha on host
immune response in chronic persistent tuberculosis: possible role for limiting
pathology. *Infection and Immunity*, 69:1847–1855, 2001.

[Moller 1988]   G. Moller. Do suppressor T-cells exist? *Scandinavian J. Immunol.*,
27(3):247–250, 1988.

[Moodley et al. 2000]   Y. P. Moodley et al. Correlation of CD4 : CD8 ratio and tumour necrosis factor (TNF)alpha levels in induced sputum with bronchoalveolar
lavage fluid in pulmonary sarcoidosis. *Thorax*, 55(8):696–699, 2000.

[Moore et al. 2002]   C. B. Moore, M. John, I. R. James, F. T. Christiansen, C. S.
Witt, and S. A. Mallal. Evidence of HIV-1 adaptation to HLA-restricted immune
responses at a population level. *Science*, 296:1439–1443, 2002.

[Moreno et al. 1998]   C. Moreno, A. Mehlert, and J. Lamb. The inhibitory effects
of mycobacterial lipoarabinomannan and polysaccharides upon polyclonal and
monoclonal human T cell proliferation. *Clin. Exp. Immunol.*, 74(2):206–10,
1998.

[Mosmann & Coffman 1989]   T. R. Mosmann and R. L. Coffman. Th1-Cell And
Th2-Cell - different patterns of lymphokine secretion lead to different functional
properties. *Ann. Rev. Immunol.*, 7:145–173, 1989.

[Muller et al. 1987]   I. Muller et al. Impaired resistance to Mycobacterium tuberculosis infection after selective in vivo depletion of L3T4+ and Lyt-2+ T cells.
Infect. Immun., 55(9):2037–2041, 1987.

430   References

[Murphy et al. 2002]   B. M. Murphy, B. H. Singer, S. Anderson, and D. Kirschner. Comparing epidemic TB in demographically distinct heterogeneous populations. *Math. Biosci.*, 180:161–185, 2002.

[Murphy et al. 2003]   B. M. Murphy, B. H. Singer, and D. Kirschner. On the treatment of TB in heterogeneous populations. *J. Theor. Biol.*, 223(391-404), 2003.

[Murray 1990]   J. D. Murray. *Mathematical Biology*. Springer, corrected 2nd edition, 1990.

[Měch & Prusinkiewicz 1996]   R. Měch and P. Prusinkiewicz. Visual models of plants interacting with their environment. In *SIGGRAPH 96*, pages 397–410. ACM Press, 1996.

[Myrvik et al. 1984]   Q. Myrvik, E. Leake, and M. M. Wright. Disruption of phagosomal membranes of normal alveolar macrophages by the H37Rv strain of Mycobacterium tuberculosi. *Am. Rev. Respir. Dis*, 129:322–328, 1984.

[Napolitani et al. 2005]   G. Napolitani, A. Rinaldi, F. Bertoni, F. Sallusto, and A. Lanzavecchia. Selected toll-like receptor agonist combinations synergistically trigger a T helper type 1-polarizing program in dendritic cells. *Nat. Immunol.*, 6(8):769–776, 2005.

[Nasaroui et al. 2002]   O. Nasaroui, F. González, and D Dasgupta. The fuzzy artificial immune system: Motivations, basic concepts, and application to clustering and web profiling. In *IEEE International Conference on Fuzzy Systems*, pages 711–716, Hawaii, HI, 2002.

[Nathan et al. 1983]   C. Nathan et al. Identification of interferon-gamma as the lymphokine that activates human macrophage oxidative metabolism and antimicrobial activity. *J. Exp. Med.*, 158(670-689), 1983.

[Neal 2002]   M. Neal. An artificial immune system for continuous analysis of time-varying data. In [Timmis & Bentley 2002], pages 76–85.

[Neal 2003]   M. Neal. Meta-stable memory in an artificial immune network. In [Timmis et al. 2003], pages 168–180.

[Newborough & Stepney 2005]   R. Newborough and S. Stepney. A generic framework for population based algorithms. In [Jacob et al. 2005].

[Ngo & Marks 1992]   J. T. Ngo and J. Marks. Computational complexity of a problem in molecular-structure prediction. *Protein Eng.*, 5(4):313–321, 1992.

[Ngo et al. 1994]   J. T. Ngo, J. Marks, and M. Karplus. Computational complexity, protein structure prediction, and the levinthal paradox. In K. Merz and S. Le Grand, editors, *The Protein Folding Problem and Tertiary Structure Prediction*. Birkhauser, 1994.

[Nicosia et al. 2004]   G. Nicosia, V. Cutello, P. Bentley, and J. Timmis, editors. *Proc. of the 3rd International Conference on Artificial Immune Systems (ICARIS)*, volume 3239 of *LNCS*. Springer, 2004.

[Nicosia 2004]   G. Nicosia. *Immune Algorithms for Optimization and Protein Structure Prediction*. PhD thesis, University of Catania, 2004.

[Noss et al. 2000]   E. H. Noss, C. V. Harding, and W. H. Boom. Mycobacterium tuberculosis inhibits MHC class II antigen processing in murine bone marrow macrophages. *Cell*, 201(1):63–74, 2000.

[Nossal 1994]   G. J. Nossal. Negative selection of lymphocytes. *Cell*, 76:229–239, 1994.

[Novitsky et al. 2003]   V. Novitsky, P. Gilbert, T. Peter, M. F. McLane, S. Gaolekwe, N. Rybak, I. Thior, T. Ndung'u, R. Marlink, T. H. Lee, and

M. Essex. Association between virus-specific T-cell responses and plasma viral load in human immunodeficiency virus type 1 subtype C infection. *J. Virol.*, 77(2):882–890, January 2003.

[Nowak & May 2000]   M. A. Nowak and R. A. May. *Virus dynamics*. Oxford University Press, 2000.

[Nowak *et al.* 1992]   M. A. Nowak, K. Tarczy-Hornoch, and J. M. Austyn. The optimal number of major histocompatibility complex molecules in an individual. *Proc. Natl. Acad. Sci. USA*, 89:10896–10899, 1992.

[Ober *et al.* 1997]   C. Ober, L. R. Weitkamp, N. Cox, H. Dytch, D. Kostyu, and S. Elias. HLA and mate choice in humans. *Am. J. Hum. Genet.*, 61:497–504, 1997.

[O'Brien & Yuhki 1999]   S. J. O'Brien and N. Yuhki. Comparative genome organization of the major histocompatibility complex: lessons from the Felidae. *Immunol. Rev.*, 167:133–144, February 1999.

[Ochsenbein & Zinkernagel 2000]   A. Ochsenbein and R. Zinkernagel. Natural antibodies and complement link innate and acquired immunity. *Immunology Today*, 21(12), 12 2000.

[O'Connor 2002]   D. H. O'Connor. Acute phase cytotoxic T lymphocyte escape is a hallmark of simian immunodeficiency virus infection. *Nature Med.*, 8:493–499, 2002.

[Ohki *et al.* 1987]   H. Ohki, C. Martin, C. Corbel, Coltey, and N. Le Douarin. Tolerance induced by thymic epithelial grafts in birds. *Science*, 237:1032–1035, 1987.

[Oksendal 2003]   B. Oksendal. *Stochastic Differential Equations: An Introduction with Applications*. Springer, 2003.

[Orange *et al.* 2002]   J. S. Orange, M. S. Fassett, L. A. Koopman, J. E. Boyson, and J. L. Strominger. Viral evasion of natural killer cells. *Nat. Immunol.*, 3(11):1006–1012, 2002.

[Orme & Cooper 1999]   I. M. Orme and A. M. Cooper. Cytokine/chemokine cascades in immunity to tuberculosis. *Immunol. Today*, 20(7):307–312, 1999.

[Orme 1999a]   I. M. Orme. Beyond BCG: the potential for a more effective TB vaccine. *Mol. Med. Today*, 5(11):487–492, 1999.

[Orme 1999b]   I. M. Orme. New vaccines against tuberculosis. the status of current research. *Infect. Dis. Clin. North. Am.*, 13(1):169–185, 1999.

[Orme 1999c]   I. M. Orme. Vaccination against tuberculosis: recent progress. *Adv. Vet. Med.*, 41:135–143, 1999.

[Ott 1993]   E. Ott. *Chaos in dynamical systems*. Cambridge University Press, 1993.

[Ottaviani & Franceschi 1996]   E. Ottaviani and C. Franceschi. The Neuroimmunology of stress from invertebrates to man. *Progress in Neurobiology*, 48:421–440, 1996.

[Ottenhoff *et al.* 1998]   T. H. Ottenhoff, D. Kumararatne, and J. L. Casanova. Novel human immunodeficiencies reveal the essential role of type-I cytokines in immunity to intracellular bacteria. *Immunol. Today*, 19(11):491–494, 1998.

[Pagie & Hogeweg 2000]   L. Pagie and P. Hogeweg. Individual- and population-based diversity in restriction-modification systems. *Bull. Math. Biol.*, 62:759–774, 2000.

[Pai 2002]   R. K. Pai. Regulation of class II MHC expression in APCs: roles of types I, III, and IV class II transactivator. *J. Immunol.*, 169(3):1326–33, 2002.

[Paine & Flower 2002]   K. Paine and D. R. Flower. Bacterial bioinformatics: Pathogenesis and the genome. *Journal of Molecular and Microbial Biotechnology*, 4:357–365, 2002.

[Panka *et al.* 1988]   D. J. Panka, M. Mudgett-Huner, D. R. Parks, L. L. Peterson, L. A. Herzenberg, E. Haber, and M. N. Margolies. Variable region framework differences results in decreased or increased affinity of variant anti-digoxin antibodies. *Proc. Natl. Acad. Sci. USA*, 85:3080–3084, 1988.

[Pantophlet *et al.* 2003]   R. Pantophlet, E. O. Saphire, P. Poignard, P. W. H. I. Parren, I. A. Wilson, and D. R. Burton. Fine mapping of the interaction of neutralizing and nonneutralizing monoclonal antibodies with the CD4 binding site of human immunodeficiency virus type 1 gp120. *J. Virol.*, 77:642–658, 2003.

[Parham & Ohta 1996]   P. Parham and T. Ohta. Population biology of antigen presentation by MHC class I molecules. *Science*, 272:67–74, 1996.

[Parham *et al.* 1989a]   P. Parham, R. J. Benjamin, B. P. Chen, C. Clayberger, P. D. Ennis, A. M. Krensky, D. A. Lawlor, D. R. Littman, A. M. Norment, H. T. Orr, R. D. Salter, and J. Zemmour. Diversity of class I HLA molecules: Functional and evolutionary interactions with T cells. *Cold Spring Harbor Symp. Quant. Biol.*, 54:529–543, 1989.

[Parham *et al.* 1989b]   P. Parham, D. A. Lawlor, C. E. Lomen, and P. D. Ennis. Diversity and diversification of HLA A,B,C alleles. *J. Immunol.*, 142:3937–3950, 1989.

[Park & Deem 2004]   J. M. Park and M. W. Deem. Correlations in the T-cell response to altered peptide ligands. *Physica A*, 341:455–470, 2004.

[Parker *et al.* 1994]   K. C. Parker, M. A. Bednarek, and J. E. Coligan. Scheme for ranking potential HLA-A2 binding peptides based on independent binding of individual peptide side-chains. *J. Immunol.*, 152:163–175, 1994.

[Parks & Weigle 1980]   D. E. Parks and W. O. Weigle. Maintenance of immunologic unresponsiveness to human gamma globulin: evidence for irreversible inactivation in B lymphocytes. *J. Immunol.*, 124:1230–1236, 1980.

[Parks *et al.* 1978]   D. E. Parks, M. V. Doyle, and W. O. Weigle. Induction and mode of action of suppressor cells generated against human gamma globulin. I. an immunologic unresponsive state devoid of demonstrable suppressor cells. *J. Exp. Med.*, 148:625–638, 1978.

[Parnes 2004]   O. Parnes. From interception to incorporation: Degeneracy and promiscuous recognition as precursors of a paradigm shift in immunology. *Molecular Immunology*, 40:985–991, 2004.

[Paul *et al.* 1951]   J. R. Paul, J. T. Riordan, and J. L. Melnick. Antibodies to three different antigenic types of poliomyelitis in sera from north alaskan eskimos. *Am. J. Hyg.*, 54:275–285, 1951.

[Paul 1999]   W. E. Paul. *Fundamental Immunology*. Raven Press, 1999.

[Peitgen & Richter 1986]   H. O. Peitgen and D. H. Richter. *The Beauty of Fractals: Images of Complex Dynamical Systems*. Springer, 1986.

[Penn *et al.* 2002]   D. J. Penn, K. Damjanovich, and W. K. Potts. MHC heterozygosity confers a selective advantage against multiple-strain infections. *Proc. Natl. Acad. Sci. USA*, 99:11260–11264, 2002.

[Penn 2002]   D. J. Penn. The scent of genetic compatibility: Sexual selection and the major histocompatibility complex. *Ethology*, 108:1–21, 2002.

[Percus *et al.* 1993]   J. K. Percus, O. E. Percus, and A. S. Perelson. Predicting the size of the antibody combining region from consideration of efficient self-nonself discrimination. *Proc. Natl. Acad. Sci. USA*, 90:1691–1695, 1993.

[Perelson & Macken 1995]   A. S. Perelson and C. A. Macken. Protein evolution on partially correlated landscape. *Proc. Natl. Acad. Sci. USA*, 92:9657–9661, 1995.

[Perelson & Oster 1979]   A. S. Perelson and G. F. Oster. Theoretical studies of clonal selection: minimal antibody repertoire size and reliability of self–non-self discrimination. *J. Theor. Biol.*, 81:645–670, 1979.

[Perelson & Weisbuch 1997]   A. S. Perelson and G. Weisbuch. Immunology for physicists. *Rev. Mod. Phys.*, 69(4):1219–1267, 1997.

[Perelson et al. 1976]   A. S. Perelson, M. Mirmirani, and G. F. Oster. Optimal strategies in immunology. I. B-cell differentiation and proliferation. *J. Math. Biol.*, 3:325–367, 1976.

[Perelson 1989]   A. S. Perelson. Immune network theory. *Immunological Review*, 110:5–36, 1989.

[Perelson 2002]   A. Perelson. Modelling viral and immune system dynamics. *Nature*, 2:28–36, 2002.

[Perutz 1970]   M. F. Perutz. Stereochemistry of cooperative effects of hemoglobin. *Nature*, 228:726–739, 1970.

[Petrovsky & Brusic 2004]   N. Petrovsky and V. Brusic. Virtual models of the HLA class I antigen processing pathway. *Methods*, 34(4):429–35, 2004.

[Peyerl et al. 2004]   F. W. Peyerl, H. S. Bazick, M. H. Newberg, D. H. Barouch, J. Sodroski, and N. L. Letvin. Fitness costs limit viral escape from cytotoxic T lymphocytes at a structurally constrained epitope. *J. Virol.*, 78(24):13901–13910, 2004.

[Picca & Caton 2005]   C. C. Picca and A. J. Caton. The role of self-peptides in the development of CD4+ CD25+ regulatory cells. *Curr. Op. Immunol.*, 17:131–136, 2005.

[Pichler 2002]   W. J. Pichler. Modes of presentation of chemical neoantigens to the immune system. *Toxicology*, 181:49–54, 2002.

[Pile 1999]   K. Pile. Broadsheet number 51: HLA and disease associations. *Pathology*, 31:202–212, 1999.

[Pinch 1993]   E. R. Pinch. *Optimal Control and the Calculus of Variations*. Oxford Science Publications, OUP, 1993.

[Polani et al. 2001]   D. Polani, T. Martinetz, and J. T. Kim. An information-theoretic approach for the quantification of relevance. In *ECAL 2001*, volume 2159 of *LNCS*, pages 704–713. Springer, 2001.

[Pollastri et al. 2002]   G. Pollastri, D. Przybylski, B. Rost, and P. Baldi. Improving the prediction of protein secondary structure in three and eight classes using recurrent neural networks and profiles. *Proteins*, 47(2):228–235, 2002.

[Poole & Claman 2004]   J. A. Poole and H. N. Claman. Immunology of pregnancy. implications for the mother. *Clin. Rev. Allergy Immunol.*, 26:161–170, 2004.

[Potts et al. 1991]   W. K. Potts, C. J. Manning, and E. K. Wakeland. Mating patterns in seminatural populations of mice influenced by MHC genotype. *Nature*, 352:619–621, 1991.

[Probst-Kepper et al. 2001]   M. Probst-Kepper, V. Stroobant, R. Kridel, B. Gaugler, C. Landry, F. Brasseur, J. P. Cosyns, B. Weynand, T. Boon, and B. J van den Eynde. An alternative open reading frame of the human macrophage colony-stimulating factor gene is independently translated and codes for an antigenic peptide of 14 amino acids recognized by tumor-infiltrating CD8 T lymphocytes. *J. Exp. Med.*, 193:1189–1198, 2001.

[Probst-Kepper et al. 2004]   M. Probst-Kepper, H. J. Hecht, H. Herrmann, V. Janke, F. Ocklenburg, J. Klempnauer, B. J. van den Eynde, and S. Weiss. Conformational restraints and flexibility of 14-meric peptides in complex with HLA-B*3501. J. Immunol., 173:5610–5616, 2004.

[Profit Program]   Profit Program. http://www.bioinf.org.uk/software/.

[Prusinkiewicz & Lindenmayer 1990]   P. Prusinkiewicz and A. Lindenmayer. The Algorithmic Beauty of Plants. Springer, 1990.

[Pulendran 2005]   B. Pulendran. Variegation of the immune response with dendritic cells and pathogen recognition receptors. J. Immunol., 173:2457–2465, 2005.

[Purisima & Scheraga 1987]   E. O. Purisima and H. A. Scheraga. An approach to the multiple-minima problem in protein folding by relaxing dimensionality test on enkephalin. J. Molec. Biol., 196:697–709, 1987.

[Quick & Dautenhahn 1999]   T. Quick and K. Dautenhahn. Making embodiment measurable. In Embodied Mind / Alife Workshop, KogWis'99, 1999.

[Quick et al. 1999]   T. Quick, K. Dautenhahn, C. L. Nehaniv, and G. Roberts. On bots and bacteria: ontology independent embodiment. In ECAL'99, volume 1674 of LNAI, pages 339–343. Springer, 1999.

[Quick et al. 2000]   T. Quick, K. Dautenhahn, C. L. Nehaniv, and G. Roberts. The essence of embodiment: A framework for understanding and exploiting structural coupling between system and environment. In CASYS'99, volume 517 of AIP Conference Proceedings, pages 649–660, 2000.

[Rabow & Scheraga 1996]   A. A. Rabow and H. A. Scheraga. Improved genetic algorithm for the protein folding problem by use of cartesian combination operators. Protein Science, 5:1800–1815, 1996.

[Rajalingam et al. 1996]   R. Rajalingam et al. Polymerase chain reaction–based sequence-specific oligonucleotide hybridization analysis of HLA class II antigens in pulmonary tuberculosis: relevance to chemotherapy and disease severity. J. Infect. Dis., 173(3):669–676, 1996.

[Ramachandran & Sassiekharan 1968]   G. B. Ramachandran and V. Sassiekharan. Conformation of polypeptides and proteins. Adv. Protein Chem., 23:283–437, 1968.

[Rammensee et al. 1995]   H. G. Rammensee, T. Friede, and S. Stevanovic. MHC ligands and peptide motifs: first listing. Immunogenetics, 41:178–228., 1995.

[Rammensee et al. 1999]   H. G. Rammensee, J. Bachmann, N. P. Emmerich, OA. Bachor, and S. Stevanovic. SYFPEITHI, a database for MHC ligands and peptide motifs. Immunogenetics, 50:213–219, 1999.

[Ramsay 1972]   W. S. Ramsay. Analysis of individual leucocyte behaviour during chemotaxis. Exp. Cell. Res., 70:129–139, 1972.

[Reche & Reinherz 2003]   P. A. Reche and E. L. Reinherz. Sequence variability analysis of human class I and class II MHC molecules: functional and structural correlates of amino acid polymorphisms. J. Molec. Biol., 331(3):623–641, 2003.

[Reddehase et al. 1989]   M. J. Reddehase, J. B. Rothbard, and U. H. Koszinowski. A pentapeptide as minimal antigenic determinant for MHC class I-restricted T lymphocytes. Nature, 337:651–653, 1989.

[Reinhardt et al. 2001]   R. L. Reinhardt, A. Khoruts, R. Merica, T. Zell, and M. K. Jenkins. Visualizing the generation of memory CD4 T cells in the whole body. Nature, 410(6824):101–105, 2001.

[Remondini et al. 2003]   D. Remondini, A. Bazzani, C. Franceschi, F. Bersani, E. Verondini, and G. C . Castellani. Role of connectivity in immune and neural network models: memory development and aging. Riv. Biol, 96:225–39, 2003.

[Reusch et al. 2001]   T. B. Reusch, M. A. Haberli, P. B. Aeschlimann, and M. Milin-ski. Female sticklebacks count alleles in a strategy of sexual selection explaining MHC polymorphism. *Nature*, 414:300–302, 2001.

[Richman et al. 2003]   D. D. Richman, T. Wrin, S. J. Little, and C. J. Petropoulos. Rapid evolution of the neutralizing antibody response to HIV type 1 infection. *Proc. Natl. Acad. Sci. USA*, 100:4144–4149, 2003.

[Riegler 2002]   A. Riegler. When is a cognitive system embodied? *Cognitive Systems Research*, 3:339–348, 2002.

[Roberts et al. 1987]   S. Roberts, J. C. Cheetham, and A. R. Rees. Generation of an antibody with enhanced affinity and specificity for its antigen by protein. *Nature*, 328:731–734, 1987.

[Robinson et al. 2003]   J. Robinson, M. J. Waller, P. Parham, N. de Groot, R. Bontrop, L. J. Kennedy, P. Stoehr, and S. G. E. Marsh. IMGT/HLA and IMGT/MHC: sequence databases for the study of the major histocompatibility complex. *Nucl. Acids Res.*, 31:311, 2003.

[Robinson et al. 2005]   J. Robinson, M. J. Waller, P. Stoehr, and S. G. E. Marsh. IPD–the Immuno Polymorphism Database. *Nucl. Acids Res.*, 33:D523–D526, January 2005.

[Rodgers & Cook 2002]   J. R. Rodgers and R. G. Cook. MHC class IB molecules bridge innate and acquired immunity. *Nat. Rev. Imm.*, 5(6):459–471, 2002.

[Rognan et al. 1999]   D. Rognan, S. L. Lauemoller, A. Holm, S. Buus, and V. Tschinke. Predicting binding affinities of protein ligands from three-dimensional models: application to peptide binding to class I major histocompatibility proteins. *J. Med. Chem.*, 4(42):4650–8, 1999.

[Roitt 1997]   I. Roitt. *Essential Immunology*. Blackwell Science, 9th edition, 1997.

[Rossi & Young 2005]   M. Rossi and J. W. Young. Human dendritic cells: Potent antigen-presenting cells at the crossroads of innate and adaptive immunity. *J. Immunol.*, 175(3):1373–1381, 2005.

[Rothbard et al. 1994]   J. B. Rothbard, K. Marshall, K. J. Wilson, L. Fugger, and D. Zaller. Prediction of peptide affinity to HLA DRB1*0401. *Int. Arch. Allergy Immunol.*, 105:1–7, 1994.

[Rudikoff et al. 1982]   S. Rudikoff, A. M. Ginsti, W. D. Cook, and M. D. Scharff. Single amino-acid substitution altering antigen binding specificity. *Proc. Natl. Acad. Sci. USA*, 79:1979–1983, 1982.

[Ruppert et al. 1993]   J. Ruppert, J. Sidney, E. Celis, R. T. Kubo, H. M. Grey, and A. Sette. Prominent role of secondary anchor residues in peptide binding to HLA-A*0201 molecules. *Cell*, 74:929–937, 1993.

[Santambrogio et al. 1999]   L. Santambrogio et al. Abundant empty class II MHC molecules on the surface of immature dendritic cells. *Proc. Natl. Acad. Sci. USA*, 96(26):15050–5, 1999.

[Saper et al. 1991]   M. A. Saper, P. J. Bjorkman, and D. C. Wiley. Refined structure of the human histocompatibility antigen HLA-A2 at 2.6A resolution. *J. Molec. Biol.*, 219:277–319, 1991.

[Saunders & Cooper 2000]   B. Saunders and A. Cooper. Restraining mycobacteria: role of granulomas in mycobacterial infection. *Immunol Cell Biol*, 78:334–341, 2000.

[Saunders 1993]   P. T. Saunders. The organism as a dynamical system. In W. D. Stein and F. J. Varela, editors, *Thinking About Biology*. Addison-Wesley, 1993.

[Sawyer 1931]   W. Sawyer. The persistance of yellow fever immunity. *J. Prev. Med.*, 5:413–428, 1931.

[Saxova et al. 2003]   P. Saxova, S. Buus, S. Brunak, and C. Kesmir. Predicting proteasomal cleavage sites: a comparison of available methods. *Int. Immunol.*, 15:781–787, 2003.

[Scherer et al. 2004]   A. Scherer et al. Quantifiable cytotoxic T lymphocyte responses and HLA-related risk of progression to AIDS. *Proc. Natl. Acad. Sci. USA*, 101:12266–12270, 2004.

[Schluns & Lefrancois 2003]   K. S. Schluns and L. Lefrancois. Cytokine control of memory T-cell development and survival. *Nat. Rev. Imm.*, 3:269–279, 2003.

[Schnare et al. 2001]   M. Schnare, G. M. Barton, A. C. Holt, K. Takeda, S. Akira, and R. Medzhitov. Toll-like receptors control activation of adaptive immune responses. *Nat Immunol*, 2:947–950, 2001.

[Schomburg et al. 2002]   I. Schomburg, A. Chang, O. Hofmann, C. Ebeling, F. Ehrentreich, and D. Schomburg. BRENDA: a resource for enzyme data and metabolic information. *Trends Biochem Sc*, 27:54–56, 2002.

[Schonbach et al. 2000]   C. Schonbach, J. L. Y. Koh, X. Sheng, L. Wong, and V. Brusic. FIMM, a database of functional molecular immunology. *Nucleic Acids Res*, 28:222–224, 2000.

[Schonbach et al. 2005]   C. Schonbach, J. L. Koh, D. R. Flower, and V. Brusic. An update on the functional molecular immunology (FIMM) database. *Appl. Bioinformatics*, 4:25–31, 2005.

[Schumacher & Heemels 1990]   T. N. Schumacher and M. T. Heemels. Direct binding of peptide to empty MHC class I molecules on intact cells and in vitro. *Cell*, 62(3):563–567, 1990.

[Scollay et al. 1980]   R. G. Scollay, E. C. Butcher, and I. L. Weissman. Thymus cell migration. Quantitative aspects of cellular traffic from the thymus to the periphery in mice. *Eur. J. Immunol.*, 10:210–218, 1980.

[Scott 1999]   A. Scott. *Nonlinear Science*. Oxford, 1999.

[Searle 1980]   John Searle. Minds, brains, and programs. *Behavioral and Brain Sciences*, 3:417–424, 1980.

[Secker et al. 2003a]   A. Secker, A. Freitas, and J. Timmis. AISEC: An artificial immune system for email classification. In *Proc. of Congress on Evolutionary Computation (CEC)*, pages 131–139, 2003.

[Secker et al. 2003b]   A. Secker, A. Freitas, and J. Timmis. A danger theory inspired approach to web mining. In [Timmis et al. 2003], pages 156–167.

[Segal & Perelson 1988]   L. A. Segal and A. S. Perelson. Computations in shape-space: a new approach to immune network theory. In *Theoretical Immunology, Part two*, 1988.

[Segovia-Juarez et al. 2004]   J. L. Segovia-Juarez, S. Ganguli, and D. Kirschner. Identifying control mechanisms of granuloma formation during M. tuberculosis infection using an agent-based model. *J. Theor. Biol.*, 231(3):357–76, 2004.

[Selin et al. 1994]   L. Selin, S. Nahill, and R. Welsh. Cross-reactivities in memory cytotoxic T-lymphocyte recognition of heterologous viruses. *J. Exp. Med.*, 179:1933–1943, 1994.

[Selin et al. 2004]   L. K. Selin, M. Cornberg, M. A. Brehm, S. K. Kim, C. Calcagno, D. Ghersi, R. Puzone, F. Celada, and R. M. Welsh. CD8 memory T cells: cross-reactivity and heterologous immunity. *Semin. Immunol.*, 16:335–347, 2004.

[Selvaraj et al. 1998]   P. Selvaraj, H. Uma, A. M. Reetha, S. M. Kurian, T. Xavier, R. Prabhakar, and P. R. Narayanan. HLA antigen profile in pulmonary tuberculosis patients and their spouses. *The Indian Journal Of Medical Research*, 107:155–158, 1998.

[Serbina & Flynn 1999] N. V. Serbina and J. L. Flynn. Early emergence of CD8+ T cells primed for production of type 1 cytokines in the lungs of Mycobacterium tuberculosis-infected mice. *Infection and Immunity*, 67:3980–3988, 1999.

[Serbina et al. 2000] N. V. Serbina, C. C. Liu, C. A. Scanga, and J. L. Flynn. CD8+ CTL from lungs of mycobacterium tuberculosis-infected mice express perforin in vivo and lyse infected macrophages. *J. Immunol.*, 165(1):353–363, 2000.

[Sercarz & Maverakis 2004] Eli E. Sercarz and E. Maverakis. Recognition and function in a degenerate immune system. *Molecular Immunology*, 40:1003–1008, 2004.

[Sette & Livingston 2001] A. Sette and B. Livingston. The development of multi-epitope vaccines: Epitope identification, vaccine design and clinical evaluation. *Biologicals*, 29(3-4):271–276, 2001.

[Sette & Sidney 1998] A. Sette and J. Sidney. HLA supertypes and supermotifs: a functional perspective on HLA polymorphism. *Curr. Op. Immunol.*, 10(4):478–482, 1998.

[Sette & Sidney 1999] A. Sette and J. Sidney. Nine major HLA class I supertypes account for the vast preponderance of HLA-A and -B polymorphism. *Immunogenetics.*, 50:201–212, 1999.

[Sette et al. 1989] A. Sette, S. Buus, E. Appella, J. A. Smith, R. Chesnut, C. Miles, S. M. Colon, and H. M. Grey. Prediction of major histocompatibility complex binding regions of protein antigens by sequence pattern analysis. *Proc. Natl. Acad. Sci. USA*, 86:3296–3300, 1989.

[Sette et al. 1994a] A. Sette, J. Sidney, S. del Guercio, M. F.and Southwood, J. Ruppert, C. Dalberg, H. M. Grey, and R. T. Kubo. Peptide binding to the most frequent HLA-A class I alleles measured by quantitative molecular binding assays. *Molecular Immunology*, 31:813–822, 1994.

[Sette et al. 1994b] A. Sette, A. Vitiello, B. Reherman, P. Fowler, R. Nayersina, W. M. Kast, C. Melief, C. J. Oseroff, L. Yuan, and J. Ruppert. The relationship between class I binding affinity and immunogenicity of potential cytotoxic T cell epitopes. *J. Immunol.*, 153:5586–5592, 1994.

[Shakhnovich 1993] Shakhnovich. Ground state of random copolymers and the discrete random energy model. *J. Chem. Phys.*, 98:8174–8177, 1993.

[Shepard 1963] R. N. Shepard. Analysis of proximities as a technique for the study of information processing in man. *Human Factors*, 5:33–48, 1963.

[Shepard 1964] R. N. Shepard. Attention and the metric structure of the stimulus space. *J. Math. Psychol.*, 1:54–87, 1964.

[Shortman et al. 1991] K. Shortman, D. Vremec, and M. Egerton. The kinetics of T cell antigen receptor expression by subgroups of $CD4^+8^+$ thymocytes: delineation of $CD4^+8^+3^{2^+}$ thymocytes as post-selection intermediates leading to mature T cells. *J. Exp. Med.*, 173:323–332, 1991.

[Sidney et al. 1996a] J. Sidney, H. M. Grey, R. T. Kubo, and A. Sette. Practical, biochemical and evolutionary implications of the discovery of HLA class I supermotifs. *Immunol Today*, 17(6):261–266, 1996.

[Sidney et al. 1996b] J. Sidney, H. M. Grey, S. Southwood, E. Celis, P. A. Wentworth, M. F. del Guercio, R. T. Kubo, R. W. Chesnut, and A. Sette. Definition of an HLA-A 3-like supermotif demonstrates the overlapping peptide-binding repertoires of common HLA molecules. *Hum. Immunol.*, 45(2):79–93, 1996.

438   References

[Sidney et al. 2003]   J. Sidney, S. Southwood, V. Pasquetto, and A. Sette. Simultaneous prediction of binding capacity for multiple molecules of the HLA B44 supertype. *J. Immunol.*, 171(11):5964–5974, 2003.

[Sidney et al. 2005]   J. Sidney, S. Southwood, and A. Sette. Classification of A1- and A24-supertype molecules by analysis of their MHC-peptide binding repertoires. *Immunogenetics*, 57(6):393–408, 2005.

[Silver et al. 1992]   M. L. Silver, H. C. Guo, J. L. Strominger, and D. C. Wiley. Atomic structure of a human MHC molecule presenting an influenza virus peptide. *Nature*, 360:367–369, 1992.

[Silver et al. 1998]   R. Silver, Q. Li, and J. Ellner. Expression of virulence of mycobacterium tuberculosis within human monocytes: virulence correlates with intracellular growth and induction of tumor necrosis factor alpha but not with evasion of lymphocyte-dependent monocyte effector functions. *Infection and Immunity*, 66:1190–1199, 1998.

[Silverstein & Rose 1997]   A. Silverstein and N. Rose. On the mystique of the immunological self. *Immunol. Rev.*, 159:197–206, 1997.

[Silverstein & Rose 2000]   A. M. Silverstein and N. R. Rose. There is only one immune system! The view from immunopathology. *Seminars in Immunology*, 12(3):173–178, 2000.

[Simon 1996]   H. A. Simon. *The Sciences of the Artificial*. MIT Press, 3rd edition, 1996.

[Singer & Kirschner 2004]   B. Singer and D. E. Kirschner. Influence of backward bifurcation on interpretation on R_0 in a model of epidemic tuberculosis with reinfection. *Mathematical Biosciences and Engineering*, 1(1):81–93, 2004.

[Singer & Linderman 1990]   D. F. Singer and J. J. Linderman. The relationship between antigen concentration, antigen internalization, and antigenic complexes: modeling insights into antigen processing and presentation. *J. Cell Bio.*, 111(1):55–68, 1990.

[Singer & Linderman 1991]   D. F. Singer and J. J. Linderman. Antigen processing and presentation: how can a foreign antigen be recognized in a sea of self proteins? *J. Theor. Biol.*, 151(3):385–404, 1991.

[Singh et al. 1983]   S. P. N. Singh et al. Human leukocyte antigen (HLA)-linked control of susceptibility to pulmonary tuberculosis and association with HLA-DR types. *J. Infect. Dis.*, 148(4):676–681, 1983.

[Slade & McCallum 1992]   R. W. Slade and H. I. McCallum. Overdominant *versus* frequency-dependent selection at MHC loci. *Genetics*, 132:861–864, 1992.

[Slifka et al. 1998]   M. K. Slifka, R. Antia, and J. K. Whitmire an R. Ahmed. Humoral immunity due to long-lived plasma cells. *Immunity*, 8:363–372, 1998.

[Smith et al. 1996a]   K. J. Smith, A. J. Reid, S. W.and Harlos, A. J. McMichael, D. I. Stuart, Bell. J. I., and E. Y. Jones. Bound water structure and polymorphic amino acids act together to allow the binding of different peptides to MHC class I HLA-B53. *Immunity*, 4:215–228, 1996.

[Smith et al. 1996b]   K. J. Smith, S. W. Reid, D. I. Stuart, A. J. McMichael, E. Y. Jones, and Bell. J. I. An altered position of the alpha 2 helix of MHC class I is revealed by the crystal structure of HLA-B*3501. *Immunity*, 4:203–213, 1996.

[Smith et al. 1997]   D. J. Smith, S. Forrest, R. Hightower, and A. S. Perelson. Deriving shape space parameters from immunological data. *J. Theor. Biol.*, 189:141–150, 1997.

[Smith *et al.* 1999]   D. J. Smith, S. Forrest, D. H. Ackley, and A. S. Perelson. Variable efficacy of repeated annual influenza vaccination. *Proc. Natl. Acad. Sci. USA*, 96:14001–14006, 1999.

[Smith *et al.* 2004]   D. J. Smith, A. S. Lapedes, J. C. De Jong, T. M. Besteborer, G. F. Rimmelzwaan, A. D. M. E. Osterhaus, and R. A. M. Fouchier. Mapping the antigenic and genetic evolution of influenza virus. *Science*, 305:371–376, 2004.

[Snell 1968]   G. D. Snell. The H-2 locus of the mouse: observations and speculations concerning its comparative genetics and its polymorphism. *Folia. Biol. (Praha)*, 14:335–358, 1968.

[Snir & Otto 1998]   M. Snir and S. Otto. *MPI-The Complete Reference: The MPI Core*. MIT Press, 1998.

[Soderberg *et al.* 2005]   K. A. Soderberg, G. W. Payne, A. Sato, R. Medzhitov, S. S. Segal, and A. Iwasaki. Innate control of adaptive immunity via remodelling of lymph node feed arteriole. *Proc. Natl. Acad. Sci. USA*, 102(45):16315–16320, 2005.

[Sompayrac 2002]   L. Sompayrac. *How the immune system works*. Blackwell, 2002.

[Southwood *et al.* 1998]   S. Southwood, J. Sidney, A. Kondo, M. F. del Guercio, E. Appella, R. T. Hoffman, S. Kubo, R. W. Chesnut, H. M. Grey, and A. Sette. Several common HLA-DR types share largely overlapping peptide binding repertoires. *J. Immunol.*, 60(7):3363–73., 1998.

[Speir *et al.* 2001]   J. A. Speir, J. Stevens, E. Joly, Butcher. G. W., and I. A. Wilson. Two different, highly exposed, bulged structures for an unusually long peptide bound to rat MHC class I RT1-Aa. *Immunity*, 14:81–92, 2001.

[Spivey 1992]   J. M. Spivey. *The Z Notation: a reference manual*. Prentice Hall, 2nd edition, 1992.

[Starovasnik *et al.* 1997]   M. A. Starovasnik, A. C., Braisted, and J. A. Wells. Structural mimicry of a native protein by a minimized binding domain. *Proc. Nat. Acad. Sci. USA*, 94:10080–10085, 1997.

[Stepney *et al.* 2004]   S. Stepney, R. E. Smith, J. Timmis, and A. M. Tyrrell. Towards a conceptual framework for artificial immune systems. In [Nicosia *et al.* 2004], pages 53–64.

[Stepney *et al.* 2005a]   S. Stepney, S. Braunstein, J. Clark, A. Tyrrell, A. Adamatzky, R. Smith, T. Addis, C. Johnson, J. Timmis, P. Welch, R. Milner, and D. Partridge. Journeys in non-classical computation I: A grand challange for computing research. *International Journal of Parallel, Emergent and Distributed Systems*, 20:97–125, 2005.

[Stepney *et al.* 2005b]   S. Stepney, R. Smith, J. Timmis, A. Tyrrell, M. Neal, and A. Hone. Conceptual frameworks for artificial immune systems. *Int. J. Unconventional Computing*, 1(3):315–338, 2005.

[Stibor *et al.* 2004]   T. Stibor, K. M. Bayarou, and C. Eckert. An investigation of R-chunk detector generation on higher alphabets. In *Proc. of Genetic and Evolutionary Computation Conference (GECCO)*, LNCS, pages 299–307. Springer, 2004.

[Stibor *et al.* 2005a]   T. Stibor, P. Mohr, J. Timmis, and C. Eckert. Is negative selection appropriate for anomaly detection? In *Proc. of Genetic and Evolutionary Computation Conference (GECCO)*, LNCS. Springer, 2005.

[Stibor *et al.* 2005b]   T. Stibor, J. Timmis, and C. Eckert. A comparative study of real-valued negative selection to statistical anomaly detection techniques. In [Jacob *et al.* 2005], pages 262–275.

[Stolzenberg et al. 1997]   E. Stolzenberg, G. Anderson, M. Ackermann, R. Whitlock, and M. Zasloff. Epithelial antibiotic induced in states of disease. Proc. Natl. Acad. Sci. USA, 94:8686–8690, 1997.

[Streilein 1999]   J. W. Streilein. Regional immunity and ocular immune privilege. Chem. Immunol., 73:11–38, 1999.

[Streilein 2003]   J. W. Streilein. Ocular immune privilege: therapeutic opportunities from an experiment of nature. Nat. Rev. Imm., 3(879-889), 2003.

[Strogatz 1994]   S. H. Strogatz. Nonlinear Dynamics and Chaos: with applications to physics, biology, chemistry, and engineering. Westview Press, 1994.

[Stryhn et al. 1996]   A. Stryhn, P. S. Andersen, L. O. Pedersen, A. Svejgaard, A. Holm, C. J. Thorpe, L. Fugger, S. Buus, and J. Engberg. Shared fine specificity between T-cell receptors and an antibody recognizing a peptide major histocompatibility class I complex. Proc. Natl. Acad. Sci. USA, 93:10338–10342, 1996.

[Stuber & Dillner 1995]   G. Stuber and J. Dillner. HLA-A0201 and HLA-B7 binding peptides in the EBV-encoded EBNA-1, EBNA-2 and BZLF-1 proteins detected in the MHC class I stabilization assay. Low proportion of binding motifs for several HLA class I alleles in EBNA-1. Int Immunol, 7(4):653-63, 1995.

[Sturgill-Koszycki et al. 1994]   S. Sturgill-Koszycki et al. Lack of acidification in Mycobacterium phagosomes produced by exclusion of the vesicular proton-ATPase. Science, (678-681), 1994.

[Sturniolo et al. 1999]   T. Sturniolo, E. Bono, J. Ding, L. Raddrizzani, O. Tuereci, U. Sahin, M. Braxenthaler, F. Gallazzi, M. P. Protti, F. Sinigaglia, and J. Hammer. Generation of tissue-specific and promiscuous HLA ligand databases using DNA microarrays and virtual HLA class II matrices. Nature Biotechnology, 17:555–5561, 1999.

[Styblo et al. 1969]   K. Styblo, J. Meijer, and I. Sutherland. The transmission of tubercle bacilli: its trend in a human population. Bulletin of the International Union of Tuberculosis, 42:5–104, 1969.

[Su & Anderson 2004]   M. A. Su and M. S. Anderson. Aire: an update. Curr. Op. Immunol., 16:746–752, 2004.

[Sudo & Kamikawaji 1995]   T. Sudo and N. Kamikawaji. Differences in MHC class I self peptide repertoires among HLA-A2 subtypes. J. Immunol., 155(10):4749–4756, 1995.

[Sullivan et al. 2001]   A. D. Sullivan, J. Wigginton, and D. Kirschner. The coreceptor mutation CCR5 Delta32 influences the dynamics of HIV epidemics and is selected for by HIV. Proc. Natl. Acad. Sci. USA, 98(18):10214–10219, 2001.

[Sun & Jang 1996]   Z. Sun and B. J. Jang. Patterns and conformations commonly occurring supersecondary structures (basic motifs) in protein data bank. J. Protein Chem., 15:675–690, 1996.

[Sun et al. 1997]   Z. Sun, X. Rao, L. Peng, and D. Xu. Prediction of protein supersecondary structures based on the artificial neural network method. Protein Eng., 10:763–769, 1997.

[Surh & Sprent 1994]   C. D. Surh and J. Sprent. T-cell apoptosis detected in situ during positive and negative selection in the thymus. Nature, 372:100–103, 1994.

[Swain et al. 1983]   S. L. Swain, R. W. Dutton, R. Schwab, and J. Yamamoto. Xenogeneic human anti-mouse T cell responses are due to the activity of the same functional T cell subsets responsible for allospecific and major histocompatibility complex-restricted responses. J. Exp. Med., 157:720–729, 1983.

[Sylvester-Hvid *et al.* 2004]   C. Sylvester-Hvid, M. Nielsen, K. Lamberth, G. Roder, S. Justesen, C. Lundegaard, P. Worning, H. Thomadsen, O. Lund, S. Brunak, and S. Buus. SARS CTL vaccine candidates; HLA supertype-, genome-wide scanning and biochemical validation. *Tissue Antigens*, 63(5):395–400, 2004.

[Takahata & Nei 1990]   N. Takahata and M. Nei. Allelic genealogy under overdominant and frequency-dependent selection and polymorphism of major histocompatibility complex loci. *Genetics*, 124:967–978, 1990.

[Takahata 1995]   N. Takahata. MHC diversity and selection. *Immunol. Rev.*, 143:225–247, 1995.

[Tan *et al.* 1997]   J. Tan et al. Human alveolar T lymphocyte responses to mycobacterium tuberculosis antigens: Role for CD4+ and CD8+ cytotoxic T cells and relative resistance of alveolar macrophages to lysis. *J. Immunol.*, 159:290–297, 1997.

[Tanchot & Rocha 1995]   C. Tanchot and B. Rocha. The peripheral T-cell repertoire: Independent homeostatic regulation of virgin and activated CD8+ T-cell pools. *Eur. J. Immunol*, 25:2127–2136, 1995.

[Tascon *et al.* 1998]   R. E. Tascon et al. Protection against mycobacterium tuberculosis infection by CD8+ T cells requires the production of gamma interferon. *Infection and Immunity*, 66(2):830–4, 1998.

[Tato *et al.* 2006]   C. M. Tato, A. Laurence, and J. J. O Shea. Helper T cell differentiation enters a new era: Le roi est mort; vive le roi! *J. Exp. Med.*, 203(4):809–812, 2006.

[Tauber 2000]   A. I. Tauber. Moving beyond the immune self? *Seminars in Immunology*, 12(3):241–248, 2000.

[Tew *et al.* 1990]   J. G. Tew, M. H. Kosco, G. F. Burton, and A. K. Szakal. Follicular dendritic cells as accessory cells. *Immunol. Rev.*, pages 185–212, 1990.

[Thomas-Vaslin *et al.* 1987]   V. Thomas-Vaslin, D. Damotte, M. Coltey, N. Le Douarin, A. Coutinho, and J. Salaun. Abnormal T cell selection on nod thymic epithelium is sufficient to induce autoimmune manifestations in C57BL/6 athymic nude mice. *Proc. Natl. Acad. Sci. USA*, 94(9):4598–4603, 1987.

[Thompson & Layzell 1999]   A. Thompson and A. Layzell. Analysis of unconventional evolved electronics. *Comm. ACM*, 42(4):71–79, 1999.

[Tieri *et al.* 2005]   P. Tieri, S. Valensin, V. Latora, G. C. Castellani, M. Marchiori, D. Remondini, and C. Franceschi. Quantifying the relevance of different mediators in the human immune cell network. *Bioinformatics*, 21:1639–1643, 2005.

[Timmis & Bentley 2002]   J. Timmis and P. Bentley, editors. *Proc. of the 1st International Conference on Artificial Immune Systems (ICARIS)*. University of Kent Printing Unit, 2002.

[Timmis & Edmonds 2004]   J. Timmis and C. Edmonds. A comment on opt-AINet: An immune network algorithm for optimisation. In *Proc. of Genetic and Evolutionary Computation Conference (GECCO)*, volume 3102 of *LNCS*, pages 308–317. Springer, 2004.

[Timmis & Knight 2001]   J. Timmis and T. Knight. Artificial immune systems: Using the immune system as inspiration for data mining. In H. Abbas, A. Ruhul, A. Sarker, and S. Newton, editors, *Data Mining: A Heuristic Approach*, pages 209–230. Idea Group, 2001.

[Timmis & Neal 2001]   J. Timmis and M. Neal. A resource limited artificial immune system for data analysis. *Knowledge Based Systems*, 14(3–4):121–130, 2001.

[Timmis *et al.* 2000]   J. Timmis, M. Neal, and J. Hunt. An artificial immune system for data analysis. *Biosystems*, 55(1/3):143–150, 2000.

[Timmis *et al.* 2002]   J. Timmis, R. de Lemos, M. Ayara, and R. Duncan. Towards immune inspired fault tolerance in embedded systems. In L. Wang, J. Rajapakse, K. Fukushima, S. Lee, and X. Yao, editors, *Proceedings of 9th International Conference on Neural Information Processing*, pages 1459–1463. IEEE, 2002.

[Timmis *et al.* 2003]   J. Timmis, P. Bentley, and E. Hart, editors. *Proc. of the 2nd International Conference on Artificial Immune Systems (ICARIS)*, volume 2787 of *LNCS*. Springer, 2003.

[Timmis *et al.* 2004]   J. Timmis, C. Edmonds, and J. Kelsey. Assessing the performance of two immune inspired algorithms and a hybrid genetic algorithm for function optimisation. In *Proc. of Congress on Evolutionary Computation (CEC)*, volume 1, pages 1044–1051. IEEE, 2004.

[Timmis 2000]   J. Timmis. *Artificial Immune Systems: a novel data analysis technique inspired by the immune system* . PhD thesis, University of Wales, Aberystwyth, 2000.

[Tononi *et al.* 1999]   G. Tononi, O. Sporns, and G. M. Edelman. Measures of degeneracy and redundancy in biological networks. *Proc. Natl. Acad. Sci. USA*, 96(6):3257–3262, 1999.

[Toseland *et al.* 2005]   C. P. Toseland, D. J. Clayton, H. McSparron, S. L. Hemsley, M. J. Blythe, I. A. Paine, K. Doytchinova, P. Guan, C. K. Hattotuwagama, and D. R. Flower. Antijen: a quantitative immunology database integrating functional, thermodynamic, kinetic, biophysical, and cellular data. *Immunome Res*, 1(1):4, 2005.

[Tough & Sprent 1994]   D. Tough and J. Sprent. Turnover of naïve- and memory-phenotype cells. *J. Exp. Med.*, 179:1127–1135, 1994.

[Tough *et al.* 1996]   D. Tough, P. Borrow, and J. Sprent. Induction of bystander T-cell proliferation by viruses and type I interferon in vivo. *Science*, 272:1947–1950, 1996.

[Tracey 1997]   K. Tracey. Tumor necrosis factor. *Marcel Dekker*, pages 223–239, 1997.

[Trachtenberg *et al.* 2003]   E. Trachtenberg, B. Korber, C. Sollars, T. B. Kepler, P. T. Hraber, E. Hayes, R. Funkhouser, M. Fugate, J. Theiler, Y. S. Hsu, K. Kunstman, S. Wu, J. Phair, H. Erlich, and S. Wolinsky. Advantage of rare HLA supertype in HIV disease progression. *Nat. Med.*, 9:928–935, 2003.

[Tramontano 2006]   A. Tramontano. *Protein Structure Prediction: Concepts and Applications*. Wiley, 2006.

[Tynan *et al.* 2005a]   F. E. Tynan, N. A. Borg, J. J. Miles, T. Beddoe, D. El-Hassen, S. L. Silins, W. J. van Zuylen, A. W. Purcell, L. Kjer-Nielsen, J. McCluskey, S. R. Burrows, , and J. Rossjohn. High resolution structures of highly bulged viral epitopes bound to major histocompatibility complex class I. Implications for T-cell receptor engagement and T-cell immunodominance. *J. Biol. Chem*, 280:23900–23909, 2005.

[Tynan *et al.* 2005b]   F. E. Tynan, S. R. Burrows, A. M. Buckle, C. S. Clements, N. A. Borg, J. J. Miles, T. Beddoe, J. C. Whisstock, M. C. Wilce, S. L. Silins, J. M. Burrows, L. Kjer-Nielsen, L. Kostenko, A. W. Purcell, J. McCluskey, and J. Rossjohn. T cell receptor recognition of a 'super-bulged' major histocompatibility complex class I-bound peptide. *Nat. Immunol.*, 6:1114–1122, 2005.

[Udaka *et al.* 2001]   K. Udaka, K. H. Wiesmuller, S. Kienle, G. Jung, H. Tamamura, H. Yamagishi, K. Okumura, P. Walden, T. Suto, and T. Kawasaki. An automated prediction of MHC class I-binding peptides based on positional scanning with peptide libraries. *Immunogenetics*, 51:816–828, 2001.

[Valentine *et al.* 2004]   S. H. Valentine, S. Stepney, and I. Toyn. A Z Patterns catalogue II: definitions and laws. Technical Report YCS-2004-383, University of York, UK, 2004.

[van Boven & Weissing 2001]   M. van Boven and F. J. Weissing. Competition at the mouse T complex: rare alleles are inherently favored. *Theor. Popul. Biol.*, 60:343–358, 2001.

[van den Berg & Kiselëv 2004]   H. A. van den Berg and Y. N. Kiselëv. Expansion and contraction of the cytotoxic T lymphocyte response - an optimal control approach. *Bulletin of Mathematical Biology*, 66:1345–1369, 2004.

[van den Berg & Rand 2003]   H. A. van den Berg and D. A. Rand. Antigen presentation on MHC molecules as a diversity filter that enhances immune efficacy. *J. Theor. Biol.*, 224(2):249–267, 2003.

[van den Berg 2004]   H. A. van den Berg. Control of T-cell immunity:design principles without the wetware. Technical Report UKC/IMS/04/36, University of Kent, 2004.

[van Deventer 2001]   S. J. van Deventer. Transmembrane TNF-alpha, induction of apoptosis, and the efficacy of TNF-targeting therapies in Crohn's disease. *Gastroenterology*, 121(5):1242–6, 2001.

[van Deventer 2002]   S. J. van Deventer. Anti-tumour necrosis factor therapy in Crohn's disease: where are we now? *Gut*, 51(3):362–3, 2002.

[Varela *et al.* 1988]   F. Varela, A. Coutinho, B. Dupire, and N. Vaz. Cognitive Networks: Immune, Neural and Otherwise. *J. Theor. Imm.*, 2:359–375, 1988.

[Varela *et al.* 1991]   F. J. Varela, E. Thompson, and E. Rosch. *The Embodied Mind: cognitive science and human experience.* MIT Press, 1991.

[Varela 1981]   F. J. Varela. Autonomy and autopoiesis. In *Self Organising Systems*, pages 14–23. New York Campus Press, 1981.

[Vidal *et al.* 2000]   K. Vidal, C. Daniel, I. Vidavsky, C. A. Nelson, and P. M. Allen. Hb (64-76) epitope binds in different registers and lengths to I-Ek and I-Ak. *Mol. Immunol.*, 37:203–212, 2000.

[Vidović & Matzinger 1988]   D. Vidović and P. Matzinger. Unresponsiveness to a foreign antigen can be caused by self-tolerance. *Nature*, 336:222–225, 1988.

[Vogel *et al.* 1999]   T. U. Vogel, D. T. Evans, J. A. Urvater, D. H. O'Connor, A. L. Hughes, and D. I. Watkins. Major histocompatibility complex class I genes in primates: Coevolution with pathogens. *Immunol. Rev.*, 167:327–337, 1999.

[Von Schwedler *et al.* 2003]   U. K. Von Schwedler, K. M. Stray, J. E. Garrus, and W. I. Sundquist. Functional surfaces of the human immunodeficiency virus type 1 capsid protein. *J. Virol.*, 77(9):5439–5450, 2003.

[Von Boehmer 1994]   H. Von Boehmer. Positive selection of lymphocytes. *Cell*, 76:219–228, 1994.

[Vordermeier 1995]   H. M. Vordermeier. T-cell recognition of mycobacterial antigens. *The European Respiratory Journal*, 20:657s–667s, 1995.

[Wagner 2005]   A. Wagner. Distributed robustness versus redundancy as causes of mutational robustness. *Bioessays*, 27:176–188, 2005.

[Waldrop *et al.* 1998]   S. L. Waldrop, K. A. Davis, V. C. Maino, and L. J. Picker. Normal human CD4+ Memory T cells display broad heterogeneity in their activation threshold for cytokine synthesis. *J. Immunol.*, 161:5284–5295, 1998.

[Walker *et al.* 2002]   J. Walker, D. R. Flower, and K. Rigley. Microarrays in hematology. *Current Opinion in Hematology*, 9:23–29, 2002.

[Watkins & Salter 2005]   S. C. Watkins and R. D. Salter. Functional connectivity between immune cells mediated by tunneling nanotubules. *Immunity*, 23:309–318, 2005.

[Watkins & Timmis 2004]   A. Watkins and J. Timmis. Exploiting parallelism inherent in AIRS, an artificial immune classifier. In [Nicosia *et al.* 2004], pages 427–438.

[Watkins *et al.* 2003]   A. Watkins, B. Xintong, and A. Phadke. Parallelizing an immune-inspired algorithm for efficient pattern recognition. In *Intelligent Engineering Systems through Artificial Neural Networks: Smart Engineering System Design: Neural Networks, Fuzzy Logic, Evolutionary Programming, Complex Systems and Artificial Life*, pages 224–230. ASME Press, 2003.

[Watkins *et al.* 2004]   A. Watkins, J. Timmis, and L. Boggess. Artificial Immune Recognition System (AIRS): An Immune Inspired Supervised Machine Learning Algorithm. *Genetic Programming and Evolvable Machines*, 5(3):291–318, 2004.

[Watkins 2001]   A. Watkins. AIRS: A resource limited artificial immune classifier. Master's thesis, Mississippi State University, USA., 2001.

[Watkins 2005]   A. Watkins. *Exploiting Immunological Metaphors in the Development of Serial, Parallel and Distributed Learning Algorithms*. PhD thesis, University of Kent, UK., 2005.

[Watson & Crick 1953]   J. D. Watson and F. H. C. Crick. Genetical implications of the structure of deoxyribonucleic acid. *Nature*, 171:964–967, 1953.

[Wauben & van der Kraan 1997]   M. H. Wauben and M. van der Kraan. Definition of an extended MHC class II -peptide binding motif for the autoimmune disease-associated lewis rat RT1.BL molecule. *Int. Immunol.*, 9(2):281–290, 1997.

[Wedekind *et al.* 1995]   C. Wedekind, T. Seebeck, F. Bettens, and A. J. Paepke. MHC-dependent mate preferences in humans. *Proc. R. Soc. Lond. B. Biol. Sci.*, 260:245–249, 1995.

[Weidt *et al.* 1995]   G. Weidt, W. Deppert, S. Buchhop, H. Dralle, and F. Lehmann-Grube. Antiviral protective immunity induced by major histocompatibility complex class I molecule-restricted viral T-lymphocyte epitopes inserted in various positions in immunologically self and nonself proteins. *J. Virol.*, 69:2654–2658, 1995.

[Weinand 1990]   R.G. Weinand. Somatic mutation, affinity maturation and the antibody repertoire: a computer model. *J. Theor. Biol.*, 143:343–382, 1990.

[Weisbuch & Atlan 1988]   G. Weisbuch and H. Atlan. Control of the immune response. *Journal of Physics A: Mathematical and General*, 21(3):189–192, 1988.

[Weisbuch 1990]   G. Weisbuch. A shape space approach to the dynamics of the immune system. *J. Theor. Biol.*, 143:507–522, 1990.

[Weissing & van Boven 2001]   F. J. Weissing and M. van Boven. Selection and segregation distortion in a sex-differentiated population. *Theor. Popul. Biol.*, 60:327–341, 2001.

[Weng *et al.* 1997]   N. P. Weng, L. Granger, and R. J. Hodes. Telomere lengthening and telomerase activation during human B cell differentiation. *Proc. Natl. Acad. Sci. USA*, 94:10827–10832, 1997.

[Whisstock & Lesk 2003]   J. C. Whisstock and A. M. Lesk. Prediction of protein function from protein sequence and structure. *Q. Rev. Biophys.*, 36:307–340, 2003.

[Whitesides & Boncheva 2002]   G. M. Whitesides and M. Boncheva. Beyond molecules: Self-assembly of mesoscopic and macroscopic components. *Proc. Natl. Acad. Sci. USA*, 99(8):4769–4774, 2002.

[Wierzchon & Kuzelewska 2002]   S. Wierzchon and U. Kuzelewska. Stable clusters formation in an artificial immune system. In [Timmis & Bentley 2002], pages 68–75.

[Wigginton & Kirschner 2001]   J. E. Wigginton and D. Kirschner. A model to predict cell-mediated immune regulatory mechanisms during human infection with Mycobacterium tuberculosis. *J. Immunol.*, 166(3):1951–67, 2001.

[Wilkinson *et al.* 1999]   R. J. Wilkinson et al. Influence of polymorphism in the genes for the interleukin (IL)-1 receptor antagonist and IL-1beta on tuberculosis. *J. Exp. Med.*, 189(12):1863–74, 1999.

[Wills 1991]   C. Wills. Maintenance of multiallelic polymorphism at the MHC region. *Immunol. Rev.*, 124:165–220, 1991.

[Wilson & Garrett 2004]   W. Wilson and S. M. Garrett. Modelling immune memory for prediction and computation. In [Nicosia *et al.* 2004], pages 343–352.

[Wold & Hellberg 1987]   S. Wold and S. Hellberg, editors. *Proc. Symp. on PLS Model Building: Theory and Application*, 1987.

[Wold 1995]   S. Wold. *Chemometric Methods In Molecular Design*, chapter PLS for multivariate linear modelling, pages 195–218. 1995.

[Wolfram 1994]   S. Wolfram. *Cellular Automata and Complexity: collected papers.* Addison-Wesley, 1994.

[Wolfram 2002]   S. Wolfram. *A New Kind of Science.* Wolfram Media Incorporated, 2002.

[Wolpert & G. 1997]   D. H. Wolpert and MacReady W. G. *No Free Lunch Theorems for Optimization*, volume 1. 1997.

[World Health Organization 2001]   World Health Organization. Who report 2001: Global tuberculosis contro, 2001.

[Wu & Kabat 1970]   T. T. Wu and E. A. Kabat. An analysis of the sequences of the variable regions of bence jones proteins and myeloma light chains and their implications for antibody complementarity. *J. Exp. Med.*, 132:211–250, 1970.

[Wuensche & Lesser 1992]   A. Wuensche and M. Lesser. *The Global Dynamics of Cellular Automata.* Addison-Wesley, 1992.

[Wuensche 2002]   A. Wuensche. Finding gliders in cellular automata. In A. Adamatzky, editor, *Collision-Based Computing.* Springer, 2002.

[Wysocki *et al.* 1986]   L. Wysocki, T. Manser, and M. L. Gefter. Somatic evolution of variable region structures during an immune response. *Proc. Natl. Acad. Sci. USA*, 83:1847–1851, 1986.

[Yan *et al.* 2003]   S. S. Yan, Z. Wu, H. Zhang, G. Furtado, X. Chen, S. Yan, A. M. Schmidt, C. Brown, A. Stern, J. Lafaille, L. Chess, D. M. Stern, and H. Jiang. Suppression of experimental autoimmune encephalomyelitis by selective blockade of encephalitogenic T-cell infiltration of the central nervous system. *Nature Medicine*, 9:287–293, 2003.

[Yates *et al.* 2001]   A. Yates, C. C. W. Chan, R. E. Callard, A. J. T. George, and J. Stark. An approach to modelling in immunology. *Briefings in Bioinformatics*, 2:245–257, 2001.

[Yeager & Hughes 1999]   M. Yeager and A. L. Hughes. Evolution of the mammalian MHC: natural selection, recombination, and convergent evolution. *Immunol. Rev.*, 167:45–58, 1999.

[Young *et al.* 1994]   A. C. Young, W. Zhang, J. C. Sacchettini, and S. G. Nathenson. The three-dimensional structure of H-2Db at 2.4 A resolution: implications for antigen-determinant selection. *Cell*, 76:39–50, 1994.

[Young 2001]   D. Young. *Computational Chemistry: A Practical Guide for Applying Techniques to Real World Problems.* Wiley Inter-Science, 2001.

[Yu *et al.* 2002]   K. Yu, N. Petrovsky, C. Schonbach, J. Koh, and V. Brusic. Methods for prediction of peptide binding to MHC molecules: a comparative study. *Mol Med,* 8:137–148, 2002.

[Yusim *et al.* 2002]   K. Yusim, C. Kesmir, B. Gaschen, M. M. Addo, M. Altfeld, S. Brunak, A. Chigaev, V. Detours, and B. T. Korber. Clustering patterns of cytotoxic T-lymphocyte epitopes in human immunodeficiency virus type 1 (HIV-1) proteins reveal imprints of immune evasion on HIV-1 global variation. *J. Virol.,* 76:8757–8768, 2002.

[Zachary *et al.* 1996]   A. Zachary et al. The frequencies of HLA alleles and haplotypes and their distribution among donors and renal patients in the UNOS registry. *Transplantation,* 62(2):272–283, 1996.

[Zanetti & Croft 2001]   M. Zanetti and M. Croft. Immunological memory. In *Encyclopedia of Life Sciences.* 2001.

[Zarling *et al.* 2001]   A. L. Zarling, S. B. Ficarro, F. M. White, J. Shabanowitz, D. F. Hunt, and V. H. Engelhard. Phosphorylated peptides are naturally processed and presented by major histocompatibility complex class I molecules in vivo. *J. Exp. Med.,* 192:1755–1762, 2001.

[Zeng *et al.* 2005]   R. Zeng et al. Synergy of IL-21 and IL-15 in regulating CD8 T cell expansion and function. *J. Exp. Med.,* 201:139–148, 2005.

[Zhang *et al.* 1992]   W. Zhang, A. C. Young, M. Imarai, S. G. Nathenson, and J. L. Sacchettini. Crystal structure of the major histocompatibility complex class I H-2Kb molecule containing a single viral peptide: implications for peptide binding and T-cell receptor recognition. *Proc. Natl. Acad. Sci. USA,* 89:8403–8407, 1992.

[Zhang *et al.* 1993]   Q. J. Zhang, R. Gavioli, G. Klein, and M. G. Masucci. An HLA-A11-specific motif in nonamer peptides derived from viral and cellular proteins. *Proc. Natl. Acad. Sci. USA,* 90:2217–2221, 1993.

[Zhang *et al.* 1998]   C. Zhang, A. Anderson, and C. DeLisi. Structural principles that govern the peptide-binding motifs of class I MHC molecules. *J. Molec. Biol.,* 281(5):929–947, 1998.

[Zhao *et al.* 1998]   Z. S. Zhao, F. Granucci, L. Yeh, P. A. Schaffer, and H. Cantor. Molecular mimicry by herpes simplex virus-type 1: Autoimmune disease after viral infection. *Science,* 279:1344–1347, 1998.

[Zinkernagel & Hengartner 2004]   R. M. Zinkernagel and H. Hengartner. On immunity against infections and vaccines: Credo 2004. *Scand. J. Immunol,* 60:9–13, 2004.

[Zinkernagel *et al.* 1996]   R. M. Zinkernagel, M. F. Bachmann, T. M. Kündig, S. Oehen, H. Pirchet, and H. Hengartner. On immunological memory. *Ann. Rev. Immunol.,* 14:333–367, 1996.

[Zinkernagel 1986]   R. M. Zinkernagel. Biological role of major transplantation antigens in T cell self-recognition. *Experientia,* 42(9):970–2, 1986.

[Zinkernagel 2002]   R. M. Zinkernagel. On differences between immunity and immunological memory. *Curr. Opin. Immunol.,* (14):523–536, 2002.

Index

Printed in the United States of America.